Introduction
Development

For my mother, who taught me so much

Introduction to Language Development

Shelia M. Kennison

Oklahoma State University

Los Angeles | London | New Delhi
Singapore | Washington DC

Los Angeles | London | New Delhi
Singapore | Washington DC

FOR INFORMATION:

SAGE Publications, Inc.
2455 Teller Road
Thousand Oaks, California 91320
E-mail: order@sagepub.com

SAGE Publications Ltd.
1 Oliver's Yard
55 City Road
London EC1Y 1SP
United Kingdom

SAGE Publications India Pvt. Ltd.
B 1/I 1 Mohan Cooperative Industrial Area
Mathura Road, New Delhi 110 044
India

SAGE Publications Asia-Pacific Pte. Ltd.
3 Church Street
#10-04 Samsung Hub
Singapore 049483

Acquisitions Editor: Reid Hester
Editorial Development Manager: Eve Oettinger
Editorial Assistant: Sarita Sarak
Production Editor: Brittany Bauhaus
Copy Editor: Megan Markanich
Typesetter: C&M Digitals (P) Ltd.
Proofreader: Scott Oney
Indexer: Kathy Paparchontis
Cover Designer: Anthony Paular
Marketing Manager: Shari Countryman
Permissions Editor: Karen Ehrmann

Printed in the United States of America

Library of Congress Cataloging-in-Publication Data

Kennison, Shelia M.

Introduction to language development / Shelia M. Kennison, Oklahoma State University.

pages cm.

ISBN 978-1-4129-9606-8 (pbk. : alk. paper)
ISBN 978-1-4522-5629-0 (web pdf)

1. Language acquisition. 2. Language awareness—Children. I. Title.

P118.K46 2013
401′.93—dc23 2013008133

This book is printed on acid-free paper.

SFI label applies to text stock

13 14 15 16 17 10 9 8 7 6 5 4 3 2 1

Brief Contents

Detailed Contents

3 The First Twenty-Four Months 57

7 Life With More Than One Language 177

About the Author

Shelia M. Kennison is a professor of psychology at Oklahoma State University. She has taught a variety of courses in psychology, including language development, statistics, and research design. She earned her MS and PhD in Cognitive Psychology from the University of Massachusetts at Amherst and her bachelor's degree in linguistics and psychology from Harvard University. Her research on language comprehension has received funding from the National Science Foundation, the Fulbright Foundation, and Psi Chi. She has published research articles in numerous journals, including *Cognition*, *Journal of Experimental Psychology: Learning, Memory, and Cognition*, *Journal of Memory and Language*, and *Discourse Processes*. Her research includes studies of language processing in English as well as other languages, including Spanish, German, Japanese, Chinese, and Finnish.

Preface

Today, there are between 6,000 and 7,000 languages in the world. The characteristics of these languages vary widely. However, by the age of 4, most children are producing and comprehending their language's most complex sentence structures. My motivation for writing this textbook comes from my personal fascination with human language and language development. Among the many fascinating facts is that an infant born today could master any language in the world, regardless of the language's characteristics, if the infant were raised in an environment where the language was used regularly. Infants born to parents who speak Swahili will come to speak and understand Swahili. Infants born to parents who speak Finnish will come to speak and understand Finnish. Infants born to parents who speak both Urdu and Hindi will come to speak both Urdu and Hindi. My parents happened to speak American English, a variety unique to the Appalachian Mountains of West Virginia, so that it is what I learned.

In this book, you will learn about the research related to language development and about the theoretical perspectives that attempt to explain how language development occurs. You will review the core empirical findings that inform us about how infants develop physically, cognitively, and socially on the road to becoming competent users of their language(s). You will also learn about the research methodologies that are routinely used to study language development. The book contains 12 chapters, each providing an in-depth perspective on a central topic in language development. In Chapter 1, you will learn about the nature of language and current methodologies for studying language development. In Chapter 2, you will explore the biological basis of language, including how different parts of the brain are involved in language behavior and how genes are involved in some language disorders. Chapter 3 describes the development of language occurring up to the first twenty-four months of life. Chapter 4 focuses on children's grammatical development. Chapter 5 provides readers with an introduction into research on the organization of knowledge of words and word meanings. In Chapter 6, you will learn about the social aspects of language and about how children learn to become competent communicators. In Chapter 7, you will learn about bilingualism. Chapter 8 examines the role of culture in language processing and development. Chapters 9 and 10 will describe what is known about language production and language comprehension, respectively. Chapter 11 describes language in the school years and contains discussions of how children learn to read as well as discussions of how language-related delays and disorders are diagnosed and treated. Last, Chapter 12 focuses on language as it changes across the life span.

Each chapter of the book contains pedagogical features that are designed to enhance the learning experience. At the end of each chapter, there are key terms, review questions, recommended reading, recommended films, and suggested class projects. In addition, each chapter contains special features that appear as text boxes. The text boxes have three recurring themes: (1) the diversity of human languages, (2) extraordinary individuals, and (3) research discoveries.

The book has been developed for undergraduate and graduate courses in language development and/or the psychology of language. Such courses are offered in numerous departments, such as psychology, education, human development, and communication sciences and disorders, as well as English, linguistics, and modern languages. An effort has been made to include a broad coverage of the relevant topics in order for the book to meet the needs of students in these multiple disciplines.

For additional ancillary resources, please visit the companion website at **www.sagepub.com/kennison**.

Acknowledgments

I would like to extend my sincerest thanks to those who helped make this book a reality. When one writes a book, it may appear that it is done by one person, but it is certainly a group effort. I thank my SAGE editors, Chris Cardone and Reid Hester, as well as Eve Oettinger and Sarita Sarak, who helped me at each stage in the process as the book was being developed, reviewed, and revised. I would also like to thank Rachel Messer, MS, a graduate student at Oklahoma State University, who read and commented on multiple versions of each chapter and who has taught the Language Development course with me over the years. I also would like to thank those educators at Nuttall Middle School and Midland Trail High School in Fayette County, West Virginia, who played an important part in inspiring me to pursue my interests in science and language. I sincerely thank Jo Davison, Claude McGraw, Randall Patterson, Joel Davis, Alma Burr, Helen Nuckols, Suzanne Skaggs, Scott Wilson, and James Workman. I also would like to thank my academic mentors Peter C. Gordon and Charles Clifton, Jr. Last, I thank my husband, Larry Liggett, who has made it possible for me to devote myself to this endeavor over the past 3 years.

The authors and SAGE would like to acknowledge the contributions of the following reviewers:

Caitlin Cole, *University of Minnesota: Twin Cities*

Herbert L. Colston, *University of Wisconsin–Parkside*

Justin Coran, *University of Florida*

Wind Cowles, *University of Florida*

Priscilla Davis, *University of Alabama at Birmingham*

Amy L. Franklin, *University of Texas at San Antonio*

Joseph Galasso, *California State University, Northridge*

Susan A. Gelman, *University of Michigan*

Tilbe Goksun, *University of Pennsylvania*

Carla Hudson Kam, *University of British Columbia*

Usha Laksmanan, *Southern Illinois University*

Mary Lou Gutierrez, *Walden University*

Max Louwerse, *University of Memphis*

David Ludden, *Lindsey Wilson College*

Mandy Maguire, *University of Texas at Dallas*

Daniel S. McConnell, *University of Central Florida*

Meghan Moran, *San Diego State University*

Judith Olson, *Bemidji State University*

Seyda Ozcaliskan, *Georgia State University*

Cathy Quenin, *Nazareth College of Rochester*

Joan Sereno, *University of Kansas*

Lauren Shapiro Crane, *Wittenberg University*

L. Kathryn Sharp, *East Tennessee State University*

Erik C. Tracy, *Ohio State University*

Lydia E. Volaitis, *Northeastern University*

Larry D. Williams *North Carolina Central University*

CHAPTER 1

LANGUAGE AND LANGUAGE DEVELOPMENT

Children all over the world have the remarkable ability to learn any language that they hear on a daily basis. This remarkable ability remains one of the most intriguing topics in science. A child born today can learn any of the 6,000 languages spoken in the world today. These languages vary a great deal in terms of how they sound, how the words are ordered within sentences, and how new words can be created. A child born in France will acquire French. A child born in Beijing will acquire Mandarin. A child born in Brazil will acquire Portuguese. Regardless of where the child is born, by the age of 5, the child will have mastered the basics of his or her native language(s). They are able to carry out everyday conversations, producing and comprehending both simple and complex sentence structures. By the age of 5, children may know as many as 4,000 to 5,000 words. In this book, you will learn how language development occurs in English as well as in the other languages of the world. In this chapter, you will learn about the major milestones of language development, the elements of language, the broad range of human languages, the major theoretical approaches to language development, and different types of methods that are used to study language development.

▲ **Photo 1.1** Children from around the world acquire language similarly, regardless of the language that is learned. Have you ever considered how children in different countries learn language?

The Nature of Language and Language Development

What Are the Major Milestones of Language Development?

Despite the many ways in which the lives of children around the world vary, the major milestones of language development are remarkably similar. Table 1.1 provides a timeline of the major milestones. Children come into the world able to make a limited number of types of vocalizations. Mostly, they cry. By the age of 8 weeks, they begin **cooing** or producing elongated **vowel** sounds. Between the ages of 4 and 6 months, infants begin to practice the sounds or phonology of their language. They start producing nonsense sequences of sounds (e.g., *bababa* or *gabudu*). The term ***babbling*** is commonly used to describe the infant's nonsensical speech sounds. Usually, by the age of 10 to 12 months, the first word emerges. However, even before infants produce their first spoken word, they have begun to understand the words and phrases spoken by others. By the age of 24 months, they are producing multiword utterances. By the age of 48 months, they are producing and comprehending syntactically complex sentences.

Children continue to fine-tune their language production throughout childhood. The fine-tuning of pronunciation may extend into late childhood. Many children experience difficulty with specific phonemes, such as /s/ or /f/. Most of the difficulty with pronunciation and fluency is resolved either with short-term speech therapy or without any intervention

(Sander, 1972). Children must also learn the social norms of language or **pragmatic rules.** For example, children must learn how to choose the appropriate word and tone for the particular social setting (Ninio & Snow, 1996). For the young child, they begin to learn how and when to be polite when making requests to parents, teachers, and other powerful others. Older children must learn when it is and is not appropriate to use slang or informal forms of language. For example, children might learn the expression *hey, dude* on the playground, but they will find that addressing a grandparent or a teacher in this way might be viewed as inappropriate.

Table 1.1 Timeline of Language Milestones From Birth to Forty-Eight Months

Milestone	Average Age
Crying	Birth
Cooing	6 to 8 weeks
Babbling	4 to 6 months
First word	10 to 12 months
Multiple word utterances	18 to 24 months
Complex sentences	30 to 48 months

Source: Adapted from Brown (1973).

As you begin your study of language development, you may find the topic to be unexpectedly complex. To understand language development, one must not only understand the basic aspects of human language that are learned and how knowledge of language is stored in memory but also appreciate the changes that occur throughout human development that might influence how language is learned and processed. The empirical studies that have been conducted to shed light on how language development occurs rely on increasingly technical methodologies. The ways in which researchers make inferences about what infants know and do not know may be difficult to grasp initially. As you become more familiar with the strategies that researchers use to test hypotheses about language development, you may come to believe, as I do, that although language development is a challenging topic to study, it is an important one that has relevance in all of our lives.

Understanding the Range of Human Languages

An important question in the study of language development is whether language development occurs the same way regardless of the language that is being acquired. Some might assume that because there are big differences across the world's languages, the process of acquiring each language would differ as well. A first step in understanding how differences across languages may be related to how language is acquired is learning more about the basic differences that exist across languages. Today, there are only about 6,000 languages still spoken. Many of these languages are spoken by millions of speakers. The top 10 most widely spoken languages in the world are displayed in Table 1.2.

Of the approximately 6,000 languages that exist in the world today, quite a few are on the verge of extinction with very few speakers. Table 1.3 displays the 16 languages with just one documented native speaker remaining. Austin and Sallaback (2011) estimated that by 2100, between 50% and 90% of the world's languages will be lost. Languages become

Table 1.2 Top Ten World Languages

Language	Millions of Speakers
1. Chinese	1,213
2. Spanish	329
3. English	328
4. Arabic	221
5. Hindi	182
6. Bengali	181
7. Portuguese	178
8. Russian	144
9. Japanese	122
10. German	93

Source: Adapted from Lewis (2009).

vulnerable to extinction when they are not learned fully by the next generation and the language is used primarily in the home. Within a couple of decades, children may not learn the language at all. Those in the family who know the language are grandparents and other elders. With each generation, if not enough young people learn the language fluently, the language will eventually be lost.

When one encounters a language that one does not know, it is a natural response to find the language unusual—perhaps even exotic. Many of the world's languages are spoken by groups of people who live lifestyles that are very different from yours and mine. These speakers may live in remote regions of the world that are cut off from industrialized society (Lewis, 2009). For example, there are hundreds of languages spoken by the indigenous tribes in the Brazilian rainforest. These

Table 1.3 Selection of the World's Most Endangered Languages

Language	Location	Number of Speakers
1. Apiaka	Brazil	1
2. Chana	Argentina	1
3. Dampa	Indonesia	1
4. Diahoi	Brazil	1
5. Kaixana	Brazil	1
6. Laua	Papua New Guinea	1
7. Patwin	United States	1
8. Pazeh	Taiwan	1
9. Pemono	Venezuela	1
10. Taje	Indonesia	1
11. Taushiro	Peru	1
12. Tolowa	United States	1
13. Volow	Vanuatu	1
14. Wintu-Nomlaki	United States	1
15. Yaghan	Chile	1
16. Yarawi	Papua New Guinea	1

Source: Adapted from Comrie, Matthews, and Polinsky (2003).

tribes rely on a hunting and gathering lifestyle. Despite how different the lives of these people may be from those in the industrialized world, it would be a mistake to assume that the languages spoken by such tribes are more primitive than other languages.

▲ Photo 1.2 Ferdinand de Saussure compared human language to chess. What aspect of language would be the rules? What would be the pieces?

The Swiss philosopher and language scholar Ferdinand de Saussure (1916/1977) argued convincingly that all human languages are comparably complex. He compared human language to a chess game. Just as a chess game has rules, so too does each human language. The words or vocabulary or **lexicon** of a language are the chess pieces. These are manipulated following the rules of the language or **grammar.** Because the different languages of the world operate similarly with their own vocabularies manipulated by their own distinct grammars, no human language is more or less primitive than any other. The task of every child is to learn the vocabulary and the rules of their language and become skilled at playing the language game.

What Are the Elements of Language?

Since the work of Saussure, our understanding of the range of human languages has grown. Researchers agree that all human languages can be described as having two major parts: (1) the vocabulary or lexicon and (2) the complete set of rules or grammar that must be followed in order to produce acceptable sentences in the language. Today, there is also a greater awareness of how the languages of the world differ in terms of vocabulary and grammar rules. Vocabulary differences across languages are relatively easy to appreciate. Different languages use different words to refer to the same concepts (e.g., *book* vs. the German *buch* or the French *livre*). Some languages have a larger number of vocabulary items than others; languages with a long tradition of writing and existing in cultures with technical occupations (e.g., law, medicine, and science) tend to have larger vocabularies than languages that are not written and existing in cultures without highly technical occupations. Differences in grammar across languages are more challenging to grasp. Grammar involves different types of rules—some applying to word order, some applying to the sounds of language, some applying to the formation of new words, and some applying to meaning. In the remainder of this section, you will learn more about each of these and how the languages of the world vary.

Syntactic Rules

Every language grammar involves rules about how words are to be ordered within sentences. Such rules are called **syntactic rules,** or syntax. Most sentences have subjects, verbs, and objects. Verbs express actions or states of being. Subjects are typically the doer of actions. Objects are typically the entities that are affected by the action. The abbreviations SVO, SOV, OVS, etc., are used to describe the ordering of constituents in different languages. In the English sentence *The cow kicked the horse,* the subject *cow* precedes the verb *kicked*, and the object *horse* follows the verb, so English is described as an SVO language. In contrast, sentences in Japanese typically follow an SOV word order, the verb coming last (i.e., *cow horse kicked*). Across all the world's languages, one can find languages having all possible orders, suggesting that there is not an order that is impossible to learn (Comrie, 1989; Croft, 2002). Table 1.4 displays the percentage of languages classified in each of the six possible word orders.

Morphological Rules

Language grammar also includes rules about how new words can be formed. Such rules are called **morphological rules,** or morphology. The term ***morpheme*** is used to refer to the smallest unit of meaning within a language. Some words involve only one morpheme (e.g., *chair, dog,* and *cloud*). Other words are composed of multiple morphemes. For example, the word *renter* contains two morphemes—(1) *rent,* the root verb *rent*, and (2) *-er,* a suffix that can be added to a verb to form a noun. The word *joyousness* contains three morphemes—(1) *joy*, the root noun; (2) *-ous*, a suffix that can be added to a noun to form an adjective; and (3) *-ness*, a suffix that can be added to an adjective to form a noun.

Speakers of English apply morphological rules often without consciously realizing that there are rules involved—for example, when creating noun compounds, such as *pancake syrup* and *flower arrangement*. The first noun in an English compound cannot be plural. Speakers avoid the compounds *pancakes syrup* and *flowers arrangement* (Alegre & Gordon, 1996), despite the fact that syrup is usually poured on more than one pancake and flower arrangements contain more than one flower. If you ask a native speaker of English why the phrase *pancake syrup* sounds better than *pancakes syrup,* he or she is not likely to know; it just does.

The languages of the world vary in morphological rules and also the prevalence of multimorphemic words (Comrie, 1989; Croft, 2002; Greenberg, 1974). Some languages, such as English, typically have a small number of morphemes per word. Such languages have been called **isolating languages.** In the English sentence *The girl is a good athlete,* each word

Table 1.4 Estimated Percentages of Languages Classified in Each of the Six Word Order Types and Percentages of Languages That Are Unclassified

Source	SOV	SVO	VSO	VOS	OVS	OSV	Unclassified
Greenberg (1966)	37.0	43.0	20.0	0.0	0.0	0.0	—
Ruhlen (1975)	51.5	35.6	10.5	2.1	0.0	0.2	—
Tomlin (1986)	44.8	41.8	9.2	3.0	0.0	0.0	0.0
Mallinson & Blake (1981)	41.0	35.0	9.0	2.0	1.0	1.0	11.0

contains just one morpheme. Chinese and Vietnamese are also isolating languages. Other languages typically have many more morphemes per word. Such languages are called **synthetic languages.** Hungarian, Finnish, and some languages that are indigenous to the Americas provide extreme cases of synthetic languages. In these languages, there are many morphemes per word. For example, the longest word in Finnish is *lentokonesuihkuturbiinimoottoriapumekaanikkoaliupseerioppilas;* and it means technical warrant officer trainee specialized in aircraft jet engines. The longest uncompounded word is *epaejaerjestelmaellistyttaemaettoemyydellaensaekaeaen* and means even with their lack of ability to disorganize.

Synthetic languages tend to have more flexible word order than isolating languages. The morphemes within a word in a synthetic language may indicate not only the gender of the word and whether the word is singular or plural, but may also indicate whether the word is the subject or object of the sentence. Finnish and Hungarian are examples of free word order languages. The major constituents of the sentence (i.e., subject, verb, and object) can occur in any order without a change in meaning. In contrast, words in isolating languages, such as English, tend to have fixed word order. The subject, object, and verb must appear in particular positions within the sentence. The following examples are Finnish sentences composed of the same words in the six possible orders; each order means *the fish ate the worm*.

a. Kala söi madon
fish ate worm

b. Madon söi kala
worm ate fish

c. Söi kala madon
ate fish worm

d. Söi madon kala
ate worm-acc fish-nom

e. Kala madon söi
fish worm ate

f. Madon kala söi
worm fish ate

Phonological Rules

A language's grammar also includes rules about how the sounds of a word are combined to form words and sentences. These rules are called **phonological rules,** or phonology. The term ***phoneme*** is used to refer to the smallest unit of sound within a language. Languages vary widely in how many different phonemes they have. In American English, there are 40 phonemes, 15 vowels and 25 **consonants** (Roach, 2000). Some languages have fewer phonemes. For example, Rotokas is a language spoken in Papua New Guinea, an island north of Australia. It has only 11 phonemes, five vowels and six consonants (Robinson, 2006). The International Phonetic Association developed a system to represent the phonemes of the world's languages. As of 2008, about 107 distinct letters have been identified. It is remarkable that the typical, normally developing child has the capacity to learn all of these possible phonemes (International Phonetic Association, 1999). They learn the phonemes that are used in their presence on a daily basis.

Some human languages are produced without any speech sounds at all. They use gestures formed with the hands. The various signed languages that exist around the world are used by those who are hearing impaired or belong to families and communities where there are hearing impaired individuals. The manual signing in signed languages is analogous to the sounds in spoken language. In the United States, the most familiar of the signed languages is **American Sign Language** (ASL). Text Box 1.1 describes the characteristics of ASL as well as its historical origins.

Text Box 1.1 Diversity of Human Languages: American Sign Language

In the United States, deaf and hearing impaired individuals routinely use ASL to communicate (Padden & Humphries, 1988). ASL is a language distinct from any other language, spoken or signed. Some may assume that it is merely a system that is used to translate English into a series of hand gestures. This could not be further from the case. ASL has its own grammar and its own syntactical and morphological rules. It has its own vocabulary or lexicon, which consists of manual gestures involving either one or both hands. ASL dates back to about 1817 in the United States when Thomas Hopkins Gallaudet established a school for the deaf in Connecticut. He had traveled to France and learned about Old French Sign Language. Today, ASL shares about 60% of its vocabulary with French Sign Language; however, sign languages in other countries have little overlap with ASL or each other. Consequently, users of signed languages from different countries would not be able to understand one another. Also, users of the same signed language who reside in different parts of the same country may be able to detect differences in each other's language use that could be considered a **dialect** or variant of the same language. Users of signed languages who have learned the language from early childhood can also detect dysfluencies in users who learned signed language relatively late in life and who perform differently than those who are fluent. Finger spelling, as shown in Photo 1.3, is used primarily for spelling names or when one would like to emphasize a word. Some words in ASL are iconic in that they have the appearance of the concept to which they refer. For example, the sign for *tree* involves an arm held perpendicular to the floor with the hand and figures moving as a tree might move in the wind. However, the formation of most signs has no physical relationship with the meaning of the word to which the sign refers (Klima & Bellugi, 1979).

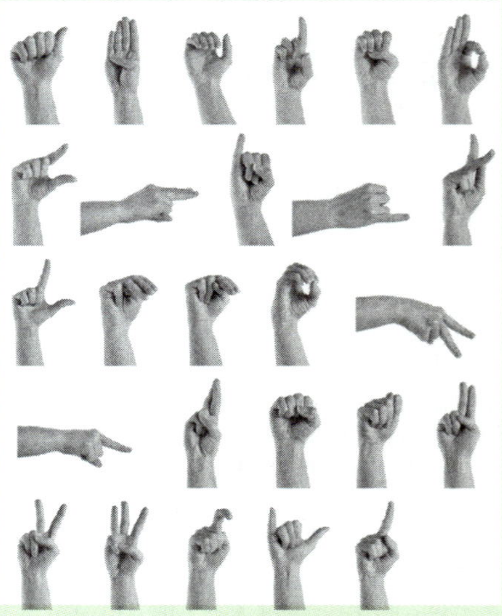

▲ **Photo 1.3** In ASL, one uses finger spelling for names or to emphasize words. The signs in the photo represent the letters of the English alphabet. Can you spell your name using ASL?

Phonological rules in a language may also result in predictable changes in how sounds are articulated in particular environments (Hayes, 2009). In English, one phonological rule is the *flap rule*. The phoneme /t/ is pronounced differently depending on what sounds are around it in a word. The phoneme /t/ as in the word *rent* is pronounced as a /t/ in the word *renter*, as it is when the root *wilt* is used to make the word *wilted* and the word *blurt* is used to make the word *blurted*. However, the way the /t/ is pronounced changes to more of a /d/

sound when *write* becomes *writer, boat* becomes *boating,* and *loot* becomes *looted.* Although most speakers of English are completely unaware that the flap rule is being applied, the evidence is in our speech. In the case of /t/ the sound is articulated more similarly to /d/ when a vowel occurs before and after it in a word.

Semantic Rules

Grammar rules pertaining to the meaning of words and sentences are called **semantic rules.** One of Chomsky's most famous examples—*Green ideas sleep furiously*—illustrated that sentences that are well formed syntactically can be meaningless (Chomsky, 1957). Speakers of the same language interpret words and sentences similarly. For example, in English, one understands the word *coincidence* similarly, despite not thinking too much about it or looking it up in the dictionary. If you ask a speaker of English to tell you the definition of the word, after a while, she might be able to give you something similar to this: *The co-occurrence of two low probability events that are related in meaning in some way that is important to the observer*. I once witnessed my 8-year-old niece use the word completely correctly while we were out shopping in a small town and had run into a family friend multiple times that day in different locations. I was quite sure that no one in the family had schooled my niece on the proper use of the word, and the word was not likely to have been on her vocabulary lists at school.

Semantic rules are also involved when we interpret the sentences *I can't dance* and *I can't draw* as meaning *I have no dancing ability* and *I am not good at drawing,* rather than *I am physically unable to move around as music plays* or *I am unable to hold a pencil or pen and draw lines on a page.* Also consider this exchange: *Do you want to go to lunch? No, I already ate.* Semantic rules explain why all speakers of English interpret the sentence *I already ate,* as meaning *I already ate some sort of food* rather than *I already ate a shoe or a brick.* Other semantic rules involve how we interpret the meaning of sentences in which words are omitted. In the sentence *Tyrone can play the trombone, and Cecilia can too,* readers generally settle on the interpretation that both Tyrone and Cecilia can play the trombone.

Understanding the rules of one's own language on a conscious, analytical level can be especially difficult, because when we learned these rules as children, we did so in an unconscious, unanalytical way. Fortunately, there are scholars who analyze the patterns found in language for the purpose of discovering the rules that may be at work. These scholars are called linguists; their field of study is **linguistics,** which is the scientific study of language and languages. Their work is similar to, yet different from, that of psycholinguists, who study how language is used in everyday life, in speaking, listening, writing, etc. Their field of study is called **psycholinguistics.** What is known about how the languages of the world differ has been discovered by linguists.

Why Are Some Human Languages So Similar?

Since the 18th century, linguists have compared languages and noted differences and similarities. Similarities and differences across languages can be the result of language history. Language tends to change over time (Lass, 1997). Imagine that a group of people speaking the same language decided to split into two groups with one group setting off from the village to relocate. Imagine that 100 years pass. Over those 100 years, the language spoken by the two groups would have changed. The changes occurring within each group could be very different. If the speakers of the two groups were once again reunited, each might find the language spoken by the other to be different from their own language. They might view the other language as a strange dialect, somewhat similar to their own language but different.

Evidence of language change can be found even within a language. Many words in English have undergone changes in meanings over time (Barnhart, 1988). For example, the word *worm* used to mean serpent or dragon but came to refer to a much less threatening creature. The word *deer* originally meant animal but came to refer to one particular type of animal. The word *clown* referred to a country peasant before it referred to a funny or foolish person.

When one finds two languages spoken by different groups of people who may live relatively far apart, it might be the case that the similarities that exist between the two languages can be explained by the two languages being historically linked. They may have descended from the same ancestor language. The term ***language family*** is used to describe a group of languages that descend from the same ancestor language. For example, the **Romance languages** are a language family. They include Spanish, French, Italian, Portuguese, and Romanian, as well as many others. The ancestor language that they all share is Latin. There are similarities in vocabulary and grammar rules among these languages. Table 1.5 displays vocabulary items for five modern Romance languages.

English belongs to the West Germanic language family, which also includes German, Dutch, Yiddish, and others. The similarities in vocabulary for these West Germanic languages can be seen in Table 1.6.

Comparing the words in these two language families, one definitely finds some striking similarities. For example, the words seem to start with similar sounds and have similar syllable structure. It is not a coincidence; rather, the forms of the words in each language are directly related to the form of the equivalent word in the ancestor language. The Romance languages and the West Germanic Languages both belong to the larger **Indo-European language** family, which means that one would expect to find some similarities between these types of languages, such as the basic SVO word order. The diagram in Table 1.7 illustrates the wide variety of languages that belong to this language family.

Table 1.5 Translations for the Words *Mother, Father,* and *Night* in Five Romance Languages

Language	Mother	Father	Night
Spanish	Madre	Padre	Noche
Italian	Madre	Padre	Notte
French	Mere	Père	Nuit
Portuguese	Mãe	Pai	Noite
Romanian	Mamă	Tată	Noapte

Note: Examples are from http://translate.google.com/

Table 1.6 Translations for the Words *Mother, Father,* and *Night* in Four West Germanic Languages

Language	Mother	Father	Night
English	Mother	Father	Night
German	Mutter	Vater	Nacht
Dutch	Moeder	Vader	Nacht
Yiddish	Muter	Foter	Nacht

Note: Examples are from http://translate.google.com/

Table 1.7 The Indo-European Language Family

Indo-European

Italic	Germanic	Greek	Balto-Slavic	Indo-Iranian	Armenian	Celtic
Latin	German	Slavic	Baltic	Iranian	Indic	Irish
	Dutch	Russian	Lithuanian		Sanskrit	Welsh
Spanish	English	Bulgarian	Latvian	Persian	Hindi	Scottish Gaelic
French	Danish	Serbo-Croatian		Kurdish	Urdu	Breton
Italian	Swedish	Polish		Pashto	Bengali	
Portuguese	Icelandic	Czech		Punjabi		
Romanian						

Source: Adapted from Comrie and colleagues (2003).

The First Language

There is no consensus about the question of whether all human languages descended from a single ancestor language or whether there were multiple ancestor languages. One can find evidence supporting each possibility. Recent research involving DNA provides the most compelling evidence so far to support the view that all of the languages spoken in the world today descend from a single group of individuals. The research traced the female ancestry of humans today to a single ancestor who lived approximately 160,000 years ago. This ancestor has been referred to as Mitochondrial Eve. The term *mitochondria* refers to the part of human cells inherited only from one's mother. This research analyzing human mitochondrial DNA suggests that approximately 70,000 years ago, the world's population dwindled to around 15,000 individuals (Dawkins, 2004). There is geological evidence to support the idea of a historical dwindling of the human population. Geological data indicate that a massive eruption of the Toba volcano in Indonesia occurred around this time and could have caused extensive environmental changes (Chesner, Westgate, Rose, Drake, & Deino, 1991), including in a 10-year volcanic winter that would have likely reduced the availability of food and caused a reduction in the population (Robock et al., 2009).

Linguists who are skeptical that all languages can be traced back to the same ancient language ancestor point to evidence showing that there is a small number of languages that appear to be so different from other languages that they do not fit into the family tree of languages. Such languages are called **language isolates.** The Basque language is the most common example of a language isolate. Basque is spoken in the Pyrenees Mountains in southwestern France and northeastern Spain (Comrie et al., 2003). Thus far, attempts to link Basque historically to other language families have not been successful. However, it has been suggested that the Basque may descend from a language that was spoken in Western Europe prior to the influx of Indo-European speakers in that region (Trask, 1996). It is possible that Basque descended from the same language as Indo-European and other language families, but there are no historical records enabling that connection to be confirmed.

Research may not be able to definitely answer the question of whether there was only one original human language from which all human languages descended. Recent studies have used DNA testing of populations of people to determine whether speakers of similar languages are also more genetically similar than groups who speak less similar languages (Behar et al., 2012). Computer models have been used to try to estimate the probable geographical locations of ancient ancestor languages (Bouckaert et al., 2012). However, if it could be proven that new human languages can be created without having a direct connection to any other language, then there would be support for the view that multiple ancestor languages could have existed in the past. Text Box 1.2 describes recent research investigating a new signed language by a community of deaf children in Nicaragua. Many researchers believe that the language is an example of a completely new human language.

Text Box 1.2 Research Discovery: Nicaraguan Sign Language

Some researchers believe that strong evidence against the view that all languages descend from the same ancestor language has been found in Nicaragua. Researchers claim that a group of deaf children have developed their own language (Kegl, Senghas, & Coppola, 1999). In 1980, deaf children in Nicaragua began to attend school in Managua. The school emphasized spoken language and taught only finger spelling of words. Children did not progress in developing spoken language skills at school; however, outside of school, children began to use more and more gestures with one another. In 1986, teachers realized that students were communicating with one another with signed language, while they were unable to communicate with them. The Nicaraguan Ministry of Education invited Judy Kiegl, an ASL expert from MIT, to analyze the students' language. She concluded that a new language had been created, one with the regularity of grammatical structure found in other human languages. A more recent example is a case of sign language developed by deaf children in Bedouin communities in the al-Sayyid Bedouin tribe in the Negev desert of southern Israel (Senghas, 2005).

The debate about the historical origins of human language may seem to be relatively unimportant when considering the topic of language development. The topic is related to this broader question: What is the nature of human language? This question is relevant when theories of language and language development are considered.

Theories of Language Development

Multiple theories of language development have been proposed. Each of the theories has its strengths and weaknesses. Thus far, there is no consensus that one does a better job than the others in accounting for the research results that have been obtained. Consequently, all share the same weakness; currently, they do not provide a complete account of how language development occurs. In this book, we will focus on four theories: (1) the behaviorist approach, (2) the generative approach, (3) the statistical learning approach, and (4) the social-interactionist approach.

What Is the Behaviorist Approach to Language Development?

For the first half of the 20th century, the field of psychology was dominated by behaviorism, the school of psychology that studied only observable behaviors and claimed that all knowledge was acquired through learning (Baum, 2005); thus, no knowledge was believed to be innate. Early research showed how animal behavior could be learned through classical conditioning (Pavlov, 1927/1960). In studies with dogs, Pavlov showed that if a bell was rung each time food was presented to the animal, then after several repetitions, the dog would come to salivate in anticipation of the food anytime the bell was rung, even when the food was not presented. The dog had come to associate the stimulus of the bell with the response of salivation.

Around the same time, John Watson (1913), who established behaviorism with his manifesto *Psychology as the Behaviorist Views It*, showed that humans also come to produce behaviors learned through classical conditioning. The study is known as *The Little Albert Experiment* (Watson & Raynor, 1920). A child, *Albert* (not his real name), was placed on a mattress with a white lab rat. He showed no fear of the animal. Later, each time the child touched the rat, Watson or Raynor struck a pipe with a hammer, making a loud sound that startled the child. Albert cried with fear. After several repetitions, Albert had come to produce the fear response anytime the rat was presented.

In later behaviorist developments, the American psychologist B. F. Skinner (1953) showed that frequency of voluntary behaviors could be affected by environmental forces. His theory of behavior is referred to as operant conditioning. He showed voluntary behaviors or operant responses would occur more frequently when reinforced by something applied in the environment. For example, if a pigeon pressed a lever and a food pellet was delivered, the pigeon would tend to press the lever more. The food pellet reinforced the behavior of pressing the lever, and lever pressing increased. Skinner also showed that voluntary behaviors would occur less frequently when a punishment was delivered. If a pigeon pressed a lever and then received an electric shock, the pigeon would tend to press the lever less often, if at all. The electric shock punished the behavior of pressing the lever, and lever pressing decreased.

In 1957, Skinner published *Verbal Behavior* in which he explained his view of how language behavior was brought about through learning processes. He emphasized the roles of imitation and reinforcement in language learning. While the details of the theory are complex, the core notion was that each utterance that a child produced or comprehended was produced by a prior learning episode in which that particular response had voluntarily occurred at some point and was reinforced by an external force. Caregivers are the most likely environmental reinforcers. When an infant produces a vocalization, a mom and dad may reinforce the infant with a physical caress or a pleasing smile.

B. F. Skinner probably had not foreseen that his book would mark a turning point in the field of psychology, but, indeed, it did. By the 1960s, behaviorism was losing steam as psychology's most popular approach to important questions. When Noam Chomsky published a critical review of B. F. Skinner's (1957) *Verbal Behavior* in 1959, it received a lot of attention. Many point to the review as one of the events that led to the cognitive revolution in psychology, when researchers began to turn their attention to previously ignored topics related to unobservable behaviors, such as human thinking, memory, and language (Mandler, 2007).

What Is the Generative Approach to Language Development?

Noam Chomsky is recognized as the founder of the generative approach to language and language development. He is a linguist at the Massachusetts Institute of Technology (MIT), a position he also held in 1959. In the 1959 review, Chomsky asserted "that the insights that have been achieved in the laboratories of the reinforcement theorist, though quite genuine, can be applied to complex human behavior only in the most gross and superficial way" (p. 143). He argued that when children acquire a language they acquire knowledge of the grammar of the language. "One who has mastered the language [can] distinguish sentences from non-sentences, . . . understand new sentences (in part), [and] . . . note certain ambiguities" (p. 57). He pointed out that one who has learned a language is constantly coming across new sentences, which are not familiar because of some prior learning episode, and can understand them without difficulty.

In 1965, Chomsky published *Aspects of the Theory of Syntax,* in which he described his theory that children are born predisposed to acquire language. This view is referred to as **nativism,** or the nativist approach. It assumes that nature, rather than nurture, holds the key to the mystery of language development. He initially proposed the existence of a **language acquisition device** (LAD), an organ of the brain responsible for language development. He argued that the innate knowledge contains **universal grammar** (UG), the rules shared by all human languages. UG explained why all infants, regardless of where they are born in the world and regardless of the languages to which they are exposed, acquire language relatively effortlessly. Later, the LAD was abandoned as the mechanism of language development in favor of a process in which language features or parameters are turned on or off during development in response to language input experienced in the environment. Ultimately, the rules that apply to the language being acquired are turned on or selected. The rules that apply to all of the other possible human languages are turned off or not selected.

In terms of the nature–nurture debate, which is well known in psychology, the generative approach represents an extreme *nature* approach to language. In contrast, the behaviorist

approach to language represents an extreme *nurture* approach. In the 1960s and 1970s, the generative approach gained a great deal of attention and support because its emphasis on innateness was compatible with others' observations. For example, the view is compatible with the view that the human mind is composed of specialized cognitive modules (Fodor, 1983), a view that is referred to as **modularity.**

The generative approach is also consistent with the view that language in humans may be subject to developmental critical periods as are behaviors in other species. Penfield and Roberts (1959) were the first to propose that humans, like other animals, experience a critical period for learning. In animals, critical periods have been found for birds. The term *imprinting* has been used to describe how a newly hatched bird, such as a gosling or duckling, will follow the first animal encountered (Spalding, 1873). Songbirds experience a window of time when learning the song of their species is possible; if a bird is not exposed to the song during the critical period, the song will not be learned (Nootebohm, 1969). Hubel and Wiesel (1963) demonstrated that there is a critical period for the development of vision in cats between birth and 3 months of age.

In the 1960s, Eric Lenneberg (1964, 1967) promoted the idea that in humans, language was a species-typical behavior. He pointed to the fact that language appears across all human societies and follows a set pattern of development across languages. He also believed that the complexity in human language could not be learned. The view has come to be called the **critical period hypothesis.** He proposed that if language is not learned during the critical period, it cannot be learned. Lenneberg's ideas were embraced by advocates of the generative approach. In particular, Steve Pinker (1994) named one of his most popular books *The Language Instinct,* which emphasized the evidence that language is biologically based and innate in ways that are similar to the instinctual behaviors of other animal species. Two types of evidence are used to support the view that there is a critical period for language. The first involves second language acquisition. Mastering a second language is more likely if one begins learning the second language during childhood versus after puberty (Singleton, 2005). The second involves cases of individuals who fail to learn a first language in childhood—usually because of adverse environmental experiences, such as severe neglect and/or abandonment by parents. The term ***feral children*** has been used to refer to such cases. Text Box 1.3 describes how the cases of feral children have provided support for the critical period hypothesis.

What Is the Statistical Learning Approach?

The statistical learning approach to language development asserts that language is learned in the same way that all human learning occurs (Saffran, 2003; Saffran, Aslin, & Newport, 1996; Seidenberg & McClelland, 1989). Advocates of this approach do not believe that there is anything about human language that makes it special or different from other human skills, such as tying our shoes or learning math. The same general cognitive mechanisms are involved in learning all skills. The emphasis on learning resembles the behaviorist approach in its extreme *nurture* approach to language (i.e., nothing is innate, everything is learned). The two approaches differ in the mechanisms presumed to be involved in the learning. Unlike the behaviorist approach, which presumed that conditioning principles led to language acquisition, the statistical learning approach presumes that human learning

Text Box 1.3 Extraordinary Individuals: Feral Children

The strongest evidence for a critical period for language includes cases of children who have passed puberty without adequate exposure to human language from caregivers. Perhaps, the best-known case of a feral child is Victor, the Wild Boy of Aveyron, who was discovered living in the forest near Saint-Sernin-sur-Rance, France in 1797 (Lane, 1976). His story was told in Louis Truffaut's (1970) film *L'Enfant Sauvage* (The Wild Child). A second famous case is that of a 13-year-old American girl who was rescued from an abusive home in which she lived in 1973. She lived much of her life up to the time of her rescue tied to a potty chair. In publicized accounts of her story, she was referred to as Genie (not her real name). Despite years of language instruction, neither Victor nor Genie was able to gain grammatical fluency in a language. Each was able to acquire a sizable vocabulary; however, they failed to master the syntactic and morphological rules of the language.

Unfortunately, there continue to be children who experience physical and social deprivation during early childhood who provide new case studies relevant in the debate of whether there is a critical period for language. In 1991, the case of Oxana Malaya received publicity (Grice, 2006). She was an 8-year-old girl living with a pack of dogs in the Ukraine. Her parents were addicted to drugs and refused to allow her to live indoors. When she was discovered, she displayed many behaviors observed in dogs, such as barking and growling. She was taken to a clinic in Odessa where she was cared for, studied, and taught human behaviors, including language. She has been unsuccessful in mastering the grammatical structure of language. She currently lives in a home for intellectually disabled adults.

involves sensitivity to statistical regularities in our experiences, including perceptual experiences. From birth, humans are exposed to massive amounts of information from which statistical information can be extracted and used to develop knowledge structures and processing strategies.

Advocates of the statistical learning approach have used computer models to demonstrate that language is learnable. Early examples of computer models were called parallel distributed processing (PDP) models (Elman et al., 1996; Rumelhart, McClelland, & the PDP Research Group, 1986). They are also referred to as "neural networks" or "connectionist models" (Friedenberg & Silverman, 2005). The logic of this approach is that if computer models can be created that learn either natural languages or simpler artificial languages in a manner that is similar to how children learn language then there is no need to assume that some knowledge of language has to be innate in children in order for language learning to occur. Studies involving computer models generally focus on a single aspect or pattern of language and attempt to show how a computer can eventually learn to reproduce the pattern with many learning trials and feedback regarding the correct target.

What Is the Social-Interactionist Approach?

The social-interactionist approach to language development is similar to the behaviorist and statistical learning approaches in that it also emphasizes experience as the primary way that children acquire language. Unlike the behaviorist and statistical learning approaches, the social-interactionist approach focuses on the role that social context and social relationships play in development. The approach is most closely associated with Vygotsky (1978, 1987). The process of learning for the child involves assistance from a knowledgeable other, such as a parent or other adult. The adult–child dyad engages in learning activities that may progress from simple skills to more complex skills over time. The adult is viewed as playing a central role in teaching the child language through interaction. However, an emphasis is placed on the function of language, rather than on the grammatical structure of language.

In the chapters ahead, you will find that there are aspects of language development that fit well with each of the major theoretical approaches. As you progress through this book, it will be important to consider each piece of evidence and to determine which theory the evidence supports. This type of critical evaluation of the evidence is exactly what language development researchers do in their everyday work and is what will ultimately lead to the development of a singular, comprehensive, and accurate theory.

Methods Used in the Study of Language Development

The modern study of language development is carried out by scientists working in the interdisciplinary field of cognitive science (Mandler, 2007). The field of cognitive science includes a diverse group of disciplines, including psychology, linguistics, neuroscience, computer science, and philosophy (Friedenberg & Silverman, 2005). Linguists are those who study the structure, function, and history of languages. Psychologists are those who study mind and behavior. Psycholinguists are those psychologists who study how language is used in daily life. Neuroscientists study how biological processes bring about thinking and intelligence in humans and animals. Computer scientists attempt to create machines with humanlike intelligence and also try to simulate human intelligence with computer models. Philosophers explore the relationship between the mind and body and the nature of knowledge. As the field of cognitive science becomes more complex through the use of technologically advanced equipment, such as devices that can image the brain during processing, one can find more and more interdisciplinary research collaborations.

Research in language development has relied on a variety of methods over the years. The earliest methods were purely observational. The changes that occurred in children's bodies and abilities were noted as they occurred. Observational methods also included wide-scale efforts to quantify the variation in children's rate of growth and development of skills. Starting in the 1970s, researchers began conducting experiments to determine what children can perceive and remember. The experimental techniques that have been developed now represent the mainstream in language development research.

What Observational Methods Are Used to Study Language Development?

Baby Biographies

Many are surprised to learn that one of the first researchers to focus on child development was none other than Charles Darwin (1809–1892), the naturalist most famous for finding and describing the evidence for natural selection. He wrote the first **baby biography,** which is a detailed diary of observations of a child's behavior. Darwin published the baby biography for his oldest child in 1877. It was titled *Biographical Sketch of an Infant.* His detailed descriptions of the child's changing abilities over the first 3 years of life are compelling to read even today. He described the following: "When our infant was only 4 months old, I thought he tried to imitate sounds; but I may have deceived myself, for I was not thoroughly convinced that he did so until he was 10 months old" (Darwin, 1877, p. 293). Darwin also describes how at 12 months of age, the child invented a word for food: *mum.*

The Swiss developmental psychologist Jean Piaget (1896–1980) and his wife, Valentine, also created baby biographies for their three children during their first 3 years of life (Piaget, 1957). Piaget is well known for being a pioneer in cognitive development, which will be discussed in greater detail in a later chapter. When writing the baby biographies, he would manipulate the environment and record how the children would respond. He conducted hundreds of manipulations and carefully recorded how the children responded. Ultimately, the observations that he recorded in the baby biographies provided key data for his later theory about how children's thinking changes over time. One of Piaget's observations was that children develop an understanding that objects no longer in view still exist (i.e., object permanence). Young children, who do not yet have object permanence, will not search for an object when it is taken and hidden from view. Older children with object permanence will actively search for an interesting object that was taken away and hidden.

The topic of baby biographies would not be complete without a discussion of the contribution of Milicent Shinn, who was the first woman to receive a PhD at the University of California in 1898 at the age of 40. In 1900, she published the book *The Biography of a Baby,* which contains 2 years' worth of observations of her niece Ruth's physical and emotional development. In developmental psychology, her book still remains on the list of classic readings. In 1908, she published *The Development of the Senses in the First Three Years of Childhood.* After receiving her doctorate, there were few opportunities for female researchers. She devoted herself to caring for her aging parents and later to educating her brother's children.

Gesell's Normative Studies

In the 1920s, researchers began to conduct large-scale surveys detailing the characteristics and abilities of children. Such surveys are called normative studies because the large samples enable researchers to distinguish characteristics and abilities that are frequent or normal from those that are infrequent or abnormal. A pioneer in this effort was Arnold Gesell, who published numerous books documenting the milestones of child development. In 1925, he published *The Mental Growth of the Preschool Child.* Gesell's goal was to provide both parents and teachers with information that might allow them to conclude that a child was performing similarly to what is typical for children of that age (Gesell, 1948). Some of the dimensions on which children were measured included motor control, such as coordination, as well as standing, walking, and drawing; language, such as vocalization ability, speaking, and listening; personal–social abilities, which included general social behavior and personality; and adaptive behaviors, which involved how well children could adjust to changes made in the physical environment where they were tested.

In order for normative data to be most useful, they should be drawn from large samples. Small samples may prevent a researcher from observing unusual cases of development. For example, if an abnormality occurs in only 1% of children, then one would need to survey more than 100 children to have a good chance of observing just one instance of the abnormality. If a researcher would like to have multiple cases of a particular abnormality, then a much larger sample size is needed. One of the earliest large-scale normative studies was conducted by Templin (1957). She interviewed 480 children between the ages of 3 and 8 and documented their pronunciation ability, listening ability, and vocabulary knowledge. The results showed that children from higher socioeconomic levels performed better than children from lower socioeconomic families. Unlike prior research (McCarthy, 1930), she did not find performance differences for girls and boys.

Normative studies continue to be conducted. Because the words that typical children learn might vary across decade, it is important to conduct normative studies of vocabulary knowledge among each generation of children. It is also important to conduct normative studies of children who share important characteristics that might influence their development and/or language acquisition. Collecting local norms can also be valuable, as there may be local, cultural factors that make large-scale, national norms less reliable in predicting performance. Collecting norms for special populations is also valuable. Researchers document the typical performance for children who are blind (Brambring, 2007) or children who have been adopted from a foreign country (Glennen, 2007), among others.

Today, there are normative data for just about anything that you could measure about a child. At birth, parents are likely to be told how their infant's weight, body length, and head circumference compare with those of other infants. The information is typically provided in the form of a percentile, which reflects the percentage of others that one is heavier than, longer than, etc. When an infant is described as in the 95th percentile for weight, he or she weighs more than 95% of other infants of the same age. When an infant is described as in the 10th percentile for length/height, then he or she is shorter than 90% of other infants the same age, because he or she is only longer/taller than 10% of other infants the same age. At each doctor's visit, parents might be updated regarding the extent to which their child's characteristics are similar to most children of that age. When children do not reach milestones, such as walking and talking, on time, there may be some concern that the child is not developing normally.

Peabody Picture Vocabulary Test

In 1959, the husband–wife team of Lloyd M. Dunn and Leota M. Dunn developed a vocabulary measure. They were specialists in special education. The task involves showing the child an array of four pictures and asking the child to point to the picture that refers to a target word. The Peabody Picture Vocabulary Test (PPVT) has undergone extensive study and revision over the years (Dunn & Dunn, 2007). It can be used for children as young as 2. It can also be used with adults, even those of advanced age. Those educators and clinicians who work with special populations also can use it to assess the vocabulary knowledge of children with various developmental syndromes and disorders. Because of its reliance on picture identification and spoken targets, it is not useful for the testing of individuals who are blind or deaf.

MacArthur–Bates Communicative Development Inventory

Language researchers recognized the need to assess a broader range of language skills in children than could be assessed in vocabulary tests. In 1993, researchers at the University of Southern California published research describing the Communicative Development Inventory (CDI), which assessed not only the development of vocabulary but also the development of grammar and gestures (Fenson et al., 1994). The original CDI was designed for the testing of children between the ages of 8 months and 2.5 years. Recent updates of the CDI provide information about the typical abilities of children between the ages of 8 months and 6 years. Included in the CDI are a variety of subscales, each designed to assess a different aspect of development and/or communication. The subscales include assessments for expressive language, comprehension of language, letter knowledge, number knowledge, social skills, and motor skills, as well as others.

Child Language Data Exchange System

The Child Language Data Exchange System (CHILDES) is an Internet-based repository for child language samples (MacWhinney, 2000). The system was created by Brian MacWhinney and Catherine Snow in 1984. Researchers and parents around the world are invited to submit transcripts and recordings of child language. The sample submitted may then be analyzed by researchers anywhere in the world. Currently, there are language samples from over 20 languages available for analysis. Samples from special populations, such as children with developmental syndromes, can also be submitted. CHILDES has been a major resource for research on child language. So far, there are over 3,100 research publications that rely on CHILDES data. The topics that are covered include the entire range of topics that you will find covered in this book including phonological development, development of morphology and syntax, and the use of figurative language.

What Experimental Methods Are Used to Assess Children's Cognition?

In the infant's first year of life, the infant does not make a very good interview subject. Finding out what infants know is not an easy task. Researchers now have a wide variety of techniques that have proven useful in discovering what is occurring in the minds of infants. Starting in the 1970s, researchers have relied on a small set of experimental techniques to infer what infants know and how they process aspects of their environment. These techniques include (1) the **head-turn technique**; (2) the **habituation paradigm**; and (3) the **preferential looking paradigm**.

Head-Turn Technique

Among the first used is the head-turn technique (Eilers, Wilson, & Moore, 1977). It was shown that infants could be trained to respond a certain way, such as to turn their heads. Infants will typically show interest in a toy, such as a monkey. The toy is used as a reward, being shown to the infant only when an infant produces a desired response, such as turn of the head when a different sound occurs in a series, such as /pa/, /pa/, /pa/, /ba/. Once the infant is trained to respond when a different sound is presented, the researcher can play any combination of sounds to the infant in order to determine which sounds the infant can discriminate (i.e., what sounds are perceived as different types of sounds to the infant). The training is essentially conditioning; sometimes the technique is referred to as the conditioned head-turn technique. The technique is most commonly used with infants between the ages of 5 and 18 months but can be used with younger infants as well as older children and adults (Werker, Polka, & Pegg, 1997).

Habituation Paradigm

Other techniques take advantage of a phenomenon known as habituation. People and animals tend to respond less and less when an object or a sound is repeated. In research with infants, when researchers present an object or sound to an infant, physiological responses, such as heart rate and sucking, tend to increase. As the object or sound is presented again and again, the physiological responses decrease, presumably because the infant has become accustomed to the stimulus and is growing bored. Because of the predictability of habituation processes, researchers can determine whether an infant treats an object or sound as new or as familiar by observing how the infant responds to the object or sound over time. A familiar object or sound will be responded to with less intensity than a new object or sound. Researchers can also establish whether infants experience any two objects or two sounds as different from one another. When infants treat an object as something new, they experience dishabituation, which typically results in greater interest in the object and increased responding.

In research conducted in the late 1960s and throughout the 1970s, numerous studies measured the rate of high-amplitude sucking to determine infants' ability to respond to objects or sounds (see Jusczyk, 2000, for review). Using this technique, a special pacifier is used with which the sucking rate of the infant can be recorded. The infant is monitored to identify the initial rate of sucking. When a stimulus is presented, the infant will begin to suck faster. As the infant habituates to the stimulus, the sucking rate will slow, returning to the baseline rate. A second stimulus is presented. If sucking rate increases, then the researcher can conclude that the infant considers the second stimulus to be something new, which means that the infant appreciates that the second stimulus is different from the first.

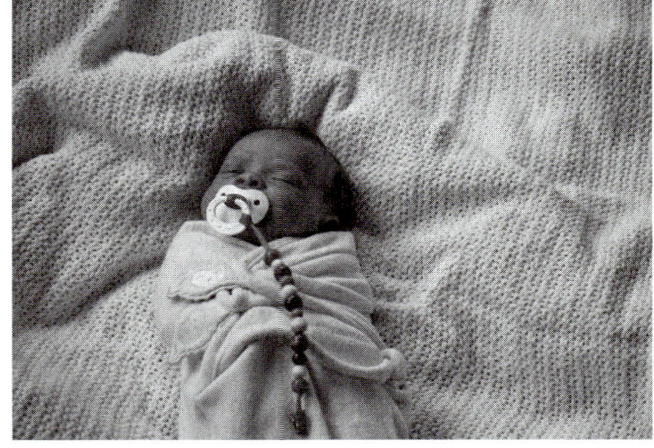

▲ Photo 1.4 Research with very young infants measures their sucking behavior with modified pacifiers that can record sucking rate. Which type of processing in infants can be studied by measuring sucking rate?

However, if the sucking rate does not increase, the researcher can conclude that the infant considers the second stimulus to be the same as the first stimulus.

Infants' pattern of responding during habituation led to researchers documenting convincingly that infants are forming memories of their environment very early in life. Rovee-Collier and colleagues investigated infants' memories by measuring the rate of an infant's kicking when lying in a crib (Rovee-Collier, 1997; Rovee-Collier, Hayne, & Colombo, 2001). In their studies, infants were placed on their back in a crib. One end of a ribbon was tied to their ankle, and the other end was tied to a mobile positioned above the crib. If the infant kicked, the mobile would move. Infants' kicking rate showed the pattern of habituation. When infants were placed into a crib decorated with a new mobile and other new and unfamiliar accessories, infants kicked more than when they were placed into a crib they had visited before. A series of careful studies varied the delay between experiencing a specific crib and also the number of familiar accessories that were in the crib. They showed that the rate of kicking was directly related to how different the crib was from the last testing session. The studies proved what many parents might have already guessed: Even in the first months of life, infants are forming memories of what is going on around them.

Preferential Looking Paradigm

Infants can also reveal what they know by how they look around their environment. Roberta Golinkoff, Kathy Hirsh-Pasek, and colleagues (Golinkoff, Hirsh-Pasek, Cauley, & Gordon, 1987; Hirsh-Pasek & Golinkoff, 1996) used a technique called the preferential looking paradigm, which involves having an infant seated on a caregiver's lap and facing two television screens. During the initial phase, the infant is shown pictures of two objects one at a time (i.e., displayed on both television screens). The infant's looking time is measured to establish a baseline level of looking. During the test phase, objects appear simultaneously, one on each television screen. When the objects appear, the child hears the word describing one of the objects. Researchers observe how long the infant looks at each object. Infants may be videotaped and their looking time coded at a later time or observers can be positioned behind a wall containing the television screens. If the infant looks longer at the object that is named, it is concluded that infant knows the word or associates the word and object as more similar than the word and the other object. The paradigm has also been used to show that infants who are just beginning to produce their first words understand key aspects of sentence meaning.

Summary

Regardless of where a child is born and what language the parents speak, the child acquires language rapidly, mastering the basics between 36 and 48 months. In order to appreciate fully the mystery that is language development, one must understand how the languages of the world differ. One must keep in mind that the similarities and differences that exist across the world's languages are related to the historical facts of where languages started and how they changed over time. The major theories of language development differ with regard to whether language is viewed as aided by innate knowledge or learned completely through direct experience with the environment and caregivers. As the field of language development has grown over the past four decades, researchers have developed a variety of creative methods to study children's language ability and their cognitive processing. Although researchers working today do not agree about how exactly language develops, all would agree that the mystery of language development is worth solving.

KEY TERMS

American Sign Language (ASL)
babbling
baby biography
consonants
cooing
critical period hypothesis
dialect
feral children
grammar
habituation paradigm
head-turn technique
Indo-European language
isolating languages
language acquisition device (LAD)
language family
language isolates
lexicon
linguistics
modularity
morpheme
morphological rules
nativism
phoneme
phonological rules
pragmatic rules
preferential looking paradigm
psycholinguistics
Romance languages
semantic rules
syntactic rules
synthetic languages
universal grammar (UG)
vowel

REVIEW QUESTIONS

1. Approximately how many human languages are still spoken in the world today? How has that number changed over the past 100 years?
2. What is grammar? What are the different types of language rules that are involved in a language's grammar?
3. Discuss Saussure's view that human languages can be described as a set of rules, similar to games such as chess.
4. How are similarities between languages related to the history of those languages?
5. What are two important ways in which the languages of world differ?
6. What are the major theories about how language develops?

7. What is the LAD? Who proposed its existence, and what role was it viewed as playing in language development?
8. How does the behaviorist approach explain how language acquisition occurs?
9. What is the critical period hypothesis? What evidence supports the critical period hypothesis?
10. What is a feral child? How are cases of feral children used as evidence for or against different theories of language development?
11. How does the statistical learning approach explain how language acquisition occurs?
12. What is a baby biography? Who was the first researcher to create a baby biography?
13. What are normative studies of child development? Why are they conducted?
14. How is the PPVT conducted? What is the test used to measure?
15. What are some of the methodologies that have been used to study language development?
16. How are the preferential looking paradigm and the head-turn technique used to study what infants know?
17. What is habituation, and how is it used by researchers who study what infants know?

RECOMMENDED READING

Chomsky, N. (1986). *Knowledge of language: Its nature, origin, and use.* New York: Praeger.

Chomsky, N. (2006). *Language and mind.* Cambridge, UK: Cambridge University Press.

Comrie, B., Matthews, S., & Polinksy, M. (Eds.). (2003). *The atlas of languages: The origin and development of languages throughout the world.* New York: Checkmark Books.

Curtiss, S. (1977). *Genie: A psycholinguistic study of a modern-day "wild child."* New York: Academic Press.

Janda, R. D., & Joseph, B. D. (Eds.). (2004). *The handbook of historical linguistics.* London: Blackwell.

Labov, W., Ash, S., & Boberg, C. (2006). *The atlas of North American English.* Berlin: Mouton-de Gruyter.

Lane, H. (1976). *The wild boy of Aveyron.* Cambridge, MA: Harvard University Press.

McWhorter, J. (2009). *Our magnificent bastard tongue: The untold history of English.* New York: Gotham.

Pinker, S. (2007). *The language instinct.* New York: Harper Perennial Modern Classics.

Trask, R. L. (1996). *The History of Basque.* New York: Routledge.

RECOMMENDED FILMS

Apted, M. (Director). (1994). *Nell* [Motion picture]. United States: Fox Home Entertainment.

Garmon, L. (Producer). (2007). *Secret of the wild child* [DVD]. *NOVA.* Boston: WGBH Educational Foundation. Available from http://www.pbs.org.

Lebrun, D. (Writer/Director/Producer), & Guthrie, R. (Producer). (2008). *Cracking the Maya code* [Television series]. *NOVA.* Boston: WGBH Educational Foundation. Available from www.pbs.org.

Truffaut, F. (Director). (1990). *L'enfant sauvage* [Motion picture]. France: MGM.

SUGGESTIONS FOR CLASS PROJECTS

1. Language change can be observed even within the span of several decades. Most libraries now have archives of newspaper writing from the past. Find an example of newspaper writing from 50, 100, or 150 years ago, and compare it with a similar type of contemporary newspaper article. In groups or individually, prepare an analysis of how the contemporary and historic articles differ in terms of word usage, sentence structure, and other matters related to writing style.

2. The relatedness of languages within the same language family can be easily observed in their vocabulary. Generate a list of 20 words that are likely learned within the first 5 years of life. Create a table in which you list the translations for those words in three languages from the same language family. Consult dictionaries that provide English translations. There are many searchable dictionaries on the Internet now with which one can translate most words in English into numerous other languages. Some suggested language combinations are any of the Romance languages, any of the Germanic languages, and any of the Balto-Slavic languages.

3. Explore the numerous morphological rules that exist in English. For the given list of suffixes (e.g., *-ion*, *-ful,* and *-ish*) or prefixes (e.g., *re-*, *un-*, *mis-*, and *de-*), generate 5, 10, or 15 examples of words that contain the morpheme. State the rule for the use of the morpheme (e.g., add *-er* to a verb to form a noun, which means the doer of the action denoted by the verb).

For additional ancillary resources, please visit the companion website at www.sagepub.com/kennison.

- Video Links
- Audio Links
- Web Resources
- Internet Activities
- Flashcards
- Web Quizzes

CHAPTER 2

THE BIOLOGICAL BASIS OF LANGUAGE

As long as there have been people, there has likely been an understanding that the contents of the head play an important role in the ability to speak and to understand language. From the historical record, we know that even the ancient Egyptians recognized that wounds to the head could result in problems with speech. An early historical account is contained in the Edwin Smith Papyrus, which dates back to the 16th century BC. It provides an account of an individual who sustained a wound to the temple and who was left unable to speak (Minagar, Ragheb, & Kelley, 2003). Writings from ancient Greece also refer to cases of language impairment following brain injury. In an account provided by Valerius Maximum (ca. 30 CE), a man from Athens who was identified as a scholar "lost his memory of letters" following a blow to the head (Benton & Joynt, 1960). In this chapter, we will learn about how our biological systems bring about language function. We will begin our discussion in the 1800s with studies of individuals who suffered from **aphasia,** which refers to any language impairment caused by brain injury. We will also examine how researchers investigated the functions of brain areas in the first half of the 20th century. Our discussion will include an in-depth look at the numerous brain-imaging technologies that are now used in both medical and research settings to map the brain. These techniques have produced a detailed understanding about the many types of aphasia that can occur. We will then look at recent genetics research, which has identified inherited language disorders and the corresponding genetic mutations. The research suggests that DNA may carry, at least in part, the building blocks of human language. We will end our discussion with a consideration of how language might have evolved.

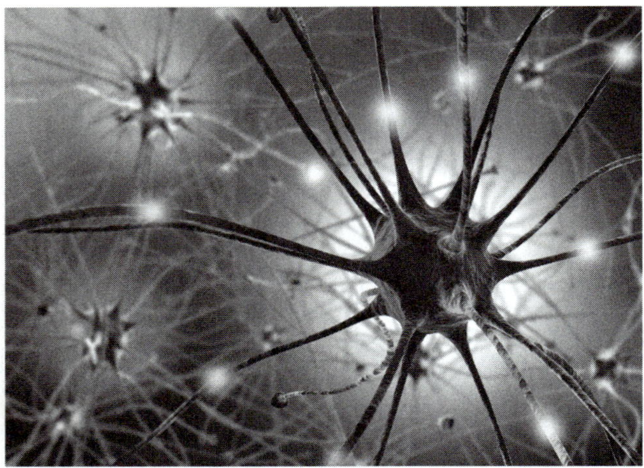

▲ Photo 2.1 This artistic representation illustrates the interconnected neurons. Have you ever thought about how many neurons would be involved in your saying *hello* to a friend?

Historical Studies of Brain and Language

What Is the Localization Hypothesis?

Today, it is generally accepted that specific brain regions are involved in specific functions (Thompson, 2000). The **localization hypothesis** refers to this view that specific brain locations perform specific functions. Among the first who believed in the localization hypothesis were the ancient Greeks. Aristotle believed that mental functions could be traced to the heart (Aristotle & Barnes, 1971). By the 1800s, many came to believe that the brain played a central role in behavior, specifically human personality. The phrenologists believed that specific locations in the brain were associated with specific personality traits, and when a brain area was larger than typical, the person was thought to possess more of that characteristic. They believed that when the brain area was particularly large, it would protrude, causing a bump to form on the skull. By the late 1800s, phrenology had fallen out of fashion. The view that specific brain locations in humans were responsible for specific human functions did not completely disappear; rather, it became popular again in the 20th century in part due to the famous case of Phineas Gage. Text Box 2.1 describes this extraordinary individual's inadvertent contribution to the study of brain science.

Text Box 2.1 Extraordinary Individuals: Phineas Gage

Much was learned about the localization of function in the brain from the case of Phineas Gage (1823–1860), who was a foreman for the Rutland and Burlington Railroad Company in Cavendish, Vermont. On September 13, 1848, he was placing an explosive powder into a large rock so that the rock could be blasted apart and removed. He was using a long metal rod to tamp down the powder. The powder accidentally ignited. The force of the explosion drove the rod into his head. The rod, which was 1.25 inches in diameter, was found 80 feet away. It entered underneath Gage's left cheekbone and exited out of the top of his skull. Eyewitness reports state that Gage was able to speak within several minutes of the event and even sat up during the wagon ride home. Gage recovered slowly, living in his parents' home. After his recovery, he had no pain in his head but reported having an unusual feeling (McMillan, 2002). It has been frequently reported that he experienced a change in personality, although it has been claimed that accounts of Gage's personality changes after the injury were greatly exaggerated (McMillan, 2002). Friends of Gage were quoted as saying that the man they met after the accident was not the Gage they had known. He had been changed by the event. Careful examination of Gage's story and also of his skull, which is now on permanent display at Harvard Medical School's Warren Anatomical Museum, suggest that the rod likely damaged Gage's frontal lobes (Damasio, Grabowski, Frank, Galaburda, & Damasio, 1994). The frontal lobes are now known to play a central role in coordinating and controlling emotional impulses. Despite his injury, Gage was able to work after his injury. He held down a job in the livery stable of the Dartmouth Hotel in Dartmouth, New Hampshire, and also cared for horses in Valparaiso, Chile. He worked until his health began to decline in 1860. He began having uncontrollable seizures. He went to live with his mother, where he resided until his death later that year.

▲ Photo 2.2 Phineas Gage lived almost 12 years after his life-altering accident, which provided insight into the functioning of the frontal lobe. How does your frontal lobe control your emotional behavior?

Did Aphasia Studies Support the Localization Hypothesis?

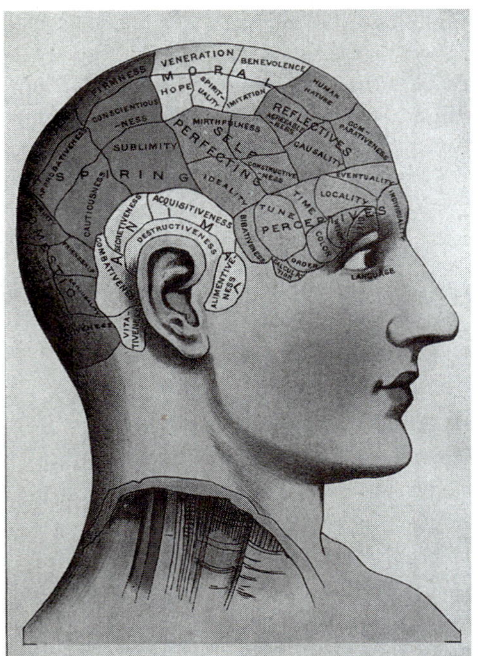

▲ Photo 2.3 The phrenologists identified areas of the head presumed to be associated with specific personality traits. If one had a bump in the area, he or she was believed to possess more of that personality trait.

Broca's Aphasia

By the mid-1880s, new evidence existed about the localization hypothesis. The evidence had nothing to do with the topic of human personality and bumps on the head. The evidence involved language ability among hospitalized patients in the suburbs of Paris, France. The physician and anthropologist Pierre Paul Broca worked in a hospital that also functioned as an orphanage, prison, and lunatic asylum (Schiller, 1998). Often, he cared for patients who were sent to the hospital because they had no other place to go. During his work there, he encountered a number of patients who had difficulty speaking. One of his patients was named Leborgne. He struggled to speak, but could produce only one word. That word was *Tan*. Broca kept careful notes about the patients' symptoms. As the patients passed away one by one, he was able to conduct postmortem autopsies on the patients, examining their brains and discovering the locations of any brain damage. His careful investigation yielded the discovery that the patients who had experienced speaking difficulties before death had damage to the same regions of the brain. The region was in the left hemisphere in the posterior inferior frontal gyrus. Broca concluded that this region was a language center. Even today, we call this region Broca's area. Figure 2.1 shows the location of Broca's area.

The patients with damage to Broca's area had similar problems when speaking. They had difficulty articulating words—if they could speak at all. Their speech lacked syntactic words, such as articles, prepositions, or helping verbs (e.g., *Eat bread now* instead of *I want to eat some bread now*). Researchers refer to this way of communicating as telegraphic speech, because it is shortened speech that still conveys meaning, similar to short communications that were once sent by telegraph machines. **Broca's aphasia** is the term used today to describe the type of aphasia characterized by difficulty in speaking. Those interested in learning more of Broca's work can view the brains of some of Broca's patients in the Musée Dupuytren in Paris.

Wernicke's Aphasia

Broca's results had a major impact when he published them in 1861 (Broca, 1861/1950). The results made a big impression on one particular person, working about 700 miles away in Berlin. His name was Karl Wernicke (pronounced with a /v/ as in /Ver-ni-kee/).

Figure 2.1 Broca's and Wernicke's areas are located in the left hemisphere. Broca's area is located in the frontal region. Wernicke's area is located in the posterior temporal region

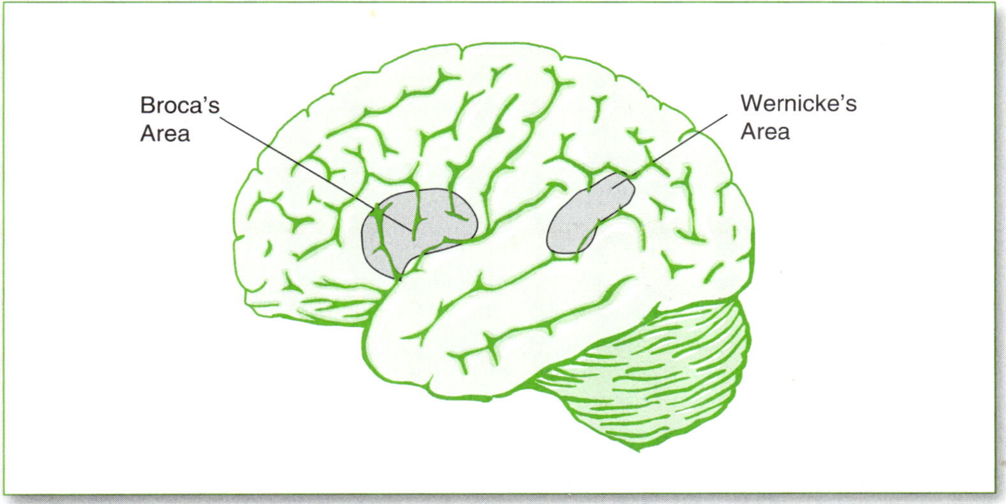

Source: Retrieved from http://en.wikipedia.org/wiki/Broca's_area

Wernicke was a physician and anatomist who also worked with patients affected by language impairments. He began observing the symptoms among his patients and soon found a patient who had an impairment quite different from those reported by Broca (Wernicke, 1874/1908). The patient had suffered a stroke. He was able to speak but understood very little of what others said to him. When the patient passed away, Wernicke carefully examined his brain in the manner similar to Broca. Wernicke found a damaged portion in an area located in the posterior region of the left superior temporal gyrus. This region became known as Wernicke's area. Figure 2.1 shows the location of Wernicke's area. The term ***Wernicke's aphasia*** is still used today to describe the type of aphasia characterized by difficulty comprehending the speech of others. Patients have fluent but sometimes nonsensical speech. Their speech typically contains syntactic words but lacks content words. Sometimes, their speech contains combinations of morphemes that are not actual words. The technical term for an invented word that is produced during speech is **neologism**.

What Is Meant by the Lateralization of Language?

The aphasia studies of the 19th century provided strong evidence that there were two locations in the brain involved in language. Both locations were in the left hemisphere of the brain. Those working in the field of anatomy often use the word *lateral* to describe something located on or at the side of the body. Consequently, it has come to be the norm that language processes are described as *lateralized* in the brain. In the 20th century, researchers have discovered that the two hemispheres are specialized in their function (Kolb & Whishaw, 2003). The left hemisphere is dominant in language processing and processing involving sequencing. The right hemisphere is dominant in spatial processing, emotional processing, the processing of faces, and the ability to appreciate and to create music. The **corpus callosum** connects the two hemispheres. One might view the corpus callosum as the communication highway between the two hemispheres. It is a bundle of nerve fibers through which all information transferred from one hemisphere to the other passes.

Wada Testing

A variety of techniques have been used to investigate the functions of the left and right hemispheres. Some of the most compelling studies used an anesthetic to paralyze one hemisphere of the brain at a time. The technique was called **Wada testing** after the neurologist Juhn Atsushi Wada, who developed the procedure (Wada, 1949, 1997). In a series of studies, he showed that language function was disrupted when the left hemisphere was paralyzed with an anesthetic (Branch, Milner, & Rasmussen, 1964; Wada & Rasmussen, 1960). The procedure was carried out on patients who were scheduled and prepped for brain surgery. It is routine, even today, for such patients to be alert, despite the fact that their brain is exposed because of the removal of part of the skull. Wada observed that when the left hemisphere was anesthetized, a patient who was communicating would abruptly stop. When the anesthetic wore off, usually in 5 to 10 minutes, speech functions returned (Bradshaw & Nettleton, 1983).

Electrical Brain Stimulation

Additional evidence of the lateralization of language was obtained in studies using **electrical brain stimulation** (EBS). Wilder Penfield pioneered the procedure to treat epilepsy by destroying brain cells to reduce or eliminate epileptic seizures (Penfield & Roberts, 1959). He also studied the function of brain regions by administering electrical currents directly to brain tissue and observing the effects. Studies were conducted on alert patients who were awaiting necessary brain surgery (Penfield, 1958). Even today, EBS is routinely used during pre-surgery evaluations of patients with epilepsy or conditions requiring the removal of brain tissue. In the studies by Penfield, electrodes delivering small amounts of electricity were placed in various locations, and the resulting behaviors were recorded. Penfield exhaustively documented patients' responses when different regions of the brain were stimulated (Jasper & Penfield, 1954; Penfield, 1958, 1972). His work has been a tremendous source of knowledge about the lateralization of the brain as well as on the organization of both hemispheres.

Split-Brain Patients

Of all the research on the lateralization of language in the brain, no research is more intriguing that the studies involving split-brain patients. Split-brain patients are those who have undergone surgery to disconnect the two hemispheres of the brain, as a treatment for severe epileptic seizures (Gazzaniga, 2005). During a seizure, the misfiring of electrical activity in the brain would begin at a location and then spread to affect more and more areas of the brain, causing the seizure to worsen. In some cases, seizures can be so severe and long lasting that they risk the lives of those experiencing them. It is for the most severe cases of seizures that splitting the brain is one of the recommended treatments. The hemispheres are disconnected through the severing of the corpus callosum. By cutting the corpus callosum, the effects of a seizure can be restricted to one hemisphere of the brain. The misfiring electrical activity does not spread to affect the opposite hemisphere; thus, the impact of the seizure is minimized.

When the surgery is successful in reducing seizures, the split-brain patient's quality of life is improved. However, research has revealed subtle effects of the surgery on language function (Sperry, 1964). Patients were interviewed in a laboratory setting and asked to perform a variety of tasks. Patients were shown pictures that were presented in either the left side or right side of the visual field. Because of the anatomy of the visual system, information presented to the rightmost part of the visual field is processed by the left hemisphere only, and information presented to the leftmost part of the visual field is processed by the right hemisphere only (Beaumont, 1983). The study showed that when a split-brain patient viewed a picture in the left visual field (i.e., right hemisphere), he could not name the picture. Pictures viewed in the right visual field (i.e., left hemisphere) could be named.

Further investigation showed that although pictures presented in the left visual field (i.e., right hemisphere) could not be named, participants saw them and understood what they were. Participants could draw what they saw and revealed knowledge of the object's identity by pointing to pictures of similar objects. The results suggested that in order for information to be named, the left hemisphere must be involved in the processing of the stimulus. The implications of the research are that in neurologically intact individuals (i.e., those with no known brain injury), information presented only to the right hemisphere can be named because the information is relayed to the left hemisphere, where a spoken response can be formulated. In the split-brain patient, a spoken response is not possible in this case because the fibers that allow information to be relayed from one hemisphere to the other have been disconnected.

Divided Visual Field Paradigm

Research investigating the lateralization of the brain has also been conducted on individuals without existing neurological conditions. The studies that used Wada testing and EBS were conducted on patients with existing neurological conditions. Consequently, they can be viewed as studies involving only neurologically unusual or atypical participants. The data obtained in such studies may not reflect results for individuals without a neurological condition. In studies with neurologically typical individuals, researchers used a technique to measure how quickly an individual could respond to a word or picture presented to either the left or right hemisphere of the brain (Mishkin & Foroays, 1952). The technique was called the divided visual field (DVF) paradigm. Studies conducted by Mishkin and Foroays and by many researchers since then have shown that words are processed faster by the left hemisphere than the right hemisphere. The left hemisphere advantage for word processing has been replicated many times by researchers (Iaccino, 1993). The effect has also been observed in languages other than English, such as Arabic and Hebrew (Faust, Kravets, & Babkoff, 1993).

Dichotic Listening Paradigm

Researchers have also conducted studies on listening to investigate how the two hemispheres process sounds. In these studies, participants listened to sounds through a special set of headphones. The headphones could present sounds to either one or the other or to both ears. The technique was called the dichotic listening paradigm. Researchers have found that participants generally respond faster to words when they are presented to the right ear (Kimura, 1961; Studdert-Kennedy, Shankweiler, & Pisoni, 1972). The finding has been called the **right ear advantage** (REA) for language. The REA may occur because the neural pathways from the ears are stronger to the opposite hemisphere (Kimura, 1961). Other studies have shown that there is a left ear advantage for processing music and nonlinguistic sounds, such as dogs barking or other environmental sounds (Bryden, 1988).

Mapping the Brain

Over the 20th century, the medical field developed a variety of technologies that can be used to produce detailed pictures of the brain. The equipment used for brain mapping is expensive and requires teams of professionals to maintain and to operate it. These instruments are housed inside hospitals or other large medical or research facilities. Over the past two decades, teams of researchers have been able to use brain-mapping techniques to explore the relationship between brain locations and brain function. In the past 30 years, the rate at which we are learning about the brain has increased dramatically. We live in an exciting era, because the coming decades promise to yield an astonishing amount of new discoveries about the role of the brain in all behaviors, including language processing. Students in college or graduate school now may be able to get hands-on research experience with brain-imaging equipment.

What Techniques Are Used to Image the Brain?

Brain-mapping techniques generally provide two types of information about the brain: (1) structural or anatomical information about the major parts or regions of the brain and (2) functional information about which anatomical regions of the brain are active during processing. The functional information provided can also indicate the level of activity in the region. These techniques are used both in medicine to investigate the brains of patients who have experienced brain damage and in research involving individuals who are neurologically intact. In many studies, researchers will compare the abilities and brain processing of patients with a specific impairment to that of neurologically intact controls who do not have the specific impairment but who are similar in sex, age, and educational background.

X-Ray Imaging

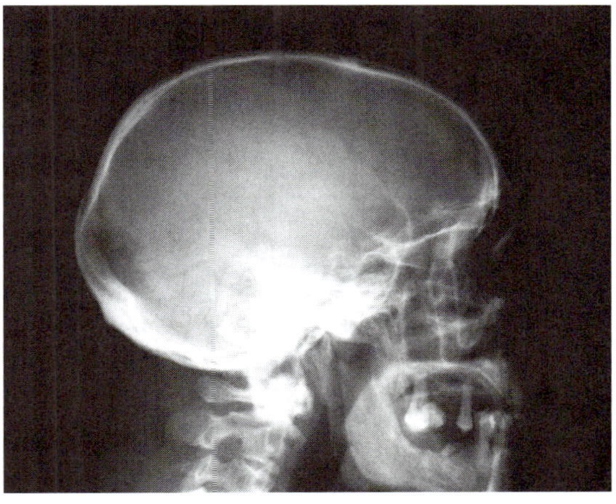

▲ Photo 2.4 X-rays of the head do not provide detailed images of the different regions. Can you identify any part of the head and face in this image?

The earliest techniques of brain (and body) imaging provided only structural or anatomical information about the body. For example, X-ray imaging was the first technology used to take pictures of the body, including the head. The discovery of the X-ray occurred in 1895 by Wilhelm Conrad Röntgen (Evens, 1995). It is interesting to note that his discovery was accidental. He was working with a cathode ray generator, which accidentally created an image of his wife's hand, which showed the bones of her hand and ring that she was wearing. The X-ray quickly came to be used to scan the body for medical purposes. One of the disadvantages of the X-ray is that the procedure exposes one to small amounts of radiation, which is not harmful, in isolation; however, those experiencing numerous, small exposures to radiation across a lifetime may be at risk for developing cancers and health problems (Berrington de González et al., 2009).

Computed Axial Tomography Scan

One of the most common scans conducted in medical settings is computed axial tomography, or the CAT scan. CAT scans also provide structural information about a scanned object, but the detail is much better than scans obtained from X-rays. Unlike X-rays, which provide two-dimensional images of a scanned object, CAT scans provide three-dimensional images. This is done through the collection of a series of two-dimensional views of the object and computer software that processes those images (Herman, 2009). Like X-rays, CAT scans produce images that do not provide fine-grained detail about small regions of the body or brain. Unfortunately, CAT scans also expose one to a small amount of radiation. Consequently, they are not widely used for research purposes.

Positron-Emission Tomography Scan

Some brain-mapping techniques can provide both structural and functional information. For example, positron-emission tomography (PET) scans provide information about blood flow in the brain. The technique was first used for medical purposes to determine whether regions were damaged and not receiving normal blood flow (Toga & Mazziotta, 2000). Researchers now also use the technique to investigate blood flow patterns in the brain when individuals perform specific cognitive tasks, such as visual perception or word processing. As a research tool, it is among the more invasive techniques. The procedure requires that a radioactive isotope be injected into the bloodstream. Commonly used isotopes are glucose or oxygen. As the isotope decays rapidly, a detector, which is positioned just above the head, records the location of the isotope in the brain. In regions of the body where there is greater blood flow, greater amounts of the isotope can be detected. Within several minutes, the isotope completely decays and leaves the body. During this time, the detector records the location of the isotope. Computer software uses this information to create colorful maps of active brain areas. The key disadvantage of using PET scans in research is that they cannot be used with all types of research subjects. PET scans are not generally used with children and pregnant women. Recruiting volunteers can be difficult and costly; typical PET scan research studies pay participants some nominal amount for the inconvenience of participating (e.g., $100 per hour or more).

Magnetic Resonance Imaging

An increasingly familiar brain-imaging technique is magnetic resonance imaging (MRI). The technique is frequently used by physicians to obtain high-resolution images of the body. Among your family and friends, there may be people who have received MRI scans for medical purposes. As the name implies, the images are created with help of a magnet. The machine directs a strong magnetic field toward the individual. Atoms in the body are changed by the magnetization. The machine then uses radio frequency fields to systematically alter the alignment of this magnetization (Toga & Mazziotta, 2000). Despite the complexity of this process, the images that can be produced are remarkable. MRIs are used to examine injured ankles and knees as well as to scan the body for tumors, large or small. In this way, the MRI provides excellent structural information about the body. One advantage of MRI over the X-ray and the CAT scan is that it does not expose the individual to radiation. However, because MRI involves magnetization, individuals who have metal in their bodies from existing wounds, prior surgical implants, or industrial work such as welding or painting cannot be scanned. Undergoing a scan when there is metal in the body can result in extreme pain and serious injury, because the magnet will cause the metal fragments to move.

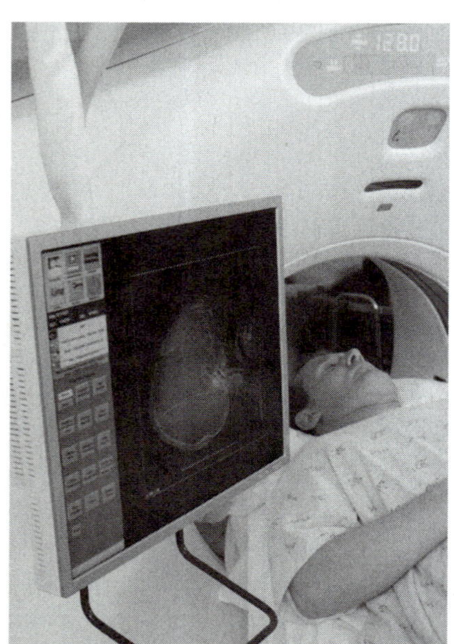

▲ Photo 2.5 MRI scanners are routinely used in medical settings and are increasingly used in research. Do you know anyone who has undergone an MRI scan?

Functional Magnetic Resonance Imaging

Researchers who are interested in the function of brain regions can use a variant of the MRI referred to as functional magnetic resonance imaging (fMRI). Like the MRI, fMRI uses a magnetic field and radio frequency to manipulate the cells of the body. Unlike MRI, fMRI provides information about the changes in blood flow and blood oxygenation in the body. Researchers can design brief tasks for an individual to perform during scanning.

By recording during the time that a person performs a task, researchers can determine whether there are some regions of the brain that are more activated than others evidenced by the greater blood flow and blood oxygenation in those areas. The tasks that are typically performed last several minutes. The image that is produced reflects the total processing occurring during that period. Consequently, fMRI cannot provide good information about the time course of processing.

Electroencephalography

Researchers interested in the time course of processing often use electroencephalography (EEG). Researchers using EEG measure the electrical activity emitted by the brain and measurable from the scalp. Researchers may place small electrodes individually on the scalp or have a test subject wear an electrode cap, which can hold a large number of electrodes in place on the scalp. In studies, researchers typically present a stimulus, such as a picture or a sound, and examine the changes in brain electrical activity that follow. This procedure is referred to as an event-related potential (ERP) or evoked brain potential. Research has shown that there are bursts of electrical activity produced by the brain during perceptual and cognitive processing. Increasingly, researchers use the pattern of electrical activity across the scalp to infer where in the brain the electrical activity has been generated (Toga & Mazziotta, 2000). However, tracing electrical activity back to specific brain regions is quite difficult. Consequently, when researchers use EEG to study cognitive processing, they usually are most interested in the time course of processing and/or the extent to which processing appears to be involving the left versus right hemisphere or frontal versus posterior region of the brain. With EEG, it is difficult to infer where inside the brain electrical activity has originated.

What Types of Aphasia Have Been Identified?

Since Broca's and Wernicke's studies in the 1800s, a great deal has been learned about aphasia and its causes. There are approximately 1 million people in the United States with aphasia (National Aphasia Association, n.d.). Each year, approximately 80,000 to 112,000 new cases of aphasia are diagnosed in the United States (Chapey, 1994). Eighty-five percent of cases of aphasia result from stroke, which occurs when there is a blockage of blood flow in the brain or bleeding in the brain. Other causes of aphasia include traumatic brain injury, brain tumors, diseases causing degeneration of the brain, and surgical procedures. Patients with aphasia rarely have damage to identical regions of the brain and rarely experiences identical symptoms. Consequently, those involved in treating individuals with aphasia must treat patients on a case-by-case basis.

Researchers have broadly classified varieties of aphasia into three types (Kolb & Whishaw, 2003; Lass, 1988): (1) **expressive aphasia,** which is characterized by difficulties producing language (e.g., speaking, writing, signing), as in Broca's aphasia; (2) **receptive aphasia,** which is characterized by difficulties understanding language (i.e., listening and reading), as in Wernicke's aphasia; and (3) **pure aphasia**, which involves problems affecting just one aspect of language, such as reading, writing, or listening. It is important to note that those who have aphasia may also have memory impairments, attentional deficits, and perceptual problems (Kolb & Whishaw, 2003). The symptoms that a particular person experiences will be determined by the specific location of brain injury.

Expressive Aphasia

Besides Broca's aphasia, other forms of expressive aphasia include **agrammatism**, **conduction aphasia**, **transcortical motor aphasia**, and **anomic aphasia**. Agrammatism is a variety of expressive aphasia in which one produces only telegraphic speech. Agrammatism can occur following damage to one or more subareas within Broca's area. Brain-imaging studies have found that Broca's area contains three distinct regions. The most posterior of these regions is the **pars opercularis**, which has been shown to play a role in phonological processing (Gold & Buckner, 2002; Nixon, Lazarova, Hodinott-Hill, Gough, & Passingham, 2004) and syntactic processing (Hagoort, 2005). The most frontal region within Broca's region is the **pars orbitalis**, which is believed to play a role in semantic processing (Hagoort, 2005). The region between the pars opercularis and pars orbitalis is the **pars triangularis**, which has been shown to play a role in semantic processing (Devlin, Matthews, & Rushworth, 2003; Gold & Buckner, 2002) and syntactic processing (Hagoort, 2005).

Conduction aphasia and transcortical motor aphasia generally occur following brain damage outside of Broca's area, specifically regions connected to Wernicke's area. The symptoms of conduction aphasia include poor word repetition; good comprehension; and fluent, nonsensical speech. Patients with conduction aphasia typically have damage to the **arcuate fasciculus**, which connects Wernicke's area to the premotor area, which is involved in controlling movements. Figure 2.2 displays the location of the arcuate fasciculus in relation to Broca's and Wernicke's areas. The symptoms of transcortical motor aphasia are characterized by impaired speech, but good comprehension. Utterances are composed of one or two words and take a great deal of effort to produce. Word repetition is possible, despite articulation difficulty. Patients with transcortical motor aphasia typically have damage to the region behind Wernicke's area. This area is described as the temporal-occipital-parietal junction.

Expressive aphasias can also occur when brain damage affects regions relatively far away from Broca's and Wernicke's areas. For example, the symptoms of anomic aphasia include word-finding difficulty during speaking with good comprehension. The speaker experiences a great deal of frustration because the words that cannot be produced at a given time may be familiar words that are commonly used. However, the speaker fails to be able to produce them when speaking. Recent research indicates that anomic aphasia is caused by damage to the left parietal lobe of the brain (Fridriksson, Kjartansson, Morgan, Hjaltason, & Magnusdottir, 2010). Different types of anomic aphasia have been observed. Patients may show difficulty producing only one type of word, such as verbs, or words related to a particular meaning, such as types of tools.

Receptive Aphasia

Besides Wernicke's aphasia, other forms of receptive aphasia include **transcortical sensory aphasia** and **pure word deafness** (Kolb & Whishaw, 2003). The symptoms of transcortical sensory aphasia include poor comprehension but fluent, grammatical speech and good word repetition. This type of aphasia is often seen in patients with damage

Figure 2.2 The arcuate fasciculus is located in an area called the temporal-occipital-parietal junction

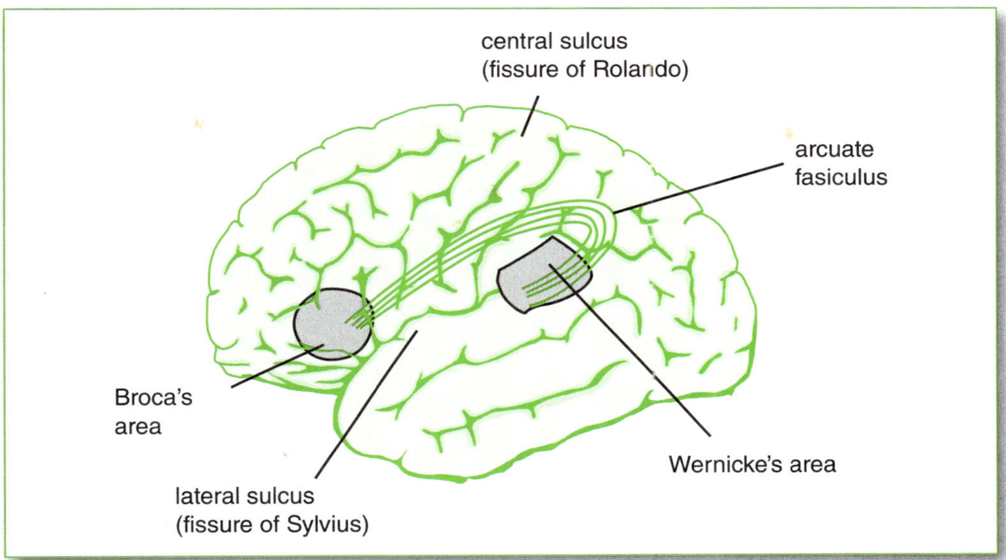

posterior to Wernicke's area. Pure word deafness is characterized by the inability to comprehend spoken language despite adequate hearing and also retaining the ability to speak, read, and write. Patients with pure word deafness typically have damage to both hemispheres in the posterior superior temporal lobes or damage to the region that connects these areas.

In rare cases, patients display symptoms of both expressive and receptive aphasia. Examples include **global aphasia, mixed transcortical aphasia,** and **primary progressive aphasia.** Patients with global aphasia lose their ability to speak and comprehend language. Their ability to repeat words and phrases is poor. Often patients with global aphasia are mute and unable to respond to the speech of others. Those with global aphasia generally have extensive damage to the perisylvian area, which includes Broca's and Wernicke's areas and the area in between (Goodglass & Kaplan, 1972). This type of damage can occur following carbon monoxide poisoning (Fabbro, 1999). The symptoms of mixed transcortical aphasia include the ability to repeat words and phrases but poor language production and comprehension. Those with mixed transcortical aphasia typically have damage to the regions surrounding but not including Broca's area, Wernicke's area, and the arcuate fasciculus. Primary progressive aphasia is a form of dementia caused by degenerative disease. Patients have difficulty finding words, reading, writing, and understanding speech. As the disease progresses, the difficulties become more and more severe. Symptoms generally occur before the age of 65, often before age 50, and sometimes as early as in the 40s.

Pure Aphasia

Among the most compelling types of language deficit following brain damage are pure aphasias in which an individual experiences a deficit in only one area of language use. Patients may be able to carry out most language tasks normally but may be completely unable to perform one type of task. For example, **alexia** is the inability to read (Damasio, 1977). A person who could read before a brain injury loses the ability to read after a brain injury. Generally, individuals are unable to recognize individual letters or to identify words from print. Sometimes, the disorder is referred to as *word blindness.* The person is able to speak normally and to comprehend spoken language. Alexia can occur with or without symptoms of other types of aphasia (Geschwind, 1965). Sometimes alexia occurs with **agraphia,** which is the inability to write. Alexia typically occurs when one experiences damage in the areas of the medial or inferior **occipital lobe.** Caramazza and Hillis (1991) reported an interesting case of two individuals who exhibit difficulty when producing verbs both in writing and in speaking. They were able to produce other types of words. Rapp and Caramazza (1995) later reported a case of an individual with aphasia who had difficulty writing verbs but relatively little problem producing verbs orally. These case studies suggest that the brain circuitry that controls language involves distinct processes for the various outputs for language (i.e., writing, speaking) and for the different types of words (e.g., nouns, verbs).

One of the rarest forms of aphasia is **crossed aphasia.** As the name implies, crossed aphasia refers to aphasia resulting from damage to the right hemisphere, instead of the left hemisphere, in a right-handed individual (Bramwell, 1899). Among cases of aphasia following stroke, crossed aphasia occurs between 0.5% and 2% of the time (Zangwell, 1967). Approximately 70% of cases appear to have right hemispheres that are mirror images of the typical left hemisphere, as the symptoms would be predictable if one assumed that the organization typical of a left hemisphere was applied to the right hemisphere (Alexander, Fischette, & Fischer, 1989). Thirty percent of cases were found to be anomalous, indicating that the pattern of aphasia in the right hemisphere was not predictable using the typical organization of the left hemisphere as a guide.

What Has Brain Imaging Revealed About Language Development?

Since Chomsky's proposal that there is a language acquisition device (LAD) that accounts for the rapid learning of language by all children, many have wondered whether the brain contains one or more regions dedicated solely to the acquisition of human language. Thus far, brain-imaging research has not found an LAD or multiple LADs or an LAD circuit. Nevertheless, brain-imaging research has shown that in the adult brain, many regions work together to achieve the production and comprehension of language. New research is exploring how infant brains respond to speech sounds (Dehaene-Lambertz, Dehaene, & Hertz-Pannier, 2002; Dehaene-Lambertz, Hertz-Pannier, & Dubois, 2006). One fMRI study showed that when infants listened to language, activity was greater in their left hemisphere language regions than in corresponding regions of their right hemispheres (Dehaene-Lambertz et al., 2002). A second fMRI study suggested that as the child matures, important changes may occur in how the multiple language centers of the brain are connected (Dehaene-Lambertz et al., 2006).

Planum Temporale

Dehaene-Lambertz and her colleagues (2010) have also proposed that one particular brain region may be important in language processing in infants. The region is called the **planum temporale,** which is a triangular region within Wernicke's area. In an fMRI study, they found that infants as young as 2 months old showed differences in activity levels in the planum temporale when listening to language versus music. One might consider the planum temporale to be a good candidate for as the brain's language learning center; however, it is unlikely to be the region that accounts for why humans have human language and other animals have different types of communication systems. In the brains of chimpanzees and other great apes, there are Broca's and Wernicke's regions and also a planum temporale (Gannon, Holloway, Broadfield, & Braun, 1998). As in human brains, the planum temporale is asymmetric, with the left side being larger than the right. However, as discussed later in this Chapter, chimpanzees and other great apes have, thus far, failed to acquire human language to the level of using both a vocabulary and a grammatical system.

There is an exciting journey ahead as researchers continue trying to understand the role of the brain in the mystery of language development. In the coming decades, researchers are likely to continue to get closer to understanding what it is about human brains that makes each child an expert language learner from birth. It is difficult to estimate how quickly the mystery will be solved. The rate of scientific discovery has definitely increased over the past several decades. There is hope that within our lifetimes, many of the key elements of the mystery will be identified. An important aspect of these discoveries will be greater understanding about the role that genetics play in how our brains and bodies developed. Another important aspect will be how one's experience with human languages is related to how the brain processes language. Text Box 2.2 discusses recent research showing that deaf signers, hearing signers, and hearing nonsigners use different parts of their brains when performing the same task.

Genetics and Language

Over the past several decades, knowledge about human genetics and the genetic makeup of other animals has increased dramatically. The genetic code is essentially the set of instructions that determine how development occurs. By understanding the genetic code, it may be possible to understand how the development of the brain and the rest of the body may occur abnormally. A thorough understanding of how the genetic code controls the development of the brain will ultimately lead to advances in the diagnosis and treatment of brain injuries. The long-term possibility is that researchers can learn why the brain

Text Box 2.2 Research Discovery: Imaging the Brains of Bimodal Bilinguals

Recent research has shed some light on the question of whether a person who knows more than one language uses the same areas of his or her brain in language processing. This research investigated brain processing in **bimodal bilinguals,** who are individuals who know both a spoken language and a signed language. Emmorey and McCullough (2009) conducted an fMRI study in which they compared the brain activity for three groups of individuals: (1) those who knew English as well as American Sign Language (ASL), (2) those who were deaf individuals and knew ASL, and (3) those who spoke English but did not know ASL. They recorded brain activity as each group viewed pictures of someone using ASL and also pictures of faces displaying a specific emotion (e.g., fear, anger, disgust). They found clear evidence of differences in brain processing between the bimodal bilinguals and deaf signers and also between the bimodal bilinguals and speakers of English who did not know ASL. The brain differences for bimodal bilinguals and deaf users of ASL were located in the superior temporal sulcus when they were processing signed language. The brain differences for the deaf signers and those who did not know ASL were found in the left hemisphere during the processing of emotional faces. The results demonstrated that both deafness and knowledge of ASL were associated with which locations were active in the brain during visual processing of pictures depicting a use of language and pictures in which language was not involved. Figure 2.3 displays the brain scans for the three groups of participants.

Figure 2.3 The brain scans of deaf participants who were signers, hearing signers, and hearing nonsigners revealed that different regions of the brain were involved as the three groups of participants performed the same task

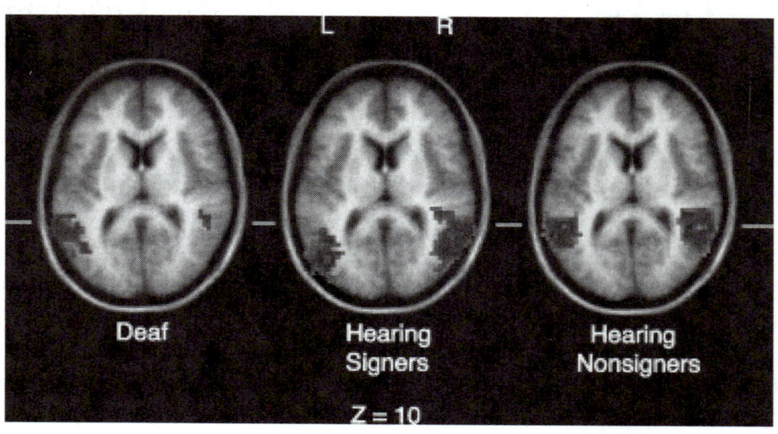

develops to have the organization that it does and why the brain repairs and reorganizes itself the way that it does following an injury.

The Human Genome Project was launched in 1990 and aimed to map the entire human genome (Human Genome Project, n.d.). The project has yielded an abundance of knowledge about human genetics, but much remains to be discovered. Since the early years of genetic research, it has been known that among all humans, regardless of ethnicity or location in the world, 99.9% of the genome is the same. The genetic material of all living things is coded in deoxyribonucleic acid, or DNA. The discovery of DNA by Watson and Crick (Watson & Crick, 1953) and Franklin (Crick, 1961/2003) led to the subsequent understanding of the building blocks of life. Human DNA is composed of 23 pairs of chromosomes, containing a total of 20,000 to 25,000 genes. Children receive half of their DNA from each parent. Sons receive a Y-chromosome from their father and an X-chromosome from their mother. Daughters receive an X chromosome from each parent. Research has yet to discover the function for half of the known human genes. Some of the most important advancements include the identification of genetic markers for inherited diseases, such as certain cancers and muscle diseases, as well as language disorders.

Is Language in Our DNA?

Specific Language Impairment

In the 1990s, researchers identified a family affected by an inherited language disorder, now referred to as **specific language impairment** (SLI) (Gopnik, 1990; Gopnik & Crago, 1991). The family members with SLI have IQs in the normal range but difficulty with grammar, specifically the proper use of syntactic rules. They make frequent mistakes with plural nouns (e.g., saying *two chair* instead of *two chairs*) and past tense verbs (e.g., saying *yesterday I cry* instead of *yesterday I cried*). The affected family members also have severe difficulty producing language because of oral apraxia, which involves difficulty moving the jaw, tongue, lips, and mouth during speaking. In the research literature, the family has been referred to as the KE family. There are 30 family members representing three generations. Sixteen of the 30 are affected by the disorder. The pattern of inheritance is autosomal dominant, which means that one can inherit the disorder when only one parent's DNA contains the mutation.

The prevalence of SLI has been estimated to affect 7.4% of kindergarten children in the United States. One's risk of having SLI is strongly influenced by one's genetic relatedness to someone who has it, which would be expected for an inherited disorder (Tomblin, 1996). If one's first-degree family members (e.g., parents, siblings, offspring) have SLI, then one's chance of also having it is 21%. If one's identical twin (who shares the same DNA) has SLI, one's chance of also having it is 71%. Some children with SLI may be late talkers, which are children whose first words are not produced until 18 to 24 months if not later. Some children with SLI will go on to have problems reading and may be diagnosed with **dyslexia,** which is characterized by difficulty learning to read despite having an IQ in the normal range. There is no cure for SLI; however, researchers have found that children with SLI can benefit from interventions that are designed to strengthen their grammatical skills (Kot & Law, 1995). Unfortunately, most children and adults with SLI do not receive a diagnosis (Rice, 1997) because educators and physicians are still unfamiliar with the disorder.

FOXP2

Attempts to locate the genetic marker for SLI have been successful (Fischer, 2006; Vargha-Khadem, Gadian, Copp, & Mishkin, 2005). Using DNA from multiple families affected by SLI, researchers have located a mutation on gene 7 that regulates the protein called **FOXP2.** There was a great deal of excitement about the discovery of the link between SLI and FOXP2; however, it remains to be determined how a normal gene 7 contributes to normal language in most people and how the mutation in gene 7 contributes to abnormal language in those with SLI. A recent brain-imaging study compared the brain scans of individuals with SLI, those from the same family without SLI, and a group of unrelated control participants without SLI. The researchers found that several brain areas differed in size for individuals with SLI and those without SLI (Watkins et al., 2002). For example, the left **caudate nucleus** was smaller in those with SLI. The caudate nucleus is part of the basal ganglia, which has been involved in motor disorders, such as Parkinson's disease and Huntington's disease. In individuals with SLI, the size of the left caudate nucleus was related to the individuals' score on tests of oral praxis or ability to move the jaw, tongue, and lips during speaking.

Studies involving comparison of twins and SLI suggest that the central language deficits in SLI are nearly 100% genetic (Bishop, North, & Donlan, 1995). Other language disorders can be inherited, but the likelihood of inheritance is lower than for SLI. Research on dyslexia has shown that some forms have a genetic basis. Twin studies of reading disability have shown a rate of heritability between 44% and 77% (DeFries, Fulker, & LaBuda, 1987). Thus far, four genes have been identified as involved in dyslexia (Gibson & Gruen, 2008). There is much still to be learned regarding how the vast human genome works together to produce the range of language abilities that most of us are able to develop. Given how quickly new knowledge is being discovered about the genetic basis of human health and human abilities, we may find the genetic basis of language problems to be treatable through genetic therapies. Certainly, this sounds a bit like science fiction. However, we are living in an era where science fiction is becoming science fact.

When Did Human Language Emerge?

No one knows for sure when human language first appeared on Earth, just as no one knows exactly when the first humans existed. Our estimates are based on the historical record, which dates back to the invention of writing. The earliest examples of writing date back to 3500 BC in Mesopotamia, which corresponds to modern-day Iraq and parts of Syria, Turkey, and Iran (Bottéro, 1992). Scholars have found examples of writing in China dating back as early as 1200 BC (Boltz, 1986). Recent excavations in Mexico have found evidence of the earliest known writing in the Western Hemisphere, which dates back to 650 BC (Pohl, Pope, & Von Nagy, 2002). Researchers believe that writing may have been invented for the purpose of keeping track of commercial transactions. For example, one might need to record that a certain number of domesticated animals were traded for a certain number of barrels of wine (Bottéro, 1992). As trade increased among communities of people, it became impossible to keep track of all transactions using memory alone.

The historical record provides only a minuscule window into human history. Spoken language is likely to have existed since the emergence of modern *Homo sapiens*. It has been estimated that *Homo sapiens* first appeared on Earth about 100,000 years ago (Tallerman, 2005). It is not known whether human language existed at this time. Today, the closest relative to modern humans is the chimpanzee, which shares about 96% of its genome with modern humans. Despite the genetic closeness, researchers tend to agree that was quite a long time ago when the family tree for modern humans and chimpanzees branched. The common ancestor shared by modern humans and chimpanzees existed over 6 million years ago (McHenry, 2009). It is important to keep in mind that the fossil record, like the historical record, is, by its very nature, incomplete. The family tree is created from the fossils that are known. As new fossils are discovered, changes to the family tree can be made.

The fossil record provides good detail about European early modern humans (EEMH) or Cro-Magnons. In current research, the term *Cro-Magnon* is rarely used by researchers. It has been replaced with the label EEMH. The EEMH lived in Europe as early as 45,000 years ago. They were similar in anatomy to modern humans, having a somewhat stockier build and larger cranial capacity (McHenry, 2009). They had wider faces than modern humans but a similar high forehead. They walked with an upright posture and stood about 5.5 feet tall. They hunted big game, such as mammoth, reindeer, and the now-extinct cave bear. They used flint tools and created textiles from spinning and dying flax fibers. Fossils have been found in southwestern France where cave paintings from the same era have been discovered (Wild et al., 2005). These cave paintings suggest EEMH culture created symbolic art. Fossils have been found in areas with multiple burial sites, and individuals appear to have been buried with objects such as tools or necklaces. The fact that they buried the dead might indicate that they had a culture that engaged in ritual and ceremony; however, it is also possible that they understood the need to bury the dead for reasons related to preventing the spread of disease.

The EEMH coexisted with *Homo neanderthalensis* (or Neanderthals). Genetic analysis of fossils suggests that they shared a common ancestor about 660,000 years ago (McHenry, 2009). Neanderthals lived between 400,000 and 30,000 years ago. They were about the same height as modern humans (i.e., 5.5 feet), but their arms and hands were likely much stronger. Their cranial capacity was as large as modern humans and perhaps somewhat larger. They hunted big game and used stone tools. They were likely primarily carnivorous. However, a recent discovery of a fossil with evidence of cooked vegetable in the teeth indicates that the diet included both meat and vegetables. There is controversy regarding the extent to which Neanderthals engaged in symbolic art

(Mithun, 2006). Most researchers believe that Neanderthal culture was much less involved in the creation of symbolic art. Mithun (2006) suggested in the book *The Singing Neanderthals* that Neanderthals may have appreciated music and song. Some of the fossil sites contained flutes carved out of the bones of birds. Additional concrete support for the proposal is lacking. Mithuen's book provides a compelling picture of what might be true. It remains to be seen whether future fossil discoveries will provide any new evidence for his proposal.

The question of whether Neanderthals and EEMH could speak in a manner similar to modern humans has been made somewhat clearer. The examination of the fossils, particularly the shape of the face and neck, suggest that the spoken language, like that used by modern humans, probably did not exist among Neanderthals; however, it might have existed among the EEMH (Lieberman, 2007). In order to produce rapid sequences of consonant and vowel syllables that are involved in human speech, one needs to have a larynx that sits relatively low in the throat and a skull whose base is arched. Infants are born with a larynx that sits high in the throat. The larynx descends by the fourth month of life (Lieberman, McCarthy, Hiiemae, & Palmer, 2001). In chimpanzees and other apes, the larynx is high, and the base of the skull is flat. This configuration severely limits the range of sounds that can be produced (Lieberman, 2007). It is impossible for apes to produce the rapid changing sequences of consonants and vowels that make up spoken human language. The fossil record shows that Neanderthal skulls lack the arch of modern humans but are somewhat more arched than chimpanzees and other apes. Some have argued that Neanderthals would have been able to produce a greater range of speech sound than chimpanzees and other apes but nowhere near the range of modern humans (Lieberman, 2007).

Perhaps some of the most compelling research on the differences among modern humans, EEMH, and Neanderthals has focused on genetic analyses. Studies have focused specifically on the evolution of the FOXP2 gene, which is involved in SLI in humans. Variants of the FOXP2 gene occur in a variety of animals, including mice, songbirds, and chimpanzees (Fisher & Scharff, 2009). In those animals, the gene is also involved in the production of vocalization. The FOXP2 gene of chimpanzees differs in a relatively small way from the FOXP2 gene of humans (i.e., in only two amino acids, Enard et al., 2002). Neanderthals had the same version of the FOXP2 gene as modern humans (Krause et al., 2007).

In 2013, Bowers, Perez-Pouchoulen, Edwards, and McCarthy suggested that FOXP2 may play a role in sex differences in communication that are observed across species. In humans, females tend to be more verbal than boys (Tannen, 2001). Bowers and colleagues (2013) found that male rat pups vocalized more than female rat pups. Analysis of the rat pups' brains revealed that there was more of a FOXP2 protein in the brains of male rat pups than female rat pups. The results suggested that the amount of vocalization was related to the FOXP2 protein. In a follow-up study, they compared the amount of the FOXP2 protein in the brains of human boys and girls and found that the brain tissue of girls contained more than that of boys. While much of these FOXP2 findings are quite provocative, it remains to be determined what exactly they mean. Given the rapid advances that are being made in genetic research, it is possible that the coming decades will bring new insight into the question of how and when language began.

How Did Language Evolve?

The mystery of how human language emerged may never be understood completely. Many scholars from a variety of disciplines (e.g., linguistics, anthropology, and cognitive science, among others) have considered this question. Generally, two types of theories have been proposed: (1) human language evolved slowly, progressing from an early, less complex system to the modern grammatical system, and (2) human language abruptly appeared on earth in *Homo sapiens* with most, if not all, of the modern complexity intact from the beginning.

Those who believe that language evolved slowly suggest that the shape of the vocal tract, the position of the larynx, the size of the spinal cord, and the size of the brain evolved slowly over the past 2 million years. The nature of communication among human ancestors is presumed to have evolved accordingly, resulting in modern human language. There have been numerous ideas about what type of utterances were the first used by *Homo sapiens*. For example, the bow-wow theory of language evolution suggests that human language began with the use of utterances whose sound resembles the objects to which they refer (Aitchison, 1998). These types of utterances are called **onomatopoeia** (e.g., *bow wow, buzz, quack*). The pooh-pooh theory suggests that humans' first words were exclamations, such as *ouch*, *oh*, and *mmm*. The heave-ho theory suggests that the first utterances were rhythmic chants or grunts, occurring during physical labor. None of the theories adequately explain how ancient humans made the leap from individual utterances to grammatical sentences.

Those who believe that language emerged abruptly have proposed a genetic mutation (Crow, 2002; Klein & Edgar, 2002). There are several variants on the view, but they essentially share the idea that language emerged in a single individual affected by the genetic mutation. From that single individual, the mutation spread, and more and more individuals shared the new, unusual way of communicating. Klein and Edgar (2002) argued that it is within the last 50,000 years that symbolic thought and the resulting cultural advances have flourished. They proposed that the innovation of human language is responsible for these advances. Desalles (2007) suggested that human language, once it appeared in humans, became a desirable characteristic, suggesting that those having language would have likely been more successful in the competition for mates. His proposal is influenced by research showing that in some species of birds, altruism plays a similar role in mate competition.

The many possibilities regarding how language evolved will likely never be fully proven or disproven. The scientific community will be eagerly awaiting the next fossil discovery and the next new report about insights gained from existing fossils. It is important to keep in mind the different possibilities as you continue to learn about how language develops in the individual. Children begin life with a vocal tract with a larynx high in the throat similar to a chimpanzee. When the larynx descends, the production of complex syllables becomes possible. At that point in life, the child has vocal abilities that surpass that of both EEMH and Neanderthals.

Communication in Nonhuman Primates and Other Animals

Is Human Language Different From Communication in Other Animals?

One might think that all forms of communication are the same. Birds tweet. Wolves howl. Dolphins squeak and whistle. Humans talk. However, human language differs from the communication systems in other species in a number of important ways. These differences were pointed out in the 1960s by the linguist Charles Hockett. He identified three characteristics that distinguish human language from forms of animal communication (Hockett, 1966). First, in all human languages, there is a way to refer to past and future events. While animals appear to be limited to communicating about events occurring in the here and now, humans can talk about something that has already occurred or something that has not yet occurred. This characteristic is referred to as the **principle of displacement.** Any event referred to in an utterance can be displaced in time from the utterance itself. Most, if not all, communication systems in nonhuman animals have no way to refer to events occurring in the past or future. Your dog may bark to signal the food dish is empty, but your dog cannot bark to communicate that the food dish will be empty tomorrow or that yesterday's dinner was particularly tasty.

A second characteristic that distinguishes human language from communication systems in other species is the **principle of productivity,** or the fact that it is possible to create an unlimited number of new messages using existing sounds or words. Animal communication systems are generally not productive. For many animals, there is a one-to-one correspondence between a vocalization and its meaning. For example, Vervet monkeys (*Chlorocebus pygerythrus*), which are native to Africa, have 36 distinct vocalizations, each with a unique message (Estes, 1992; Seyfarth & Cheney, 1992). Some of the calls are used to signal the presence of a predator (Seyfarth, Cheney, & Marler, 1980a, 1980b). One call signals a leopard. A different call signals a snake. Yet another call signals an eagle. Other vocalizations are referred to as *wrrs* and *chutters* and are used in social situations with other monkeys (Tomasello & Call, 1997).

A third important characteristic of human language is that there is rarely a connection between the meaning conveyed by an utterance and the utterance's form—how it sounds or is written. Because of the arbitrariness between form and function in human language, there is no limit to the meanings that can be conveyed by a human language. Researchers refer to this characteristic as the **principle of arbitrariness.** Communication among bees provides an example of a communication system in which the relationship between form and function is not arbitrary. When a bee returns to the hive after locating a food source, it will communicate to others about the food source by carrying out a dance. The moves carried out in this dance communicate the distance and the direction to the food source (Riley, Greggers, Smith, Reynolds, & Menzel, 2005; von Frisch, 1967). In this way, the features of the dance map correspond directly to the message being conveyed.

In human languages, there are a few exceptions to the principle of arbitrariness. For some words, the relationship between the form of the word (i.e., how it sounds or how it looks) and the meaning is not arbitrary, such as cases of onomatopoeia. While all languages have words of this type, there are differences in the form of the words used in different languages. English speakers describe rain failing by saying *drip drop*. French speakers say *plic plic*. Italian speakers say *plin plin*. Russian speakers say *kap kap*. English roosters says *cock-a-doodle-doo,* but French roosters say *cocorico*. The fact that such words have different forms in different languages illustrates humans' ability to associate meaning with arbitrary sound sequences.

Can Nonhuman Animals Learn Human Language?

Over the past century, there have been numerous attempts to teach human language to a variety of nonhuman species, including chimpanzees, gorillas, bonobos, parrots, and dolphins. These attempts have yet to produce nonhuman animals with the range of language skills that most 5-year-olds possess. The lack of complete success supports the view that human language is unique to humans and likely part of our biological endowment. Nevertheless, these efforts have led to some remarkable successes. If a future attempt can show that nonhuman animals can acquire the entire system of a human language, the demonstration would be strong evidence against the nativist view, because it could not be the case that human language is hardwired in humans. A demonstration that a nonhuman animal can fully master human language would be strong evidence in favor of the view that language is one of many cognitive skills that can be acquired through general learning principles by humans and nonhumans.

Gua

Two of the early attempts to teach chimpanzees a human language were conducted by scholars who attempted to raise a chimpanzee in the same manner as they would a human child. They tested the idea that if raised in a human environment, the infant chimpanzee would come to understand and to use human language. The first attempt was carried out by Winthrop Kellogg, a professor in Indiana, and his wife, Luella. They raised a chimpanzee named Gua from infancy alongside their infant son, Donald (Kellogg & Kellogg, 1933). The experiment ended after 9 months. The chimpanzee learned to use a cup and a spoon, learned to shake hands, but failed to learn to speak. A second attempt was made in 1951. Keith and Catherine Hayes began raising a chimpanzee named Viki and teaching her human language. They raised her as they would a human child, as the Kelloggs had raised Gua. The Hayeses recognized that Viki might have trouble articulating human speech, so they provided Viki with speech therapy. As with Gua, Viki was unable to learn to speak. In 2 years, Viki learned only two words.

Washoe

In the late 1960s, Allen and Beatrix Gardner began to work with a chimpanzee named Washoe. Washoe was a female born in 1965 in West Africa. She was captured by the United States Air Force for possible use in the space program. The Gardners recognized that chimpanzees do not have the ideal vocal architecture for human speech sounds. They decided to teach her ASL. They taught Washoe ASL in a manner similar to how children are taught. Washoe demonstrated great success in learning signs: 350 in all. Much has been made over the fact that Washoe demonstrated the principle of productivity, producing the combination of signs *waterbird* when seeing a swan. There were also reports that Washoe taught ASL to her adopted son Louis (Fouts, Fouts, & Van Cantfort, 1989). Washoe never mastered the ordering of words in sentences. She used sentences such as the following to refer to the same circumstance: *Roger tickle Washoe* and *Washoe tickle Roger* (Brown, 1970); thus, it was concluded that she failed to master the grammatical structure of language. It has been pointed out that when responding to questions, Washoe used the appropriate part of speech. For example, when asked *Where is the box?* or *When is dinner?* Washoe responded using a noun, specifying a location and a time, respectively (Gardner & Gardner, 1975).

Nim Chimpsky

In the 1970s, Herbert Terrace, a professor at Columbia University, began his own project to teach a chimpanzee ASL. Terrace aimed to disprove Chomsky's theories and also to improve on aspects of the Gardners' project, which he thought was flawed. He named the chimpanzee Nim Chimpsky, a reference to Noam Chomsky, the linguist. The chimpanzee came to be called Nim. Nim was able to learn 125 signs but, like Washoe, never seemed to master the word order rules in ASL. Terrace reported that Nim's longest utterance was "Give orange me give eat orange me eat orange give me eat orange give me you" (Terrace, 1979). One can infer quite easily Nim's desire in this utterance; however, Nim's ordering of words appears random.

▲ **Photo 2.6** Bonobos are endangered. Their native habitat in the Democratic Republic of Congo is disappearing rapidly.

Koko

One of the longest-lasting nonhuman ASL learning studies is that involving a gorilla named Koko. Koko is a western lowland gorilla. The project has lasted more than 30 years (The Gorilla Foundation, n.d.). In that time, Francine "Penny" Patterson, her trainer, has worked closely with Koko. It is estimated that Koko knows about 1,000 signs and understands approximately 2,000 spoken English words. There have been reports that Koko, like Washoe, produced novel combinations of signs, such as *waterbird.* However, like Gua, Viki, and Nim, Koko has not mastered the systematic use of word order when forming sentences. Nevertheless, Koko is a remarkable animal, being one of the only captive animals that has had a pet of her own. Over the years, Koko has cared for several cats.

Kanzi

The most impressive example of a nonhuman primate learning human language has involved a bonobo (*Pan paniscus*) or pygmy chimpanzee named Kanzi (Greenfield

& Savage-Rumbaugh, 1991; Savage-Rumbaugh, Shanker, & Taylor, 1998). Kanzi was born in the Yerkes Field Station at Emory University in Atlanta, Georgia. He was cared for by a dominant female, who had actually stolen him away from his birth mother soon after birth. The dominant female was selected to take part in a language learning study involving a keyboard containing lexigrams (i.e., symbols representing words). Kanzi was always taken along. Despite the fact that Kanzi appeared not to be paying much attention to the activities occurring with his adoptive mother, later he showed remarkable mastery of the lessons. He has been described as understanding simple spoken sentences and responding appropriately with the keyboard of lexigrams (Savage-Rumbaugh & Lewin, 1994). Today, Kanzi lives at the Great Ape Trust in Des Moines, Iowa, and understands approximately 3,000 spoken English words. When the word is pronounced, he can point to the appropriate symbol on the keyboard (Raffaele, 2006).

▲ Photo 2.7 Bottlenose dolphins are found in oceans around the world and are frequently featured performers in aquariums and zoos. Do you think that one day it will be possible for humans to have conversations with dolphins?

Dolphin Studies

Remarkably, one can find evidence that species other than primates also can learn to understand simple sentences. Studies with bottlenose dolphins (*Tursiops truncates*) have shown that there are dolphins that can be trained to understand simple artificial languages, having a set of systematic grammatical rules (Herman, 1980; Herman, Richards, & Wolz, 1984). For the studies, the researchers created artificial languages, involving either sounds generated by a computer or arm–hand gestures performed by humans. The artificial languages were created to have word order rules. For both types of artificial languages (i.e., acoustic and visual), dolphins were able to carry out an action indicated by the sentence with accuracy well above chance, such as *fetch frisbee* or *toss frisbee (in) basket*. They were also able to comprehend novel sentences and novel sentence structures very well. The studies focused only on the comprehension abilities of dolphins and did not attempt to teach dolphins to produce language forms. Another species that has shown an ability to communicate with humans is the African gray parrot. Text Box 2.3 describes the research that was conducted with Alex, probably the most famous African gray parrot in the world.

Text Box 2.3 Extraordinary Individuals: Alex, the African Gray Parrot

▲ Photo 2.8 This is a typical African gray parrot. Have you ever interacted with a talking parrot?

Alex (1976–2007) was an African gray parrot (*Psittacus erithacus*) that became a celebrity, of sorts, among those interested in language development. His owner and trainer Irene Pepperberg worked with him for over 30 years and found that he was able to master a wide range of cognitive skills, including listening comprehension, which up to that point were considered to be only possible for primates. Alex had a vocabulary of 150 words, learned to recognize on sight 50 objects, and could distinguish different quantities of objects, involving sets of 6 or fewer ("Alex the African Grey," 2007). The technique that Pepperberg used with Alex is described as the model/rival technique. One trainer made a request of a second trainer, who performed the task with Alex observing. When the second trainer performed correctly, the first trainer rewarded the person with positive attention or with a treat. When the second trainer performed incorrectly, no reward was given. Then, Alex was given an opportunity to perform a similar task. By carefully changing the words and objects involved in the tasks, it was shown that Alex was not merely mimicking previously observed behaviors, but, in fact, he could interpret the spoken requests, such as show me the yellow square.

Summary and Theoretical Implications

Research has shown that there are specific locations in the brain involved in language production and language comprehension. Since the mid-1800s, researchers have attempted to locate these brain areas using a variety of techniques, beginning with postmortem autopsy of patients with aphasia and continuing today with modern brain-imaging techniques. Modern brain-mapping techniques can provide both structural and functional information about the brain. Using these techniques, researchers have identified many types of aphasia. Recent advances in genetic analysis have shown that there are language impairments that can be inherited and that are associated with specific genes. Studies of brain anatomy and genetics of humans and nonhuman primates provide clues to how language might have evolved. Efforts to teach nonhuman primates human languages have generally been unsuccessful. These results are expected because of the differences in vocal tracts and brain areas of human versus nonhuman primates.

In terms of the four theoretical approaches to language development, the evidence regarding the biological basis for language is most consistent with the generative approach because research has shown that there are specific brain areas associated with language processing and because specific genes appear to be involved in the occurrence of language-related disorders. The fact that nonhuman primates and other animals have not been completely successful in mastering human languages is also support for the view that language may be unique to humans and learnable only by humans. On the other hand, advocates of the statistical learning approach could argue that the biological facts regarding language do not rule out the possibility that language learning occurs in the same manner as other types of learning. Advocates of this perspective might say that having multiple language areas in the brain, rather than a singular area, is unexpected if language is processed in a modular fashion, as claimed by the generative approach. One might expect the language module to be a single area or regions that are closer together in the brain. The biological facts of language provide little support for either the behaviorist or social-interactionist views of language development. The fact that there are language disorders linked to genes is inconsistent with the behaviorist view that language develops due solely to learning experiences.

KEY TERMS

agrammatism
agraphia
alexia
anomic aphasia
aphasia
arcuate fasciculus
bimodal bilinguals
Broca's aphasia
caudate nucleus
conduction aphasia
corpus callosum
crossed aphasia
dyslexia
electrical brain stimulation (EBS)
expressive aphasia
FOXP2
global aphasia
localization hypothesis
mixed transcortical aphasia
neologism
occipital lobe
onomatopoeia
pars opercularis
pars orbitalis
pars triangularis
planum temporale
primary progressive aphasia
principle of arbitrariness
principle of displacement
principle of productivity
pure aphasia
pure word deafness
receptive aphasia
right ear advantage (REA)
specific language impairment (SLI)
transcortical motor aphasia
transcortical sensory aphasia
Wada testing
Wernicke's aphasia

REVIEW QUESTIONS

1. What is the localization hypothesis? In history, who has believed the hypothesis?
2. What evidence was instrumental in convincing people that the phrenologists' assumptions about the brain were wrong?
3. What are the symptoms of Broca's aphasia? What region of the brain is typically damaged in individuals with Broca's aphasia?
4. What are the symptoms of Wernicke's aphasia? What region of the brain is typically damaged in individuals with Wernicke's aphasia?
5. When researchers say that the human brain is lateralized for language, what do they mean?
6. What is Wada testing? What did studies using this procedure show about language and the brain?
7. What is EBS? What did studies using this procedure show about language and the brain?
8. What evidence is there that people differ in the extent to which their brains are lateralized for language?
9. What are the two broad categories of brain-imaging techniques? Provide an example of each.
10. Describe what is involved in the following brain-imaging techniques from the patient or participant's point of view: PET, fMRI, and EEG?
11. Besides Broca's aphasia, identify and describe two other types of expressive aphasia.
12. Besides Wernicke's aphasia, identify and describe two other types of receptive aphasia.
13. Identify the three subregions within Broca's area. What type of processing is associated with each region?
14. What are the symptoms of agrammatism? What area of the brain is typically damaged in someone who suffers from agrammatism?
15. What brain regions have been found to be responsible for language acquisition?
16. What is the best evidence that DNA plays a role in typical language development?
17. What are the symptoms of SLI?
18. What does the fossil record suggest about the language abilities of Cro-Magnons?
19. What does the fossil record suggest about the language abilities of Neanderthals?
20. What theories about the evolution of language have been proposed?
21. How do human languages differ from the forms of communication in other species?
22. Discuss the attempts that have been made to teach chimpanzees, gorillas, and bonobos human language. How successful have these attempts been?

23. What did Washoe demonstrate about the ability of chimpanzees to learn human language?

24. What has the case of Kanzi revealed about the ability of nonhuman animals to learn human language?

25. Discuss the attempts that have been made to teach dolphins human language. How successful have these attempts been?

RECOMMENDED READING

Battro, A. M. (2001). *Half a brain is enough: The story of Nico.* Cambridge, UK: Cambridge University Press.

Dawkins, R. (2004). *The ancestor's tale, a pilgrimage to the dawn of life.* Boston: Houghton Mifflin Company.

Gazzaniga, M. S. (2005). Forty-five years of split-brain research and still going strong. *Nature Reviews Neuroscience, 6*(8), 653–659.

Hess, E. (2008). *Nim Chimpsky: The chimp who would be human*. New York: Bantam.

Kellogg, W. N., & Kellogg, L. A. (1933). *The ape and the child: A comparative study of the environmental influence upon early behavior*. New York: Hafner Publishing Co.

Lewis, J. (1981). *Something hidden: A biography of Wilder Penfield*. New York: Doubleday.

McMillan, M. (2002). *An odd kind of fame: Stories of Phineas Gage.* Cambridge, MA: MIT Press.

Mithen, S. (2006). *The singing Neanderthals: The origins of music, language, mind, and body.* Cambridge, MA: Harvard University Press.

Pepperberg, I. (2009). *Alex and me: How a scientist and a parrot discovered a hidden world of animal intelligence—and formed a deep bond in the process*. New York: Harper Paperbacks.

Savage-Rumbaugh, S., & Lewin, R. (1994). *Kanzi: The ape at the brink of the human mind*. Hoboken, NJ: Wiley.

Sebeok, T. A., & Umiker-Sebeok, J. (1980). *Speaking of apes: A critical anthology of two-way communication with man.* New York: Plenum Press.

Taylor, J. B. (2008). *My stroke of insight: A brain scientist's personal journey*. New York: Viking.

Terrance, H. S. (1979). *Nim: A chimpanzee who learned sign language.* New York: Knopf.

RECOMMENDED FILMS

Apsell, P. S. (Executive producer). (1993). *Stranger in the mirror* [Television series]. *NOVA*. Boston: WGBH Educational Foundation.

Apsell, P. S. (Producer). (1997). *In search of the first language* [Television series]. Boston: WGBH Foundation.

Grubin, D. (Director). (2002). *The secret life of the brain* [DVD]. Available from www.pbs.org.

Rubin, J. (Director/Writer/Producer). (2008). *Ape genius* [Television series]. *NOVA*. Boston: WGBH Foundation. Available from www.pbs.org.

Schroeder, B. (Director). (2007). *Koko: The talking gorilla*. United States: Criterion.

Spinks, S. (Director & Producer). (2002). *Inside the teenage brain* [Television series]. *Frontline.* Boston: WGBH Foundation. Available from www.pbs.org.

White, J. B. (Producer). (2009). *Becoming human.* [Television series]. *NOVA.* Boston: WGBH Foundation. Available from www.pbs.org.

SUGGESTIONS FOR CLASS PROJECTS

1. Investigate the extent to which the left hemisphere is dominant for language by observing the ear preference for cell phone users. Most people display a REA for language and tend to prefer to hold a telephone or cell phone to their right ear when listening. Carry out naturalistic observation in public areas where lots of people are likely to be talking on their phones. Work individually or in teams. Record the sex of the person that is observed using a cell phone and which ear the cell phone is near when the person is talking.

2. Investigate the relationship between sex, handedness, and ear preference for telephone use in a survey. Use the Oldfield (1971) handedness questionnaire as the core of the survey, and add in questions about the respondent's sex (i.e., male vs. female) and ear preference when using a telephone (i.e., left vs. right). Analyze the data on your own, or your instructor can compile data from all students and analyze it during class time.

3. Make a study of examples of onomatopoeia across languages. Generate 20 English words that are examples of onomatopoeia. In a dictionary, the origin of the word will be listed as onomatopoeia. Use searchable dictionaries on the Internet to find the translations of these words in three other languages. You will find that there are differences in onomatopoetic words across languages. Speculate about why such differences exist.

For additional ancillary resources, please visit the companion website at www.sagepub.com/kennison.

- Video Links
- Audio Links
- Web Resources
- Internet Activities
- Flashcards
- Web Quizzes

CHAPTER 3

THE FIRST TWENTY-FOUR MONTHS

The Infant at Birth

▲ Photo 3.1 The infant is wearing headphones. At what age do you think that infants typically start listening closely to speech sounds and music?

According to the view that the infant comes into the world as a *tabula rasa*, or blank slate, one would expect the infant to come into the world knowing nothing. Research suggests that this extreme view is incorrect. Children come into the world with some pretty amazing inborn behaviors. For example, they come equipped with reflexes, which are the result of innate mechanisms. Table 3.1 describes some of these reflexes. At birth, all infants display the stepping reflex, which occurs when infants' feet come in contact with the ground or a hard surface. Infants lift their feet and make a walking motion. Infants are not strong enough to support their weight on their legs and feet; thus, walking is not actually possible by the infant. By the third month of life, the stepping reflex disappears. All infants also display the crawling reflex, which occurs when infants are placed on their stomachs. They move their arms and legs in a crawling motion. The reflex disappears by the age of 2 months. Later in development, most infants go on to crawl before they walk; however, some

Table 3.1 Some of the Reflexes Observed Universally in Newborns

Reflex	Description	Disappears
Sucking reflex	When roof of infant's mouth is touched, lips will close and sucking will occur	About 2 months
Crawling reflex	When placed on stomach, infant's legs will make a crawling motion	About 2 months
Startle reflex	Infant will cry when losing support and falling or when a loud sound is heard	About 3 months
Stepping reflex	When feet touch the ground or a surface, infant will lift feet and make a walking motion	About 3 months
Babinski reflex	When side of foot is touched, toes will fan out and foot will turn inward	About 4 months
Rooting reflex	When cheek is touched, infant will turn in the direction of the touch	About 4 months
Palmar grasp	When palm is touched by a finger, infant will grasp the finger	About 6 months
Gag reflex	Prevents choking	Does not disappear
Right reflex	Head will be lifted when placed on the stomach to clear nose and mouth	Does not disappear

Source: Levine and Munsch (2011).

infants skip the crawling phase altogether. Those infants who do crawl devise numerous ways to achieve it (Arlitt, 1946).

Table 3.2 Timeline of Selected Language Milestones From Birth to Twenty-Four Months

Milestone	Average Age
Facial imitating	At birth
Social smiling	6 weeks
Cooing	6 to 8 weeks
Babbling	4 to 6 months
Gesturing	8 to 10 months
Saying first words	10 to 12 months

Source: Adapted from Brown (1973).

Noam Chomsky, the founder of the generative approach, has compared learning language to learning to walk (Chomsky, 1959). No one teaches a child to walk; however, the physically healthy and normally developed children will eventually discover their ability to do so. Most parents will vividly recall their child's first step. In many cases, children will spend days or weeks pulling themselves up and standing. After some experimenting, the child is off and walking, taking the first few wobbly steps. Similarly, Chomsky claimed that the infant comes into the world with the physical mechanisms in place necessary for language to emerge (Searchinger, Male, & Wright, 2005). Table 3.2 displays the language development milestones for the first 24 months.

Infants are, in fact, remarkably capable at birth and soon after birth. Many of their abilities appear to be related to, at least in part, inborn knowledge. The remainder of this section will review infants' ability to imitate, to see, and to hear. Infants' hearing ability is particularly well developed, as they are able to perceive the entire range of possible human speech sounds.

Imitation

Infants are well equipped at birth to imitate the facial expressions of others. Being able to imitate facial expressions can aid in learning how to produce speech sounds. In their now classic study, Meltzoff and Moore (1983) showed that newborns who are only minutes old can imitate facial expressions of strangers. They showed that infants who had viewed only one or two human faces in the minutes following birth could imitate facial expressions. The logistics of how the study was done is almost as interesting as the study's results. The research team received permission from a group of expecting mothers prior to the delivery of their infants. The research team made sure that they were in the hospital when the mothers arrived in labor. Soon after the delivery, the researchers were allowed a brief time with the infant. During testing, a member of the research team made a series of facial movements and videotaped the responses of the newborn. These videotapes showed that infants had no problem matching the researcher's facial expression. The photos displayed in Figure 3.1 were taken during a similar study carried out on infants who were between 12 and 21 days old (Meltzoff & Moore, 1977). The three facial expressions that were successfully imitated were (1) the tongue protruding, (2) the mouth opened in a circle, and (3) the lips protruding.

The fact that infants can imitate others so soon after birth suggests that imitation requires little or no learning. Today, researchers believe that there are specialized cells

Figure 3.1 Infants were able to imitate a stranger's facial expressions. The stranger in the photo is the researcher Andrew Meltzoff

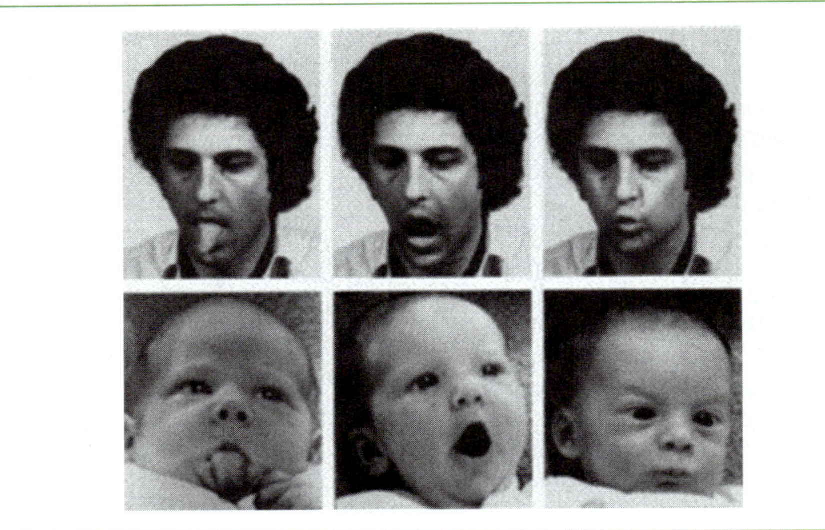

or neurons in the brain that may explain the ability of humans to imitate others so easily. These cells are called **mirror neurons.** They were first reported in 1992 by neuroscientists studying macaque monkeys (Di Pellegrino, Fadiga, Fogassi, Gallese, & Rizzolatti, 1992). The researchers discovered specialized neurons in the brain that responded when the animal observed another performing an act and when the animal itself performed the act, such as putting something to its mouth. The research on the functioning of mirror neurons in animals and humans is in its beginning stages. There are many questions that remain unanswered. For now, their existence provides an intriguing possible explanation for infants' imitative skills. Infants demonstrate other extraordinary abilities as well. Text Box 3.1 describes the research that discovered yet another way in which infants are extraordinary: They can add and subtract.

The fact that newborn infants can imitate faces is even more amazing when one considers the fact that at birth, the infant's visual system is not fully developed (Vital-Durand, Atkinson, & Braddick, 1996). At birth, their vision is about 20/400, which for an adult would be legally blind. At birth, infants cannot see color, only shades of gray, and their eyes cannot focus on objects that are near them. Parents may notice that sometimes infants' eyes do not appear to be moving together. It takes several months for infants' eyes to become well coordinated. By the age of 6 months, infants' vision is typically 20/25, which is almost as good as an adult with perfect vision (i.e., 20/20). Between birth and 6 months, infants can develop their visual skills faster if they experience an interesting variety of visual stimuli. The American Optometric Association (AOA) (2011) has

Text Box 3.1 Extraordinary Individuals: Infants Can Add and Subtract?

Long before infants can say the numbers *one, two, three,* they show an understanding of adding and subtracting. In the experiments conducted originally by Karen Wynn (1992) and replicated by several others (Simon, Hespos, & Rochat, 1995; Uller, Carey, Huntley-Fenner, & Klatt, 1999), infants as young as 5 months of age were seated on a caregiver's lap in view of a small stage on which a toy mouse was placed. A screen was lifted between the infant and the stage. The infant then saw a research assistant place a second toy mouse behind the screen. The screen was then lowered, allowing the infant to see the stage. On some trials, the stage held the two toys. On some trials, the stage held just one toy. The researcher measured how long the infant looked at the stage when the screen was lowered. They found that the infants looked longer when they saw only one toy on the stage, indicating that the outcome violated an expectation of seeing two toys. Figure 3.2 displays the sequence of events that occurred in the study. The researchers also carried out trials in which infants' subtraction ability was tested. Two toys were initially placed on the stage. The screen was then raised, and the researcher removed one of the toys. When the screen was lowered, either one or two toys were present. Infants, again, looked longer when the unexpected outcome occurred. These and other similar results demonstrate that children appreciate the individuality of small numbers of objects and use that knowledge when evaluating what they see.

Figure 3.2 Experimental manipulations, testing infants' ability to add and to subtract

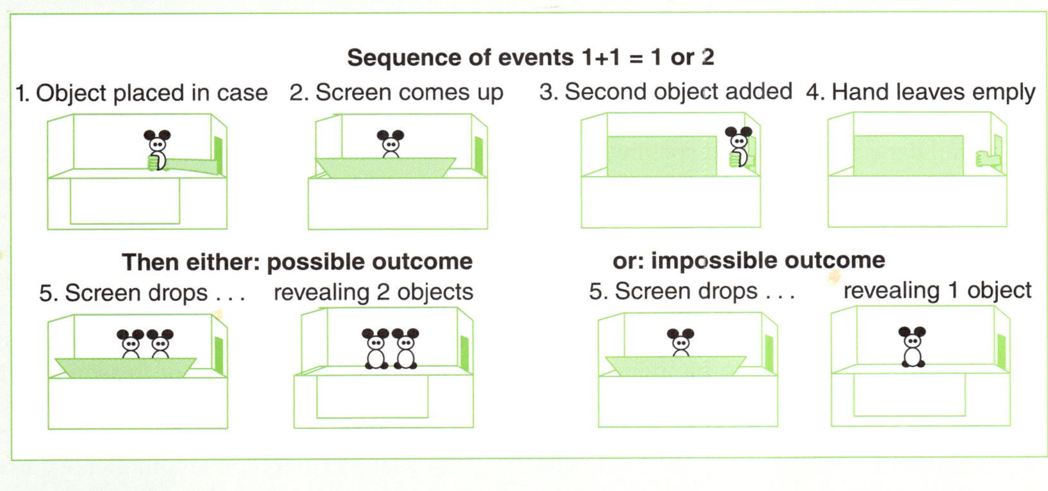

Source: Wynn (1992).

recommended that parents routinely change their infants' environment by changing the location of the crib. Many decorations for infants' rooms and mobiles now come with vibrant colors and interesting geometric shapes. These decorations may be helpful as infants learn how to see.

Hearing

At birth, the hearing of infants is particularly well developed. There is some reduction in hearing due to the fact that the middle ear may contain fluid. As the fluid dissipates, hearing improves. Hospitals routinely test newborns' hearing before they are allowed to leave the hospital (Northern & Downs, 2002). Two tests are routinely used. The first test is the **otoacoustic emissions test,** or OAE. A small earphone and microphone are placed in the infant's ear canal. Sounds are played. When an infant hears normally, an echo is produced. If an infant is hearing impaired, then no echo will be observed. The second test is a test of **auditory brainstem response,** or ABR. Small electrodes are placed on the infant's head, and the electrical activity is recorded as the infant listens to sounds. Before these techniques were developed, deaf or hearing impaired infants were diagnosed much later when they were discovered to be unable to respond to environmental sounds, such as their parents calling their names.

Even with the fluid in the middle ear, infants are quite sophisticated hearers. Newborns are able to distinguish speech sounds from all the world's languages—even the speech sounds that do not occur in the language(s) spoken in their environment (Eimas, Siqueland, Jusczyk, & Vigorito, 1977; Jusczyk, 1999). Infants' perceptual abilities are quite extraordinary, considering that it has been estimated that across the world's languages, there are about 600 consonants and 200 vowels (Ladefoged, 2001). Infants appear to come into the world equipped to handle any speech sound that they might hear. During the first year of life, infants come to learn how to produce the speech sounds that they hear regularly.

The phonemes that infants of English-speaking parents hear and learn to produce are displayed in Table 3.3 along with their phonetic symbols. In English, there is a loose relationship between spelling and pronunciation, which can make the discussion of phonemes more challenging for speakers of English than speakers of languages in which the relationship between sound and spelling is unambiguous. In English, the spelling of a word may only partially indicate how the word is to be pronounced. Some letters, such as *c,* are associated with more than one pronunciation. The *c* in *car* is pronounced as a /k/, but the *c* in *receive* is pronounced as /s/. The word **grapheme** is used to refer to the spelling of a word or phoneme. In English, graphemes are letters. In other languages, such as Chinese, graphemes may be either an entire symbol or a portion of a symbol (e.g., 鸟 is the symbol *bird*). English is likely the most complicated language to spell, because the spelling-to-sound or **grapheme-to-phoneme correspondences** are so variable. One reason is that English has been a written language since before the 10th century. Many words with unusual grapheme-to-phoneme correspondences are borrowed by other languages or represent a period of time in the English language that had grapheme-to-phoneme correspondences that were different from those that are productive today. The poem "Our Strange Lingo" in Table 3.4 illustrates the complexities of English spelling.

Table 3.3 Phonetic Symbols for Consonants and Vowels in American English

Consonants				Vowels	
Top	/t/	Choke	/ tʃ /	Eat	/ iː /
Dig	/d/	Shake	/ ʃ /	Ate	/ eI /
Pot	/p/	Joy	/dʃ /	Shoot	/u:/
Bat	/b/	Measure	/ʒ/	Goat	/ o ʊ /
Mat	/m/	Wife	/w/	Rat	/æ/
Nap	/n/	Yes	/y/	Cut	/ Λ /
Fun	/f/	Sing	/ŋ/	Kite	/ a Λ /
Van	/v/	Uh-oh	/ʔ/	Boy	/Λj/
Sit	/s/	Look	/l/	Lit	/I/
Zoo	/z/	Ram	/r/	Set	/Λ/
Kiss	/k/			Foot	/Λ/
Game	/g/			Caught	/Λ:/
Those	/ ð/			Pot	/a/
Think	/ θ/			Uh	/au/
Home	/h/			Pout	/aw/

Source: Glucksberg and Danks (1975).

In the now-classic study that showed that infants can distinguish phonemes that do not occur in their native language, Eimas and colleagues (1977) utilized the head-turn technique, which you learned about in Chapter 1. Infants were trained to turn their heads when they heard a target speech sound. A variety of speech sounds were played. Of interest in this study and similar studies was the extent to which infants could distinguish pairs of speech sounds that differed in a minimal way, such as /p/ versus /b/, /t/ versus /d/, and /k/ versus /g/. These pairs of consonants differ only in one feature—**voicing,** or the presence of vibration in the vocal cords. The consonants /b/, /d/, and /g/ are all voiced consonants because they are produced with voicing. The consonants /p/, /t/, and /k/ are voiceless consonants because they are produced without voicing.

Categorical Perception

Infants' ability to distinguish speech sounds is described as **categorical perception,** which means that the hearer perceives most sounds as one phoneme or the other despite the fact that the sounds vary systematically along an acoustic continuum. For example, the sounds /pa/ and /ba/ are perceived by English speakers as referring to different speech sounds—/p/ as in *pat* and /b/ as in *bat*. Acoustically, there is only a small difference between the two

Table 3.4 The Spelling Oddities of English Are Wonderfully Manipulated in a Poem

Our Strange Lingo

Our Strange Lingo When the English tongue we speak.
Why is break not rhymed with freak?
Will you tell me why it's true
We say sew but likewise few?
And the maker of the verse,
Cannot rhyme his horse with worse?
Beard is not the same as heard
Cord is different from word.
Cow is cow but low is low
Shoe is never rhymed with foe.
Think of hose, dose, and lose
And think of goose and yet with choose
Think of comb, tomb and bomb,
Doll and roll or home and some.
Since pay is rhymed with say
Why not paid with said I pray?
Think of blood, food and good.
Mould is not pronounced like could.
Wherefore done, but gone and lone—
Is there any reason known?
To sum up all, it seems to me
Sound and letters don't agree.

Source: Cromer (1902).

sounds. The difference is found in the amount of silence that occurs right after the sound begins. Both sounds begin with a closing of the lips. During the production of a /b/ in English, the vocal cords vibrate, causing there to be only a small amount of silence following the beginning of articulation. The amount of time between the beginning of articulation for a consonant and when the vocal cords begin to vibrate is called **voice-onset time.** During the production of a /p/, the vocal cords do not vibrate, and there is a longer amount of silence following the beginning of articulation. If there are 25 milliseconds or more of silence, the sound is generally perceived as /p/; otherwise, it is perceived as /b/ (Wood, 1976). Figure 3.3 displays the relationship between amount of voice-onset time for the /p/ to /b/ continuum and the percentage of /b/ to /p/ responses.

Consonants can vary in terms of other features, besides voicing. They can differ in terms of **place of articulation** or where in the vocal tract the airflow is interrupted. For example, some consonants are created when the airflow is interrupted at the lips. These consonants, which include /p/, /b/, and /m/, are referred to as **labial consonants.** Consonants can also differ in terms of **manner of articulation,** or the extent to which the airflow is completely or only partially stopped during the production of the sound. Consonants that involve a complete interruption of airflow are called **stop consonants,** as in /p/, /b/, /t/, /d/, /g/, and /k/. In contrast, the consonants /f/ and /v/ are produced when the air is permitted to flow between the lips and teeth. Consonants produced when airflow is partially stopped between any two places of articulation (e.g., lips, teeth, or palate) are classified as **fricatives.** Figure 3.4 displays the human speech tract. During the production of speech, the airflow can be interrupted at many points along the route from the larynx to the lips.

Later research showed that the ability to distinguish speech sounds is not unique to human infants. Other animals, such as chinchillas, can also do this (Kuhl, 1987). Both infants and chinchillas (and likely other animals) demonstrate categorical perception. Unlike chinchillas, by the end of the first year of life, infants lose the ability to distinguish those speech sounds that do not occur in their environment (Jusczyk, 1999). One of the earliest studies by Werker and Tees (1984) showed that infants gradually lose the ability to perceive

Figure 3.3 Relationship between voice-onset time and perception of /b/ to /p/

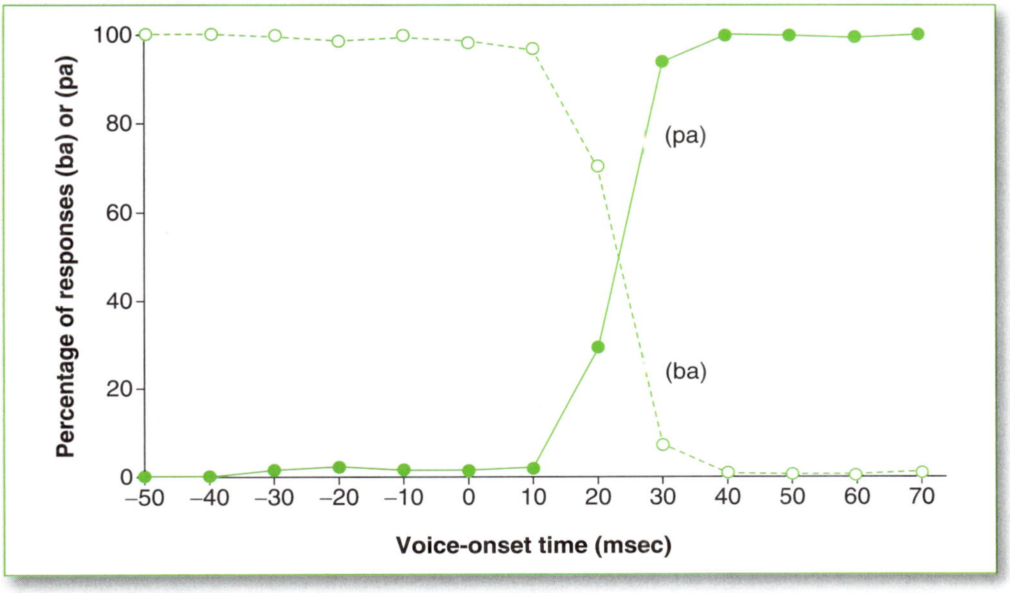

Percentage of responses (ba) or (pa)

Voice-onset time (msec)

(pa)

(ba)

Source: Adapted from Wood, C. C. (1976).

phonemes that they do not hear in their environments. In the study, infants of English-speaking mothers heard pairs of consonants that varied in voicing. Some of the consonant pairs never occurred in English. They were consonants that occur in Hindi and also in Inslekepmx, a language that is spoken in British Columbia. The results showed that all infants between 6 and 8 months of age could distinguish these phonemes, but few infants between the ages of 10 and 12 months could. In follow-up studies with infants whose mothers spoke Hindi, a language spoken in India, all of the 10- to 12-month-old infants could distinguish the sounds. These results and others like them demonstrate that the infants' perceptual abilities rapidly tune to their environment. Over time, infants lose the ability to distinguish contrasts that they do not hear. Recent research has shown that Japanese infants can distinguish the English phonemes /l/ and /r/ early in life but gradually lose that ability between 6 and 12 months of age (Kuhl et al., 2006). American infants raised in English-speaking environments have also been studied. They also lose the ability to distinguish phonemes that do not occur in English between the ages of 6 and 12 months (Kuhl, 2007).

Auditory Adaptation

Research by Patricia Kuhl and colleagues (Kuhl, Conboy, Padden, Nelson, & Pruitt, 2005) has shown that there may be a relationship between how quickly infants' perceptual abilities adapt to the language(s) in their environment and their later language development. In two

Figure 3.4 The diagram depicts the parts of the human speech tract. During the production of speech sounds, the airflow can be interrupted at numerous points from the larynx to the lips

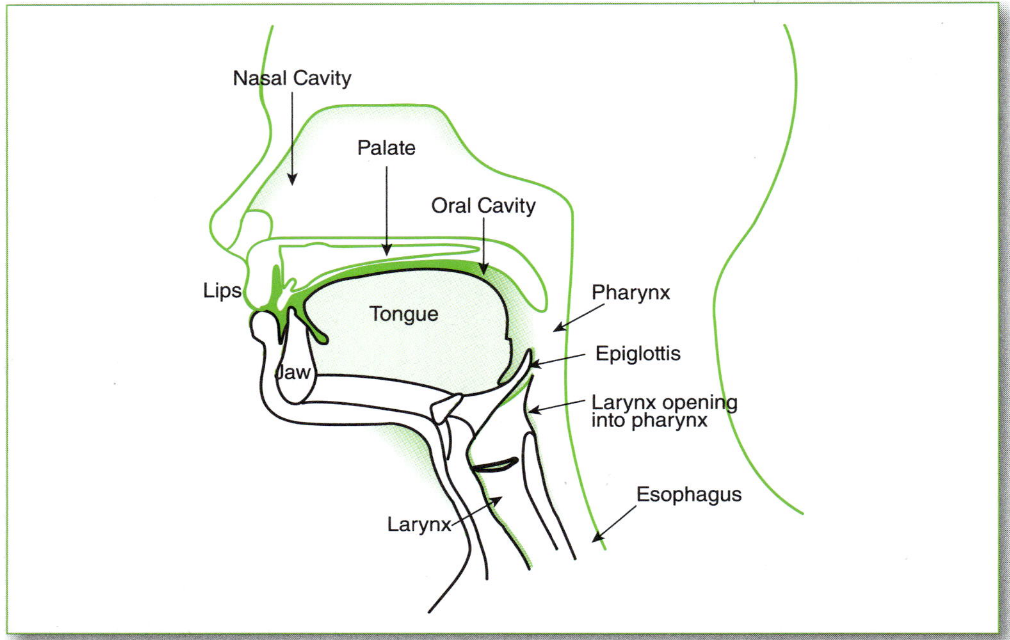

Source: Retrieved from http://en.wikipedia.org/wiki/Motor_theory_of_speech_perception

studies so far, they found that infants who demonstrated better performance when perceiving contrasts in their native language than when perceiving contrasts from other languages showed better performance in word comprehension later. They measured performance at 14, 18, 24, and 30 months (Kuhl et al., 2005). One implication of this research is that infants may, one day, be tested relatively early in life in order to predict their later word-learning aptitude. Although this type of intervention may take many years to develop, the hope is that infants showing delays in tuning into their native language sounds could be assisted through special learning sessions to improve their perceptual abilities.

When Does Learning Begin?

Although the imitative and perceptual abilities of infants demonstrate that infants are far from *blank slates* at the time of birth, infants have a great deal to learn after they enter the world. The knowledge that they gain becomes part of their memories, first starting out as a sensory memory, such as a sound or visual image, before becoming a long-term memory. One of the amazing facts about infants is that they begin forming long-term memories even before they are born. The best evidence that infants form long-term memories before birth comes from studies investigating infants' memories for sound. We know that infants can hear before they are born, because expectant mothers sometimes report feeling fetal movement when there is a loud sound in the environment. Studies have shown that fetuses only 26 weeks old (6.5 months) can respond to environmental sounds (Kisilevsky, Muir, & Low, 1992).

A substantial body of research has shown that infants' experiences before birth are remembered and can influence their behavior after birth. Research by Spence and DeCasper (1987) found that infants prefer to listen to actual human speech as compared with speech that has been altered by a computer (Spence & DeCasper, 1987). They also prefer to listen to regular speech versus speech that is played in reversed order (Peña et al., 2003). Studies have also shown that infants can distinguish their mother's voice, which they have heard through the womb during pregnancy, from the voices of unfamiliar others (DeCasper & Fifer, 1980; Kisilevsky et al., 2003). Fetuses are likely to form their best auditory memories of the mother's voice, because it is the only sound that is transmitted from within the mother's body (Petitjean, 1989). Infants can also distinguish between auditory samples of their mother's language and samples of other languages (Mehler et al., 1988). In a study by Moon, Cooper, and Fifer (1993), 2-day-old infants demonstrated a preference for their native language. These studies and others like them do not mean that infants understand a particular language; rather, the sound and rhythm of the language are familiar to them due to their prior exposure to the language.

Studies with infants have also demonstrated that they form memories of specific auditory events that occur before birth. In a particularly clever study, it was shown that infants formed memories for chunks of specific language that they heard before birth. DeCasper and Spence (1986) recruited a group of pregnant women to read a part of a story every day during the last 6 weeks of their pregnancies. When the infants were born, the researchers compared infants' responses to the previously heard passage and responses to a passage that had never been read by the mother. The results showed that the infants responded differently to the familiar passage than the novel one. In contrast, a control group of infants whose mothers did not read the target passage during pregnancy responded to the target and novel passages similarly.

▲ Photo 3.2 After 26 weeks' gestation, the fetus can hear sounds occurring outside of the womb. Do you think it would be useful for expectant parents to talk to their unborn baby?

Types of Memory

In both children and adults, there are multiple types of memories that can be formed (Ashcraft

& Radvansky, 2009). Memories for specific events that one experiences are called episodic memories (also called autobiographical memory). For adults, episodic memories might include birthday parties, experiences with family and friends, and notable occasions, such as graduating from high school. The memory contains details of the episode as it was experienced. In contrast, memory for factual knowledge, such as the knowledge learned in school, is called **semantic memory.** One's knowledge about the meanings of words would be included in semantic memory. Most of the research on long-term memory has been conducted on adults. In that work, researchers refer to memory that we know we have and usually can talk about as explicit memory. Memory that we may not know we have stored is called implicit memory. An example of implicit memory would be memory for how to ride a bike or memory about how to punch in numbers on a phone keypad to call a particular person. These memories for how to perform specific tasks are examples of procedural memory.

The exact nature of the memories that infants first form is unclear. It is also not clear how infant memory develops over time. The nature of the memories formed soon after birth may differ from the memories formed later in life, such as when the child is learning facts in school. What is clear is that memories begin to form not only before birth but soon after birth as well. Swain, Zelazo, and Clifton (1993) conducted a study with infants only 24 hours old and showed that they demonstrated long-term memory for speech sounds. They first tested the infants when they were 24 hours old and had them listen to the speech sounds. They tested the infants 24 hours later, using a variation of the head-turn procedure. The infants responded differently to sounds that they had heard the previous day than to unfamiliar sounds. Prior to that study, there was no evidence that infants formed memories that lasted longer than a few minutes.

The Earliest Memories

While we know that infants form memories before birth and soon after, there is still a great deal that we do not know about how memory develops in the first 24 months of life and beyond. Most adults are not able to report many memories for early childhood. The term ***childhood amnesia*** has been used to describe the fact that most of our childhood experiences are forgotten. Early research has suggested that most of the early childhood memories remembered by adults occurred between 2 and 3 years of age (Usher & Neisser, 1993). More recent research has attempted to identify the age at which the childhood amnesia wanes, which is the age at which one has memories for life events. The age is estimated to be between 3 and 5 years (Multhaup, Johnson, & Tetirick, 2005). The causes of childhood amnesia are not known. It may be that memories formed before one is able to verbalize one's experience are not stored in memory very well or are not easily retrieved from memory. Another possibility is that the changes occurring in the child's brain due to growth and maturation make it difficult to access memories formed early in life. It may be a combination of these possibilities as well as causes not yet discovered.

Most, if not all, of the memories formed in the first year of life are not remembered later in life. It is only through carefully designed studies that researchers have been able to obtain evidence that memories are formed. Mandel, Jusczyk, and Pisoni (1995) showed that infants 4.5 months of age preferred listening to their own names. Infants show a preference for listening to familiar words, such as cup and spoon, versus unfamiliar words by the age of 7.5 months (Jusczyk & Aslin, 1995). Infants who are 10 months old are able to recognize words that they have heard before in a laboratory environment (Jusczyk, 1997) (see Table 3.5).

Table 3.5 Milestones of Lexical Development

Milestone	Age
Perceiving speech	6 to 8 months
Recognizing familiar speech sequences	8 to 10 months
Preferring familiar words	10 to 12 months
Learning to associate words with objects	12 to 14 months

Source: Based on Werker and Tees (1999).

What Role Do Adults Play in Infants' Language Development?

When one takes an informal poll of friends and family and asks how they think children learn language, one tends to find that the most frequently cited answer is that parents teach children to talk. The idea that development is influenced by the social environment, which is a central assumption of the social-interactionist approach to language development, was influenced by Lev Vygotsky (1896–1934), a prolific researcher and theorist. He was born in Orsha, in modern-day Belarus. He emphasized the role of social processes in child development, particularly language development (Lee & Smagorinsky, 2000). He coined the term *zone of proximal development* to describe those tasks that children can carry out themselves and those tasks that children can complete with some help from adults. He viewed the zone of proximal development as a continuum with the lower part of the range including the activities children can carry out independently and the upper part of the range including the more complex activities that children can accomplish with assistance from another. He believed strongly that play was a critical activity in development, enabling a child to engage in abstraction, a key element of cognitive development. He also recognized the strong connection between language and thought. He pointed out the relationship between oral speech and inner speech, which refers to the voice in our minds that we can sometimes experience when we are thinking to ourselves. He believed that language, specifically oral language, is first used by the child for social interaction and occurs when thought is turned into speech. In contrast, inner speech develops as a way to direct one's own behavior and occurs when one turns speech into thought. Perhaps, his most significant contribution is the idea that children acquire knowledge through social interactions, which inherently involve cultural knowledge. The process of *internalization* is used as children learn how to carry out activities in the context of the social environment or culture. He wrote extensively, publishing six volumes of work over a 10-year period. Considering his early death at the age of 34 from tuberculosis, one cannot help but wonder what he could have accomplished had he lived longer.

Today, many researchers find the work of Vygotsky relevant, particularly when considering how infants acquire language. Certainly, parents and other caregivers play some role in children's learning; however, the important question is what exactly their role is. Those who believe that language development occurs as it does because of innate mechanisms believe that the role of parents and caregivers in language learning is overstated (Pinker, 2007). There are ways in which adults interact with children that may be beneficial to children—either in their survival or their language learning or both.

Infant-Directed Speech (Motherese)

When most adults talk to infants, they adopt a way of talking that is different from how they would talk to another adult or teenager. They may use a high-pitched voice and an exaggerated intonation. When speaking to infants and small children, adults typically use short sentences and repeat key words (e.g., Hey there. Look at you! You are a happy baby. Yes, you are. Who's the happy baby? You are!). In English, the phrase *baby talk* is used to describe this manner of speaking. Among language development researchers, the terms ***motherese*** (Newport, Gleitman, & Gleitman, 1977) and ***infant-directed speech*** are used.

Careful examination of samples of motherese shows that it provides the infant with excellent examples of vowel sounds. The vowel sounds produced in motherese are typically stretched out, produced over a longer duration, and also particularly good prototypes for the vowel (Kuhl, 1999). Motherese may also help infants start to identify word boundaries in speech (Thiessen, Hill, & Saffran, 2007). The research investigating how infants use motherese supports the **phonological bootstrapping hypothesis,** which claims that language learning is aided by the infants' analysis of the characteristics of the speech that they hear (Morgan & Demuth, 1996). The more specific **prosodic bootstrapping hypothesis** suggests that the prosody or melody of the speech stream provides the infants with information about language that aids them during acquisition. The prosody of speech may provide information that infants can use to identify the beginnings and endings of words as well as the beginnings and endings of clauses and phrases. Furthermore, infants may use information about stress patterns within words to identify the beginnings and endings of syllables. There is evidence that infants are sensitive to the different types of adult speech (i.e., motherese vs. speech used between adults). Research has shown that infants actually prefer to listen to motherese over regular, adult-directed speech (Cooper & Aslin, 1990; Fernald, 1985).

▲ Photo 3.3 Around the world, the speech used to address infants is acoustically different from the speech used to address adults. Do you speak differently when you talk to a small child versus one of your peers?

Motherese Across Cultures

Motherese appears to be used all around the world in many cultures (Bryant & Barrett, 2007; Fernald et al., 1989; Grieser & Kuhl, 1988). The ability of infants and adults to identify the meaning of the motherese may be a product of natural selection, serving in some way to promote the survival of our species (Fernald, 1982). Darwin was the first to suggest that the expression of human emotion is the same across all humans (Darwin, 1872/2007). In 1971, Ekman and Friesen obtained convincing evidence for Darwin's view. The six universal expressions that all people can easily identify are sadness, happiness, anger, fear, surprise, and disgust. Fernald (1989) showed that adults listening only to the intonation of motherese utterances can distinguish those expressing approval from those expressing prohibition. In a recent study, Bryant and Barrett (2007) showed that samples of motherese can be readily distinguished from adult-directed speech by those who do not speak the language. They played American English samples of motherese and adult-directed speech to Shuar adults. The Shuar people are indigenous to Ecuador. The motherese or adult-directed speech expressed one of four meanings: (1) approval, (2) comfort, (3) attention, and (4) prohibition. The Shuar adults were highly accurate in distinguishing motherese from adult-directed speech. They were accurate in distinguishing the intentional content of the four categories of motherese but were less able to distinguish them in adult-directed speech.

If motherese is necessary for language development to occur, then one would expect that in cultures in which motherese or child-directed speech occurs infrequently, children would have trouble learning language. In fact, there are cultures where children are not generally spoken to before they can talk, and those children learn language in a timely fashion. The example of such a culture is the !Kung San people who live in the Kalahari Desert in Angola, Botswana, and Nambia. The !Kung San live in small groups of 10 to 30 people in a semipermanent camp located near a water source. They speak the !Kung language, which is spoken using click phonemes that do not occur in English or other Indo-European languages. The main character in the film *The Gods Must Be Crazy* is !Kung San. For thousands of years, they have followed a hunting and gathering lifestyle, eating meat provided by the men who hunt and roots, nuts, and berries that women gather. Women who give birth do so alone away from the camp. They return to work within hours after delivery. Giving birth in this manner gives women a sense of pride. The child-rearing practices of the !Kung San are particularly compelling and informative on the topic of language development. Children are not spoken to until they can talk (Konner, 2002; Pinker, 2007). This is not out of indifference to children. To the contrary, mothers are attentive caregivers. They carry their infants close to their bodies about 90% of the day and respond immediately to their crying. Studies have shown that !Kung San mothers respond to their crying infants more rapidly than mothers in the United States. The fact that !Kung San children acquire language fully and in a similar time frame as other children in the world has been cited as evidence that caregivers play a limited role in language development.

The evidence that child-directed speech is not the magic bullet in explaining language acquisition should not be taken to imply that parents and caregivers play no role in language development. Recent research suggests that infants' categorical perception can be influenced by the social environment (Kuhl, 2007). In a study in which American infants reared in

English-speaking environments were taught Mandarin speech sounds, Kuhl and colleagues (Kuhl, Tsao, & Liu, 2003) found that infants learned when a person or tutor was in the room but not when the tutor's image was delivered on a television screen or when the infant only heard the sounds without the presence or image of the tutor. The results are among the first to demonstrate that the presence of a person strongly facilitated infants' learning. The authors suggest that the presence of a person can increase the infants' attention during learning, leading infants to make better use of the acoustic characteristics of the speech sounds.

Motherese in Signed Languages

Research has shown that motherese is not just a phenomenon observed in spoken languages. Mothers who communicate with their infants using signed languages also display different signing behaviors with their infants than with adults. Masataka (1992) recorded interactions of eight mothers who were deaf and users of Japanese Sign Language as their first language when they were interacting either with their deaf infants or with a deaf adult. The results showed that mothers signed differently with infants than adults. When they signed to their infants, they signed more slowly, repeated signs, and used exaggerated movements in the execution of the signs. In a follow-up study, Masataka (1992) showed that when presented with signing that was intended for infants and signing that was intended for adults, deaf infants preferred the former.

First Communications

When Do Infants Begin Communicating?

For the first weeks of life, infants spend most of their time sleeping, around 16 to 18 hours a day. They cry when they are hungry or physically uncomfortable. They will also make noises when they breathe and eat. Such sounds are called **vegetative sounds.** Before the age of 6 weeks, the only sounds that infants typically make are cries and vegetative sounds. After the age of 6 weeks, infants begin to produce cooing, which consists of pleasant-sounding vocalizations, usually involving one elongated vowel (e.g., *oooooo* or *aaahhh*). Between 6 weeks and 12 months, children go from smiling to producing first words. By the middle of the second year, children typically have developed a vocabulary of several hundred words or more. The remainder of this section provides information about each of the important milestones that children experience during this time.

Social Smile

It's not until around 6 weeks of age that infants begin to show signs that they are recognizing others. Showing a sign of recognition can be considered a form of nonverbal communication. Around this time, when Mom or Dad enters the room, the infant might smile and appear pleased. This behavior is called the **social smile.** When infants begin smiling in response to a caregiver's presence, there is usually a clear sense that the infant recognizes the person and is happy that the person is there. Research shows that an infant does not need to see in order to display social smiles, as blind infants also produce social smiles, but less often than sighted infants (Rogers & Puchalski, 1986) The blind infant may recognize the presence of caregivers through the other senses, such as the sounds of a voice.

When infants older than 12 weeks do not take interest in others, make eye contact, or display social smiles, there is some cause for concern. Such infants should be evaluated by a pediatrician because of the possibility of **autism** or some other developmental disorder. Autism is characterized by impaired social functioning. Autism is difficult to diagnose, particularly in very young infants. Children may not receive a definitive diagnosis until the age of 3. Infants with autism may also show delays in cooing and may not produce laughter, which emerges around 16 weeks. Because laughter requires quite a lot of muscle control, children with neuromuscular problems may also not produce laughter at the typical point in development (Gallagher, Jens, & O'Donnell, 1983).

Babbling

The first speech sounds that infants produce occur when they begin **babbling,** which involves the production of nonsensical sounds and sound sequences. When one listens closely to the young infants' productions, one finds examples of vocal play in which the infant produces a range of sounds. Some of the sounds are speech sounds; some are not, such as blowing a raspberry, which involves extending the tongue and blowing so that the lip vibrates against the tongue. As the infant comes to produce primarily speech sounds, vocal play turns into the first stage of babbling. Babbling involves the repetition of sounds in a sequence that usually involves consonant–vowel combinations (Oller, 1980; Stark, 1980).

There are two stages of babbling. Initially, the infant repeats the same consonant and vowel, such as *bababa* or *gagaga*. This stage is typically first observed when the infant is between 4 and 6 months old. This type of babbling is referred to as **canonical babbling.** By 11 to 12 months, infants will produce sequences of speech sounds involving different consonants and vowels, such as *bagadabaga* or *mabadagama* The second stage of babbling is called **variegated babbling.** Around this time, infants will also produce intonation during babbling. The most easily recognizable intonation involves that used when one forms a question. The pitch of the voice rises at the end of the question. For declarative sentences, the pitch of the voice tends to fall at the end of the statement. Examinations of variegated babbling have found that infants' babbling increasingly takes on characteristics of the language(s) of the home. Brown (1958) referred to this phenomenon as **babbling drift.** By the time infants are about 10 months old, it is possible to distinguish the babbling of infants who have been raised in different language environments (Boysson-Bardies, 1993). Infants are likely to be able to imitate the speech sounds of those in their environment. Laboratory studies have shown that infants can be trained to imitate vocalizations by around 5 months of age (Kuhl & Meltzoff, 1982).

Babbling is not physically possible at birth. When an infant is born, the larynx, or voice box, is high in the throat, which allows the infant to breathe and eat at the same time. With a high larynx, one can produce vowel sounds but cannot articulate speech in which there are alternative consonant and vowel combinations. In the first 2 months of life, the infant's larynx descends, after which the infant is more capable of producing the types of syllables found in speech. With a descended larynx, the infant is also at a greater risk of having food enter the windpipe, causing choking. Some researchers have written about the fact that the adaptive value of human language must have been quite great because it comes with a clear risk—one that can sometimes cause death.

Infants who are deaf or have hearing loss will babble; however, there are some differences between their babbling and the babbling of hearing infants (Stoel-Gammon, 1988). Deaf infants may utilize a smaller set of speech sounds than hearing children, including more sounds produced with the lips (i.e., labials), such as /p/, /b/, and /m/; nasal sounds, such as /m/, /n/, and /ŋ/; as well as sounds produced using a partial stoppage of the airflow during articulation (i.e., fricatives), such as /f/, /v/, /s/, and /z/. Deaf infants who are taught sign language babble with their hands. They produce a variety of nonsensical gestures involving the hands and arms (Naeve Velguth, 1996; Petitto & Marentette, 1991).

Despite the fact that language is unique to humans, babbling is not. Intriguing research by Elowson, Snowdon, and Lazaro-Perea (1998) has shown that babbling occurs in nonhuman primates. They studied pygmy marmosets, which are native to the Amazon jungle of South America. Adult marmosets have numerous calls that are used to signal to other marmosets about the presence of dangers. Observations made in a captive colony of marmosets showed that young marmosets produced seemingly random sequences of adultlike vocalizations in the absence of any dangers. The vocalizations appeared to increase interactions with the adult marmoset caregivers. The authors suggested that babbling in human infants may similarly serve to increase interactions between infants and caregivers, which may be adaptive because it may increase the likelihood of infant survival.

First Gestures

Smiling and vocalization are not the only ways in which infants can communicate with others. They can also use gestures. For example, infants learn to indicate *no* by moving their head side to side. They learn to indicate *yes* by moving their head up and down. They learn to wave goodbye to others or blow kisses to show affection. They will eventually learn that a shrug of the shoulders means I don't know. They may indicate that they want to be picked up by stretching out their arms toward someone. They might hold out an object to show or give it to another.

By the age of 11 to 12 months, infants point (Leung & Rheingold, 1981). The extent to which the pointing is carried out for the purposes of communication is in the eye of the beholder. Parents are likely quite good at interpreting the gestures and sounds produced by their infant. Researchers have spent quite a lot of time investigating what infants' gestures actually mean.

Infants' gestures and some of their vocalizations toward the end of the first year of life are likely produced for the purposes of communication. The critical feature of communication is intentionality on the part of the speaker. Austin (1962) proposed three stages in the development of communication, which provide some distinctions that are useful when thinking about infants' gestures. These are displayed in Table 3.6. Gestures and utterances without intention represent the infants' early productions and would be classified as belonging to the perlocutionary stage. The illocutionary stage would include communications having intentionality but falling short of being adultlike productions. This stage likely includes most of infants' first use of gestures with or without accompanying words.

Research has shown that not all of infants' pointing is for communication. Infants will point when no one else is with them (Delgado, Gómez, & Sarriá, 2009). By the 13th month of life, infants use pointing to direct the attention of others (Tomasello, Carpenter, & Liszkowski, 2007). Research conducted by Goldin-Meadow and colleagues has shown that

Table 3.6 Three Stages of Development of Communication

Stage		Age
Stage 1: Perlocutionary	No communicative intent	Birth to 10 months
Stage 2: Illocutionary	Has communicative intent Does not use adultlike forms	10 to 12 months
Stage 3: Locutionary	Has communicative intent Uses adultlike forms	12 months and older

Source: Based on Austin (1962).

early pointing behavior is related to later vocabulary development (Iverson & Goldin-Meadow, 2005; Özçaliskan & Goldin-Meadow, 2006). They have also shown that the earlier that children use gestures with words, the earlier they are likely to produce two-word utterances (Goldin-Meadow & Butcher, 2003; Iverson, Capirci, Volterra, & Goldin-Meadow, 2008; Iverson & Goldin-Meadow, 2005) Infants' ability to use gestures has led many new parents of hearing, pre-verbal infants to consider teaching their infant a small set of gestures to facilitate communication. Textbox 3.2 describes the popularity of baby sign language.

Text Box 3.2 Diversity of Human Languages: Baby Sign Language

Many new parents have taught sign language to their hearing infants. Because infants can learn to gesture before they can learn to speak, some have suggested that using *baby signs* could facilitate communication between preverbal infants and caregivers. Some of the signs that are taught include signs for *eat, more, drink, sleep,* and *all done,* among others. Early studies of the use of baby signs in hearing infants suggested that hearing infants can easily learn to use gestures for communications (Acredolo, Goodwyn, Horobin, & Emmons, 1999; Goodwyn, Acredolo, & Brown, 2000). The studies also suggested that there are several advantages experienced by infants who learned and used signs early in life as compared with other infants. They went on to develop larger vocabularies, showed enhanced mental development, and exhibited fewer behavioral problems, such as tantrums. Not all researchers agree about the benefits of baby sign language. An examination of the research supporting the benefits of baby sign language found that the studies from which the results were obtained were methodologically weak (Johnston, Durieux-Smith, & Bloom, 2005). The conclusion, then, is that the scientific evidence does not fully support the claims that infants who use baby signs will experience long-term cognitive advantages over infants who do not use baby signs. Nevertheless, there is no study to suggest that using baby signs causes harm. Most of the parents who use baby signs with their infants find them easy to use and also useful. Parents may want to explore the possibility of using baby signs with their infants and determine for themselves whether the experience is beneficial.

When Do Children Produce Their First Words?

Healthy infants will begin producing their first words at 10 to 12 months of age and will continue to add to their lexicons throughout their lives. Researchers have closely examined how infants begin the journey of building their lexicons. This research has found that initially there is quite a lot of initial variability. Some infants' first words are words of their own invention. They have not heard the word used by others in the way that they use the word. The terms ***protowords*** and ***idiomorph*** are used to refer to such inventions. For example, an infant may refer to a favorite food, such as cereal, as *mum mum*. With such examples, one might speculate that the word has come from sounds made during mealtime, such as *mmm*. The child might refer to a bed as *godi*. Parents may never figure out where the invented word's origin. The invented words may become part of funny family stories retold even when the infant is an adult. Some infants may create names for caregivers that only they use. Some families will come to use the invented name for the person, because it comes to be used so often and is used in such a positive, sweet context.

The First Ten Words

Infants' first 10 words can be quite variable. Harris, Barrett, Jones, & Brookes (1988) compared the first words of four children from the time that they were 6 months old until they were 24 months old. Among the first 10 words, *mommy* was used by three of the four children. Two of the four children produced *no, go, here, more, there, hello, bye-bye, teddy, moo,* and *shoe* or *shoes.* The remainder of the words produced by only one of the four children included *quack, buzz, boo, teddy, ball, wee, down, bee, choo-choo, doggy, car, brum, woof, baby, yes,* and a name. First words may be truly referential words, because they are used across many contexts and are generally used to refer to objects of that type. Often first words are context bound words, or words that are used only in a particular situation in which the objects being described are present. For example, a child may use the word *teddy* first only when the soft toy is present.

Only a few studies have investigated the extent to which children's first 10 words differ across languages. In a study of 32 infants reared with German as the first (and only) language, Kauschke and Hofmeister (2002) observed that the first 10 words included relational predicates (e.g., *oben* [up] and *wieder* [again]) and social terms (e.g., *hallo* [hi] and *nein* [no]) when the infants were between 13 and 15 months old. Later on, infants produced more nouns, verbs, and other types of words. In a large-scale comparison of infants learning English, Mandarin, or Cantonese, Tardif et al. (2008) compared the first 10 words learned between the ages of 8 to 16 months. The results revealed some similarities and some differences across languages. Words for people, sounds, and sound effects (e.g., *vroom*) were included in the earliest words learned by each group of children. The groups of children differed in terms of how early nouns and verbs appeared. English-speaking children produced far more nouns that verbs. Mandarin-speaking children produced far more verbs than nouns. Cantonese children produced a comparable number of nouns and verbs.

The First Fifty Words

As children become more skilled word learners and their vocabularies grow larger, they tend to learn more nouns than any other type of word. When researchers have closely examined the first fifty words that English-speaking children know, they have found that

40% to 60% of the words are nouns (Bates et al., 1994; Dromi, 1987; Goldin-Meadow, Seligman, & Gelman, 1976). Only about 3% of first-learned words are verbs (Caselli et al., 1995). Some have suggested that the nouns may not be universally dominant in children's early vocabularies. In some languages, the advantage of nouns over verbs is not as large as in English. In some languages, verbs may be more salient than they are in English. Tardif, Gelman, and Xu (1999) provided evidence for the salience of verbs in a study involving Mandarin-speaking children. Choi and Gopnik (1995) reported similar results from a study involving Korean-speaking children. No study has found that verbs are more frequent than nouns in children's initial vocabularies (Waxman & Lidz, 2006).

The Emergence of Words

How Do Children Build Their Lexicons?

Researchers describe knowledge of vocabulary as the mental dictionary or lexicon. Children's lexicons grow one word at a time. By the age of 18, one usually has learned about 100,000 words, involving about 2,000 root words and words related to each root (e.g., for root word *write,* the related words could be *writes, wrote, writing, written, writer,* or *unwritten*) (Nation & Waring, 1997). A 5-year-old usually knows between 4,000 and 5,000 word families. On average, children in the United States learn approximately 1,000 additional word families each year. For anyone who has tried to learn a word a day to study for a standardized test, such as the ACT or SAT, one knows how hard it is for the teenager or adult to keep up with that pace.

The Word Spurt

Somehow, children are able to acquire an amazing number of new words without seeming to try. Around 18 months of age, the number of words that children know increases rapidly. The vocabulary may triple in size in a matter of only a few weeks. This has been referred to as the **word spurt.** There is individual variation in when a child will experience the word spurt. Figure 3.5 displays the vocabulary sizes for four children from 14 months to 22 months.

As children acquire new words, they may not use all of the words correctly. All children appear to make the same types of errors—a fact which suggests that the errors are likely to result from something general about human cognition. The first type of error is called an **overextension.** This error occurs when children use a word more broadly than an adult would. For example, a child may learn the word *dog* but use it to refer not only to dogs but also to cats, cows, horses, and birds. Eventually, the child will come to understand that a dog is a particular kind of animal and that other animals are called by other names. A second type of error is called an **underextension,** which occurs when children use a word in a more limited way than an adult would. For example, a child may use the word *hat* to refer to a particular hat but does not use the word to refer to any example of a hat. Research by Kay and Anglin (1982) showed that overextensions are less common than underextensions. Anglin (1986) also observed in his daughter a third type of error, which he called an overlap. His daughter would use a word that was underextended in one way but overextended in another. For example, she would use *brella* to refer to open umbrellas but also use it to refer to kites.

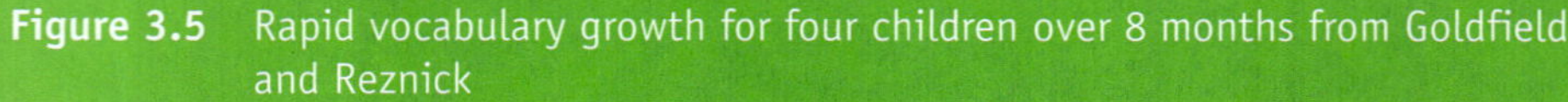

Figure 3.5 Rapid vocabulary growth for four children over 8 months from Goldfield and Reznick

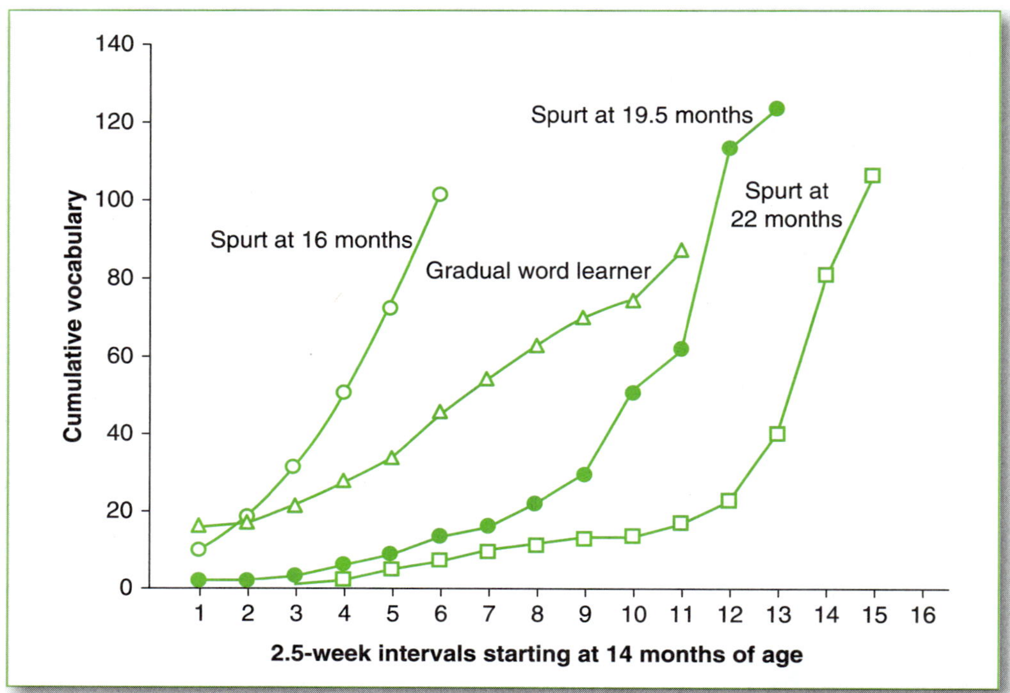

Source: Goldfield and Reznick (1990).

Fast Mapping

The apparent ease with which children can learn new words can be explained by the process of **fast mapping,** which refers to children's ability to learn new words after one exposure (Carey, 1978). Fast mapping was convincingly demonstrated in a study in which children between 3 and 4 years of age were in a room with two trays: one blue and one a shade of olive green (Carey & Bartlett, 1978). The interviewer asked the child to *Get me the chromium tray, not the blue tray, the chromium one.* Children had no problem retrieving the olive colored tray. They demonstrated some knowledge of the word *chromium* even after 6 weeks. Halberda (2003) has shown that fast mapping occurs in 17-month-old infants. Golinkoff, Hirsh-Pasek, Bailey, and Wenger (1992) demonstrated that 30-month-old infants could acquire six new words via fast mapping in one experimental session.

The Original Word Game

▲ Photo 3.4 Children learn many new words by asking adults, *What's that?* This strategy has been called the *original word game.* Have you spent time with a toddler who asks a lot of these types of questions?

The social context is also important as children learn new words. As children grow more verbal in the second year of life, they aid in their own word learning by requesting information from parents and others in their environments. Children will ask, What's that? Anyone who has been around a toddler of this age is likely to make note of how exhausting it can be when a child asks what's that? throughout the day. Usually, adults will provide them with labels for every object. Some have referred to this interactive word-learning activity as the **original word game** (Brown, 1958). When playing the original word game with children, adults tend to provide labels that they would not provide to an adult asking the same question. For example, an adult unfamiliar with life in the countryside may point to a crow and ask *What's that?* The friend would likely say, *That's a crow.* If a child asked the same question, the adult is likely to say, *That's a bird.* Adults tend to provide children with the **basic level category** for objects, such as the label *bird* rather than the label for a particular type of bird (Brown, 1958, 1973).

Research also shows that infants who are too young to ask *What's that*? will use information from the social context when learning new words. Eighteen-month-olds, for example, will fail to learn new words if adults do not make it clear that the infant should pay attention and learn the new word (Baldwin et al., 1996). Baldwin and colleagues (1996) suggested that this is a sensible strategy on the part of infants, because half of parents' speaking during a day is not focused on the child and word-learning. Briganti and Cohen (2011) showed that 18-month-old infants learned words when viewing a picture of a person pointing at an object; 14-month-old infants did not. The implication is that infants' ability to use social cues in word learning develops after 14 months of age. For some individuals, the social context is more important in language learning than for others. Text Box 3.3 describes the life of Helen Keller, who learned language despite being blind and deaf from the age of 19 months. She was able to achieve so much because of one particularly dedicated teacher.

Text Box 3.3 Extraordinary Individuals: Helen Keller

▲ Photo 3.5 Anne Sullivan with Helen soon after Sullivan's arrival in Alabama. Have you seen any of the films about Helen Keller's life?

Helen Keller (1880–1968) is best known for learning to speak, read, and write despite becoming blind and deaf at the age of 19 months. You may be familiar with her story, because her life served as the basis for the film *The Miracle Worker* (Penn, 1962; Tass, 2001). Helen was born in 1880 in Tuscumbia, Alabama. Her father, Arthur Keller, was a newspaper editor and had served as an army captain. When Helen was 6 years old, the family hired Anne Sullivan to be Helen's teacher. Sullivan was only 20 years old, partially blind herself, and had attended the Perkins Institute for the Blind in South Boston, Massachusetts. Her relationship with Helen continued to the time of her death in 1935. When Sullivan arrived in Tuscumbia in 1887, she attempted to teach Helen American Sign Language (ASL). She started by teaching her finger spelling. Initially, progress was slow because Helen did not understand what a sign was or what a word was. In her autobiography *A Story of My Life* (Keller, Shattuck, & Herrmann, 2004), she described the moment, when she was almost 7 years old, and suddenly realized what a word was. The day was life changing for her. On that particular day, Sullivan saw that Helen did not understand the lesson. In an attempt to add context to the words that were being taught, she took her to the well house and pumped water onto her hands and then finger spelled w-a-t-e-r in the palm of her hand. Keller described that event in her autobiography:

▲ Photo 3.6 Helen when she was a young woman.

> Suddenly I felt a misty consciousness as of something forgotten—a thrill of returning thought; and somehow the mystery of language was revealed to me. I knew then that *w-a-t-e-r* meant the wonderful cool something that was flowing over my hand. (Keller et al., 2004, p. 16)

In May 1888, Helen began studying at the Perkins Institution for the Blind with Anne Sullivan by her side. In 1890, she embarked on her biggest challenge yet: learning to speak. For one who cannot hear, learning to speak is particularly challenging. She was taught by Sarah Fuller, the principal of Horace Mann School for the Deaf. Fuller allowed Helen to learn how to make speech sounds and syllables by using her hands to touch Fuller's face and mouth during speaking. Fuller also allowed Helen to explore the inside of Fuller's mouth during speaking, so that Helen could understand how the position of the lips, tongue, and throat changed during articulation. When Helen was 24 years old, she graduated from Radcliffe College in Cambridge, Massachusetts. She traveled the world, speaking about her life and advocating for the rights of the disabled. She was a prolific writer, publishing 12 books. In 1964, she received the Presidential Medal of Freedom from President Lyndon B. Johnson.

What Strategies Do Children Use When Learning New Words?

Whole Object Bias

When learning new words, children all over the world appear to make the same basic assumptions about words and their possible labels. In this sense, they are biased learners. For example, a child points to a giraffe and asks, *What's that?* If mom says, *It's a giraffe,* the child will assume that the word is the name for the whole object rather than a part of the object or some characteristics of the object (e.g., color). Imagine that a child sees a helicopter and asks, *What's that?* Dad says, *It's a helicopter.* The child is not likely to assume that the word refers to the rotating blade; rather, the word will be understood as referring to the whole object. This bias is referred to as the **whole object bias** when acquiring new words.

Mutual Exclusivity Principle

The second type of bias that children demonstrate is that they tend to assume that every object has one, and only one, label or name. This bias is called the **mutual exclusivity principle.** For example, if a child knows that the word for bucket is *bucket,* the child will assume that it is not referred to using another word. Markman and Wachtel (1988) demonstrated children's use of the mutual exclusivity principle in a series of experiments in which 3-year-olds were shown pairs of objects. One of the objects in the pair was a familiar object (e.g., banana, cow, spoon). The second object was unfamiliar (e.g., a pair of tongs). They were then told, Show me the plunk. *Plunk,* or another made-up word, was used. Children were significantly more likely to show the interviewer the unfamiliar object than the familiar one. Markman and Wachtel concluded that children reasoned that the plunk could not be the banana, because a banana is a banana; bananas cannot be called bananas and plunks.

Taxonomic Bias

The third bias that children demonstrate is that they tend to assume that a word refers to a type of an object or whole category rather than a specific example of the category. For example, children will assume the word *cat* does not refer to just Grandma's kitty but is a word that is used to describe all kitties. This has been referred to as the **taxonomy bias** (Markman & Hutchinson, 1984). Children as young as 18 months of age have been found to use these three strategies in acquiring new words (Markman, 1990).

How Do Children Differ in Word Learning?

Late Talkers

Some children begin using their first words much later than other children. The term ***late talker*** is used to describe a child who is 18 to 24 months of age or older who has a vocabulary of 50 words or fewer (Kelly, 1998). Studies that have followed late talkers to determine the extent that they continue to experience problems with language suggest that some late talkers will do as well as their same-age peers. A recent study by Rescorla (2009) assessed the language ability of 17-year-olds who had received a diagnosis of language delay between the ages of 24 and 31 months. The results showed they performed in the normal range on language and reading tasks but had lower performance on vocabulary and verbal memory tasks as compared with peers from the same socioeconomic background.

Pronunciation Problems

Children may also struggle with the clear and precise production of phonemes for many years. Typical children may not master adultlike pronunciation until they are 8 years old (Ferguson, Menn, & Stoel-Gammon, 1992). Children may have difficulty producing /f/ and /v/ until 4 years of age (Sander, 1972). It is not until the age of 5 years that children can produce /r/, /l/, and /s/. The most difficult sounds to master are /θ/ as used in *thigh*; /ð/ as in *the*; and /ʒ/, as in *measure*. A small percentage of children may have difficulty producing some speech sounds even after age 8. Table 3.7 displays the average ages at which English phonemes are typically mastered. Most schools now provide speech therapy for such children. Although parents sometimes are reluctant to place their children in speech therapy out of fear that doing so will be stigmatizing, many adults who received speech therapy as children can report that the experience was not traumatizing and was helpful. My own experience with speech therapy occurred when I was about 9 years old and in the fourth grade. I received just a few lessons about how to pronounce /s/ and /f/. I vividly recall the day that I happened to notice how my speech therapist was placing her teeth on her lip to form the /f/ sound. My way of pronouncing the /f/ was not quite the same. Making that

Table 3.7 Average Ages at Which English Phonemes Are Typically Mastered

Age	Phonemes
Between 18 months and 3 years	/p/, /m/, /h/, /n/, /w/
Between 18 months and 4 years	/b/
Between 2 and 4 years	/k/, /g/, /d/
Between 2 and 6 years	/t/, /ŋ/
Between 2.5 and 4 years	/f/, /y/
Between 3 and 6 years	/r/, /l/
Between 3 and 8 years	/s/
Between 3.5 and 7 years	/tʃ/, /ʃ/
Between 4 and 7 years	/z/
Between 4.5 and 7 years	/θ/
Between 5 and 8 years	/ð/
Between 6 and 8.5 years	/ʒ/

Source: Based on Sander (1972).

connection was all that I needed. Soon, I was on my way of making clearer /f/ and /s/ sounds, and the speech therapy sessions ended.

Summary and Theoretical Implications

Infants are born with some extraordinary abilities, which position them well to begin to learn language. They are able to imitate facial expressions and to perceive the full range of speech sounds that exist in all the world's languages. The fact that infants appear equipped with useful abilities at birth is most consistent with nativist views of language development and least consistent with the view of the behaviorists that all knowledge, including language, is acquired through specific learning experiences.

By the end of the first year of life, infants lose the ability to distinguish the speech sounds that they do not hear regularly. Starting at 4 months, infants begin communicating with others, first through babbling, then by gesturing, and finally through producing words. Between 12 and 24 months, infants go from knowing few or no words to knowing 100 or more words. During this time of rapid vocabulary growth, infants rely on multiple strategies to learn new words, including asking adults for the names of objects. These facts are consistent with each of the major theories of language development and do not help us determine which theory of language development is better than the others. The behaviorist approach can explain the learning of vocabulary through imitating, classical conditioning, and operant conditioning. The social-interactionist approach can explain infants' learning of speech sounds and vocabulary through learning experiences that occur between caregivers and children, and the statistical learning approach can explain the learning of speech sounds and vocabulary through the infant relying on basic cognitive learning mechanisms that are not unique to language. Advocates of the generative approach have proposed that word-learning strategies may be innate, which can account for why language development proceeds relatively quickly for all children, regardless of the language that they are learning, but they have not yet provided detail regarding how innate mechanisms of language learning interact with the environment to bring about the changes that are observed in infants, ability to perceive and produce speech sounds.

Last, the chapter described how parents and caregivers play an important role in the health and well-being of infants; however, their role in language development is not yet clear. There are some cultures in which the adults do not routinely talk to preverbal children, yet those children acquire language normally. Recent laboratory studies suggest that the mere presence of a person in the room when language is being used may result in the infant paying greater attention to language, which may result in increased learning. The overall picture that language development does not critically depend on caregiver input is consistent with both the generative and statistical learning approaches and least consistent with the behaviorist approach, as parents are likely to be the largest source of language experience and language learning opportunities for the child. The results that show that language learning may be facilitated in infants by the presence of a person is consistent with the social-interactionist view; however, the evidence obtained so far suggests that social factors play less of a role in language development than advocates of this view claim.

KEY TERMS

auditory brainstem response (ABR)

autism

babbling

babbling drift

basic level category

canonical babbling

categorical perception

childhood amnesia

fast mapping

fricatives

grapheme

grapheme-to-phoneme correspondences

idiomorph

infant-directed speech

labial consonant

late talker

manner of articulation

mirror neurons

motherese

mutual exclusivity principle

original word game

otoacoustic emissions test (OAE)

overextension

phonological bootstrapping hypothesis

place of articulation

prosodic bootstrapping hypothesis

protowords

semantic memory

social smile

stop consonants

taxonomy bias

underextension

variegated babbling

vegetative sounds

voicing

voice-onset time

whole object bias

word spurt

REVIEW QUESTIONS

1. Describe infants' visual abilities at birth. How do they compare with the visual abilities of adults?
2. Describe infants' hearing abilities at birth. How is infant hearing tested in newborns?
3. What are newborn reflexes? How long are newborns able to produce the reflexes?
4. Discuss the evidence showing that memories of sounds are formed before birth.
5. How does an infant's ability to distinguish speech sounds change during the first year of life?
6. What evidence is there that the perceptual abilities of infants are influenced by the social environment?
7. What evidence is there that infants can form long-term memories?
8. At what ages do infants begin to recognize familiar words' speech sequences, prefer listening to familiar words, and learn to associate words with objects?
9. What are the different types of babbling that all children produce? In what order are they generally observed chronologically?

10. What evidence is there that babbling occurs in nonhuman species?
11. Who are the !Kung San people, and what do their child-rearing practices suggest about how children learn language?
12. Do deaf mothers who use sign language use motherese with their infants? What evidence is there that deaf mothers and infants use and experience motherese?
13. What were the challenges that Helen Keller faced in learning language?
14. What are the three stages in the development of communication?
15. Discuss infants' pointing behavior and the extent to which pointing is used for communication.
16. Describe the types of words that children typically learn first.
17. Explain when the word spurt occurs for typical, healthy children, and discuss possible causes for the word spurt.
18. What evidence is there that children use social cues in learning new words?
19. What biases do children exhibit in learning new words?
20. What is the original word game? Who plays it, and what role does it play in lexical development?
21. What is a late talker? What has research shown about the long-term outcomes for children who start out as late talkers?

RECOMMENDED READING

Acredolo, L. P., Goodwyn, S. W., & Abrams, D. (2002). *Baby signs.* New York: McGraw-Hill.

Agin, M. C., Feng, L. F., & Nicholl, M. (2004). *Late talker: What to do if you child isn't talking yet*. New York: St. Martin's.

Brown, R. (1973). *A first language: The early stages*. Cambridge, MA: Harvard University Press.

Gopnik, A., Meltzoff, A. N., & Kuhl, P. K. (2000). *The scientist in the crib: Minds, brains and how children learn.* New York: Harper Paperbacks.

Jusczyk, P. W. (1997). *The discovery of spoken language.* Cambridge, MA: MIT Press.

Keller, H., Shattuck, R., & Herrmann, D. (2004). *The story of my life: The restored classic.* New York: W. W. Norton.

Konner, M. (2002). *The tangled wing: Biological constraints on the human spirit*. New York: Times Books.

Ladefoged, P., & Maddieson, I. (1996). *The sounds of the world's languages*. Oxford, UK: Blackwell.

Pullum, G. K. (1991). *The great Eskimo vocabulary hoax and other irreverent essays on the study of language.* Chicago: University of Chicago Press.

Tomasello, M. (2008). *Origins of human communication*. Cambridge, MA: MIT Press.

RECOMMENDED FILMS

Apted, M. (Director). (1994). *Nell* [Motion picture]. United States: Fox Home Entertainment.
Aronson, J. (2002). *The sound and the fury* [Motion picture]. United States: New Video Group.
Gary, D., & Hott, L. R. (Directors). (2007). *Through deaf eyes* [DVD]. Available from www.pbs.org.
Haines, R. (Director). (1986). *Children of a lesser god* [Motion picture]. United States: Paramount.
Scarl, H. (Producer/Director). (2011). *See what I am saying* [Documentary]. United States: New Video Group.
Searchinger, G., Male, M., & Wright, M. (Writers). (2005). *Human language series* [DVD]. United States: Equinox Films/Ways of Knowing Inc.
Tass, N. (Director). (2001). *The miracle worker* [Motion picture]. United States: Walt Disney Video.
Uys, J. (Director). (2004). *The gods must be crazy* [DVD]. United States: Sony Pictures.

SUGGESTIONS FOR CLASS PROJECTS

1. Because of the irregular spelling-to-sound correspondences in English, English-speaking students find it somewhat challenging to pick out the sounds of common words. You will be provided with a list of 20 commonly used words, involving a combination of regularly and irregularly spelled words. Provide phonetic transcriptions for each word using the symbols displayed in Table 3.3. You may find other systems for transcribing the sounds in other sources. Only use the symbols in Table 3.3.

2. Carry out observations of children under the age of 20 months. Observe the children's speaking, and make note of the quality of the articulation of the different speech sounds. Children are able to produce some speech sounds better than others at that young age. If possible, observe the speech of up to three children and compare the results.

3. Visit the Child Language Data Exchange System (CHILDES) website (http://childes.psy.cmu.edu/). Choose a transcript of a conversation for a child between the ages of 12 and 18 months. Analyze the extent to which the child's speech does or does not involve nouns to a greater extent than other types of words.

For additional ancillary resources, please visit the companion website at www.sagepub.com/kennison.

- Video Links
- Audio Links
- Web Resources
- Internet Activities
- Flashcards
- Web Quizzes

CHAPTER 4

GRAMMATICAL DEVELOPMENT

By the time most children are between 3 and 4 years of age, they are producing complex sentences. Their sentences contain grammatical morphemes, such as **function words** (e.g., *the, was,* and *and*) and suffixes (e.g., plural *-s*, past tense *-ed*). Some of their sentences will even contain multiple clauses (e.g., *I like cookies* and *I like swimming*). During this time, most parents are quite busy, keeping up with their toddlers, making sure that they are getting good nutrition, making sure that the their clothes are still fitting well, and making sure that they keep out of harm's way. Parents are so busy that they may not even notice changes in the toddler's grammatical development. Despite the popular notion that parents and caregivers teach children language, there is little evidence that children's grammatical development is substantially influenced by parental intervention. When one closely examines the daily lives of parents and toddlers, one finds that there is little or no time each day dedicated to teaching the toddler the grammatical rules of language. Toddlers appear to figure out the rules of their language with little conscious effort. In this chapter, you will learn about how children acquire the grammatical rules of their language. You will learn about the grammatical errors that children make and what these errors tell us about how children are learning grammar. You will also learn about how grammatical development occurs in children from populations with developmental disorders, including deafness, and children who have experienced a hemispherectomy.

▲ Photo 4.1 Three children are having snacks. Have you ever overheard the conversations of children around this age?

Beginnings of Grammar

When Do Infants Begin to Show Signs of Grammar?

The beginning of grammatical development occurs long before children are producing multiple-word utterances. Researchers distinguish infants' **receptive language ability,** which is their ability to understand language, from their **productive language ability,** which is their ability to produce language. Recent research shows that there is evidence that infants are figuring out the syntactic relations of their language before their productive language ability is well developed. Around the time that infants have one or two words in their vocabulary, most are already beginning to figure out the word order rules of their language. Evidence for this was obtained in a particularly clever set of studies by Golinkoff and colleagues (Golinkoff, Hirsh-Pasek, Cauley, & Gordon, 1987; Hirsh-Pasek &

Golinkoff, 1991). Golinkoff and colleagues (1987) used the preferential looking paradigm with infants who were 13 to 15 months old. Infants sat on the lap of a caregiver in front of two screens, each displaying a different video. Both videos showed Big Bird and Cookie Monster. In one video, Big Bird was shown tickling Cookie Monster, and in the other video, Cookie Monster was shown tickling Big Bird. Over loudspeakers, the infant heard the following sentence: Big Bird is tickling Cookie Monster. An analysis of infants' looking preferences toward the video screen showed that infants looked longer at the screen that matched the sentence that they heard. The authors concluded that the infants had already learned that in English the first noun in a sentence is usually the **agent,** or performer, of the action, and the second noun in a sentence is most usually the object that is being affected by the action. In a second investigation, Hirsh-Pasek and Golinkoff (1991) found similar results in a study where infants were shown videos of a woman with a ball and a set of keys. In one video, she is holding the keys and kissing the ball. In the other video, she is holding the ball and kissing the keys. As the videos played, the loudspeaker played the following sentence: She's kissing the keys. An analysis of infants' looking time showed that infants looked longer at the video that matched the sentence. The results supported the view that infants comprehended how the second noun (i.e., *the keys*) was manipulated by the agent (i.e., being kissed vs. being held).

When one considers how infants go from understanding nothing in the speech being spoken around them to understanding the relationships among words to being able to produce sentences, one cannot help but view this as a monumental feat. Infants appear able to infer differences among types of words from how they sound in the speech of adults. Shi, Werker, and Morgan (1999) showed that the infants between 1 and 3 days old can distinguish between function or **closed class words** (e.g., *it, this, in, of, these,* and *some*) and content or **open class words** (e.g., *baby, table, eat, slowly,* and *happy*). Characteristics of function words and content words differ not only in length and stress pattern but also frequency of usage. Function words are typically shorter than content words and, unlike content words, are unstressed. Function words are far more frequent than most content words. Consider the paragraph in 1. The function words are shaded in gray. Shi and colleagues found that infants who are 6 months of age show a preference for content words (Shi & Werker, 2001) and that at 11 months of age, infants are able to distinguish between highly frequent function words and content words (Shi, Werker, & Cutler, 2006).

1. Once upon a time, there was a prince and a princess, and they lived in a castle. The castle was located in a beautiful valley. All the people who lived in the valley liked the prince and princess, because they were kind to all the people. Once a year, the prince and princess threw a big party and invited everyone in the valley to attend. There was lots of food and music. Everyone had a very nice time.

For many infants in the world, their language environment contains not just one language but two. Further, the two languages may not always be used separately. Text Box 4.1 describes code-switching, which is commonly used by bilinguals. Infants reared in environments in which code-switching routinely occurs have not been shown to experience any language delays or problems acquiring language.

Text Box 4.1 Diversity of Human Languages: Code-Switching

In homes in which more than one language is spoken, one can frequently hear examples of code-switching or switching from one language to another in quick succession. The reasons that people code-switch are many. One may have difficulty finding an appropriate word or expression in the language in which one is having a conversation. If the listener is also bilingual, the speaker can insert words from their mutual other language and get the intended message across. Some researchers have speculated that some code-switching may happen because words from the second language become mentally available more quickly than the corresponding word in the other language. For infants who are reared in bilingual environments, one can easily appreciate how difficult it is for the infant to realize that there are different languages with different grammars being used. In a home in which bilingual parents and family members engage in code-switching, the challenge for the infant is even greater. There are times when words and phrases from each language end up becoming part of utterances in the other language. Despite the confusion that this scenario suggests, children raised in bilingual environments in which code-switching occurs become fluent speakers of both languages. In the southwest region of the United States, one might hear utterances containing both English and Spanish, or Spanglish. Among bilinguals who know Chinese and English, one can hear Chinglish. In the Tamil-Nadu area of South India, one can hear the mixing of Tamil and English or Tanglish. In cases in which code-switching becomes prevalent in a community or society, it is possible that children rarely hear examples of only one language or the other. They acquire the mixture of languages as a native language. Linguists refer to such native languages as mixed languages (Bakker & Mous, 1994). An example of a mixed language is mednyj Aleut, which is a mixture of Russian and Aleut. The language has retained the verbs from Aleut, verbs but has replaced the Aleut verb endings with their Russian equivalents.

Bootstrapping

The work of Shi and colleagues illustrates a process that has been referred to as bootstrapping, which was first introduced in the context of language acquisition by Pinker (1984) to describe how children acquire their language by using relatively simplistic sources of information to acquire relatively complex knowledge. In Chapter 3, we learned about phonological and prosodic bootstrapping, which referred to the possibility that children may learn new words because the stress pattern and sounds of adult speech provide cues that signal word beginnings and word endings. The term ***semantic bootstrapping*** has been used to describe how children might acquire the knowledge of the syntactic rules of their language from their acquisition of the meanings of words (Pinker, 1984). Others have proposed that children's knowledge of grammar can help them learn new words. This process has been described as **syntactic bootstrapping** (Fisher, 1994; Gleitman & Gleitman, 1992). It is possible that children use both strategies, perhaps relying on one strategy earlier in development than the other strategy.

Semantic bootstrapping may be particularly useful early in the second year of life. For example, consider children's one-word utterances. Some of these utterances appear to be intended to communicate more than a single word's worth of meaning. Children may want to communicate a semantic relationship, but they are physically only able to articulate one word. For example, a toddler may pick up a favorite video and say *watch* to Mom or Dad. They might also combine their one-word utterance with a gesture. The toddler might give Mom or Dad an empty sippy cup and say, *more*. After being taken for a ride on Mom or Dad's shoulders, the toddler might outstretch her arms and say, *again*. Mom and Dad and even family friends can likely infer what the toddler is trying to express.

Holophrases

The term ***holophrase*** (Dore, 1975) has been used to describe these types of one-word utterances. Greenfield and Smith (1976) analyzed children's holophrases and identified nine relations that were expressed in a single word. Examples are provided in Table 4.1. A single word may be used to imply an action, which would be expressed as a verb in adult speech or something that would be the doer of an action or agent, which would be expressed as a subject noun phrase in adult speech. A single word may also be something that is acted upon by an agent; the term ***patient*** is used to refer to entities that are acted upon in sentences. Patients are usually expressed as an object noun phrase. One-word utterances may also specify a location, where an action occurred or where a patient was relocated, or an action.

Two-Word Utterances

Just a few months later, most infants are producing two-word utterances. By this time, they are already demonstrating knowledge of basic word order. Research has shown that children's two-word utterances adhere to the basic word order of their language. Brown (1973) collected two-word utterances of English-speaking children and found that a child might

Table 4.1 Semantic Relations in One-Word Speech

Word	Context	Relation
Kitty	Looking at pet	Naming
Mama	Looking at cookie longingly	Violation
Dada	Hearing someone come in	Agent
Ball	Having just thrown it	Action
Down	Having just thrown something down	State of object
Cracker	Pointing to the snack drawer	Associated object
Grandma	Upon seeing Grandma's empty bed	Possessor
Bowl	Putting grapes in a bowl	Location

Source: Based on Greenfield and Smith (1976).

Table 4.2 Sample Two-Word Utterances and Their Relational Meanings

Example	Relational Meaning
Mommy throw	Agent + action
Throw ball	Action + patient
Mommy ball	Agent + patient
Sit swing	Action + location
Ball bucket	Patient + location
My ball	Possessor + possession
Ball small	Patient + attribute
This ball	Demonstrative + patient

Source: Based on Brown (1973).

say *throw ball, eat cookie*, or *sleep bed*. In these cases, the children correctly ordered the words in terms of the word order of English (i.e., agent-action or action-patient or agent-patient). Table 4.2 displays additional examples of two-word utterances with their relational meanings. Brown further noted that English-speaking children do not produce two-word utterances that violate the word order of English, such as *eat kitty* to mean that the kitty is eating or *ball throw* to mean that Daddy should throw the ball. It is worth pointing out that most parents and caregivers would not likely notice this interesting regularity in children's speech. Usually, adults are most focused on interpreting children's intention, rather than noting specific word choice or word order.

Utterances of Three or More Words

As children begin to produce utterances of three or more words, they are on the road to figuring out how to express their intentions as sentences. In order to understand how children may begin to apply the grammatical knowledge that they are acquiring in the production of sentences, it is important to understand how language researchers describe the grammatical structure of sentences. Since the publication of Noam Chomsky's book *Syntactic Structures* (1957), the traditional way in which linguists have represented the syntax of a language has been in terms of **phrase structure rules.** A simplified set of phrase structure rules for English is displayed in 2. The first phrase structure rule is that in English, a sentence (S) must contain at least one noun phrase (NP) and one verb phrase (VP), as in *the dog barked*. The second phrase structure rule is that a verb phrase must contain a verb but can optionally contain an adverb (ADV) (e.g., *slowly*) and a noun phrase (NP), which would serve as an object, as in *the dog suddenly bit the mailman*. Noun phrases include a determiner (e.g., *the* or *a*) and a noun but may optionally contain an adjective phrase (ADJ) (e.g., *tall*) or a prepositional phrase (PP). A prepositional phrase (PP) must contain a preposition (e.g., *on, in,* or *above*) and a noun phrase (NP), as in *the dog suddenly bit the mailman on the porch*. Table 4.3 provides examples of tree diagrams, which provide a graphical representation of the phrase structure of sentences.

2. a. S → NP VP
   b. VP → (ADV) V (NP)
   c. NP → Det (ADJ) N (PP)
   d. PP → P NP

Knowing the phrase structure rules of the language is just one part of learning how to construct grammatical sentences. Another important type of information that children must acquire is information about how specific words can and cannot be used. For example, some verbs can be used with or without a direct object, such as *sing* as in 3. When a verb is used with a direct object, it is described as a **transitive verb.** When a verb is used without a direct object, it is described as an **intransitive verb.** Some verbs can only be used intransitively, such as *bark* in Table 4.3 and *die* as in *the guest died.* Such verbs are described as obligatorily intransitive. Some verbs must be used transitively, such as *bit* in Table 4.3 and *bought* as in *Logan bought the boat.* The sentence *Logan bought yesterday* is not grammatically correct. Such verbs are described as obligatorily transitive. Some obligatorily transitive verbs also require that a location phrase be used, such as *put* as in *Tom put the book on the table.* If the location phrase is not specified, the sentence is not grammatically correct; language researchers use an asterisk to indicate that a sentence is ungrammatical (e.g., **Tom put the book*).

Table 4.3 Example Tree Diagrams of English Sentences

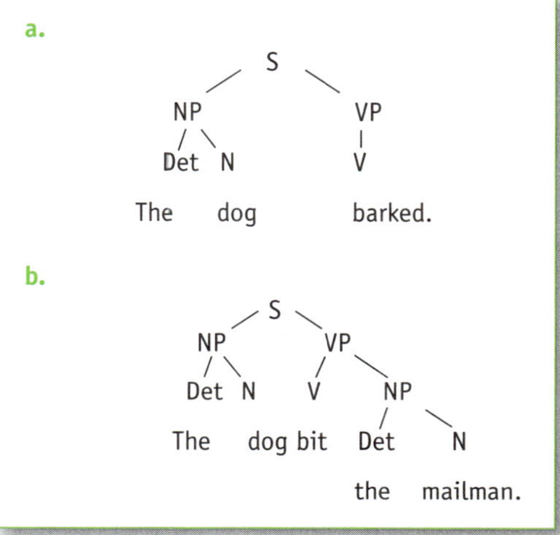

3. a. Miriam sang beautifully. — Intransitive
   b. Miriam sang the hymn beautifully. — Transitive

Our knowledge about how specific words must be used is generally viewed as being stored in memory as part of the word's lexical information (MacDonald, Pearlmutter, & Seidenberg, 1994). Lexical information can include a variety of details about word usage—not only information about how verbs are used. Nouns are also associated with lexical information. One way in which nouns differ is their grammatical number, which refers to whether they are grammatically singular or plural. In English, nouns must agree in number with the verb in a sentence, as shown in 4.

4. a. The cookie was delicious. — Singular noun
   b. The cookie tastes great. — Singular noun
   c. The cookies were delicious. — Plural noun
   d. The cookies taste great. — Plural noun

Singular nouns refer to an individual object; plural nouns refer to more than one item. In English, some nouns that have a singular meaning are grammatically plural. For example, the words *scissors* and *pliers* refer to individual tools; however, when used in a sentence, each noun must be used with a plural verb, as shown in 5a and 5b. There are also nouns that refer to a group involving multiple individuals but are grammatically singular, as in 5c and 5d. In English, there are relatively few grammatical distinctions across nouns. In languages such as Spanish, French, and German, nouns are associated with gender as well as number. In some languages, there are many more grammatical distinctions for nouns. Text Box 4.2 describes the noun classes in Fulani, which is a Bantu language spoken in Africa. There are 22 grammatically distinct noun classes.

5. a. The scissors were rusty.
   b. The pliers were inexpensive.
   c. The family was saddened by the news.
   d. The flock was headed south for the winter.

Text Box 4.2 Diversity of Human Languages: Noun Classes in Fulani

Table 4.4 Fulani Noun Classes With Examples

Stem	Suffix	English Translation
Laam	*-be*	Chiefs
Loo	*-de*	Storage-pots
Bow	*-di*	Mosquitoes
Laam	*-do*	Chief
Biraa	*-dam*	Milk
Nood	*-a*	Crocodile
Njar	*-am*	A drink
Les	*-di*	Country
Yiit	*-e*	Fire
Nyor	*-go*	Cover-mat
Taador	*-gol*	Girdle
Nyala	*-hol*	Calf
Ngas	*-ka*	Hole
daN	*-ki*	Grass shelter
loo	*-nde*	Storage-pot
?en	*-ndu*	Breast
Dem	*-ngal*	Tongue
hottoll	*-o*	Cotton

Source: Comrie, Matthews, and Polinsky (1996).

In English, nouns can be either singular or plural, as in *book* or *books*. In Spanish, nouns can be singular or plural and also masculine or feminine, as in *el amigo* and *los amigos*, which are the masculine forms, or *la amiga* and *las amigas*, which are the feminine forms. Some languages, such as German, have three genders: (1) masculine, (2) feminine, and (3) neuter. In other language families, there is much more complex morphological markings on nouns. For example, Fulani, a language spoken around the world with roots in West Africa, has about 25 classes of nouns (Arnott, 1970). Dialects of Fulani are spoken by millions of people throughout Africa. There are about 25 distinct noun classes, which means that a noun stem could be marked in 1 of 25 suffixes. Some of the noun classes are similar to classes found in Indo-European languages, such as plural, masculine, and feminine. Other classes are human, animate, and inanimate, among many others. Table 4.4 displays 18 of the most commonly used noun classes in Fulani.

How Is Syntactic Development Measured?

Mean Length Utterance

Infants' syntactic development progresses rapidly. Their utterances become longer and more complex. In 1973, Roger Brown published his seminal work on children's language development. One of his important innovations was a method of measuring syntactic development. The method that he created remains widely used today. He coined the term ***mean length utterance*** (MLU) to refer to the average number of morphemes produced per utterance. One calculates MLU by adding up the total number of morphemes produced in a conversation and dividing it by the total number of utterances produced by the child. This sounds rather straightforward. However, determining where an utterance begins and ends was challenging. Some morphemes occur as individual words. Such morphemes are referred to as **free morphemes.** In contrast, **bound morphemes** are those morphemes that must be attached to another word or morpheme in order to be used. Examples of bound morphemes are the tense markers on verbs (e.g., *-ed, -ing,* and *-s*), the plural marker *-s* for nouns, and various suffixes (e.g., *-ish, -ful,* and *-ly*) and prefixes (e.g., *-un* and *-re*). Consequently, in some cases the number of morphemes in an utterance will be more than the number of words in the utterance. Brown also had to decide what to do about repeated words and exclamations (e.g., *oh* or *um*). He decided not to code exclamations; however, repeated words were counted. Brown first used MLU in the analysis of language produced by three children over a period of several months. Table 4.5 displays a brief excerpt of a child's conversation. The number of morphemes is indicated for each utterance.

Table 4.5 Example of Child's Utterances With Mean Length Utterance Computed Following Brown

Utterance	Number of Morphemes
Adult: Where's mommy?	3
Child: At home.	2
Adult: She's not here?	4
Child: Yeah.	1
Child: At home.	2
Adult: Home?	1
Child: Yeah.	1
Child: At home.	2
Adult: Oh, Mommy's home.	3
Child: Oh . . . my . . . my mommy's not here.	6
Adult: Mommy's home.	3
Child: That's where my mom is at.	7
Child: At . . . at Dad's house.	5

Source: Based on Brown (1973); Child Language Data Exchange System (CHILDES) (MacWhinney, 2000).

In Brown's (1973) research, he found that children's MLU increased with age. Figure 4.1 displays the MLU measured multiple times for three children. For each child, MLU increased steadily over time. Brown proposed that children's syntactic development occurred in five stages. Each of the five stages is associated with a range of MLU. In the first stage, children's MLU is between 1 and 2. Stages 2 through 5 have MLUs that increase by 0.5. Children in the second stage have an MLU between 2.0 to 2.5. Children in the third stage have an MLU between 2.5 and 3.0. Children in the fourth stage have an MLU between 3.0 and 3.5. Children in the fifth stage have

Figure 4.1 The graph displays the relationship between mean length utterance and age for three children. For each child, mean length utterance increases over time

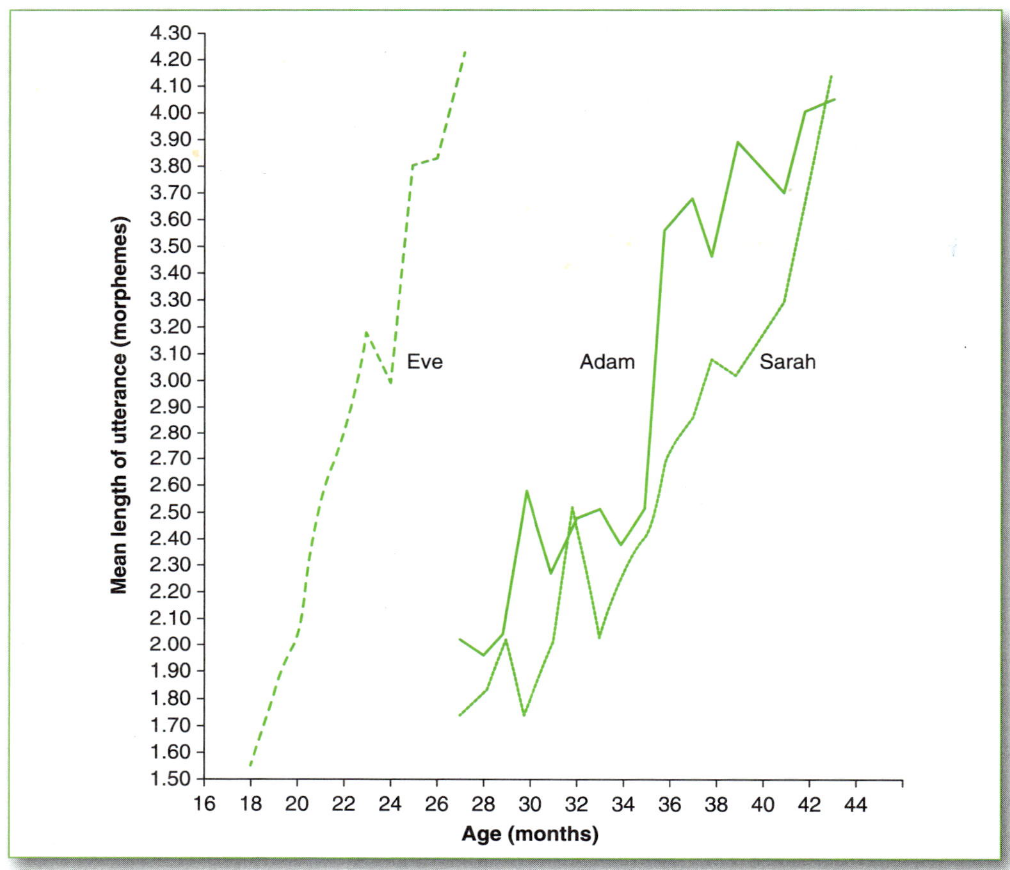

Source: Brown (1973).

an MLU between 3.5 and 4.0. While the stages end at an MLU of 4.0, children can produce, on average, utterances longer than four morphemes. Consequently, the stages of syntactic development are most useful when describing the speech of children with an MLU between 1 and 4.

The longitudinal data that Brown (1973) obtained from the three children also provided information about how the grammatical morphemes of English are acquired. As you learned in Chapter 1, morphemes are the smallest unit of meaning in the language. In English, there are suffixes that are added to verbs to change the meaning of the verb in systematic ways, such as causing a change in tense. Table 4.6 displays some example verb tenses. There is also a suffix that is added to nouns to form the plural; however, in English, there are quite a few nouns that have irregular plural forms. Examples are displayed in Table 4.7.

Table 4.6 Examples of Regular and Irregular Verb Tenses

Regular Verbs

Present Tense	Present Progressive	Past Tense	Past Participle
Talk	Talking	Talked	Talked
Smile	Smiling	Smiled	Smiled
Kiss	Kissing	Kissed	Kissed
Laugh	Laughing	Laughed	Laughed

Irregular Verbs

Present Tense	Present Progressive	Past Tense	Past Participle
Eat	Eating	Ate	Eaten
Break	Breaking	Broke	Broken
Throw	Throwing	Threw	Thrown
Bring	Bringing	Brought	Brought

Source: Dictionary.com.

Table 4.7 Examples of Regular and Irregular Plural Nouns in English

Regular		Irregular	
Singular	**Plural**	**Singular**	**Plural**
Dog	Dogs	Ox	Oxen
Table	Tables	Sheep	Sheep
Hat	Hats	Knife	Knives
Apple	Apples	Goose	Geese
Girl	Girls	Man	Men

Source: Dictionary.com.

Brown's (1973) study of children's syntactic development showed that morphemes appear to be acquired in a particular order (see also Cazden, 1968). They analyzed the utterances of four children, making note of when they appeared to have mastered differently. Mastery was defined as using the morpheme 90% of the time in contexts in which the morpheme was obligatory (i.e., not using it resulted in an ungrammatical utterance). Of the 14 morphemes that Brown identified, the children acquired the present progressive first, as in *I driving*. Prepositions were acquired next, and then plurals. Table 4.8 displays the 14 morphemes in their order of acquisition with examples. As you review the table, you might find the term ***copula*** unfamiliar. It refers to a word that serves to link a subject and a predicate as in *Nora is smart* or *Nora is going to college*. Of the 14, the contractible copula and contractible auxiliary verb forms are the last to be mastered. In 1973, de Villiers and de Villiers published a similar study. Their study involved 21 children who were between 16 and 40 months of age. Their results were remarkably similar to those obtained by Brown (1973).

Index of Productive Syntax

Many researchers still use MLU; however, another method was created by Scarborough (1989). It is called the **Index of Productive Syntax** (IPSyn). A child's score on the IPSyn is calculated using a transcript of 100 of their utterances. The researcher categorizes the child's utterances in terms of their syntactic structures (e.g., noun phrases, verb phrases, questions, negative sentences). For each category of structure, two different examples from the child's utterances are coded. One point is given for each of the different types of structures used by the child. The child's IPSyn score is the total number of points obtained. The advantage of IPSyn over MLU is that IPSyn can be used to distinguish children whose MLUs are above 4.0. Studies have shown that there are strong correlations between children's scores on the IPSyn and their MLU (Scarborough, Rescorla, Tager-Flusberg, Fowler, & Sudhalter, 1991).

Table 4.8 Typical Order of Acquisition of Grammatical Morphemes in American English

Order	Morpheme	Example
1.	Present progressive	I driving
2–3.	Prepositions	in, on
4.	Plural	balls
5.	Irregular past tense	broke, fell, threw
6.	Possessive	Daddy's chair
7.	Uncontractible copula	This is hot
8.	Articles	a, the
9.	Regular past tense	She walked
10.	Third person present tense, regular	He works
11.	Third person present tense, irregular	She does
12.	Uncontractible auxiliary	The horse is winning
13.	Contractible copula	He's a clown
14.	Contractible auxiliary	She's drinking

Source: Brown (1973).

How Do Children Acquire Morphemes?

Researchers have disagreed about the reasons that English-speaking children acquire morphemes in the order that they do. Brown (1973) explored the possibility that children's acquisition of morphemes was influenced by the frequency with which the morphemes occurred in the environment. One might expect that the morphemes that are being produced the most in the environment and being heard the most by children would be the first ones that children would produce. For example, Brown (1973) observed that articles (e.g., *the* and *a*) were the most frequent morphemes. In contrast, the present progressive *-ing* was much less frequent. However, in children's speech, the present progressive was always mastered first and the articles were mastered much later (i.e., after seven other morphemes). Brown (1973) concluded that there was no direct relationship between the order in which children acquired the morphemes and the frequency with which they experienced the morphemes. Newport, Gleitman, and Gleitman (1977) investigated the relationship between parents' frequency of usage of auxiliaries and children's acquisition and found no evidence that children acquire frequent morphemes earlier than less frequent ones. Moerk (1980, 1981) reported data in which he claimed to have observed a relationship between frequency of usage and children's acquisition of morphemes. Aspects of the methodology were subsequently questioned (Pinker, 1981).

Brown (1973) entertained the hypothesis that linguistic complexity of the morpheme, rather than frequency of usage, determined the order with which morphemes are acquired by children. He analyzed morphemes as having two types of complexity: (1) **semantic complexity,** which refers to the meaning or relation expressed, and (2) **syntactic complexity,** which refers to the complexity of the phrase structure used. For example, some morphemes involve both semantic and syntactic complexity, such as third person present tense (e.g., *works* as in *he works*). The morpheme *-s* codes the number (i.e., plural vs. singular) of the subject noun and also tense (i.e., present vs. past) for the verb. In contrast, the plural *-s* codes only the number for the noun. By Brown's (1973) analysis, the third person present tense *-s* is more complex than the plural *-s*. As Table 4.8 shows, it is acquired later than the plural *-s*. Despite the intuitive appeal of the complexity explanation, Brown (1973) was unable to determine which type of complexity was the determining factor. Because the English language involves fewer grammatical morphemes than other languages, it may be the case that English does not provide the ideal conditions for studying the acquisition of morphemes.

Learning Rules

Since Chomksy's (1959) review of Skinner's book *Verbal Behavior* (1957), researchers have recognized the fact that learning language is about learning language rules rather than simply memorizing every word one hears. When one considers how children acquire the grammatical morphemes of language, one might wonder whether children are learning to use words individually or learning to use morphological rules. For example, a child might say, "Mama was singing in car." The word *singing* contains the suffix *-ing*, which can be added to verb stems to express the present progressive. One possible way in which children might come to learn to use the word *singing* is to learn it as a whole. An alternative possibility is that children learn a new rule for how to form a particular type of verb, and they learn that they can apply the rule to any verb of that type. In accordance with this view, the child would learn that *-ing* is a morpheme that can be added to any present tense verb. Strong evidence for rule learning was obtained by Jean Berko (1958). Text Box 4.3 provides more detail about her important discovery.

Text Box 4.3 Research Discovery: Toddlers Learn Grammatical Rules

Figure 4.2 The image that is displayed is a sample of the stimuli used in Berko's wug test

Source: Berko (1958).

In 1958, Berko conducted a now-classic experiment in which she demonstrated that children, indeed, learn rules. Children as young as 3 years old were shown a picture of a bird and told, "This is a wug." In a second picture, there were now two birds. The interviewer said, "Now there are two of them. There are two _____." Figure 4.2 displays sample materials from Berko's (1958) study. Children could correctly produce the plural form of wug, which is wugs, despite the fact that they had never heard the plural form before. The task has come to be known as the wug test. The results of the study showed that children were able to produce a new verb form for a word that they had encountered for the first time, showing that they, indeed, applied a rule to the new word.

Additional evidence that children's acquisition of grammatical word endings involves the learning of rules comes from the errors that they make during this time of their life. Most children will go through a phase in which they apply a grammatical rule to a case for which the rule does not apply. These errors are called **overregularization errors** (Cazden, 1968; Ervin, 1964; Slobin, 1973). For example, the past tense in English is typically formed by adding an *-ed* to the verb stem, as in *walked* and *talked*. Children will produce overregularized forms, such as **goed* and **broked* (the asterisks indicate ungrammatical words). Remarkably, early on in children's language development, they will routinely produce correct versions of the same irregular verbs, which later, they get wrong. When the past tense form is learned, children will produce overgeneralized forms. Children will come to sort out which verbs are regular and whose past tense is formed using the suffix *-ed* and which verbs are irregular.

Pinker (1999) has argued that mastering the past tense in English involves two processes. Mastering the form for regular verbs involves learning the rule. Producing a regular past tense form involves the application of the rule. In contrast, mastering past tense for irregular verbs involves storing in memory the irregular past tense form. Producing an irregular past tense form involves retrieving it from memory. Evidence for the existence of these two mechanisms has been obtained in studies with adults with different types of brain damage. Ullman and colleagues (1997) reasoned that the left frontal area of the brain has been found to be involved in the application of language rules and the left temporal medial area has been found to be involved in the retrieval of stored lexical items. He hypothesized that individuals with damage to the left frontal region would be expected to have difficulty producing regular verbs but not irregular verbs. In contrast, individuals with damage to the left posterior region would be expected to have difficulty producing irregular verbs but not regular forms. He obtained evidence supporting this hypothesis in a study in which he compared the verb processing of four groups of individuals: (1) adults with posterior aphasia, (2) adults with Alzheimer's, disease, (3) adults with Parkinson's disease, and (4) adults with Huntington's disease. The former two groups performed better on regular verbs than on irregular verbs. The latter two groups performed better on irregular verbs than regular verbs.

Of the bound morphemes, such as the word endings on verbs and nouns, there are two types: (1) **inflectional morphemes** and (2) **derivational morphemes.** Inflectional morphemes, when added to a word, do not change the part of speech (e.g., noun, verb, adjective). For example, when the plural *-s* is added to a noun, the resulting word is also a noun. In contrast, derivational morphemes, when added to a word, change the part of speech or substantially change their meaning. For example, when the suffix *-ment* is added to a verb, such as *acknowledge,* the resulting word is a noun, as in *acknowledgment*. Most of the research on the acquisition of morphemes has focused on inflectional morphemes. Very little research has been conducted on the acquisition of derivational morphemes, despite the fact that there are numerous derivational morphemes in English and other languages and their usage is quite frequent. Table 4.9 displays some common derivational morphemes in English.

▲ Photo 4.2 The children are collaborating during play. What types of sentences do you imagine the children using during this type of activity?

Table 4.9 Sample of Derivational Morphemes in English

Morpheme	Rule	Examples
-ly	Adjective + *-ly* = Adverb	Slowly, personally
-er	Verb + *-er* = Noun	Jumper, listener
-ment	Verb + *-ment* = Noun	Acknowledgment, entanglement
-ness	Adjective + *-ness* = Noun	Happiness, contrariness
-ify	Noun + *-ify* = Verb	Glorify, horrify
-able	Verb + *-able* = Adjective	Washable, drinkable
-ize	Adjective + *-ize* = Verb	Privatize, modernize
-al	Verb + *-al* = Noun	Denial, removal
-ance	Verb + *-ance* = Noun	Deliverance, perseverance
-less	Noun + *-less* = Adjective	Childless, heartless

Source: Dictionary.com.

Syntactic Bootstrapping

There is a great deal of lexical information that must be acquired in order for children to become grammatically competent. Some lexical knowledge may even involve groups of words. For example, in English there are many verbs that are used with prepositions, which serve not as the beginning of a prepositional phrase but as a verb particle, as shown in the examples in 6. Some verb particle constructions can be used both with a direct object occurring after the particle, as in 6b, and with the object occurring before the particle, as in 6c. In this case, the particle is separable from the verb. In other verb particle constructions, the particle is not separable from the verb, as in 7a. Still other verb particle combinations must be used such that the particle is separated from the verb, as in 7c. In the examples, an asterisk is used to indicate that the usage is awkward and might be ungrammatical for some speakers of English.

6. a. John threw the trash into the Dumpster. — Prepositional phrase
   b. John threw out the trash. — Particle
   c. John threw the trash out. — Separable particle

7. a. The gasoline gave off fumes.
   b. *The gasoline gave fumes off.
   c. *They couldn't tell apart the twins.
   d. They couldn't tell the twins apart.

The child who has knowledge of some verbs and the ways in which they are used appears to be able to use that knowledge to infer the meanings of new verbs. For example, if a child hears Mom say, "You are verb-ing the kitty," then the child can use information from the context to infer the verb meaning. The verb may be *holding* or *petting*. If a child hears Mom say, "The kitty is verb-ing at me," then the verb may be *looking*. Evidence for this type of syntactic bootstrapping has been obtained in a series of studies by Naigles and colleagues (Lee & Naigles, 2008; Naigles, 1990; Naigles & Swensen, 2007). Naigles (1990) showed that when 2-year-olds were presented with a sentence containing a novel verb, the children's interpretation of the verb depended on the syntactic structure in which the verb appeared.

The Production of Complex Sentences

The rapid changes that occur in children's speech toward the end of the second year and throughout the third year may not get much attention. Certainly, parents likely notice that children are able to communicate better; however, parents may not be keenly aware of the subtle changes occurring in their children's word order from week to week. The research that has been conducted on how children master the production of complex sentences shows that they typically produce forms of the complex sentence that are unlike adult language early on and only later come to produce consistently the structures that would be considered grammatically correct for the adult speaker. The research on how children produce complex sentences has focused on a small set of constructions, which we will discuss in this section. These include **negative sentences,** *wh*-questions, **passive sentences,** sentences containing **coordinated clauses,** and sentences containing relative clauses.

How Do Children Master Complex Sentences?

Negative Sentences

Children learn the word *no* relatively early. As you learned in Chapter 2, for some children, the word *no* is among the first 10 words that they are able to produce. However, producing negative sentences comes much later. Consider the examples of negative and declarative sentences in 8. Klima and Bellugi (1966) observed that children's production of negative sentences seemed to involve multiple stages of development. They claimed that some children go through an initial stage in which they form a negative sentence by producing the word *no* before an affirmative sentence as in *no eat cookie* or *no kitty bite.* In the second stage, the negative word *no* occurs in the middle of the sentence, in front of the verb, as in *kitty no bite* or *you no eat.* The third stage involves the use of a negative contraction, as in *kitty doesn't bite* or *you don't eat.* Typically, children will produce the negative contraction before they produce the corresponding affirmative form (i.e., *kitty does bite* or *you do eat*). Some researchers have questioned whether children actually produced the first type of negative sentence (Bloom, 1970). Evidence for such sentences was later found by de Villiers and de Villiers (1985) in the speech of their son.

8. The store had a lot of bananas. — Declarative sentence

   The store didn't have a lot of bananas. — Negative sentence

Hiramatsu (2003) investigated more complex negative sentences that were produced by children between the ages of 3 and 5 years. She found that children produced ungrammatical sentences, such as * *What did the Smurf didn't buy?* (the asterisk indicates that the sentence is ungrammatical). She tested the same children in a grammaticality judgment task. The task used a procedure similar to that used by McDaniel and Cairns (1996). One experimenter was the storyteller, and one experimenter manipulated a puppet named LuLu, who was described as speaking *moon talk*. The storyteller explained to the child that LuLu was still learning English and she needed help. So the child could help by telling LuLu whether her sentences were right or wrong. The results showed that children judged the sentences that they had produced as incorrect. Hiramatsu (2003) concluded that children's productions may not represent perfectly their knowledge about language.

Questions

Most children do not correctly produce questions until they are between 24 and 36 months. Children first correctly produce questions that require *yes* or *no* answers, as in *Do you want a cookie?* Other questions involve the use of a **wh-word** (e.g., *which, where,* and *when*), as shown in 9. Wooten, Merkin, Hood, and Bloom (1979) conducted a longitudinal study of children's language development and observed that questions including *what, where,* and *who* are produced earlier than questions including *when, how,* and *why.* One explanation for this pattern of development is that sentences containing *what, where,* and *who* are easier to understand. Support for this explanation was obtained by Winxemer (1981) in a study that investigated children's ability to comprehend different types of questions.

9. a. Which cookie do you want?
   b. Where do you want to sit?
   c. When do you want to go to the park?

Klima and Bellugi (1966) observed that before children produce adultlike *wh*-questions, they produce earlier ungrammatical versions. Initially, children produce shortened sentences containing the *wh*-word, but not the auxiliary verb as in *What that?* and *Where kitty go?* Later, children will include the auxiliary verb but may not properly order it before the subject noun as is *What kitty is eating?* and *Where kitty is going?* In order to produce the adultlike question, children must learn to order the auxiliary verb before the subject noun.

Passive Sentences

In adult speech, one observes both active and passive sentences. Examples are provided in 10. In the **active sentence**, the agent of the action is the subject, and the patient of the action is the object. In contrast, in the passive sentence, the patient is in subject position, and the agent of the action is expressed in a prepositional phrase. In one of the few studies that have investigated children's productions of passive sentences, Horgan (1978) showed pictures to children between the ages of 2 and 13. The children were asked to produce sentences describing the pictures. She found that children produced full passives, which included the prepositional phrase containing the agent noun, more often that they produced **truncated passives,** which did not specify the agent. According to Horgan (1978), children view the regular and short passives as separate, unrelated constructions.

10. Clarence visited Thelma on Sunday. Active sentence
    Thelma was visited by Clarence on Sunday. Passive sentence

Bever (1970) investigated how children between the ages of 2 and 5 comprehended passive sentences. He found that children between the ages of 3.5 and 4 understood passive sentences better than children who were between 4 and 5. In a later study, Maratsos (1974) obtained similar results, showing that children between 3 and 3.5 years of age understood passives better than children between the ages of 3.5 and 4 years. Bever's (1970) interpretation of the data is that children initially comprehend passive correctly but later come to adopt a comprehension strategy in which subject nouns are always interpreted as agents. He suggests that when the children use the strategy for passive sentences, they overuse the strategy. He suggests that this overgeneralizing of their comprehension strategy is similar to children's overregularization errors.

Other languages also have passive constructions. Research conducted with two other languages has shown that passives may be acquired earlier in those languages than in English. Allen and Crago (1996) found that children learning Inuktitut, a language spoken in Eastern Canada, use passives regularly by the age of 3. Demuth (1990) investigated how children learn passives in Sesotho, a language spoken in Africa. The results showed that passives were used regularly when children were 2 years old. In these languages, passives appear to be used more frequently in adult language than in English and in other languages in which passives are acquired relatively late (e.g., by 5 years of age in German, de Villiers, 1984, and by 8 years of age in Hebrew, Berman, 1985).

English-speaking children do appear to have knowledge of how to form passives, despite the fact that they rarely produce them. In a study by Pinker, Lebeaux, and Frost (1987), children between the ages of 3 and 4 were taught how to produce active sentences with verbs that they had never heard before. The verbs were created just for use in the study, as *The boy pilked the dog*. Children were then asked, "What happened to the dog?" Children were able to produce the short passive form, which they had not heard before, as in *The dog was pilked*. Brooks and Tomosello (1999) conducted a similar study and showed that children between the ages of 2 and 4 could produce both full and short passive forms for verbs that they had never heard before.

Coordinated Clauses

Among complex sentences are those that contain coordinated clauses. In such sentences, one or more clauses are connected by a conjunction, such as *and, but,* and *or.* Examples of coordinated sentences are provided in 11. Bloom, Lahey, Hood, Lifter, and Fiess (1980) investigated children's acquisition of coordinated sentences. Despite quite a bit of variability across children's utterances, some general tendencies were observed. The results showed that children's early uses of coordinated clauses differ from adult sentences and also differ from the children's later usages. Children's first coordinated clauses appear to have little semantic relationship, as in 12a. Later on, children produce coordinated clauses in which the conjunction *and* appears to be used to express a temporal relation, as in 12b.

Still later, the conjunction *and* is used to indicate a causal relationship between the two clauses, as in 12c.

11. a. James loved Tammy's cooking, and his favorite meal was turkey casserole.
    b. Clarence saw the movie, but he did not find it very interesting.
    c. Sometimes, Tim played softball on Sundays, or he watched sports on TV.

12. a. I can swim and you can swim too.
    b. I am going home and take my bath.
    c. I was outside too long and I was cold.

Complement Clauses

When children are around 3 years of age, they typically are producing some sentences containing a **complement clause,** which is a structure that expresses, at minimum, the second of two subject–verb relations in the sentence. In complements, the subject or verb may not be explicitly stated; it may be inferable. Examples are provided in 13. Reich (1986) investigated children's acquisition of sentences containing complements and found that they are produced around the time that children have an MLU between 3.5 and 4.0. Children are likely to produce complements that appear in subject position of the sentence, as the example in 13c, later than other forms.

13. a. I tried to go outside.
    b. I pushed the boy down.
    c. That you liked the kitty surprised me.

How Do Children Use Pronouns?

Pronouns (e.g., *he, she, himself, herself, it,* and *they*) are among the most frequently used words in English. They are highly frequent in most languages as function words. All languages appear to have words or morphemes that function as pronouns (Ariel, 1990). From the point of view of a speaker, a pronoun is the preferred way to refer to someone or something that has already been introduced into the discourse. As children learn to use pronouns correctly, they must come to appreciate some pretty subtle grammatical regularities. Chomsky (1981) pointed out that in English, there are predictable syntactic relationships between pronouns and their antecedents. Consider the examples in 14. In 14a, the pronoun *she* refers to the antecedent *Mary.* The antecedent of a pronoun does not appear in the same clause as the pronoun. This is true for all pronouns. A pronoun is not interpreted as referring to the same person as a noun when that noun is within the same clause, as shown in 14d. In contrast, antecedents for reflexive pronouns, such as *himself, herself, themselves,* and others, are always within the same clause, as in 14b. A sentence is ungrammatical if an antecedent of a reflexive pronoun occurs in a different clause as the reflexive pronoun, as shown in 14c. In the examples, an asterisk is used to indicate that the usage is ungrammatical for some speakers of English.

14. a. Because Mary was tired, she fell asleep quickly.
    b. Mary rewarded herself with an ice cream sundae.
    c. *Because Mary was tired, herself fell asleep quickly.
    d. Mary rewarded her with an ice cream sundae.

Pronouns are among the first words that children learn (Fenson et al., 1994). Chien and Wexler (1990) tested a group of 120 children between the ages of 2.5 and 6.5 years. One of the tasks was the Simon Says game. They used a girl puppet named Kitty and a boy puppet named Snoopy. Each child was asked to perform an action spoken by either Kitty or Snoopy. The children then heard sentences similar to those shown in 15. The researchers found that children's production demonstrated an understanding that an antecedent for a pronoun cannot occur within the same clause. If a child understood the grammatical rules of pronoun use, then female children hearing sentences similar to 15a should point toward themselves. When they hear sentences similar to 15b, they should point toward Kitty. When they heard sentences similar to 15c, they should point toward Snoopy. Children over 5.5 years old correctly interpreted reflexive pronouns but still made errors when interpreting pronouns.

15. a. Kitty says that Lori should point to herself.
    b. Kitty says that Lori should point to her.
    c. Snoopy says that Lori should point to him.

Children's delay in mastering the use of pronouns has been observed in other studies involving English-speaking children (Grimshaw & Rosen, 1990) and also involving children acquiring other languages, such as Dutch (Philip & Coopmans, 1996).

Grammatical Development in Special Populations

Those researchers who believe that language is a skill that is learned just as all other cognitive skills are learned might expect there to be a relationship between general cognitive ability and language ability—specifically a positive correlation. For those with normal or above average cognitive ability, one would expect normal and above average language ability, respectively. For those with below average cognitive ability, one would expect below average language ability. In contrast, those researchers who believe that language is a human skill that is distinct from other types of cognitive skills would predict that language ability would not always be related to cognitive ability.

How Are Language Ability and Cognitive Ability Related?

One of the most common assumptions that people have about language ability is that it is always correlated with cognitive ability. In other words, individuals with high IQs have above average language ability, and individuals with low IQs have below average language ability. Researchers who believe that language is a product of a general cognitive processing mechanism would expect that there would be a strong correlation between individuals' cognitive abilities and their language abilities (Bates, 1994). In contrast, researchers who believe that language ability stems from processing mechanisms that are separate from those involved in other forms of cognitive processing would not expect language and cognitive processing always to be correlated (Bellugi, Marks, Bihrle, & Sabo, 1988); rather, they would expect there to be evidence that the two processing systems are distinct and separate.

In neuropsychology, researchers often want to know whether two mental processes are distinct and separate, in terms of how the brain carries out the processing. They have developed a methodology for this purpose. The term ***single dissociation*** refers to circumstances in which one is studying two processes, such as production of function words and the production of content words, and a factor influences one process but not the other. For example, damage to one area of the brain might cause disruption in one's production of function words, but not in producing content words. The existence of the single dissociation indicates that the two processes are somewhat separable in terms of how the brain carries out the two types of processing. The term ***double dissociation*** is used to describe the demonstration that there is also a factor that affects the second process but not the first—for example, if there was also a factor that affected the processing of content words but not the processing of function words. The existence of a double dissociation provides researchers with insight into how related two processes or two processing systems are (Baddeley, 2003).

Chatterbox Syndrome

There have been intriguing case studies of individuals with severe cognitive deficits who demonstrated near normal language ability. For example, Cromer (1994) described an individual who displayed **chatterbox syndrome.** The person had hydrocephalus, a condition that involves fluid building up inside the brain. As the region containing the fluid expands, the surrounded brain tissue is displaced and often damaged. Despite a severe intellectual deficit, the person was highly verbal, talking so much as to merit being called a chatterbox. Table 4.10 provides an excerpt of a conversation between D.H., a woman with chatterbox syndrome, and a researcher. D.H.'s IQ is 44. Despite D.H.'s fluency and apparently normal grammatical abilities, she was generally unable to perform simple cognitive tasks, such as ordering three pictures to form a coherent story of an event.

Down Syndrome

More often, when there are significant cognitive deficits, there are also language deficits. Language may be delayed to a comparable extent as cognition, or it may be delayed to a greater extent. There are over 200 genetic syndromes that can result in children with **mental retardation** (MR) (Moser, 1992), which is generally defined as an IQ under 70 (i.e., two standard deviations below the mean of 100). In recent years, the term ***intellectual disability*** is more commonly used. It is becoming more commonly used because there is such a negative stigma associated with MR. The most common disorder associated with intellectual disability is **Down syndrome.** It is caused by a genetic error that occurs during development. When the DNA from the egg and sperm joins, an extra copy of chromosome 21 becomes part of the embryo's DNA. The term ***trisomy 21*** is another name for Down syndrome.

Table 4.10 Conversation Between Researcher and D.H., a Woman With Chatterbox Syndrome and an IQ of Forty-Four

Researcher:	So how long have you been here then?
D.H.:	Two and a half years.
Researcher:	Uh-huh.
D.H.:	And Dad's getting fed up with moving around. He thinks it's time that I settle down—to school, which is fair enough. To him, it . . . he feels it's going to ruin my whole life if I don't settle down sometime.
Research:	Uh-huh.
D.H.:	So I'm going to have to, at some point settle down, somewhere.
Researcher:	Uh-huh.
D.H.:	Somehow, Mum didn't mind me moving about, but Dad objected to it because he knew it was bothering me and it was bothering my schoolwork.

Source: Cromer (1994).

Down syndrome affects approximately 1 in 730 births. The chance of having a child with Down syndrome increases with the age of the mother. For women who conceive over the age of 40, the chance of giving birth to a Down syndrome child is about 1 in 100 (Hook, 1982). The rates of Down syndrome are similar for boys and girls. For individuals with Down syndrome, there can be a great deal of variability in severity, which is directly related to how many copies of the extra chromosome are created during development. Individuals with Down syndrome have almond shaped eyes with epicanthic skin folds on the corners of the eyes, low muscle tone, a small chin, a flat and broad face, a flat nasal bridge, short neck, a short fingers, a small mouth, and a protruding tongue.

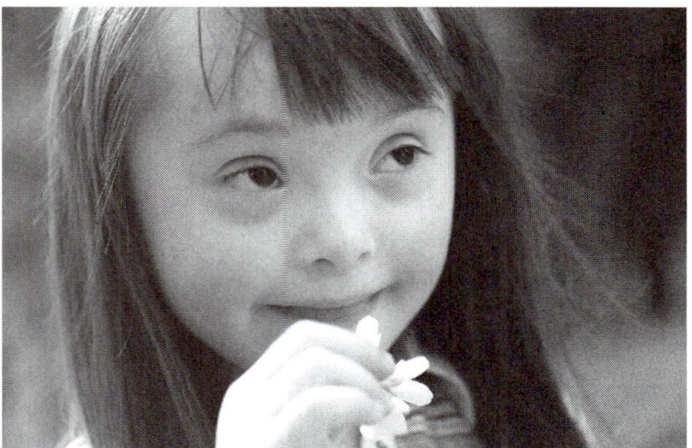

▲ Photo 4.3 A young girl with Down syndrome smells a flower. Have you had an opportunity to interact with anyone with Down syndrome?

The average IQ of individuals with Down syndrome is about 50; 100 is the average of neurologically typical individuals. The delays observed in language skills are greater than other cognitive skills. Receptive language skills are generally better than productive language skills. Syntactic development is severely delayed (Price et al., 2008). The language deficit is greater than the cognitive delay. Oliver and Buckley (1994) found that infants with Down syndrome learned vocabulary similar to typically developing infants; however, their development was delayed up to 18 months. In the study, the Down syndrome infants did not show the word spurt pattern of rapid vocabulary growth that typically developing infants often show. Fowler (1990) found evidence for grammatical deficits in Down syndrome. Grammatical suffixes, such as verb tenses, are commonly omitted (Laws & Bishop, 2003).

Fragile X

The second most common disorder that results in intellectual disability is **fragile X,** or Martin–Bell syndrome (also called Escalante's syndrome) (Hagerman & Silverman, 1996). The cause of fragile X is genetic. The genetic abnormality occurs on the X chromosome. Males are more likely to be affected than women. A father who carries the genetic abnormality will pass it on to his daughters but not to his sons. A mother who carries the genetic abnormality experiences a 50% chance of passing it on to her child. Fragile X has a prevalence of 1 in about 3,600 births for males, and 1 in about 4,000 to 6,000 births for females. The prevalence is similar across ethnic groups. Those born with fragile X will share similar physical features, such as large ears, an elongated face, high-arched palate, low muscle tone, flat feet, and soft skin.

The typical IQ for males with fragile X is between 40 and 60; 100 is the average IQ. A small percentage of males will have an IQ above 70. Females with fragile X may be mildly affected and have an IQ in the lower part of the normal range. Many individuals with fragile X will have symptoms of autism, such as gaze aversion and repetitive behaviors. When one receives a diagnosis of autism for a child, it is generally recommended that a genetic test for fragile X be obtained. Studies of language in individuals with fragile X have shown that there are deficits observed in speech development, vocabulary, grammar, and the social aspects of language use (Price et al., 2008).

Specific Language Impairment

Further evidence for some dissociation between language ability and cognitive ability is the disorder known as specific language impairment (SLI), which you first learned about in Chapter 2. SLI is a disorder in which language ability is more impaired than cognitive ability. Consequently, SLI provides evidence for a single dissociation between language ability and cognitive ability. In order to be diagnosed with SLI, one must have a nonverbal IQ of 80 or above and no other cognitive or neurological problems (American Psychiatric Association, 2000). SLI affects approximately 7% of children (Leonard, 1998). SLI appears to be more common in boys than in girls (Robinson, 1991). Research has found that children who went on to be diagnosed with SLI first exhibited symptoms when they began using multiple-word utterances (Hick, Joseph, Conti-Ramsden, Serratrice, & Faragher, 2002). The symptoms of SLI have been found to vary a great deal across children; however, researchers have not yet been able to agree on specific subtypes of SLI (Bishop, 2004). Some children with SLI demonstrate grammatical deficits, specifically difficulty producing correct morphological suffixes on verbs. They also make frequent lexical errors. One of the most common errors is the omission of the plural suffix *-s* as in *two big dogs* and the omission of the third person singular verb suffix *-s* as in *the boy jumps*.

Among SLI researchers, there are different opinions regarding why children with SLI make the errors that they make. Tallal and colleagues (Tallal & Piercy, 1978; Tallal et al., 1996) have argued that SLI is the result of a phonological processing deficit. Children may fail to perceive speech sounds occurring at the ends of words. Leonard (1998) suggested that the underlying perceptual problem is particularly problematic in learning noun and verb suffixes because there are often phonological changes to the word when a suffix is added. For example, when one adds the plural *-s* to nouns, the resulting plural is sometimes pronounced as an /s/ as in *socks* and sometimes pronounced as a /z/ as in *dogs*. Rice and colleagues have argued that SLI occurs because of a deficit in mechanisms involved in representing grammatical knowledge or in processing grammatical information (Rice, 1997; Rice & Wexler, 1996; see also van der Lely, 1998; van der Lely & Christian, 2000).

Williams Syndrome

In the 1980s, language researchers reported the existence of a congenital disorder characterized by cognitive deficits combined with what appeared to be near normal language ability. The disorder was called **Williams syndrome.** The research community immediately took notice because Williams syndrome appeared to be another example of a single dissociation between language and cognition. More important, when considered with SLI, it appeared to provide strong evidence for a double dissociation between language and cognition. It is a rare developmental disorder affecting approximately 1 in 7,500 to 1 in 20,000 births. The syndrome is caused by the deletion of the region q11.23 of chromosome 7; the deleted region involves about 25 genes (Morris, Lenhoff, & Wang, 2006). People with Williams syndrome have distinctive facial features, often described as elfin. They have a low nasal bridge and a cheerful demeanor, and they are interpersonally engaging. Some individuals with Williams syndrome may also have exceedingly well developed abilities or **savantism** in music or art. In the 1980s, Ursula Bellugi and colleagues described the abilities of individuals with Williams syndrome, concluding that the syndrome provided evidence that language develops independently from other cognitive abilities (Bellugi, Marks, Bihrle, & Sabo, 1988). They described the syndrome as being characterized by low IQ (i.e., an average of about 60, well below the average of 100) and near normal language ability (Bellugi, Lichtenberger, Mills, Galaburda, & Korenberg, 1999).

In recent research, this view of Williams syndrome has been challenged. Mervis and Becerra (2007) analyzed cognitive and linguistic performance for 306 children with Williams syndrome. They argued that the previous claims that individuals with Williams syndrome have severe intellectual disability combined with near normal language ability are overstated. They found that the average IQ was about 71, and the distribution of IQ scores approximated a normal distribution. They noted that performance on spatial processing tasks was particularly poor, which served to decrease overall IQ scores. They also found that language ability did not exceed the cognitive ability of the individual; rather, language ability was comparable to cognitive ability. They point out that most children with Williams syndrome have difficulties in the pragmatic use of language and could benefit from language therapies.

How Does Grammar Develop in Deaf Children?

Each year in the United States, many infants are born profoundly deaf. The prevalence of congenital deafness is estimated at 1 to 6 in 1,000 births (Kemper & Downs, 2000). The cochlear implant is an electronic device that enables a previously deaf individual to process sound. The first electronic devices that were designed to help the deaf to hear were invented in the 1950s (Clark, Tong, & Patrick, 2000). The first commercially available cochlear implant was sold in 1972 by House 3M. According to the Food and Drug Administration, there have been over 188,000 people worldwide who have received a cochlear implant, including about 30,000 children (National Institute on Deafness and Other Communication Disorders, 2011). Language outcomes are best for children who receive their implants before the age of 4 (Nevins & Chute, 1997). Infants as young as 12 months old can receive an implant (Spencer & Marschark, 2003).

Deaf children without cochlear implants experience difficulty in acquiring grammar (Bollard, Chute, Popp, & Parisier, 1999). Using a variant of the wug test (Berko, 1958), Cooper (1967) showed that profoundly deaf children performed behind their same-age hearing peers in language performance. Presnell (1973) found that grammatical development, in particular, occurs more slowly for profoundly deaf children than for hearing children. Elfenbein, Hardin-Jones, and Davis (1994) tested a group of 5- to 7-year-old children with hearing loss in the mild to severe range. The results showed that the children made errors on verb forms, irregular noun plurals, possessives (e.g., *Ricky's shirt* or *Jose's car*), and comparatives (e.g., *smaller* or *prouder*). Although there are few studies on the language development of deaf children after receiving a cochlear implant, those studies have suggested that language development is facilitated following implantation (Hammes et al., 2002; Svirsky, Robbins, Kirk, Pisoni, & Miyamoto, 2000). However, grammatical development for deaf children with cochlear implants still lags behind their same-age peers (Nikolopoulos, Dyar, Archbold, & O'Donoghue, 2004; Schorr, Roth, & Fox, 2008).

Language delays in deaf children most likely occur because most deaf children are born to hearing parents and are not taught sign language until relatively late in childhood (Quigley & King, 1982). Only 5% to 10% of deaf children are born to deaf parents (Meier & Newport, 1990). For those children who are born deaf to hearing parents and who are not provided with exposure to American Sign Language (ASL) or some other signed language, there is the possibility that they will miss the critical period for learning language and, like feral children, face difficulties learning grammar. Holt and Svirsky (2008) claimed that after the age of 3.5, there is a decline in children's ability to acquire language.

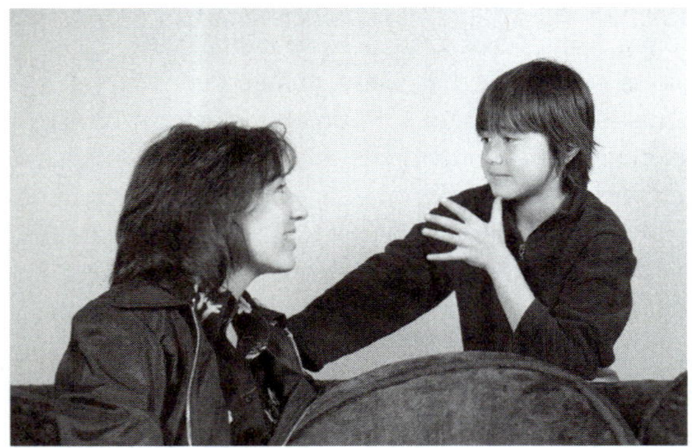

▲ Photo 4.4 A son communicates with his mother using American Sign Language (ASL). For many deaf children, ASL is the first language acquired during childhood. Do you know anyone who is deaf or who has a deaf child?

Among those involved in deaf communities, cochlear implants are not always viewed in a positive light. Some individuals believe that cochlear implants may ultimately lead to the loss of deafness and the well-organized and active communities that have existed for the deaf and their families (Padden & Humphries, 1988). Because deaf individuals who use a signed language live fulfilling, normal lives, many in the deaf community to do not view deafness as a disability that should be cured. The use of sign language is preferred and valued over the use of oral language. There is a perception that receiving a cochlear implant will increase the likelihood that the deaf individual will come to rely primarily on oral language for communication.

How Does Grammar Develop With Only One Hemisphere?

For some children who suffer from chronic, severe seizures, the treatment of last resort is a hemispherectomy, the removal of one of the hemispheres of the brain. The treatment would only be recommended when one has very little quality of life. Case studies have shown that young children who have a hemispherectomy do remarkably well. Because of brain plasticity, the functions of the missing hemisphere are often taken over by the remaining hemisphere. Cases of children who have their left hemispheres removed are particularly interesting to language researchers. Because the left hemisphere has been found to play a dominant role in language processing, one must wonder whether children who lose their left hemisphere early in life would be able to acquire their language's grammar.

Research conducted by Susan Curtiss and colleagues (Curtiss & de Bode, 2003; Curtiss & Schaeffer, 2005) investigated the grammatical development of children who had hemispherectomies. Curtiss and de Bode (2003) obtained speech samples of eight children who had experienced a left hemispherectomy and compared the samples with similar samples obtained from children with similar MLUs. The results indicated that the grammatical characteristics of the language of the two groups of children were remarkably similar. The authors concluded that the grammar produced by the intact right hemisphere appears normal. Curtiss and Schaeffer (2005) compared speech samples of 10 children who had experienced either a left or right hemispherectomy with similar samples from children with similar MLUs. They again found that children who had experienced a left hemispherectomy could develop normal grammatical function; however, they did find that those children made slightly more grammatical errors than children who had experienced a right hemispherectomy. Pulsifer and colleagues (2004) reported long-term outcomes for 71 individuals who had undergone hemispherectomy. Among the data available were IQ scores before and after surgery. Their results indicated little change in IQ as a result of surgery. For 34 of the 53 individuals for whom scores were available, IQ changed on average less than 15 points. IQ decreased for 11 individuals following surgery and increased for 8 individuals.

Summary and Theoretical Implications

The development of grammar certainly begins before children are able to produce multiple-word utterances. Studies using the preferential looking paradigm have shown that infants with only one or two words in their vocabulary are already beginning to understand what types of words precede other types of words in sentences. When infants begin producing two-word utterances, the ordering of the words in these utterances follows the word order norms of the language. Over the next year, children begin to acquire the syntactic morphemes of their language. The order in which morphemes are acquired appears to be determined by the cognitive complexity of the relationship represented by the morpheme. Research shows that children learn grammatical rules, rather than memorizing individual word forms that they hear. As they learn grammar rules, they routinely apply them incorrectly, producing forms that adults never produce. These errors are called overregularizations. Children also commonly make errors when learning how to produce negative sentences and questions, as well as sentences containing coordinated clauses, complement clauses, and pronouns.

Because grammatical development occurs rapidly and without direct instruction on the part of parents, the behaviorist and social-interactionist views of language development have difficulty accounting for such learning. On the other hand, the generative approach and the statistical learning approach to language development can each account for rapid, early learning without direct instruction. Because the statistical learning approach views language development as not involving special language-specific mechanisms, it is unclear how it would account for recent facts related to the dissociation of language and cognitive abilities in disorders. Cases in which language ability is more delayed than cognitive ability and cases in which language ability is less delayed than cognitive ability suggest that language ability and cognitive ability are brought about via different mental systems. The grammatical development of deaf individuals may be delayed, particularly if they are not exposed to a signed language very early in life. Surprisingly, children who have received a hemispherectomy go on to develop relatively normal language—even those children whose left hemisphere has been removed. Because the generative approach to language development assumes that language abilities are brought about from different mechanisms than other cognitive abilities, which are biologically based, it has the potential to incorporate the data better than the other theoretical approaches.

KEY TERMS

active sentence
agent
bound morphemes
chatterbox syndrome
closed class words
complement clause
coordinated clauses
copula
derivational morphemes
double dissociation
Down syndrome
fragile X
free morphemes
function words
holophrase
Index of Productive Syntax (IPSyn)
inflectional morphemes
intellectual disability
intransitive verb
mean length utterance (MLU)
mental retardation (MR)
negative sentences
open class words
overregularization errors
passive sentences
patient
phrase structure rules
productive language ability
receptive language ability
savantism
semantic bootstrapping
semantic complexity
single dissociation
syntactic bootstrapping
syntactic complexity
transitive verb
trisomy 21
truncated passives
wh-word
Williams syndrome

REVIEW QUESTIONS

1. What evidence is there that children understand English word order rules before they are able to produce sentences?
2. Explain how tree diagrams are used to represent the syntactic rules for a language.
3. How are children's two-word utterances evidence that children already appreciate English word order rules?
4. What evidence is there that children learn grammatical rules?
5. What is MLU? How does MLU change as children age?
6. What grammatical morphemes in English are learned first? What are the explanations for why some morphemes are acquired before others?
7. What is the wug test? What did it show about children's development of grammar?
8. What is an overregularization error? Why do children make overregularization errors? What does the existence of such errors say about the nature of grammatical development?
9. When children learn to form negative sentences, what are the different forms that they tend to produce? What is the typical sequence in which the different sentence structures are produced?
10. When children learn to form questions with wh-words, what are the different word orders that they tend to produce? What is the typical sequence in which the different sentence structures are produced?
11. How do children learn to produce passive sentences?
12. How do children begin to use complement clauses?
13. What is the difference between semantic and syntactic bootstrapping?
14. What is fragile X syndrome? What are the language and cognitive abilities of individuals with fragile x syndrome?
15. What is Down syndrome? What are the language and cognitive abilities of individuals with Down syndrome?
16. What is Williams syndrome? What are the language and cognitive abilities of individuals with Williams syndrome?
17. What is SLI? What are the language and cognitive abilities of individuals with SLI?
18. What is chatterbox syndrome? What are the language and cognitive abilities of individuals with chatterbox syndrome?
19. How does grammatical development occur in individuals who have had a hemispherectomy of the left hemisphere?
20. What does research suggest about how IQ is changed following hemispherectomy?

RECOMMENDED READING

Bates, E., Bretherton, I., & Snyder, L. (1988). *From first words to grammar: Individual differences and dissociable mechanisms.* Cambridge, UK: Cambridge University Press.

Brown, R. (1973). *A first language: The early stages*. Cambridge, MA: Harvard University Press.

Carey, S. (2011). *The origin of concepts.* Oxford, UK: Oxford University Press.

Chomsky, N. (1986). *Knowledge of language: Its nature, origin, and use.* New York: Praeger.

Morgan, J., & Demuth, K. (1996). *Signal to syntax: Bootstrapping from speech to grammar in early acquisition.* Mahwah, NJ: Erlbaum.

Slobin, D. I. (1985–1997). *A cross-linguistic study of language acquisition*. Hillsdale, NJ: Erlbaum.

Synder, W. (2007). *Child language: The parametric approach*. Oxford, UK: Oxford University Press.

RECOMMENDED FILMS

Codina, A. (Director). (2009). *Monica and David* [Documentary]. United States: Home Box Office.

Searchinger, G., Male, M., & Wright, M. (Writers). (2005). *Human language series* [DVD]. United States: Equinox Films/Ways of Knowing Inc.

SUGGESTIONS FOR CLASS PROJECTS

1. Transcripts of child speech can be viewed and downloaded at the CHILDES database (www.childes.com). Visit the database, select a transcript of a child between the ages of 20 months and 48 months, and prepare an analysis of the transcript to determine whether the child is a typical or atypical child of that age based on the grammatical structures that the child has used in the conversation. An alternative is that you could be provided with an excerpt of a conversation involving a child. Estimate the child's age, based on the types of grammatical structures that the child has used in the conversation.

2. You can obtain experience representing English sentences using the tree diagrams routinely used by linguists. The following sentences can be diagrammed using the phrase structure rules displayed in Table 4.3.

   a. The little bird drank the water.
   b. The large dog chased the skinny cat.
   c. The boy from the neighborhood helped Susan.
   d. The workers moved the boxes to the curb.
   e. The students often complained about the homework.
   f. The tree fell onto the car.
   g. The cook prepared the meal without onions.
   h. The quarterback threw an interception.

For additional ancillary resources, please visit the companion website at www.sagepub.com/kennison.

- Video Links
- Audio Links
- Web Resources
- Internet Activities
- Flashcards
- Web Quizzes

CHAPTER 5

THE LEXICON

By the time that students in the United States are ready for college, they know about 100,000 words (Nation & Waring, 1997). In order for children to accumulate that many words in a period of 18 years, they must learn an average of 10 words a day. This rate of word learning is nothing less than amazing. As most teenagers can attest, learning just one new word a day for a school assignment or to prepare for a standardized test (e.g., SAT

▲ Photo 5.1 A young boy is making words with letters. Do you remember playing with toy letters when you were a child?

or ACT) is a challenge. As you are reading this, you have access to your vast knowledge of words stored away in your memory. Each word that you read triggers an experience of understanding in your mind; you rapidly become aware of what the word is and how it fits in with the material that has preceded it on the page. Researchers use the term *mental lexicon* to refer to our vast knowledge of word meaning and word usage. When researchers use the term *mental lexicon,* they do not suggest that there is a single location in the brain where all word knowledge is stored; rather, the mental lexicon is a metaphor for how the mind might organize all that we know about the words in our language. In this chapter, you will learn about what information researchers believe is contained within the mental lexicon, the processes involved in accessing the mental lexicon, and how word knowledge is organized in memory with other types of factual knowledge.

The Mental Lexicon

What Information Is Contained in a Lexical Entry?

The contents of a traditional dictionary are likely quite familiar to you. Dictionary entries usually contain information about the word's spelling, pronunciation, part of speech (e.g., noun, verb, adverb, adjective, and preposition), and possible meanings. In some dictionaries, entries also contain information about the history of the word (also called the word's etymology). Table 5.1 displays the entries for the words *stalactite* and *stalagmite,* which both have Greek origins. These words are often confused. One useful way that I have found to remember the difference between the two words is that if you are not careful, you *might* trip over a *stalagmite*.

The information contained in the mental lexicon is believed to be somewhat similar to the information contained in the traditional dictionary; however, there are important differences. Researchers continue to study and debate the organization of the mental lexicon (Aitchison, 2003). Researchers generally agree that the mental lexicon must contain all the information about a word that one needs to use the word appropriately in speaking and in writing or to comprehend during listening and reading. Most researchers agree that the mental lexicon would specify any information that is not predictable, given the grammatical rules of the language (Aitchison, 2003). For example, in English, there is a grammatical rule that dictates how a plural noun can be formed from a singular noun (e.g., add the suffix *-s*

Table 5.1 Dictionary Entries for the Words *Stalactite* and *Stalagmite*

sta · lac · tite [stuh-lak-tahyt, stal-uhk-tahyt] noun

a deposit, usually of calcium carbonate, shaped like an icicle, hanging from the roof of a cave or the like, and formed by the dripping of percolating calcareous water. 1670–1680; Neo-Latin stalactites Greek stalakt (ós) dripping (stalag-, stem of stalássein to drip + -tos verbid suffix) + Neo-Latin -ites -ite

sta · lag · mite [stuh-lag-mahyt, stal-uhg-mahyt] noun

a deposit, usually of calcium carbonate, more or less resembling an inverted stalactite, formed on the floor of a cave or the like by the dripping of percolating calcareous water. origin 1675–1685; Neo-Latin stalagmites Greek stálagm (a) a drop (stalag-, stem of stalássein to drip + -ma noun suffix of result) + Neo-Latin -ites -ite

Source: Dictionary.com.

to a singular noun). Because there are some nouns whose plural forms are different from this rule, there is a need for the mental lexicon to specify whether the singular noun follows the rule (i.e., is a **regular word**) or does not follow the rule (i.e., is an **irregular word**). The mental lexicon must also specify the irregular plural form. Table 5.2 provides a list of some common irregular plural nouns. When one is listening or reading and encounters an incorrectly formed plural noun (e.g., **mouses* or **deers*; the asterisks indicate ungrammatical words), one presumably has discovered the error during the process of lexical access after one has accessed all the information about the word and determined that the word is irregular instead of regular and has become aware of what the correct plural form is.

Table 5.2 Examples of English Irregular Plural Nouns

Singular Form	Plural Form
Man	Men
Woman	Women
Mouse	Mice
Louse	Lice
Goose	Geese
Foot	Feet
Tooth	Teeth
Wife	Wives
Leaf	Leaves
Scarf	Scarves
Life	Lives
Knife	Knives
Moose	Moose
Deer	Deer

Source: Dictionary.com.

Irregular Verbs

In addition to irregular nouns, there are also irregular verbs, which must be specified in the mental lexicon. For example, the past tense of a regular verb is formed by adding the suffix *-ed*. Quite a few verbs in English are irregular. Their past tense forms do not follow the rule and must

Table 5.3 Examples of English Irregular Past Tense Verbs

Present Tense Form	Past Tense Form
Break	Broke
Hold	Held
Take	Took
Shake	Shook
Ring	Rang
Sing	Sang
Sit	Sat
Hit	Hit
Fly	Flew
Blow	Blew
Catch	Caught
Choose	Chose
Freeze	Froze
Breed	Bred

Source: Dictionary.com.

be learned individually, verb by verb, and stored in the lexicon. Table 5.3 provides examples of irregular past tense verbs that are common in English. As you learned in Chapter 4, as children acquire past tense forms of verbs, they typically experience a period in which they make overregularization errors, such as **braked* or **holded* (the asterisks indicate ungrammatical words). By the time children have mastered verb past tense forms, their lexicons are likely to contain adequate information about which verbs are regular versus irregular and what irregular verb forms should be used.

As you may remember from Chapter 4, there are many languages that have a greater number of noun and verb classes than we have in English. The lexicons of the speakers of those languages would contain information about all the possible types of words in their languages. For example, many languages have grammatical gender. One of these languages is Italian. Nouns in Italian carry the suffix *-a* or *-o*, indicating that the noun is feminine or masculine, respectively. The meaning of the noun is usually consistent with a feminine or masculine meaning. A small number of nouns have a meaning that is inconsistent with their gender classification, such as *pilota,* which means pilot, and *pirata,* which means pirate. These cases would be considered irregular forms in Italian and would need to be specified in the mental lexicon.

Idioms

The lexicon must also specify multiword expressions that have come to mean something different from the words contained in the expression. Such expressions are called idioms, such as *that's all she wrote,* which means something has come to an end. Although no one is quite sure what the origin of this idiom is, it is suggested that it arose during World War II when many soldiers received breakup letters from sweethearts. The expression *Dear John Letter* arose around this time also. Such letters were quite common ("Dear John Letter," n.d.). Each generation of language learners must learn idioms one by one through experience with the language. For those learning a second or third language, idioms are particularly challenging to learn. Table 5.4 displays some familiar idioms in American English. Many verbs in English are used idiomatically with prepositions (e.g., *in, on, up, out,* and *down*), such as in *Bill threw up the pizza* versus *Bill threw up the ball* and *Sara put down the performance* versus *Sara put down the box*. These have been referred to as verb–particle constructions (Fraser, 1976).

Table 5.4 Examples of Common Idioms in American English

Idiom	Meaning
To kick the bucket	To die
To throw someone under the bus	To betray
The cat has your tongue	You are shy and not talking
To be out of the woods	To be out of danger
To pull someone's leg	To be insincere or to joke
To lose one's shirt	To lose a lot of money
To spill the beans	To reveal a secret
To twist someone's arm	To persuade
To be in hot water	To be in trouble
To face the music	To confront the truth

Source: Dictionary.com.

How Are Ambiguous Words Represented in the Mental Lexicon?

When one considers the number of lexical entries that might be contained in the mental lexicon, one would expect the number of entries to equal the number of words that one knows. What exactly is a word? In fact, the way that traditional dictionaries define the word *word* differs somewhat from how psycholinguists define it. The traditional dictionary has one entry for each letter sequence that makes up a word. For example, consider the entry for the word *hit* in Table 5.5; there are many meanings, some of them more closely related to each other than others. In contrast, the mental lexicon is likely to make finer distinctions related to meaning. Words that are spelled and pronounced the same but have different meanings are believed to have distinct lexical entries. For example, when *hit* is used to refer to a murder, it is quite different from the typical usage when *hit* is used to refer to a punch or a play in a baseball game. Also, consider the word *bug*. It can be used to refer to an insect, as in *James shrieked when he felt the bug crawling on his neck*. It can also be used to refer to a small microphone that may be used for spying, as in *The FBI were able to record the bribery attempt because they placed a bug in the hotel room*. Some researchers have described this type of ambiguity as **lexical ambiguity** (Rayner & Duffy, 1986). Others have used the term ***homograph*** to describe such words, indicating that there are two different words that happen to be spelled the same. Table 5.6 provides more examples of homographs in English.

The mental lexicon must also distinguish words that can be used as more than one part of speech (e.g., noun, verb, adjective, adverb). Such words are quite similar in meaning, despite functioning in a different way in a sentence. For example, *hit* can be used as both a

Table 5.5 Dictionary Entry for the Word *Hit*

Hit [hit]
verb (used with object)

1. to deal a blow or stroke to: Hit the nail with the hammer.
2. to come against with an impact or collision, as a missile, a flying fragment, a falling body, or the like: The car hit the tree.
3. to reach with a missile, a weapon, a blow, or the like, as one throwing, shooting, or striking: Did the bullet hit him?
4. to succeed in striking: With his final shot he hit the mark.
5. Baseball. to make (a base hit): He hit a single and a home run.

verb (used without object)

6. to strike with a missile, a weapon, or the like; deal a blow or blows: The armies hit at dawn.
7. to come into collision (often followed by against, on, or upon): The door hit against the wall.
8. Slang. to kill; murder.
9. (of an internal-combustion engine) to ignite a mixture of air and fuel as intended: This jalopy is hitting on all cylinders.
10. to come or light (usually followed by upon or on): to hit on a new way.

noun

11. an impact or collision, as of one thing against another.
12. a stroke that reaches an object; blow.
13. a stroke of satire, censure, etc.: a hit at complacency.
14. Baseball. base hit.
15. Backgammon. a. game won by a player after the opponent has thrown off one or more men from the board. b. any winning game.

verb phrases

16. hit off, a. to represent or describe precisely or aptly: In his new book he hits off the American temperament with amazing insight. b. to imitate, especially in order to satirize.
17. hit on, Slang. to make persistent sexual advances to: guys who hit on girls at social events.
18. hit out, a. to deal a blow aimlessly: a child hitting out in anger and frustration. b. to make a violent verbal attack: Critics hit out at the administration's new energy policy.
19. hit up, Slang. a. to ask to borrow money from: He hit me up for ten bucks. b. to inject a narcotic drug into a vein.

Source: Dictionary.com.

Table 5.6 Examples of Common Homographs in English

Word	Meanings
Poker	A game played with cards
	A tool used during wood-burning
Table	A piece of furniture
	An illustration used in a piece of writing
Ball	A round object used in many games
	A formal celebration, usually involving dancing
Plant	A general term used to describe flora
	A factory or other manufacturing setting
Bank	A financial institution
	A raised area of land, near a river or a road
Habit	A frequent behavior
	A type of clothing worn by a nun
Scale	A device used for weighing objects
	A type of tissue found covering many fish
Horn	A musical instrument
	A bony protrusion found on some animals
Boxer	An athlete who competes in boxing
	A type of dog
Cabinet	A piece of furniture, often used to store dishes
	Members of a president's executive team

Source: Sereno, O'Donnell, and Rayner (2006).

noun and a verb. Such words are examples of **syntactic category ambiguity.** Other examples are *kick* and *drink,* as shown in 1. Other examples of words that can be used in more than one syntactic category include *silver* (noun or adjective), *to* (infinitive, preposition), and *her* (noun, adjective). Some of the cases of syntactic category ambiguity also involve a phonological change. The words are spelled the same but have different pronunciations for the different parts of speech. A word such as this is called a **heteronym.** Consider the examples in 2. In contrast, words that are spelled differently and mean different things but are pronounced the same, such as *hair* and *hare* as well as *red* and *read*, are called **homophones.**

1. John saw the kick. — Noun usage
   Mary tried the drink. — Noun usage
   Horses kick sometimes. — Verb usage
   Kids drink milk and juice. — Verb usage

2. a. The garden was used to *produce* lots of *produce*.
   b. The medic *wound* the shirt around the *wound*.
   c. When she saw the *tears* in the dress, she definitely shed some *tears*.

Beginning in elementary school, children in the United States and other English-speaking countries have the opportunity to demonstrate their knowledge of words during spelling bees. The person who correctly spells the most words wins. Text Box 5.1 describes this American tradition and provides a list of all the winning words from the National Spelling Bee.

Text Box 5.1 Extraordinary Individuals: National Spelling Bee Winners

▲ **Photo 5.2** These children are competing in a spelling bee. Did you ever take part in a spelling bee in elementary or middle school?

The first National Spelling Bee in the United States was held in Washington, D.C., in 1925; however, there are reports of spelling bees being held as far back as the mid-1700s (Scripps National Spelling Bee, n.d.-a). Those competing in the spelling bee receive a word to spell. The word is spoken aloud by a person referred to as the *pronouncer.* If the competitor correctly spells the word, he or she advances to the next round. Those who spell their word incorrectly are eliminated. The competition continues until the round in which there is only one person who correctly spells a word on that round. When a competitor receives a word to spell, the competitor is able to ask the pronouncer for information about the word, such as other pronunciations and, its part of speech, language of origin, and definition. Anyone who has watched the final round of the National Spelling Bee might conclude that asking for additional information is a useful strategy. Table 5.7 displays the winning words from 1925 to 2012.

Table 5.7 Winning Words from the National Spelling Bee Competitions From 1925 to 2012

Years	National Spelling Bee Winning Words
1925–1929	Gladiolus, cerise, luxuriance, albumen, asceticism
1930–1939	Fracas, foulard, knack, torsion, deteriorating, intelligible, interning, promiscuous, sanitarium, canonical
1940–1949	Therapy, initials, sacrilegious, semaphore, chlorophyll, psychiatry, dulcimer*

Years	National Spelling Bee Winning Words
1950–1959	Meticulosity, insouciant, vignette, soubrette, transept, crustaceology, condominium, schappe, syllepsis, catamaran
1960–1969	Eudaemonic, smaragdine, esquamulose, equipage, sycophant, eczema, ratoon, Chihuahua, abalone, interlocutory
1970–1979	Croissant, shalloon, macerate, vouchsafe, hydrophyte, incisor, narcolepsy, cambist, deification, maculature
1980–1989	Elucubrate, sarcophagus, psoriasis, Purim, luge, milieu, odontalgia, staphylococci, elegiacal, spoliator
1990–1999	Fibranne, antipyretic, lyceum, kamikaze, antediluvian, xanthosis, vivisepulture, euonym, chiaroscurist, logorrhea
2000–2009	Demarche, succedaneum, prospicience, pococurante, autochthonous, appoggiatura, Ursprache, serrefine, Guerdon, Laodicean
2010–2012	Stromuhr, cymotrichous, guetapens

Source: Scripps National Spelling Bee (n.d.-b).

*There was no competition for 1943, 1944, and 1945.

So far, we have discussed just a few of the distinctions that must be coded in the mental lexicon. There are many more distinctions proposed by researchers (Aitchison, 2003). It is generally the case that fluent language users are not aware of the distinctions that exist in their mental lexicons. They are using the lexicon to plan what they are going to say or to write and to interpret what they are hearing or reading. The processes are occurring so rapidly that usually one has little insight into how words were selected from memory for an utterance or were retrieved from memory for comprehension. The best way that researchers have to find out which distinctions are specified is reaction time experiments. Such experiments are designed to measure how quickly words can be understood or pronounced. In the remainder of this chapter, you will learn more about the techniques that researchers have used and the discoveries that they have made.

What Part of Memory Contains the Mental Lexicon?

Memory researchers generally agree that there are different types of memory. Our mental lexicon is part of our semantic memory, which refers to memory for the meanings of words and also factual knowledge (e.g., *Earth is the third planet from the sun* and *The chemical symbol for salt is NaCl*). This type of memory is different from the most familiar type of memory, which is episodic memory (also called autobiographical memory). It involves memory for events from daily life, such as memory of your last birthday celebration. Episodic memories may vary in their level of detail. Some episodic memories are highly detailed and emotionally charged, such as one's memory of a significant worldwide event, such as one's memories for September 11, 2001, or the day that Michael Jackson passed away. Such memories are called flashbulb memories. Other types of memory include (1) working memory, which is used for on-the-spot thinking and problem solving; (2) procedural memory, which involves memory for sequences (e.g., how to tie a shoe or how to ride a bike); and (3) sensory memory, which refers to the brief memory for information coming into our senses (e.g., sight, sound, touch, taste, and smell). Much of our understanding of how memory works comes from research conducted since the 1960s, when the field of psychology began to focus more on the study of mental processes (Benjamin, 2007). One of the most elegant experiments of this decade revealed the existence of visual sensory memory. Textbox 5.2 provides details about this important research discovery.

Text Box 5.2 Research Discovery: Sperling's (1960) Discovery of Visual Sensory Memory

When I am asked to name my favorite experiment, I always respond with Sperling (1960). Sperling's experiments showed me how we can study the intangible processes involved in human thinking. In his first experiment, participants were shown an array of 12 letters, displayed as three rows with 4 letters each. The array was presented briefly. When the array disappeared, participants were asked to report as many letters from the array as possible. Initial results showed that participants' performance was quite consistent; everyone reported 3 to 4 letters. Was this the extent of our visual sensory memory—between 25% and 33% of the total of what we have just seen?

Subsequent experiments proved conclusively that the amount that we remember is closer to 100% than 25% to 33%. In these experiments, participants were again shown an array of 12 letters, but this time, when the array disappeared, participants heard one of three random tones. If they heard a high tone, they were to report only the top row of letters from the array. If they heard a low tone, they were to report only the bottom row of letters from the array, and if they heard a mid tone, they were to report the middle row. In all cases, participants were able to recall 3 to 4 letters from the array. This is remarkable because participants did not know which row they would have to report from the array until the array had disappeared. The results suggested that at the time that participants heard the tone and reported the letter from the appropriate row, they had the entire array available in memory. An example of the stimuli used in the experiment is provided in 3. Many follow-up experiments provided additional details about the nature of visual sensory memory (also called iconic memory). The content of the memory (or the icon) is a verbatim copy of the visual information processed by the senses. It is available in memory for about 250 milliseconds or until it is overwritten by more sensory input. In research that followed, similar results were obtained for listening, showing that there is auditory sensory memory (also called *echoic memory*) (Goldstein, 2007). The duration of auditory sensory memory is much longer than visual sensory memory, as it lasts about 3 to 4 seconds.

3. 7 I V F
X L 5 3
B 4 W 7

The memories that are the most salient and the most important to us are those that are conscious. We are aware of them. We can talk about them and reflect on them. This type of memory is referred to as explicit memory (Goldstein, 2007). The terms *conscious memory* and *declarative memory* have also been used to refer to explicit memory. Research has revealed that we also store vast amounts of information that is unconscious. We are not aware that we have stored the information or are using the information when we are using it. This type of memory is referred to as implicit memory. The term *unconscious memory* has also been used to refer to implicit memory. Most of the semantic, episodic, and working memories that we have are explicit in nature. Sensory memory is primarily implicit. Semantic and procedural memories can involve aspects that are both explicit and implicit.

Our knowledge of words routinely involves both explicit and implicit memories (i.e., information that we know we know and other information that we know, but we do not realize that we know it or how we use it). For example, one may not know a word's meaning completely but may still be able to judge whether the word has a positive or a negative meaning (Osgood, Suci, & Tannenbaum, 1957). Consider the word *puerile.* Even one who does not know that the word means young and irresponsible may be fairly confident that the word's meaning is negative. Words whose meanings are only partially known have been called **frontier words** (Durso & Shore, 1991). One's knowledge of a word's meaning increases as one's experience with the word increases (Reichle & Perfetti, 2003).

Lexical Access

Are Words Recognized One Letter or Phoneme at a Time?

Researchers use the term *lexical access* to refer to the process of hearing or seeing a word. Imagine reading the word *artichoke.* Quickly, you become aware of what the word means (i.e., a vegetable used in Mediterranean dishes). If you are fond of artichokes, you might know more about it (e.g., it is a member of the daisy family and has a thistle-like appearance). The processes involved in lexical access are still not well understood. Researchers do know that the time that it takes for one to become aware of a word's meaning is a fraction of a second (Rayner & Pollatsek, 1989). The process occurs so rapidly that we are generally not able to describe how we do it. It occurs below the level of conscious reflection. Research has suggested that the rate of comprehension for listening is several hundred words per minute most of the time (Rayner & Pollatsek, 1989). In daily life, the process of lexical access is occurring tens of thousands of times. Only when we encounter an unfamiliar word (i.e., a word that is not yet stored in our lexicon or when we have difficulty hearing or seeing a word fully) might we become aware that we are exerting some effort to figure out what the word means.

Perceiving Words

The majority of studies that have been conducted to investigate the mental processes involved in lexical access have investigated how printed words are recognized (Rayner & Pollatsek, 1989). The primary reason that research has focused on reading words rather than hearing words is that between the 1960s and 1990s, experiments with printed words were easier to carry out on computers than were experiments involving speech. With recent advancements in computer technology, this is no longer the case; thus, there has recently been an increase in the number of studies investigating speech recognition. One of the earliest findings stemming from studies investigating lexical access of printed words was that as the number of letters in a word increases, the time needed to recognize the word also increases. This pattern has been referred to as the **word length effect.** For example, the word *pomegranate* would take longer to process than *kiwi*. In the early days of research on lexical access, researchers hypothesized that words are recognized a letter at a time (see Gough, 1972). If presented with the word *kiwi,* one would identify the letter *k,* then the letter *i,* then the letter *w,* and finally the letter *i*. Then one would consult one's lexicon for a matching letter sequence. This view predicted that the amount of time that would be needed to identify a word would be equal to the sum of the durations needed to identify each letter within the word.

Although this view of word recognition served as a useful starting point for thinking about how lexical access might occur, there is ample evidence now showing that it is incorrect. In fact, the first experiment that provided evidence against this view was conducted in the late 1800s (Cattell, 1886). This experiment showed that whole words are actually recognized faster than any single letter within a word (Cattell, 1886; Reicher, 1969). This has been referred to as the **word superiority effect.** Consequently, word recognition cannot occur as the result of letter-by-letter identification; rather, the letters in a word must be processed simultaneously or in parallel. Cattell (1886) showed participants words or single letters and asked them to repeat what they saw. He found that participants were more accurate perceiving words than letters. In 1969, Reicher replicated the experiment, improving upon the methodology in order to eliminate possible confounds. He also modified the procedure to eliminate the possibility of guessing. He presented a stimulus, which was a word, a single letter, or a sequence of letters that did not make up a word. It was followed by a pattern mask, which served to eliminate the sensory memory of the stimulus and then two probe letters, which appeared above and below the location that previously displayed the target (i.e., real word, single letter, or nonword). The participant was asked to indicate which letter had been presented. The results showed that participants were more accurate responding to probe letters when the letter had previously appeared in a word than as a single letter or in a nonword.

What Factors Influence Word Processing?

Over the past 40 years, thousands of experiments have investigated how words are recognized. One of the most robust findings in the word recognition literature is the **word frequency effect**—how often words are used in daily life influences how quickly they can be recognized. Words that are experienced more often in daily life can be recognized more rapidly than words that are experienced less often (Rayner & Pollatsek, 1989). On average, very high frequency words can be responded to about 50 milliseconds more quickly than low frequency words. Examples of high and low frequency words are displayed in Table 5.8.

One of the most common tasks that researchers use to study lexical access, in general, and the word frequency effect, specifically, is the **lexical decision task** (Balota & Chumbley, 1984). Using a computer, a researcher presents to a participant a series of letter strings (e.g., *artichoke* or *cumumber*). The researcher instructs the participant to decide as quickly as possible whether the letters make up a word. If the letters form a word, the participant should hit one key, which is designated the *YES* key; if the letters do not form a known word, the participant should hit another key, which is designated the *NO* key. The average lexical decision time for English words is about 400 to 600 milliseconds.

Table 5.8 Examples of High and Low Frequency Words

High Frequency Nouns	Low Frequency Nouns
Time	Epoch
Paper	Sheath
Police	Constable
Face	Facade
City	Hamlet
Water	Oasis
Room	Alcove
Love	Adoration
College	Convent
Body	Corpse

Source: Francis and Kučera (1982).

A second methodology that is frequently used to investigate lexical access is the recording of eye movements during reading. When readers process text in everyday life, they do not press keys on a keyboard or speak into a microphone; they merely move their eyes to the next region of the text that they are reading. Researchers can record eye movements during natural reading and determine how long readers take to process individual words. Researchers refer to movements of the eyes as saccades, which comes from the French word for the verb *to jerk* (i.e., *saccader*). The time between saccades, when the eyes are focused on some object, is called a fixation. The average time that is spent looking at a single word in a sentence is between 250 and 350 milliseconds. During reading, the average length of a saccade is between 5 and 9 letters (or character spaces). Fixations on frequently used words are generally shorter than fixations on infrequently used words (Rayner & Duffy, 1986). Words that are high frequency are skipped more often than words that are low frequency. Some of the most frequent words in the English language are the function words (e.g., *the, that, of, and, but, to*) (Francis & Kučera, 1982).

Word frequency effects have also been observed using the **word-naming task.** In such experiments, participants view a series of words and are asked to pronounce each word as quickly as possible. A microphone detects the response. The researcher can determine how long it took the participant to begin speaking after the word appeared on the computer screen. The average naming times for English words is between 500 and 600 milliseconds. Frequently used words have been found to be named more quickly than infrequently used words (Balota & Chumbley, 1985). A variant of this task is the **picture-naming task** in which participants are shown pictures and are asked to pronounce the name of the picture as quickly as possible. Pictures are named more quickly when their names are higher in frequency than pictures whose names are lower in frequency (Meschyan & Hernandez, 2002).

Word frequency effects have also been observed during listening. Researchers demonstrated such effects using the **phoneme-monitoring task** (Connine & Titone, 1996; Foss, 1969). Participants listen to a sentence or passage in order to understand the meaning. At the same time, participants listen for a target phoneme, which is specified by the experimenter. When the target phoneme is detected, participants press a designated key on a computer keyboard. Some researchers have found that participants can respond faster when the target phoneme occurs in a high frequency word than when it occurs in a low frequency word (Dupoux & Mehler, 1990; Eimas & Nygaard, 1992; Foss, 1969; Foss & Blank, 1980).

Surprisingly, there have been relatively few studies investigating the word frequency effect using brain-imaging technology. The primary reason relates to the fact that lexical access occurs so rapidly. In an ingenious study, Rayner, Inhoff, Morrison, Slowiaczek, and Bertera (1981) estimated the minimum time needed to recognize individual words in sentences. They were able to control precisely when text was visible to readers using information about the position of the readers' eyes. They were able to vary the amount of time that text was visible to readers. They found that when readers had each word that they looked at available for only 50 milliseconds, their reading rate was the same as when they had the text available at all times during reading.

The bulk of the studies investigating how lexical access occurs have focused on a small number of factors, including orthographic, phonological, and morphological information. The role of orthographic information has been noted since the 1960s. Havens and Foote (1963) found that words that shared letters with lots of other words took longer to recognize than words that shared letters with few other words. This pattern of processing is referred to as the **orthographic neighborhood effect** (see Andrews, 1989). Words that share letters are considered neighbors. The most common definition of an orthographic neighbor is a word of the same length that differs in one letter (in any position) (Coltheart, Jonasson, Davelaar, & Besner, 1977). Recent research has shown that the number of neighbors that a word has is not the only factor that predicts the word's processing time. Words that have high frequency neighbors are processed more slowly than those without high frequency neighbors (Bowers, Davis, & Hanley, 2005a, 2005b). Some researchers have argued that orthographic neighborhood effects in word processing stem from phonological processes (Adelman & Brown, 2007) occurring when one applies grapheme-to-phoneme rules during lexical access.

The Role of Phonology

More research is needed to determine the exact roles of orthographic and phonological processing in lexical access. However, there is ample evidence supporting the view that phonological processing plays a role in lexical access, even when it occurs during silent reading (Rayner & Pollatsek, 1989). The link between phonology and reading has been the focus of research since the late 1880s. The topic was discussed in Huey's (1908/1968) book titled *The Psychology and Pedagogy of Reading*. When we first learn to read, we read aloud. As our reading proficiency increases, we are taught how to read without sounding out each word. Initially, a child might engage in **subvocalization,** which involves engaging in some of the actions involved in speech without producing audible vocalizations (e.g., moving the lips or other parts of the vocal tract and hearing the sound of the word in one's mind). The most proficient readers may experience, on occasion, the phenomenon called inner speech, which is hearing a voice in one's mind while they are reading silently.

▲ Photo 5.3 A girl reads silently. When you read silently, do you ever hear a voice in your head pronouncing the words?

To the surprise of many, studies in the 1960s, 1970s, and 1980s have shown that silent reading by highly proficient readers activates the muscles typically used during speaking. The activations of muscles in the lips, tongue, chin, larynx, and throat can be measured using electromyographic (EMG) recording (Garrity, 1977). McGuigan (1971) showed that even deaf readers who were fluent in sign language experienced increased muscle activity in their articulators during language processing; their articulators were their fingers. The role of inner speech in language comprehension has been debated. There is ample evidence suggesting that it plays some role in aiding comprehension. Studies have used a concurrent articulation task in which participants were instructed to engage in vocalizations while carrying out reading or listening. The concurrent vocalization task would prevent vocalization of the materials being comprehended. These studies found reduced comprehension during concurrent vocalization versus conditions in which participants did not vocalize during reading or listening (Slowiaczek & Clifton, 1980).

Another line of research has shown that there is a role for phonology during silent reading. Sentences containing repeated phonemes, as in 4a, take longer to read than similar sentences in which phonemes are not repeated, as in 4b (Kennison, Sieck, & Briesch, 2003; McCutchen & Perfetti, 1982). This pattern of results has been referred to as the **tongue twister effect** in reading. The effect is observed even when the repeated phonemes in the sentence involve letters or graphemes that are not visually similar. Zhang and Perfetti (1993) observed a robust tongue twister effect in an experiment involving Chinese, a language in which the sound of most words is unrelated to how the words appear visually.

4. a. Tina and Todd took the two toddlers the toys.

   b. Lisa and Chad sent the four orphans the toys.

Researchers have concluded that phonological processes play a role in the initial recognition of a word (i.e., prelexical access) and also play a role after a word is recognized (i.e., postlexical access). The findings related to subvocalization and the tongue twister effect are viewed as stemming from phonological information becoming available after lexical access occurs; readers automatically become aware of the sounds of the words. The research supporting the view that the sounds of words play a role in initially recognizing them includes studies showing that words that are spelled irregularly (e.g., *one* and *choir*) take longer to pronounce than words that are spelled regularly (e.g., *fun* and *stir*) (Baron & Strawson, 1976; Seidenberg, Waters, Barnes, & Tanenhaus, 1984). In a series of experiments, Lesch and Pollatsek (1998) investigated the possibility that words are accessed through the application of grapheme-to-phoneme rules (also referred to as assembled phonology). Their results supported the view that when one reads the word *beard,* one applies the grapheme-to-phoneme rules. In doing so, one arrives at the correct pronunciation, but also arrives at a pronunciation that rhymes with *bird*, because some words in English follow that pattern (e.g., *heard*). Some researchers are not convinced that phonology is always used in lexical access (Martensen, Dijkstra, & Maris, 2005).

There is also evidence that silent readers are activating information about the syllable structure of a word (Ashby & Rayner, 2004; Carreiras, Álvarez, & de Vega, 1993). A syllable is composed of at least one vowel sound but often occurs with one or more consonants. For example, the words *Maine*, *Kansas*, *Virginia*, and *Mississippi* contain one, two, three, and four syllables, respectively. Research has found that when readers process a word that shares a syllable with a word in the preceding context, they spend less time processing the word that does not share a syllable with a word in the preceding context.

Morphological Complexity

In addition to orthographic and phonological information, morphological information also plays a role in lexical access. Research has also shown that words that are composed of multiple morphemes take longer to recognize than words composed of only one morpheme. This pattern of processing is called the **morphological complexity effect.** Taft (1981) compared how quickly participants could process prefixed words, such as *remind*, and similarly spelled words that did not contain a prefix, such as *relish*. The latter type of word was referred to as a *pseudoprefixed* word. He found that response times were longer for words with pseudoprefixes than for words with actual prefixes. The results supported the view that during word recognition, readers initially analyze words in terms of component morphemes. The mental lexicon is searched for matches for each morpheme. When matches are found, the participant understands the meaning of the word by combining the meanings obtained for the individual morphemes. Processing time was longer for the pseudoprefixed words because participants analyzed the word as involving a prefix (i.e., re + lish). When participants attempted to match the stem *-lish* in their mental lexicons, it was not found. Participants then realized that the initial analysis of the word was incorrect and that they must match the entire word (i.e., *relish*) to their lexical knowledge. In a later eye tracking study, Lima

(1987) measured eye movements as readers processed sentences containing prefixed and pseudoprefixed words. She found results similar to Taft's; fixation durations were longer on pseudoprefixed words than prefixed words.

If readers routinely analyze words in terms of their morphemes and then attempt to match the component morphemes to information stored in the mental lexicon, then one might expect that words containing frequently occurring morphemes would be recognized more quickly than words containing infrequently occurring morphemes. In fact, research has found evidence for this type of morpheme frequency effect. Hyönä and Pollatsek (1998) recorded eye movements of Finnish readers who processed multimorphemic words occurring within sentences. As you learned in Chapter 1, Finnish is a language that involves many prefixes and suffixes; consequently, words can become quite long. They found that readers took longer to process words whose first morpheme was low frequency than they took to process words whose first morpheme was higher frequency.

Because the morphemes and words in different languages would involve different distinctions, the mental lexicons of speakers of different languages would be expected to differ somewhat. The mental lexicons of speakers of some languages may involve more distinctions than those of other languages. Consider the mental lexicon of readers of Chinese. Text Box 5.3 describes the writing system of Chinese, which uses characters instead of letters. The pronunciation of the character is not predictable from how the character is written; thus, it must be learned and stored in the mental lexicon. Some characters are composed of radicals, which are subcomponents of characters that carry their own meaning. Researchers are still debating whether the mental lexicons of Chinese readers are organized by characters or by radicals.

Imageability

Word processing can also be influenced by what a word means. For example, some words refer to concrete objects (e.g., *chair*, *tree*, *necklace*). Other words refer to abstract concepts (e.g., *patriotism*, *honesty*, *willingness*). Numerous studies have shown that words with concrete meanings are recognized faster than words with abstract meanings (Binder, Westbury, McKiernan, Possing, & Medler, 2005). Paivio (1971) proposed that concrete and abstract words differ in their memory representations. For both types of words, the meaning is stored in memory as a verbal description. In contrast, the lexical entry for concrete words contains a visual image (i.e., how the object appears). Paivio referred to this as the dual-coding hypothesis. An alternative explanation for the processing advantage of concrete words over abstract words is that concrete words are related to more words in memory than are abstract words; the difference in interconnectedness within memory leads to differences in processing. This view as been called the context availability theory (Schwanenflugel, 1991).

Age of Acquisition

One of the most intriguing effects in word processing suggests that the time one takes to process a word is influenced by how old one was when one first learned the word. Words

Text Box 5.3 Diversity of Human Languages: The Chinese Writing System

Unlike English, which has individual letters or graphemes that correspond to speech signs, Chinese has characters that correspond to words. When one reads Chinese, one realizes how the symbol sounds only when realizing what the symbol means. Writing systems such as Chinese use characters instead of letters and are referred to as logographic writing systems (DeFrancis, 1989). Characters are composed of strokes. Figure 5.1 illustrates the 12 strokes that are used to create all Chinese characters. Language researchers have found that the number of strokes in a character is positively related to its processing complexity (Seidenberg, 1985).

Chinese characters vary in terms of the number of strokes that they contain. Some characters involve a single stroke. Other characters may contain up to 64 strokes (Taylor & Taylor, 1995). Most Chinese words are made up of multiple characters; however, because Chinese does not mark word boundaries, the reader must figure out how a sequence of characters should be grouped to form words. Most Chinese characters also have multiple meanings and singular and plural nouns are identical; thus, the

Figure 5.1 The twelve elemental strokes used in writing Chinese characters

task for the Chinese reader is even more complex. Some examples of complex Chinese words are provided in 5. Traditional Chinese writing is vertical, moving from the right to the left; however, one sees other variations in contemporary society. In elementary school, when Chinese children are introduced to reading Chinese characters, they must memorize more than 3,000 core characters. Throughout life, they may continue to acquire new ones. There are as many as 80,000 different characters, but some are rarely used. Also, new characters are created regularly in popular culture and by businesses creating new corporate identities. The oldest examples of Chinese script date back to 1200 to 1050 BC (Boltz, 1986).

5. a. 電腦 [电脑] diànnǎo—"electric brain" translates as *computer*
   b. 電話 [电话] diànhuà—"electric speech" translates as *telephone*
   c. 電車 [电车] diànchē—"electric vehicle" translates as *trolleybus*
   d. 電影 [电影] diànyǐng—"electric shadow" translates as *film/movie*
   e. 電路 [电路] diànlù—"electric road" translates as *electric circuit*
   f. 電視 [电视] diànshì—"electric look" translates as *TV*
   g. 電池 [电池] diànchí—"electric pool" translates as *battery*

that are acquired early in childhood (e.g., *apple* or *milk)* are recognized more rapidly than words that are acquired later in life (e.g., *apricot* or *beer*); this has been referred to as the **age of acquisition effect** (Carroll & White, 1973; Gilhooly & Gilhooly, 1979). Table 5.9 provides examples of words learned early versus later in childhood. More recent studies have found age of acquisition effects in a variety of language processing tasks, including lexical decision (Morrison & Ellis, 1995, 2000; Stadthagen-Gonzalez, Bowers, & Damian, 2004), picture naming (Ellis & Morrison, 1998; Morrison, Ellis, & Quinlan, 1992), word naming (Brown & Watson, 1987; Morrison & Ellis, 1995, 2000), silent word reading (Zevin & Seidenberg, 2002), and eye fixation durations during sentence processing (Juhasz & Rayner, 2003). Some researchers have speculated that the age of acquisition effect in word processing occurs because of word frequency, as a word that is learned early in childhood is likely to be experienced more often across one's life than a word that is learned later in life (Oldfield & Wingfield, 1965). Recent research has demonstrated that the age of acquisition effect is distinct from word frequency effects and that both influence word processing (Meschyan & Hernandez, 2002).

Table 5.9 Words for Body Parts and Food That Differ in Age of Acquisition

Learned Before Age 4	Learned After Age 7
Mouth	Groin
Nose	Ankle
Elbow	Blood
Toe	Skin
Tongue	Jaw
Thumb	Brain
Ear	Hip
Lip	Liver
Leg	Lung
Tooth	Chest
Egg	Fig
Milk	Wine
Cheese	Steak
Cookie	Cabbage
Apple	Turnip
Juice	Beer
Corn	Steak
Pea	Stew
Grape	Cranberry
Banana	Cucumber

Source: Bowers and Kennison (2011).

Theoretical Approaches to Lexical Access

What Is the Logogen Model?

Since Reicher's (1969) work, there have been numerous studies conducted, investigating the unconscious, cognitive processes involved in lexical access. This research has produced several theories. Some of these theories have also been disproven in experimental tests. Despite their shortcomings, it is still useful to review them to understand how researchers' views of lexical access have changed over the years. Of the early models of lexical access, Morton's (1969) **logogen model** strongly influenced how researchers conceptualized lexical access. Morton proposed that a logogen (from the Greek word *logos* that means word, and the Greek word *genus* that means birth) represented each morpheme or word. The logogen specifies a morpheme's characteristics, such as how it is spelled, what it means, and how it sounds. Logogens can become activated by sensory input, such as when the word is seen or heard. In order for lexical access to occur, the activation level of the logogen must reach a certain threshold. Consequently, if the sensory input is unclear or incomplete, lexical access could fail to occur. One can think of the threshold level of activation as relating to consciousness. Lexical entries or logogens may be partially activated in our unconsciousness; however, in order for them to become available to our conscious minds, there must be an adequate amount of sensory input. The notion of the threshold of activation was an abstraction; one could not easily quantify the specific amount of activation that would be enough. Morton also proposed that a logogen can be activated by the context. For example, if one heard the sentence fragment *This morning for breakfast, Mary prepared bacon and...,* the listener could anticipate that the word following *bacon and* is *eggs.* The activation of the words in the preceding context could increase the activation of the logogen for *eggs* so much that the threshold is reached even before the word *eggs* has been experienced. The two ways in which logogens could be activated (i.e., sensory input and contextual input) were viewed as two distinct systems that functioned in parallel. Also, the threshold of activation needed for lexical access to occur could vary a great deal across contexts.

Subsequent models of lexical access tended to focus on lexical access of either printed words or spoken words. Forster (1976, 1979) proposed a model of lexical access for printed words, emphasizing the role of the orthographic properties of words (i.e., how words are spelled and how similar their spellings are to the spellings of other words). The model assumed that when one encountered a printed word and searched through memory for a match, one searched through a list of words ordered by frequency of usage. Frequently experienced words would be searched before infrequently experienced words. Forster further proposed that the processes involved in lexical access were separate and independent of other processing systems, such as processes involving meaning. Unlike Morton's (1969) logogen model, which claimed that the meaning of a context can directly influence a word's lexical access, Forster's model claimed that lexical access is not influenced by contextual information.

What Is the Cohort Model?

By the 1980s, researchers had begun to investigate lexical access in spoken language. Marslen-Wilson and colleagues (Marslen-Wilson, 1987, 1990; Marslen-Wilson & Tyler, 1980) proposed the **cohort model**, which emphasized the fact that speech is experienced by the listener over time and that the mental processes involved in lexical access are continually updated during listening. According to the model, we recognize spoken words in three stages. When we experience the sensory input or sound of the word, we activate a set or cohort of possible word candidates, which share the sounds that are present in the input. For example, if one heard the sentence fragment *Susan wanted to see the tr...,* the cohort of possible word candidates would contain *tram, trap, tray, tree, tread, tress, trek, trim, trip, trot, trust,* and *truth,* among others. In the second stage, one word from the set is selected for additional processing. This selection stage can be influenced by multiple sources of information, including information from the context and word frequency. One way in which the cohort is reduced is by consulting incoming information; thus, if the target word continued as *tra*, then the cohort would be reduced to contain *tram*, *trap*, and *tray*. Last, the last remaining word in the cohort is integrated into the preceding context. Comparing the cohort model to Morton's (1969) logogen and Forster's (1976,1979) search model, one finds both similarities and differences. As in Morton's (1969) logogen model, the cohort model presumes that word candidates are activated in parallel. However, as in Forster's (1976, 1979) search model, the cohort model presumes that the initial stage of lexical activation is influenced only by sensory input.

How Are Semantic Networks Organized?

Contemporary models of lexical access may seem extremely complex compared with the early models. These models recognize that our knowledge of words involves multiple types of information—visual information related to how a word is spelled, phonological information related to how a word is pronounced, morphological information related to whether a word has a single morpheme or multiple morphemes, and also semantic information. The critical difference among models is the extent to which the process of lexical access can be initially influenced only by sensory information. Such models are referred to as **bottom-up processing** models; processing is influenced by the input from the environment. In contrast, **top-down processing** models allow for processing to be influenced by both sensory information and also information that is derived from within the system, such as information activated in memory, including expectations generated from considering the meaning of the context so far. The Forster (1976, 1979) model is an example of a bottom-up model. Morton's (1969) logogen model is an example of a top-down model. In the cohort model (Marslen-Wilson, 1987, 1990; Marslen-Wilson & Tyler, 1980), the first stage involves bottom-up processing, but the second stage involves top-down processing. Recent models of word recognition envision processing as fully top-down and interactive (McClelland & Rumelhart, 1981; Seidenberg & McClelland, 1989).

As we learned earlier in this chapter, our knowledge of words is just one part of our semantic memory. Semantic memory also includes factual knowledge that we have learned and also our general knowledge of the world that we have gained. Since the 1960s, researchers have described the organization of semantic memory as a **semantic network** (Collins & Quillian, 1969). Concepts are thought of as being represented by nodes in the network. Nodes are connected to other nodes via pathways. Concepts that are related in meaning are connected. When a node is activated in the network because it has been recognized during processing, the activation can spread along the connecting pathways to convey some of the nodes to related nodes. This way of envisioning the semantic network in our minds is certainly an oversimplification; however, it provides researchers with a useful way of thinking about the interconnectedness of knowledge in the mind. Figure 5.2 displays the hierarchical semantic network containing information about birds and fish.

Figure 5.2 Collins and Quillian's hierarchical network model of knowledge related to birds and fish

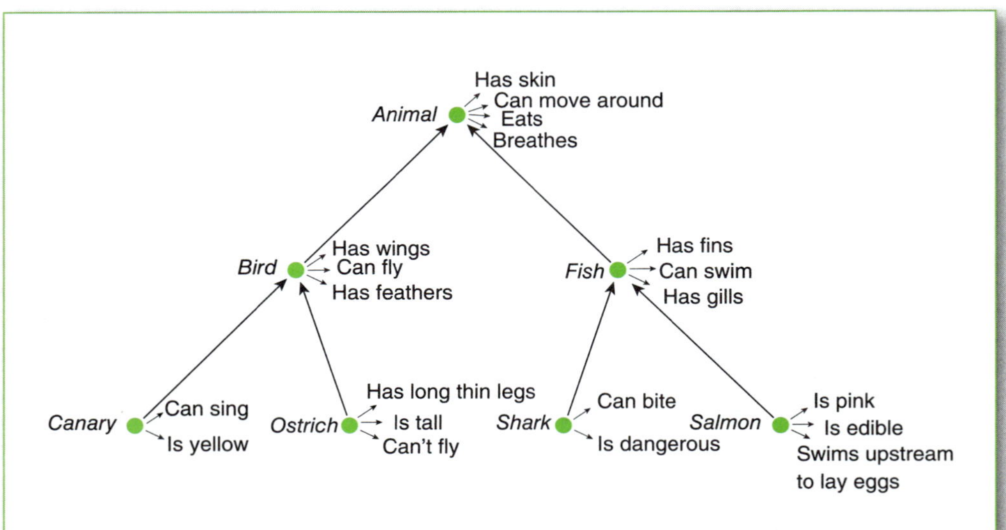

Source: Collins and Quillian (1969).

Collins and Quillian (1969) used a **semantic verification task** to test their model. Participants viewed sentences similar to those in 6. They found that the time that people took to confirm factual statements followed a predictable pattern. Participants could confirm statements faster when the information in the statement was closer in the hierarchy of the semantic network than when the information in the statement was farther away. Generally, participants could confirm 6a faster than 6b and 6c faster than 6d. This pattern of results was referred to as the category-size effect.

6. a. A canary is a bird.
   b. A canary is an animal.
   c. A bird can fly.
   d. A bird eats.

Concepts such as *bird* and *fish* are referred to as basic level categories. They represent the middle of the hierarchy. Categories above them, such as *animal*, are referred to as **superordinate categories,** and categories below them, such as *canary* and *shark*, are referred to as **subordinate categories.**

The notion that semantic memory is organized in terms of a network was embraced by researchers. The overall structure bears some resemblance to the organization of the brain; there are 80 to 120 billion neurons in the brain that communicate with one another in a neural network (Azevedo et al., 2009). Unlike nodes in the semantic network, neurons themselves do not contain information. Researchers are still trying to determine how memories are stored in the brain; however, the storage of information involves numerous neurons working together. When researchers conceptualize the semantic network, the individual nodes are likely to correspond to populations of neurons rather than individual ones.

The notion that the network was arranged in a hierarchy fell out of favor. Empirical studies observed results that could not be explained within a hierarchical network. For example, participants could judge the statement *A robin is a bird* faster than the statement *A penguin is a bird.* The pattern of results demonstrated the **typicality effect.** Typical members of categories can be verified more quickly than atypical members of categories (Rein, Goldwater, & Markman, 2010; Rips, 1975; Rothbart & Lewis, 1988). Research has shown that typicality effects can be observed in many different types of individuals, such as children (López, Gelman, Gutheil, & Smith, 1992), individuals with Alzheimer's (Smith, Rhee, Dennis, & Grossman, 2001), and individuals living in indigenous communities in Guatemala (López, Atran, Coley, Medin, & Smith, 1997).

Researchers generally agree that when a person hears a word and becomes aware of the meaning of the word, the part of the semantic network containing that information becomes activated, which is another way of saying that the information is available for processing to a greater extent than it was before the word was heard. Researchers also envision that when a word or concept is activated, there is **spreading activation** within the network. An early example of a spreading activation model was proposed by Collins and Loftus (1975), whose model claimed that knowledge was stored in an interconnected network; however, their network did not involve a hierarchical organization. Figure 5.3 provides an illustration.

A word or concept that has been activated can result in related concepts and words also becoming somewhat activated, as the activation can spread along the pathways that connect concepts. The level of activation decreases as spreading occurs; thus, concepts more closely related in the network can activate one another more than concepts more distantly related in the network. Spreading activation can explain typical performance in a word association task, which involves asking a person to say the first word that comes to mind when hearing a word. For example, if you ask someone to say the first word that comes to mind for the word *bread,* the typical response is *butter.* If you ask someone to say the first word that comes to mind for the word *shoe,* the typical response is *sock.* For the word *pickle,* one could observe a range of responses such as *sour, green, cucumber,* and *sandwich.*

Semantic Priming

The idea that our memories for words and factual knowledge are organized in a semantic network containing nodes and connections through which activation spreads during processing is consistent with empirical research showing that the time needed to process a

Figure 5.3 Collins and Loftus's spreading activation model of knowledge

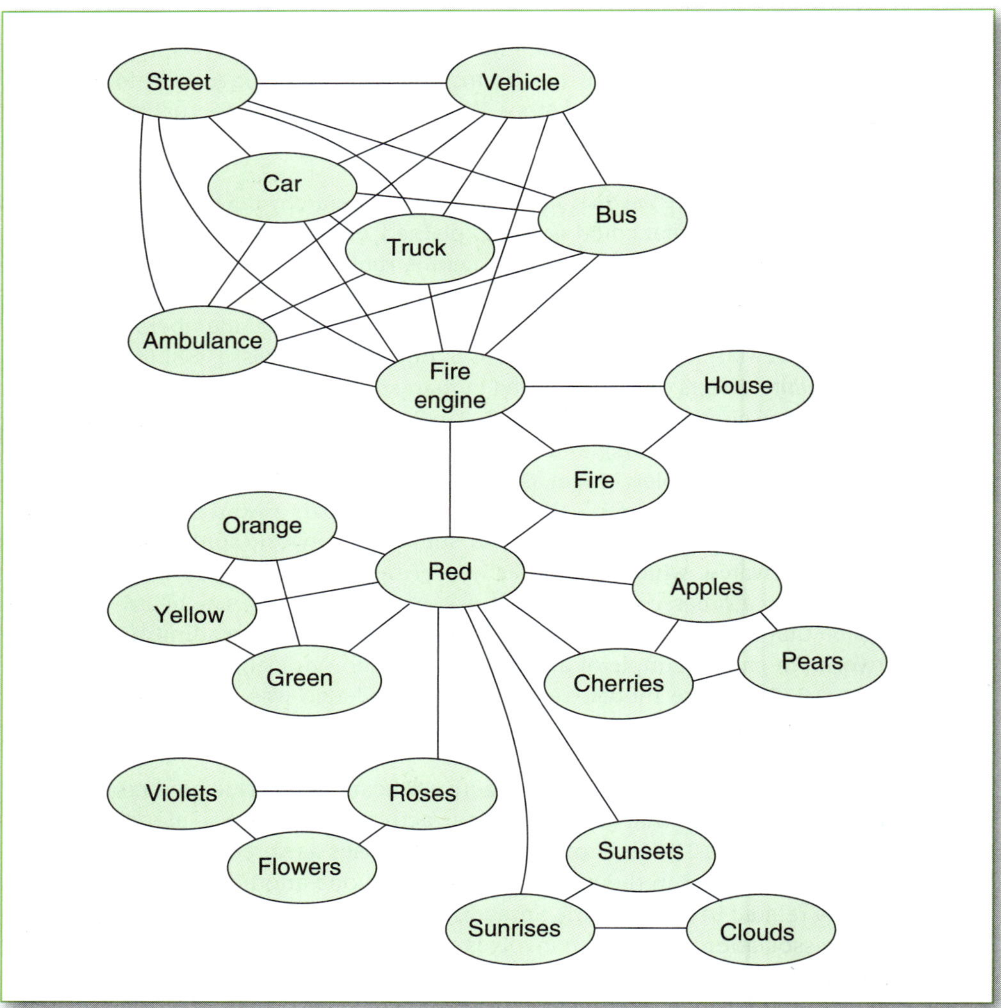

Source: Collins and Loftus (1975).

word or concept can be influenced by whether one has just processed a semantically related word or concept. This research used the **semantic priming paradigm** (Loftus, 1973; Meyer & Schvaneveldt, 1971). Researchers asked participants to perform a lexical decision task in which the order of the words was carefully manipulated. Target words, such as *butter,* were either preceded by a semantically related prime word, such as *bread,* or preceded by an unrelated control word, such as *nurse*. The results showed that participants were faster to respond to target words when they had been preceded by prime words versus control words. The **semantic priming effect** is the average difference in response time for control and prime word conditions.

Research has shown that there may be differences in semantic priming for word pairs that are strongly related in meaning (e.g., *boy, man*) and words pairs that often occur together in daily language use (e.g., *street* and *light*) (Hutchison, 2003; Lucas, 2000). The term ***semantic associate*** has been used to refer to the latter type of word pair. Attempts to identify word pairs that are semantic associates but not semantically related and vice versa have been made; however, it is difficult to find semantic associates that are not semantically similar (Hutchison, 2003) (e.g., *knife—fork, bread—butter, hockey—ice*). More research is needed to determine whether the **priming** that is observed during processing for semantically related word pairs and for word pairs that are semantic associates involves the same or different processes.

Researchers have investigated semantic priming using a variety of techniques, including lexical decision and naming tasks, as well as tasks in which participants process different stimuli through different senses or modalities. This has been called **cross modal task** (Marslen-Wilson, Tyler, Waksler, & Older, 1994; Swinney, 1979). Participants may be asked to view a single word, a sentence, or a picture. While performing a task related to the visually presented stimuli, they must carry out a secondary task involving listening, such as deciding whether the sound that they hear is a word versus a nonword. When the stimuli that participants process are semantically related, they generally respond faster than when the stimuli that participants process are unrelated. For example, Swinney (1979) investigated how listeners processed sentence pairs. The second sentence contained lexical ambiguities, such as the words *bug* or *bank*. Quickly after participants heard the target word in the sentence, a letter string was presented. Participants performed a lexical decision task. Swinney (1979) varied the semantic context of the sentence to be consistent with one of the two meanings of the ambiguous word. The visually presented word was either semantically related to one of the meanings or was unrelated to either meaning. He found that when the visually presented word appeared immediately following the ambiguous word, both meanings of the ambiguous word primed lexical decision responses. When the visually presented word appeared four syllables after the ambiguous word, only the meaning of the ambiguous word that was supported by the preceding context primed lexical decision responses.

Language researchers are particularly interested in understanding how the information about words is stored and processed differently in semantic networks than information about other types of knowledge. As we learned earlier in this chapter, our knowledge of words contains a variety of types of information (e.g., orthographic, phonological, morphological, syntactic, and semantic). Bock and Levelt (1994) proposed a network model of lexical knowledge involving three levels: (1) the conceptual level, (2) the lemma level, and (3) the **lexeme** or sound level. Figure 5.4 provides an illustration of these three levels. Important innovations in the model are the distinction between the lexeme and lemma levels and the distinction between the lemma and conceptual levels. The lemma level represents the

Figure 5.4 Bock and Levelt's lexical network model

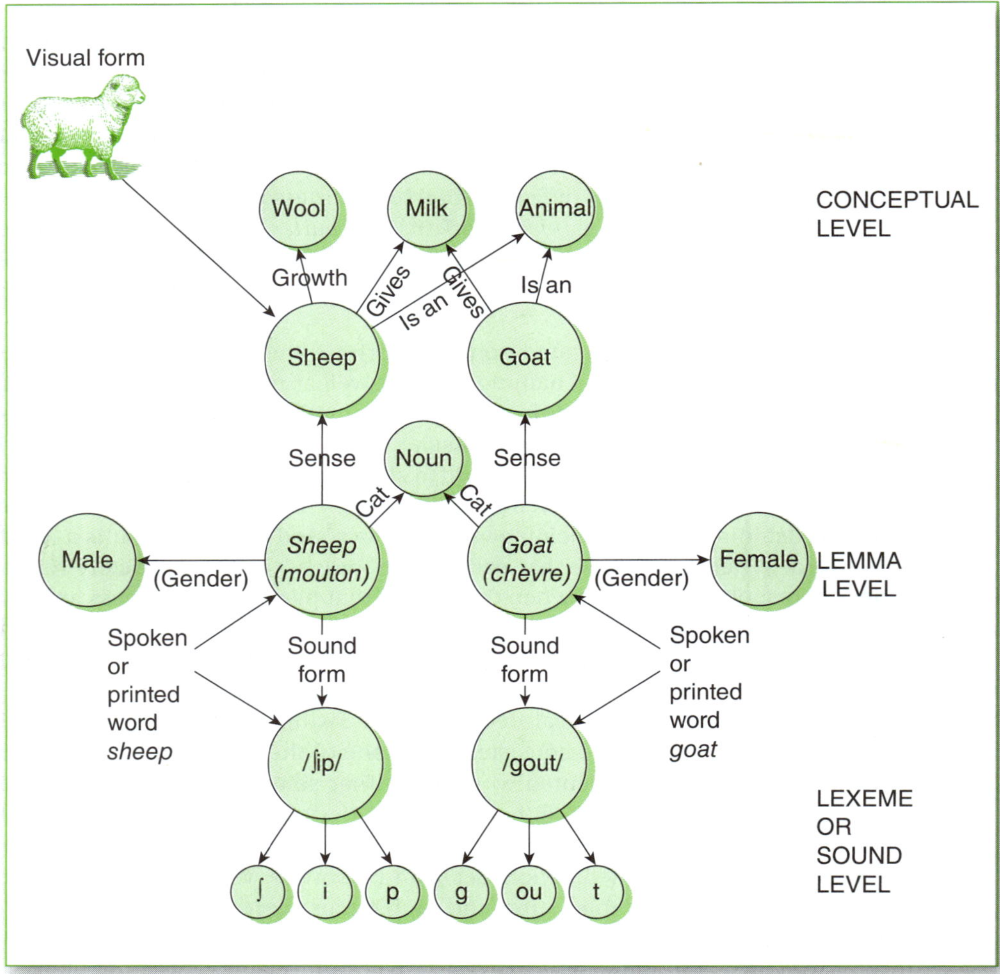

Source: Bock and Levelt (1994).

syntactic characteristics of the word, such as its part of speech and its gender. The lexeme level represents information about phonology. The lemma level is distinct from the conceptual level that contains information about the word's meaning and is also connected to concepts that are similar in meaning. It is important to note that the lexeme and conceptual levels are not directly connected.

Research with speech production has provided evidence for distinct levels of representations for phonology, grammar (i.e., morphology and syntax), and meaning

(Brown & McNeill, 1966; Schwartz, 2011). In everyday speaking, we sometimes experience difficulty finding the appropriate word. This experience has been described as a **tip-of-the-tongue state,** or TOT state. It is common for individuals who are experiencing a TOT state to be unable to produce a particular word, but they may be aware of the first phoneme or syllable. While not being able to produce the target word, they are able to generate words that sound similar. They are also able to judge similar-sounding words and determine that they are not the target word. Researchers have found that some individuals are more susceptible to experiencing TOT states than others. When speakers of languages that have gender morphology are experiencing a TOT state, they are often able to indicate the word's gender (Caramazza & Miozzo, 1997; Vigliocco, Antonini, & Garrett, 1997). Vigliocco, Vinson, Martin, and Garrett (1999) also showed a word's status as a count noun versus a mass noun (e.g., *bottle*, as in *I found two bottles* vs. *sand*, as in *I found some sand*). Older adults experience TOT states more often than young adults (Burke, MacKay, Worthley, & Wade, 1991; Dahlgren, 1998; White & Abrams, 2004). Bilingual speakers experience TOT states more often than monolingual speakers (Gollan & Silverberg, 2001).

Explanations for why TOT states occur have involved suggestions about how activation spreads through the semantic network. Burke and colleagues (Burke, 1997; Burke et al., 1991; Burke, MacKay, & James, 2000; James & Burke, 2000) proposed the Transmission Deficit Hypothesis (TDH), which claimed that one experiences a TOT state when the activation level at the phonological level of the network was insufficient. This proposal has been supported in studies demonstrating that people can recover from a TOT state if they are primed (i.e., asked to process) a phonologically similar word (Abrams, White, & Eitel, 2003; James & Burke, 2000). Abrams et al. (2003) found that recovery from TOT states occurred more often when participants were primed with a word that shared the first syllable with the target word than when they were primed with a word that shared only the initial phoneme, only the middle syllable, or only the final syllable with the target word. Prior studies have found that phonological information influences processing related to TOT states to a greater extent than does semantic information (Jones, 1989; Kennison, Fernandez, & Bowers, in press; Meyer & Bock, 1992), which generally supports a processing model involving distinct levels of representation for phonological and semantic word information.

Summary and Theoretical Implications

The process of how people recognize words has been the focus of a great deal of research. The results have shown that many factors influence how quickly words can be recognized, such as the number of letters in a word, the word's frequency of usage, how similar the word's letters are to the letters in other words, the word's phonology, the word's meaning, and when during childhood the word was learned. There have been a variety of different types of theories proposed to explain how the different types of information are used during lexical access. Researchers continue to debate whether contextual information can influence the initial lexical access of a word. As understanding about lexical access has increased, researchers have recognized that knowledge about words is but one type of information in semantic memory. The metaphor that is used to describe the organization of semantic memory is a semantic network. Research has shown that the memory representations of semantically related concepts influence one another during processing. Researchers generally agree that semantic memory is organized as a network of interconnected information through which activation can spread. When one concept is activated in the network, activation can flow along connections to activate related concepts with activation levels decreasing the farther it spreads. The semantic network likely connects information about words with world knowledge. The information about words is believed to be organized into multiple levels such that phonological information is distinct from syntactic information, which is also distinct from conceptual information.

In terms of the four theoretical approaches to language development, the research that has been conducted on lexical processing does not help us in determining which theory is most accurate. The research, thus far, is consistent with all four approaches. The learning of vocabulary could be facilitated, at least in part, by behaviorist principles involved in classical and operant conditioning. It could also be facilitated by supportive social relationships, as is the theme of the social-interactionist approach to language development. Within the generative approach to language, the issue of modularity arises in discussions of lexical processing; some advocates of the generative approach propose that syntactic processing and lexical processing are carried out via different mechanisms (i.e., modules). In contrast, the statistical learning approach to language assumes that there are no specialized modules (e.g., language vs. general cognitive processing) or language-specific submodules (e.g., lexical vs. syntactic) for language processing. You will learn more about specific models from each of these approaches when you learn about language comprehension in Chapter 10.

KEY TERMS

age of acquisition effect
bottom-up processing
cohort model
cross modal task
frontier words
heteronym
homograph
homophones
irregular word
lexeme
lexical ambiguity
lexical decision task
logogen model
morphological complexity effect
orthographic neighborhood effect
phoneme-monitoring task
picture-naming task
priming
regular word
semantic associate
semantic network
semantic priming effect
semantic priming paradigm
semantic verification task
spreading activation

subordinate categories

subvocalization

superordinate categories

syntactic category ambiguity

tip-of-the-tongue state (TOT state)

tongue twister effect

top-down processing

typicality effect

word frequency effect

word length effect

word-naming task

word superiority effect

REVIEW QUESTIONS

1. What is the mental lexicon? What information is generally viewed as being contained in the mental lexicon?
2. What is lexical ambiguity? Provide an example. Discuss how lexical access would occur for a word that involves lexical ambiguity.
3. What are homographs, heteronyms, and homophones? Provide some examples of each.
4. How does semantic memory differ from episodic and procedural memory?
5. Explain the difference between explicit and implicit memory. To what extent are sensory, episodic, semantic, and procedural memories likely to be explicit and implicit in nature?
6. What is the word length effect?
7. What is the word superiority effect? What does this effect suggest about the processes involved in word recognition?
8. What is the word frequency effect? What are some theories about why the word frequency effect occurs?
9. What is the lexical decision task? How is it used to understand the processes involved in word recognition?
10. What is the word-naming task? How is it used to understand the processes involved in word recognition?
11. What is the evidence suggesting that silent reading involves phonological processes?
12. What is the evidence suggesting that the process of lexical access for printed words is influenced by phonological processes?
13. Identify three models of lexical access. How do the models differ?
14. When researchers describe semantic memory as organized as a semantic network, what do they mean? What are the parts of the network? How are they connected? How does semantic relatedness influence the connections in the network?
15. What is a basic level category? How is it different from superordinate and subordinate categories?

16. What is the word association task? Why are most people likely to produce the word *dad* when they are cued with the word *mom*?

17. What is semantic priming? Identify three pairs of words that would likely semantically prime one another.

18. How do semantic associates differ from word pairs that prime one another?

19. What is the empirical evidence for Bock and Levelt's (1994) model of word knowledge in which there are three levels of representation: (1) the lexeme level, (2) the lemma level, and (3) the conceptual level?

20. What occurs during a TOT state? Do some individuals experience TOT states more often than others? If so, which individuals are particularly susceptible to experiencing TOT states?

RECOMMENDED READING

Austin, J. L. (1962). *How to do things with words.* Cambridge, MA: Harvard University Press.
Carey, S. (2009). *Origins of concepts.* Oxford, UK: Oxford University Press.
Lakoff, G. (1990). *Women, fire, and dangerous things.* Chicago: University of Chicago Press.
Lakoff, G. (2003). *Metaphors we live by.* Chicago: University of Chicago Press.
Murphy, G. (2002). *Big book of concepts.* Cambridge, MA: MIT Press.
Perfetti, C. (1995). *Reading ability.* New York: Oxford University Press.
Rayner, K., & Pollatsek, A. (1989). *The psychology of reading*. New York: Prentice Hall.

RECOMMENDED FILMS

Atchison, D. (Director). (2006). *Akeelah and the bee* [Motion picture]. United States: Lions Gate Films.
Blitz, J. (Director). (2004). *Spellbound* [Documentary]. United States: Columbia Tristar Homevideo.
Chaikin, E., & Petrillo, J. (Directors). (2004). *Word wars* [Documentary]. United States: Seventh Art Releasing.
Creadon, P. (Director). (2006). *Wordplay* [Documentary]. United States: IFC Films.
Peterson, S. (Director). (2003). *Scrabylon* [Documentary]. United States: Scrabylon Productions.

SUGGESTIONS FOR CLASS PROJECTS

1. You can gain experience computing the number of neighbors that exist for a list of 10 or 20 English words. A popular operational definition of an orthographic neighbor was used by Medler and Binder (2005). One counts the number of words that can be formed by changing just one letter of the word. For example, for the word *blue,* one can change the first letter and form the words *clue, flue,* and *glue.* Changing the second and third letters result in no words. Changing the last letter results in *blur.* The total number of neighbors is then four.

2. You can explore the interrelatedness of words using the word association task. Then develop a list of 10 to 20 words that refer to common objects and create a random list of the words. Ask multiple individuals to listen to each word and say or write down the first word that pops into their minds. You will find that some words have specific words that are most frequently provided. For other words, there may be more variable responses. You will be asked to discuss their results in a written assignment in which you discuss the organization of the mental lexicon.

3. The meanings of words change over time. Researching the historical changes in word meaning can provide you with insight into how the meanings of words today came to be. You will be assigned to research the histories of word meanings (also called etymologies) for a list of 10 to 20 words. The best resource for researching word histories is the *Oxford English Dictionary* (OED). There are other online resources that are free, including www.dictionary.com and www.etymology online.com.

4. Although heteronyms may seem rare, there are more than 300 in English. You may work individually or in teams to develop a list of heteronyms. Be sure to establish some rules for the task, such as whether proper names can be used, whether words can be considered without their diacritic marking (e.g., *résumé*), and whether foreign words that have been borrowed into English recently can be used.

5. You can explore differences in the frequency of usage for different words using Google. By searching for single words and noting the number of webpages that contain that word, you can get a rough measure of how often the word is being used. Using the advanced search features of Google, you can limit the searches to specific types of website (e.g., .edu, .gov, .com) or to specific websites, such as that for a specific college or newspaper. Then explore the extent to which word frequencies differ for different types of websites, such as commercial websites versus educational institutions.

For additional ancillary resources, please visit the companion website at www.sagepub.com/kennison.

- Video Links
- Audio Links
- Web Resources
- Internet Activities
- Flashcards
- Web Quizzes

CHAPTER 6

SOCIAL ASPECTS OF LANGUAGE DEVELOPMENT

Acquiring the grammatical rules of a language is a big part of language development; however, it is not the only part. Children must also acquire the social rules of language, which means that they learn to use language appropriately in different social settings (Hymes, 1966). When children have done this, they are described as having achieved **communicative competence.** Communicative competence typically develops throughout childhood and through adolescence. Many teenagers are still learning the details of how they should modify their language use across different social settings in order to achieve their particular communication goals, such as delivering information, making a joke, or making amends for some prior action. Unfortunately, for some individuals, such as those with autism and related disorders, achieving communicative competence may not be possible. For such individuals, navigating the social rules of language may be a challenge throughout life. In this chapter, you will learn about how children acquire the social rules of language. You will also learn about individuals who have difficulty with the social rules of language and about the research detailing the cognitive deficits that appear to be responsible. Last, you will learn about the characteristics of adult language use, how language use varies across different social settings and across different groups of speakers, and how children learn to produce and comprehend figurative forms of language.

▲ Photo 6.1 Two friends are talking, making eye contact. When you talk to a same-sex friend, do you make a lot of eye contact?

Development of Communication

When Do Conversations Begin?

Children begin their journey on the road to communicative competence soon after birth. Table 6.1 displays some conversational milestones and the age at which children typically master them. By the time children are between the ages of 9 and 15 months, they are usually able to interact with another person and together focus on some object or other person. This behavior has been referred to as **joint attention.** Being able to initiate and maintain episodes involving joint attention are key in the development of **conversation.** The building blocks of effective communication are established in the first year of life, as infants are introduced to **turn-taking** during interactions. In the second year, toddlers begin to demonstrate the ability to have conversations on a particular topic. They also begin to demonstrate **metalinguistic awareness,** which means that they become aware of their own language use and the extent to which their intentions have been understood by a listener. Around the age of 5 years, they become proficient at clarifying their utterances when they have been misunderstood. Those 9 years old and older typically demonstrate the regular use of polite forms, showing that they understand that their language must be modified to suit the particular social setting. Teenagers and young adults may find that they continue to fine-tune their conversational abilities, learning the varying social norms for the different social settings in their lives (e.g., classroom, home, work, friends).

Turn-Taking

Children's first accomplishment on the road to communicative competence is to learn how to have a conversation. In a conversation, two or more individuals communicate with one another about a particular topic. An essential component of a conversation is turn-taking: first one person speaks and then another. Observations of parents with infants show that even before the infant can speak, parents are encouraging turn-taking behavior. For example, Snow (1977) described that when caregivers interact with infants, the caregivers appear to take a turn and then wait for the infant to take a turn. Table 6.2 displays an interaction between a mother and her 3-month-old infant: Although the mother is not likely aware of it, she is modeling for the infant the essential element of conversation—taking turns.

Actual turn-taking by children in conversation does not emerge until much later in childhood. Before it is mastered, children will exhibit a conversational style that is quite different from that of adults. The first researcher to document the conversational behaviors of children was Piaget (1926). He observed that children will engage in turn-taking, but what they say is not directly related to the previous statement or statements. He referred to these interactions as **collective monologues,** as each child delivers a monologue, one after the other. To the casual observer, it may appear that the children are discussing the same topic. On close examination, one finds that the children's individual statements are quite separate and unrelated to the other

Table 6.1 Children's Conversational Milestones

Milestone	Approximate Age
Preverbal turn-taking	8 to 9 months
Turn-taking stabilizes	2.5 to 3.5 years
Can maintain topic in interaction with adult	2 years
Makes clarifications	2 years
Adaptation of speech style to listener	2 years
Use of early polite forms variable	2 years
Turn-taking repairs	5 years
Polite forms fully developed	9 years
Communicative competence	After 9 years

Source: Adapted from Adams (2002).

Table 6.2 Conversation Between Mom and Three-Month-Old Infant

Ann:	(smiles)
Mom:	Oh, what a nice little smile! Yes, isn't that nice? There. There's a nice little smile.
Ann:	(burp)
Mom:	What a nice wind as well! Yes, that's better isn't it? Yes.
Ann:	(vocalizes)
Mom:	Yes! There's a nice noise.

Source: Snow (1977, p. 12).

statements. For example, consider the exchange in 1. Two children are seated at a table and drawing with crayons. They each take turns making statements, but each of the children's statements is unrelated to the previous statement. Shatz and Gelman (1977) found that children as old as 4 years of age may find it difficult to sustain an actual dialogue.

1. Logan: I am drawing an airplane.
   Griffin: I need the green crayon.
   Logan: Airplanes can go fast.
   Griffin: The grass is green.
   Logan: My airplane has blue wings.
   Griffin: I am drawing a green cat.
   Logan: Airplanes fly in the sky.

Piaget claimed that young children's behavior can be explained by their egocentrism. They are locked into their point of view and do not see situations from the points of view of others. The egocentrism of children is a core element in Piaget's stages of cognitive development. Table 6.3 provides a description of these stages. The young infant who is under 2 years of age is in the sensorimotor stage and will be occupied with integrating sensory experiences with physical activities. According to Piaget, children begin to use language and images to represent the world between the ages of 2 and 7. After the age of 7, they are able to reason about concrete objects. After the age of 11, they are able to think abstractly. Being able to think abstractly is necessary for children to be able to use language creatively. As we will see later in this chapter, children acquire the ability to understand and produce creative forms of language, such as metaphors and sarcasm, relatively late in childhood.

Table 6.3 Piaget Stages of Cognitive Development

Stage	Age	Abilities
Sensorimotor	0 to 2 years	The infant progresses from instinctual reflexes to symbolic thought and coordinates sensory experiences with physical action.
Preoperational	2 to 7 years	Children use words and images to represent the world.
Concrete Operational	7 to 11 years	Children can reason logically about concrete objects and classify objects into different sets.
Formal Operational	11 to 15 years	Children can reason abstractly and logically.

Source: Piaget (1926).

What Is Private Speech?

Private Speech

A second type of conversational behavior that children demonstrate is **private speech**, which occurs when children talk to themselves. Both Piaget (1959) and Vygotsky (1934/1986) noted that children between the ages of 2 and 7 frequently engage in private speech. They often do it when they are playing alone or just before they are falling sleep. Their speech is spoken aloud, but the speech does not appear to have communicative intent, as it is not directed to others. Children appear to use private speech to guide themselves in activities and to regulate their behavior.

Children's use of private speech decreases after they enter school. In multiple studies of elementary school children, Berk (1986, 1992) found that private speech represented 20% to 60% of children's utterances. Subsequent research showed that the use of private speech in 4- to 5-year-olds was related to their performance on tasks. Children using more private speech performed better on tasks (Berk & Spuhl, 1995). Other studies have shown that children who engage in uninhibited self-talk are later more attentive when faced with a challenging task (Behrend, Rosengren, & Perlmutter, 1992; Bivens & Berk, 1990). These results suggest that parents and caregivers should not discourage children from using private speech, as it appears to facilitate children's performance on tasks. Researchers still do not understand well the mechanisms that link private speech and performance or to what extent private speech may be predictive of children's success in educational settings in general.

Speech Acts

As children become more skilled communicators, more and more of their utterances are directed toward others, each with a specific intention. Among language researchers, the term ***speech acts*** is used to describe types of utterances (Dore, 1974). Speech acts are formulated to fulfill a specific intention of a speaker, such as issuing a greeting or making a **request**. The ways in which speech acts are formulated vary across cultures. For example, the way in which one greets another person varies from culture to culture. In the United States, a handshake serves as a greeting in some formal settings. In Bangladesh, one would salute the other with one's right hand, while in Zambia, one would squeeze the thumb of the person whom one is greeting (van Patter, n.d.). The language that is used in a greeting or in other types of speech acts must be formulated according to the rules of the language.

Searle (1965) identified five types of speech acts. These are displayed in Table 6.4. As discussed in Chapter 3, infants do not typically display intentionality in their gestures and utterances until 10 months of age (Austin, 1962). Research has shown that for infants younger than 3 years of age, utterances are about objects in the environment in the form of representatives and directives (Snow, Pan, Imbens-Bailey, & Herman, 1996; Wetherby, Cain, Yonclas, & Walker, 1988). Only later are children able to produce more conceptually complex speech acts, such as commissives. McTear and Conti-Ramsden (1992) found that promising is typically observed by the time that the child is 3 to 4 years old. The extent to which children's acquisition of speech acts are related to other aspects of children's development has not been well researched. The topic certainly merits research, as educators and

Table 6.4 Five Categories of Speech Acts

Speech Act	Goal
Representatives	Conveying belief or disbelief in some proposition
Directives	Directing the listener to do something
Expressives	Expressing a psychology state, such as when making an apology or congratulating someone
Commissives	Promising the listener some future outcome
Declaratives	Stating a fact that will change the current state of affairs

Source: Searle (1965).

researchers increasingly point out that children's success in school is strongly related to children's social competence (Longoria, Page, Hubbs-Tait, & Kennison, 2009).

How Do Children Handle Communication Failures?

As children start out on the path toward communicative competence, they are not always successful in communicating their intentions. The term ***repair*** has been used to describe an utterance that is produced following a communication failure. The fact that children attempt to repair unsuccessful utterances demonstrates a rather sophisticated awareness of listeners and whether their utterance had the impact that the child intended. Children's earliest repairs are repetitions of the unsuccessful utterance. Later, children may produce a revision of their utterance (Brinton, Fujiki, Loeb, & Winkler, 1986; Gallagher, 1977). Golinkoff (1986) observed conversations between three mothers and their infants. Infants were between 12 and 19 months old and were having lunch with Mom. Golinkoff found that infants were successful in communicating their intention only 51% of time. When adults are unable to interpret children's utterances, they often follow up with a question. For example, a child might say, *Mommy toast*. Mom may be unsure about whether the child is requesting that Mommy prepare more toast for the child to eat or requesting that Mommy eat toast, too. When questions of this type are directed to infants, infants produce a response. Olsen-Fulero and Conforti (1983) observed conversations between mothers and their infants who were between the ages of 2.5 and 3 years, finding that children responded to clarification questions 77% of the time (see also Moerk, 1972). Preschool children may initially try to repair an unsuccessful utterance; however, if the repair is unsuccessful, they tend not to persist (Brinton et al., 1986).

Children's communication failures have also been investigated in laboratory studies in which researchers have created situations in which varying types of failures might occur. Gallagher (1977) tested children between the ages of 21 and 29 months in a laboratory where they were encouraged to play. During the play session, an experimenter asked the child, "What?" every 3 minutes. The question was not connected to any of the child's utterances. The results showed that children responded to the question 77% of the time. Most responses were restatements of an earlier utterance. Wilcox and Webster (1980) conducted a similar study with 17- to 24-month-old children. The researchers varied their responses to children's requests. In one condition, the researcher asked, "What?" when a child made a request. In a second condition, the researcher said, "Yes, I see it," which was not appropriate because the child's utterance was a request rather than a statement. The results showed that rather than giving up on the request, children either repeated or revised their request.

The way in which children handle communication failure appears to depend on their conversation partner. Tomasello, Conti-Ramsden, and Ewert (1990) observed children between 15 and 21 months interacting with either their fathers or mothers. They found that when fathers did not respond to their child's utterance, the child usually did not repeat the utterance or revise it. In contrast, when mothers did not respond to their child's utterance, the child usually repeated the utterance. The results also showed that when children started an interaction, the topic was less likely to be pursued by fathers than by mothers. The authors concluded that children at this age were already forming expectations about the conversational cooperativeness of individuals. Mothers were generally treated as more conversationally cooperative than fathers.

In another laboratory study, Shwe and Markman (1997) used an object-choice task in which children who were 30 months old were shown two objects and instructed to request one of them. Children could request an object by pointing, reaching, or speaking. When the child made the request, the researcher placed one object in a bucket away from the child and placed one object near the child. Half of the time the object that the child requested was placed near the child, and half of the time, it was placed in the bucket quite a distance away from the child. They found that children repeated requests more often when they did not receive the desired object than when they did. Children's second requests were frequently revised to make clearer their desire for the object. The authors concluded that children's requests are formulated to achieve communication and when misunderstandings occur, children reformulate requests to increase the likelihood of a successful communication.

When a child repairs an utterance, he must have risen to a relatively high level of communicative ability. Alexander, Wetherby, and Prizant (1997) proposed that there are three important abilities that children must have in order to accomplish a communication repair. First, they must be able to demonstrate intentionality or a desire to communicate something. Second, they must be able to realize that a communication failure has occurred, which requires that they be sensitive to the needs of their communication partner. Third, they must be effective in their own communication skills to accomplish the repair. The process involves not only being aware of one's own point of view and one's own communicative desire but also the perspective of the listener. They must monitor the situation and try to detect evidence of comprehension or confusion on the part of the listener. Without a doubt, it is a complex task.

When Do Children Master the Art of Conversation?

Gricean Maxims

Adult conversations generally are characterized by four principles proposed by Grice (1975). These principles have come to be known as the **Gricean maxims.** These are displayed in Table 6.5. Research with children has found that violations of the principle of quantity were observed often with 4-year-olds (Bishop & Adams, 1989). They often provided less or more information than was called for in the situation. In a study reported by

Table 6.5 Four Gricean Principles for Conversation

Quantity	Make your contribution as informative as is required, but do not provide too little or too much information.
Quality	Try to make your contribution one that is truthful. Do not say anything you believe to be false.
Relation	Make your contribution relevant to the aims of the ongoing conversation.
Manner	Be clear. Try to avoid obscurity, ambiguity, wordiness, and disorderliness in your use of language.

Sources: Grice (1957, 1975).

Eskritt, Whalen, and Lee (2008), 3- to 5-year-olds interacted with puppets, which could be consulted in a game to find a sticker. Prior to the game, one of the puppets interacted with the experimenter in a way that conformed to Gricean maxims. The other puppet interacted with an experimenter in a way that did not conform to Gricean maxims. The results showed that children were sensitive to violations of the relation, quality, and quantity principles, because when they sought help to find the sticker, they asked help from the former puppet more often than the latter puppet.

Conti and Camras (1984) investigated children's awareness of Gricean maxims. In the study, children heard a story with two endings. One ending violated a Gricean principle, and the other ending conformed to it. Children were asked to indicate which of the two endings was *silly.* Children between the ages of 6 and 8 were able to identify the ending that did not violate Gricean maxims. In contrast, 4-year-olds could not. Ackerman (1981) investigated children's ability to explain violations of Gricean maxims. They found that children younger than 8 or 9 years old could not give such explanations. Pellegrini, Brody, and Stoneman (1987) analyzed the speech of 2-year-olds and found that they rarely violated the Gricean maxims of *quality* and *manner.* Other studies with 2-year-olds found their speech conformed to the principles of *quantity* (Dunham, Dunham, & O'Keefe, 2000; Ferrier, Dunham, & Dunham, 2000; O'Neill, 1996).

Theory of Mind

What Is Theory of Mind?

In order for children to become effective communicators, they must be able to appreciate what their conversational partners know or do not know. Children must be able to take the perspective of another and to understand the beliefs, intentions, and knowledge of others. For example, imagine that a mom or dad asks a toddler, "Where's your teddy?" The mom or dad asks the question because the location of the teddy is not known, but more important, there is a desire to know the location. Being able to infer the knowledge and intentions of others helps one provide satisfying answers to questions. Some researchers have referred to the ability to understand the mind of another as **theory of mind.** Most children acquire theory of mind between the ages of 3 and 5 (Birch & Bloom, 2003). Before the age of 3, children are generally poor at understanding the minds of others. This might explain why a parent might discover his or her toddler hitting a smaller sibling while the sibling is shrieking in pain. When the parent directs the toddler to stop, the toddler might appear confused. Is the toddler really unaware of how his actions are affecting the sibling (i.e., causing pain)? Because of research showing that young toddlers do not understand the minds or perspectives of others, the answer to this question is likely yes.

How Is Theory of Mind Measured?

False Belief Tests

Determining whether a child does or does not have theory of mind can be challenging (Bloom & German, 2000). Researchers have developed some rather complicated tasks to detect presence of absence of theory of mind. One type of task that has been developed to test for theory of mind has been called the **false belief test** (Baron-Cohen, Leslie, & Frith, 1985; Wimmer & Perner, 1983). One example of a false belief test is the matchbox scenario. An interviewer shows a matchbox to the child. The interviewer asks, "What do you think is in the box?" The child is likely to say *matches*. The interviewer opens the box to reveal that there are M&Ms inside. The interviewer closes the box. At that time, someone else walks into the room—someone who has not observed the interaction with the matchbox. The interviewer asks the child, "What do you that he will think is in the box?" A child with theory of mind will appreciate that the person who just walked in does not know that the matchbox contains M&Ms and will understand that the person is likely to think that the matchbox contains matches. A child without theory of mind will say that the person will think that there are M&Ms in the matchbox. Individuals who fail the false belief tests have been described as suffering from a **curse of knowledge** (Camerer, Loewenstein, & Weber, 1989), which is the tendency to use one's own state of understanding to judge another's state of mind.

A second example of a false belief test is one involving two dolls—a doll named Molly and a doll named Sarah. Each doll has a basket and a marble. A test subject is introduced to the dolls by the interviewer. The interviewer tells the test subject that each doll places her marble in her own basket. Then, the interviewer says that Molly has to go. The interviewer shows the test subject Molly leaving the area and the interviewer puts her away. The interviewer then says that Sarah wants to take Molly's marble and put it in her own basket. The interviewer moves Molly's marble and places it in Sarah's basket. Then the interviewer explains that Molly is returning and she is coming to look for her marble. The interview asks the test subject, "Where will Molly look for her marble?" An individual with theory of mind will respond that Molly will look in her own basket, where she left the marble. An individual without theory of mind will answer that Molly will look for her marble in Sarah's basket.

Onishi and Baillargeon (2005) developed a nonverbal task to assess infants' theory of mind. In the task, infants watched as a researcher placed an object in one of two possible locations. Later, the researcher reached for the object. The researcher either watched the object move to another location or left the room while the object was moved. When the researcher reached into one of the two locations for the object, the infants' looking time was measured. The authors reasoned that if infants had theory of mind then they would find it natural when the researcher reached for the object where the researcher believed the object was; however, the infant would find it odd when the researcher reached for the object in a location where the researcher would not expect the object to be. The results showed that infants as young as 15 months old looked longer when the researcher reached for the object in a location where the researcher would not expect the object to be. Some researchers have wondered whether theory of mind is unique to humans. Text Box 6.1 describes recent research with chimpanzees demonstrating that theory of mind might be something that humans share with their distant primate relatives.

Text Box 6.1 Research Discovery: The Mind of the Chimpanzee

Researchers long believed that theory of mind was unique to humans (for discussion, see Premack & Woodruff, 1978). It has long been believed that nonhuman animals do not possess self-awareness or awareness of the minds of others. Some researchers, such as Jane Goodall, have questioned this assumption. Goodall (1986) pointed out that apes in their natural habitats routinely make requests of other apes, such as requests for food or help when dealing with other aggressive apes. Perhaps the most convincing evidence so far has been obtained in research by Tomasello and colleagues (Call & Tomasello, 1999; Schmelz, Call, & Tomasello, 2011). They carried out a series of studies investigating chimpanzees' theory of mind. Call and Tomasello (1999) compared the performance of human children, chimpanzees, and orangutans on a nonverbal false belief test. In the task, a researcher placed food in one of two boxes. A second researcher watched the first researcher place the food in the box. The second researcher showed the subject (i.e., child, chimpanzee, or orangutan) where the food was located by pointing to the appropriate box. In some conditions, the second researcher did not know where the food was hidden but the participant did. When the subject indicated the box that contained the food, the subject received a reward. The authors reasoned that in cases where the subject knew the food was in a box different from that indicated by the second researcher, the subject should ignore the second researcher pointing, because the subject should understand that the second researcher was operating on a false belief. The results showed that 5-year-old children performed in this manner but the chimpanzees and orangutans did not.

Schmelz et al. (2011) created a situation in which two chimpanzees searched for food on a table containing two opaque boards. One board was lying flat on the table. The other board was slanted, as though lying on top of a piece of food. They showed that chimpanzees, when at the table alone, would search for food under the slanted board. However, when they played a back-and-forth foraging game with another chimpanzee, they would avoid choosing the slanted board when the other chimpanzee had searched for the food just prior. The authors interpreted this result as indicating that the chimpanzee was able to infer that the other chimpanzee would have searched for food under the slanted board on the prior turn and would have taken the food if it were there. The fact that chimpanzees appeared to alter their foraging strategy based on the likely foraging strategy of a competitor suggests a rather sophisticated appreciation of the minds of others.

Does Theory of Mind Always Develop?

Autism

For some individuals, such as those with autism, theory of mind may not develop by the age of 5 or may not develop at all (Baron-Cohen, 2009). Autism is a developmental disorder characterized by social deficits, repetitive behaviors, and delayed language. A leading researcher, Simon Baron-Cohen, has described autism as **mind blindness** (Baron-Cohen, 1990, 2009), because individuals with autism appear to have great difficulty understanding others' intentions and perspectives. Symptoms of autism are usually observed before the child is 30 months old. There may be a lack of eye contact, aversion to physical contact, dislike of loud sounds, and lack of interest in social interactions, among others. Some children with autism display interest in mechanical objects and may repeat the speech of others in a manner that appears to be without purpose. The term *echolalia* is used to describe this type of speech. Approximately 80% of individuals with autism will have an IQ below 70. The prevalence of autism in the United States has been estimated to be approximately 1 in 88 (Centers for Disease Control and Prevention [CDC], n.d.). Autism may have multiple causes. Recent research supports a role for genetics (Abrahams & Geschwind, 2008). Other research suggests that exposure to environmental toxins may play a role (Arndt, Stodgell, & Rodier, 2005).

Onishi and Baillargeon's (2005) research with 15-month-olds suggests that it may be possible to detect deficits in theory of mind in very young children. Most children with autism are diagnosed when they are 3 years of age or older. Parents and pediatricians are likely to take a wait-and-see approach, because some children show delays in producing first words and sentences but later catch up and demonstrate normal language development. In cases in which a child may have autism, the wait-and-see approach may not be ideal, because providing therapies to children with autism as early as possible may result in their having better long-term outcomes. Onishi and Baillargeon's technique could prove to be one tool in diagnosing problems with theory of mind in children who are preverbal or producing only one or two words.

There is still a great deal to be learned about autism and how individuals with autism experience the world differently from others. It is hard to believe that our contemporary understanding of autism is only a few decades old. Text Box 6.2 describes an individual who helped raise awareness about autism and also about the remarkable talents that some autistic individuals have.

Asperger's Syndrome

Social deficits are also characteristic of **Asperger's syndrome,** a developmental disorder considered to be related to autism. In the new edition of the *Diagnostic and Statistical Manual of Mental Disorders* (DSM–V), Asperger's syndrome is listed as part of the autism spectrum, rather than a distinct disorder (American Psychiatric Association, 2013). Individuals with Asperger's syndrome display poor nonverbal communication ability, poor empathy, and physical clumsiness (Asperger, 1944). Some individuals with Asperger's develop special interests, which may occupy a lot of their time. They may discuss their special interest at length with others, in conversations that are aptly

Text Box 6.2 Extraordinary Individuals: Kim Peek (1951–2009)

The name Kim Peek may not have been widely known, if not for the 1988 film *Rain Man* (Levinson, 1988), starring Dustin Hoffman and Tom Cruise. Dustin Hoffman's character was Raymond Babbitt, the older brother of Cruise's character Charlie Babbitt. Raymond displayed many of the symptoms of autism (e.g., social awkwardness, preference for fixed schedules, avoidance of eye contact) as well as a remarkable ability to remember information viewed only a short time including numbers, names, baseball trivia, etc. The latter abilities have been referred to as savantism. Approximately half of all savants have some form of autism (Treffert, 2009). Dustin Hoffman won the Academy Award for best actor and shed light on the challenges faced by families affected by autism.

▲ Photo 6.2 Kim Peek is pictured here in 2007, where he was attending a conference. Have you seen the film *Rain Man*, which was inspired by his life story?

The story of *Rain Man* might not have been written (as it was) had it not been for the fortuitous meeting of Barry Morrow, the screenwriter of *Rain Man*, and Kim Peek in 1984. The two had met in Texas at a conference for individuals with intellectual impairments. Kim was a true savant, having the ability to read books and remember virtually everything that he read (Peek, 1996). He knew vast amounts of trivia from a wide range of subjects, including literature, geography, history, sports, and many others. Despite his astonishing memory, he had an IQ of only 87. He had lifelong problems with motor skills. He did not learn to walk until around the age of 4 and was never able to button a shirt. His set of problems and ability likely stemmed from abnormal brain development. He was born with an unusually large brain/head; a damaged cerebellum, which is the area of the brain involved in coordinating movement; and a missing corpus callosum, which is the bundle of fibers that connects the left and right hemispheres. Kim Peek lived his life with family in Murray, Utah. He passed away from a heart attack in 2009 at the age of 58.

described as *one-sided*. Although individuals with Asperger's seek out social interaction, they have trouble forming and maintaining friendships. The social norms that appear to come easily to others are confusing to those with Asperger's. They also frequently fail to perceive interpersonal cues or interpret them incorrectly, when they are perceived. IQ scores for individuals with Asperger's vary widely. Some individuals have average or above average IQs, are able to complete college, live independently, and pursue a career. Over the past several decades, awareness about Asperger's syndrome has increased. Many people who have struggled for years with social difficulties are now finding support from others who have experienced the same issues. Online communities have formed, and the term *Aspies* is often used by individuals with Asperger's syndrome to describe themselves (Aspies for Freedom, n.d.).

Agenesis of the Corpus Callosum

Autism and Asperger's syndrome are not the only disorders in which social deficits occur. Social deficits are also common in individuals with **agenesis of the corpus callosum** (ACC). ACC is a developmental disorder in which one is born without a corpus callosum, the bundle of fibers connecting the hemispheres of the brain. ACC is a rare disorder affecting between 0.03% and .07% of births (Penny, 2006). In some cases, the corpus callosum is completely missing, and in other cases, some portion of the corpus callosum is intact. With the increasing use of prenatal sonograms, parents may receive a diagnosis for the child before birth, usually after the 20th week of gestation. The corpus callosum typically is fully developed by the 18th to 20th week of gestation. ACC has been discovered in adults following a routine medical examination. Such individuals may have lived their lives up to that point unaware of having a disorder. Characteristics of ACC include memory problems, difficulty interpreting humor, difficulty in relationships, and difficulty understanding nonliteral forms of language (Paul, Van Lancker-Sidtis, Schieffer, Dietrich, & Brown, 2003).

Varying Language Across Social Setting

For the typical developing child, mastering the social norms in various social settings in society is no small feat. By the time that one has graduated high school, one has likely learned to navigate the social norms at home and at the homes of family friends and relatives. One has had to master the social norms at school, during classes, and also during athletic and social events, such as formal dances. One is likely to have had a job, where one had to learn how to deal with a boss, coworkers, and maybe even new employees for whom one had to serve as an authority figure. In these many settings, there are numerous opportunities to say or to do the wrong thing. What teenager has not made a major social gaff or *faux pas*? The social norms of various social settings are learned with experience. In some cases, the norms might be articulated by another person. This is most often done when one is experiencing a different culture and it is assumed that one would not know the norm unless explicitly told. For example, when entering a Buddhist temple, one is expected to remove one's shoes. In other cases, the norms may not be immediately clear. One may learn the norms only after violating them, so learning is through trial and error. For settings that one has experienced throughout life, the social norms may have been learned without much conscious awareness. There is a sense of knowing what would be or would not be appropriate in the setting, but one may not be able to articulate how one has come to know the norms. The term ***register*** is used to describe a particular form of language that is used in a particular social setting.

How Do Children Acquire the Politeness Register?

One of the first ways in which children learn to modify their language is using polite language or the **politeness** register. For example, polite language is used when speaking with authority figures, including parents and older relatives. It is generally used in religious settings and other settings in which individuals are expected to show respect (Brown & Levinson, 1987). Research has shown that speakers use more polite language when a listener has a higher social status (Andersen, 1996; Becker, 1986; Ervin-Tripp, 1976; Ninio & Snow, 1999) and when speakers believe that they are imposing on a listener (Axia & Baroni, 1985; Brown & Levinson, 1987; Snow, Perlmann, Berko Gleason, & Hooshyar, 1990).

Polite Requests

Parents play a major role in teaching children the social rules of language. Parents routinely teach children to say *good-bye, hello, please,* and *thank you.* Research by McTear and Conti-Ramsden (1992) found that more complex polite forms, such as promising and persuading, are not fully mastered until children are about 9 years old or older. Children's early use of polite forms has been well studied. Snow and colleagues (1990) investigated children's requests and found that the phrasing of the request varied across settings. Corsaro (1979) found that 2-year-olds used a polite form (e.g., *Can I have . . . ?*) when making a request to an adult, but used a nonpolite form (e.g., *Give me . . .*) when making a request to a peer. Research suggests that mastery of polite forms is not completed until later childhood (Bates & Silvern, 1977; Becker, 1986; Ninio & Snow, 1999; Nippold, Leonard, & Anastopoulous, 1982).

In such studies, children are instructed to interact with an interviewer or a puppet and to make requests in different kinds of ways. For example, Becker (1986) showed that 5-year-olds could modify their requests at will, either making a request in a *nice* way or in a *bossy* way. Children were asked to imagine that another child had borrowed an item. Nice requests were typically phrased using indirect forms, such as *Can you give it to me?* Bossy requests were typically phrased more directly, such as *I need it.* Bossy requests sometimes also included demanding phrases, such as *Right now, you have to,* or *really.*

Bates and colleagues (Bates, 1976; Bates & Silvern, 1977) investigated children's ability to produce polite requests. Bates (1976) tested Italian children between the ages of 3 and 6 years. Bates and Silvern (1977) tested children in the United States who were between the ages of 2 and 8 years. In both studies, children interacted with a puppet described as "the old lady." They were instructed to ask her for a piece of candy. After children asked the first time, they were told that they did a good job but that the old lady likes to hear children ask. They should ask again and do so more nicely than the first time. After the second request, children were rewarded with a piece of candy and instructed to ask again in the nicest way possible. Both studies showed that children were able to increase the politeness of their requests. Even the very young children (i.e., 2- and 3-year olds) were able to modify their requests as instructed.

Development of Polite Requests

Bates (1976) proposed that children progress through three distinct states in the development of polite requests. Stage one involves children producing direct requests or imperatives, such as *Give it to me*. This stages ends when children are around 4 years old. Stage two involves children being able to use a variety of sentence structures to make their requests; however, they are generally unable to soften their request or mask the specific aim of their request. This stage is observed in children who are 5 to 6 years old. The third and final stage is observed in children between the ages of 7 and 8; they are able to formulate requests using a variety of sentence structures and also are able to use indirect language to soften the tone of their requests.

Even very young children demonstrate that they take into account the listener when formulating their own speech. Research conducted by Ervin-Tripp and her colleagues (Ervin-Tripp & Gordon, 1985; Ervin-Tripp, Guo, & Lampert, 1990) found that children around the ages of 2.5 years old used less polite language with their fathers than with their mothers. Similar research by Kornhaber-Le Chanu (1995) with 21- to 27-month-olds found that when parents were instructed to ignore children's requests, children produced more intense gestures and speech with their mothers. The results suggest that children are sensitive to the cooperativeness of the parent.

Cultural Differences

Without a doubt, politeness is culturally defined (Brown & Levinson, 1987). The types of language use that would be considered as polite versus not polite vary across cultures. In some languages, there are different sentence structures or word forms to be used for different levels of politeness. For example, in French, the second person pronouns *vous* and *tu* are equivalent to *you* in English; however, *vous* is used when speaking with someone who is one's superior or with someone that one does not know well. Using *vous* is a way for the speaker to show respect for the listener. In contrast, the *tu* form is used with peers and is viewed as the less polite form. If one uses the *tu* form inappropriately (i.e., with a boss or with someone with whom one is not familiar), it would be considered rude. In contrast, if one has known someone for a time in a formal setting and has used the *vous* form from the beginning, then shifting to the *tu* form can communicate to the listener that the speaker views their relationship as becoming closer. Spanish and German have similar distinctions. In German, *du* and *sie* are used; *sie* is the more polite form. In Spanish, *tú* and *usted* are used; *usted* is the more polite form.

In Japanese, Korean, and Chinese, there are more elaborate distinctions between the more polite or honorific language forms and the less polite language forms (Matsumoto, 1997). Japanese is known for having a complex system of polite language. The most polite form of Japanese is *teineigo*; it is the form of Japanese that is typically used in formal television programs in Japan. One uses this form to show respect for the listener. In *teineigo* sentences, most nouns have a suffix *-desu* that is added for the polite or honorific forms of nouns and *-masu* that is added to polite forms of verbs. Prefixes *-o* or *-go*

▲ **Photo 6.3** Japanese women, one young and one older, are talking in the park. Do you use more polite speech when you speak to your elders?

are used for objects. Examples of casual and polite sentences are displayed in 2. Beginning learners of Japanese are typically taught *teineigo*.

2. a. kore wa hon da. | これは本です Casual register
   b. kore wa hon desu. | これは本である Polite register
   c. This is a book. English translation

Japanese also has two other forms of language, one referred to as *sonkeigo*, or respectful language, and another referred to as *kenjōgo*, or modest language. Sonkeigo is used to show respect toward the subject of the sentence and is used when one speaks to higher status others or customers in a business setting. Kenjōgo is used to show respect toward a non-subject of a sentence—usually the object. This form is used when one is describing activities related to a person in one's own social group, implying that the activities are used to be of service to another. Speakers use different verb forms for sonkeigo and kenjōgo sentences. A casual verb would be substituted for a more polite verb for sonkeigo and kenjōgo. Different verb forms for teineigo, sonkeigo, and kenjōgo are displayed in 3. In Japanese society, women tend to use polite forms of the language more often than men (Wetzel, 2004). Studies of language use between Japanese children and their caregivers show that honorifics are rarely used (Clancy, 1985; Cook, 1988).

3. a. do English translation
   b. する suru Causal register
   c. なさる nasaru Sonkeigo register
   d. 致す itasu Kenjōgo register
   e. します shimasu Teineigo register

What Is Genderlect?

Politeness is just one way in which the speech of men and women varies. Differences in the language use of men and women have been noted for decades (Lakoff, 1975). The term ***genderlect*** was created to describe the fact that men and women differ in their language use, specifically in the sentence structures that they use and in their choice of words (Tannen, 2001). For example, women tend to use longer, more polite sentences. In contrast, men use shorter, more direct sentences. Consider the examples in 4. Women also tend to phrase statements in the form of a question.

4. a. If you wouldn't mind, could you please pass me the napkins?
   b. Please pass the napkins.
   c. Pass the napkins.

The term *tag questions* has been used to describe these structures, as in *It sure is cold in here, isn't it?* Men's speech is more likely to include profanity, such as *damn* and *hell*

(Lakoff, 1973). Women's speech is less likely to contain profane words and more likely to include exclamations such as *Oh, dear* (Lakoff, 1973) and *Dear, me* (Farb, 1974). The greater use of profanity by men than by women was documented by Gallahorn (1971), who analyzed language used in personnel meetings involving staff in a psychiatric hospital.

Ritti (1973) found that girls tend to use expressions such as *oh* and *wow* more often than boys. The adjectives used by males and females also differ. Lakoff (1975) found that females use the words *adorable, charming, lovely,* and *divine* more than males, and males use *great, terrific,* and *neat* more often than females. In 1976, Hartman conducted a study involving older adults in Maine. She found that women used words such as *lovely, delightful, wonderful, nice, pretty, pathetic, pretty little, smartly uniformed, cute, dearest, gentle, gaily, beautifully, lovelies, very very, devoted, meek, perfectly wonderful,* and *stylish* more than men. In a study involving college students, Mehl and Pennebaker (2003) found that male college students used more profanity, anger words, articles (e.g., *the, those, these*), and long words than did female students. In contrast, female students used more *filler* words (e.g., *like, well*), more words indicating tentativeness (e.g., *would, should, could*), and more words referring to positive emotions. In a study of college students, Pennebaker and King (1999) found that women used the personal pronouns *I, me,* and *my* more often than men.

Lakoff (1975) pointed out that men's utterances are often commands (or directives), and women's utterances are requests. Hennessee and Nicholson (1972) analyzed 1,000 television commercials and found that 90% of the commands were produced by men. Soskin and John (1963) observed similar results when analyzing the transcript of a conversation of a married couple. The husband produced many more commands than the wife during the conversation. Lakoff (1975) also pointed out that women speak more tentatively than men. For example, women are more likely to tag questions onto their statements as a way to soften the directness of their statements (e.g., *It's cold, isn't it?*). Such sentences have been called tag questions.

Mulac, Bradac, and Gibbons (2001) analyzed 30 research studies that had found sex differences in language use. Across the prior studies, 16 aspects of language use had been found to differ for men and women. Applying a technique known as meta-analysis, Mulac et al. (2001) investigated the effect sizes of the observed sex differences. They observed that men used more words referring to quantity, more judgmental adjectives (e.g., *good, dumb*), more directives (e.g., *Write this down*), and more *I* references than women. In contrast, women used more intensive adverbs (e.g., *really, so*), references to emotions, uncertainty verbs (e.g., *seems to, maybe*), and negations (e.g., *not, never*) than men. They did not find sex differences in the use of tag questions (cf. Lakoff, 1975).

Laboratory studies in which men and women carry out conversations have also shown that there are gender differences in language use. Carli (1990) observed conversations between two people who were either the same or different genders. Participants chose the topic for their conversation. They were instructed to select a topic on which they held differing opinions. Participants were later asked to rate how influential their conversation partner was. The results showed that when women were talking with men, they were more tentative than men. Interestingly, when women were tentative, they were viewed as more

Text Box 6.3 Diversity of Human Languages: Gender Dialects in Yanyuwa

Yanyuwa is one of a few languages in which men and women speak different dialects of the language. Yanyuwa is an aboriginal language in northern Australia (Kirton, 1988). The language is one that has a rich morphology (e.g., prefixes and suffixes). Different morphological endings are used by male and female speakers (Bradley, 1988). The root words or word stems are the same in both dialects; however, the prefixes that are used on nouns, verbs, and pronouns differ. Examples of the men's dialect and the women's dialect in Yanyuwa are displayed in Table 6.6. Those who speak Yanyuwa do not have distinct names for the two dialects. When asked, elders are unable to provide a rationale for why men and women speak so differently (Bradley, 1988). Those living in neighboring communities, who speak other languages, perceive Yanyuwa as particularly difficult to learn because of the gender dialects. Male children start out in life using the women's dialect of the language. When they enter puberty and go through the ceremonial ritual signifying their entrance into manhood, they are expected to use the men's dialect. If they use women's dialect after that point, male elders will correct them, sometimes chastising them verbally. Men who have recounted experiencing the corrections describe them as humiliating. The only time in which it is socially acceptable for one to use the dialect of the opposite gender is when telling a story in which one is acting out a role of someone of the opposite gender or quoting verbatim something that another person said. The number of speakers of Yanyuwa is currently very small. It is in danger of becoming extinct, as the number of native speakers is decreasing.

Table 6.6 Examples of the Men's and Women's Dialects in Yanyuwa

Yanyuwa has seven classes of common nouns: male (M); female (F); masculine (MSC); feminine (FEM); food (nonmeat) (FD); arboreal (ARB); and abstract (ABS) and four cases: nominative (NOM); dative (DAT); ergative-allative (ERG/ALL) marking transitive subject and "to" a person or location; and ablative (AB).
Translation
"The short initiated man whose name is Wungkurli, went down to the sea, taking a harpoon with him for dugong or sea turtle."
Women's Dialect
Nya-ja nya-wukuthu nya-rduwarra niya-wini nya-Wungkurli kiwa-wingka This-M M-short M-initiated man his-name M-personal name he-gowayka-liya ji-wamarra-lu niwa-yirdi na-ridiridi ji-walya-wu down-wards MSC-sea-ALL he-bring ARB-harpoon MSC-dugong/turtle-DAT
Men's Dialect
Jinangu f-wukuthu f-rduwarra na-wini f-Wungkurli ka-wingka wayka-liya This short initiated man his-name personal name he-go down-wards ki-wamarra-lu na-yirdi na-ridiridi ki-walya-wu
MSC-sea-ALL he-bring ARB-harpoon MSC-dugong/turtle-DAT

Source: Bradley (1988).

influential by men but less influential by women. The language used by men was not related to how others perceived their influence.

The extent to which men and women differ in their language use varies across languages. In some languages, there may be little difference between the language used by men and women. Some languages, such as Finnish and Indonesian, do not have distinct pronouns to refer to men/boys and women/girls (i.e., *he* or *she*). Other languages have entirely different dialects that are used by men and women. Text Box 6.3 describes such a language from Australia in which men and women speak separate dialects. The differences between the versions of the language that men and women speak involve elaborate morphological and syntactic differences.

How Do Children Learn Other Social Registers?

Ways of speaking can also vary by geographic region. Linguists also use the term *dialect* to refer to a variety of a language. Languages routinely have dialects defined by geographic boundaries. In contrast, registers involve rather subtle differences in language use, which are generally accepted as grammatically correct. Although the English that is spoken in the United States was once the same as the English that was spoken in England, today one would consider American and British English as different dialects. Table 6.7 displays differences in vocabulary for American and British English.

When one speaks, one's choice of words and pronunciation are clues about where one has grown up or has lived for a long time. Certainly, some dialects are viewed as more desirable than others. In the United States and other industrialized countries, there is a version of the language that is considered the **standard dialect** of the language, which is taught in schools and in formal settings. Often, the dialects that people speak in their daily lives differ,

Table 6.7 Vocabulary Differences Between American and British English

American Word	British Word
Truck	Lorry
Apartment	Flat
Elevator	Lift
Bathroom	Loo
Police officer	Bobby
Lawyer	Solicitor (also barrister)
Pharmacist	Chemist
Wrench	Spanner
Baby stroller	Pram
Flashlight	Torch
Janitor	Porter
Parking lot	Motor park
Suspenders	Braces
Car hood	Bonnet
Car trunk	Boot
Thread	Cotton
French fries	Chips
Antennae	Aerial
Diaper	Nappy
Eraser	Rubber

either a little or a lot, from this standard dialect. In the United States, there are many regional dialects (e.g., Boston, New York, New Jersey, Maine, Martha's Vineyard, and New Orleans, among others). One of the largest dialects in the United States is **Southern American English** (SAE), which includes speakers from Virginia and other southern states on the East Coast to speakers in the eastern edge of New Mexico. SAE is characterized by differences in phonology, which is sometimes described as a *southern drawl*. Some vowels in SAE are lengthened (Nagle & Sanders, 2003). SAE also has distinctive vocabulary, such as the word *yonder* to express a distant location and the contraction *y'all* (for *you all*) to refer to second person plural. There are also expressions that you find exclusively among speakers who live or were raised in the southern United States, such as *I am fixing to wash my hair,* which means getting ready to wash my hair. In northern regions of the United States, those who speak SAE may face prejudice. In southern regions of the United States, those who do not speak SAE may be viewed as northerners or Yankees and viewed negatively.

Dialects may be associated with a particular racial or ethnic group. **African American Vernacular English** (AAVE) is one of the best-known examples (Baugh, 2000). The history of AAVE has received a great deal of attention. Some researchers have suggested that the dialect is a creole that originated from a pidgin that developed when speakers of African languages worked for English-speaking landowners and worked with English-speaking indentured servants on plantations in colonial America (Holm, 1988, 1989). Today, the characteristics of AAVE are most similar to southern dialects of American English. Without a doubt, the grammar of the dialect differs predictably from standard American English. Table 6.8 illustrates grammatical differences between sentences in standard American English and AAVE.

Slang and Jargon

The language that peers use with one another can be quite different from the language that one uses to communicate with unfamiliar others or with authority figures. Peer groups develop **slang,** which is a nonstandard form of language use serving to bond members of a group and to distinguish them from other groups. Often words originally created as slang in a particular subpopulation of a culture can come to be embraced by the standard form of the language. For example, the word *cool,* which originally meant having a low temperature, came to also mean being fashionable in the 1930s. Individuals who share the same profession or work setting also may develop specialized vocabulary that facilitates their communication and also distinguishes them from other groups. Such vocabulary is called **jargon,** which refers to specialized words and phrases common to a particular group of people, such as those working in the same profession. Military jargon has resulted in numerous new words and expressions in English, such as *jeep, barracks, brig,* and many others. Medical settings are also a common place where slang and jargon are used (Coombs, Chopra, Schenk, & Yutan, 1993). The creation and use of slang and jargon appears to are give users a distinct identity and to provide group members with a sense of belonging.

Table 6.8 Examples of Sentences From Standard American English and African American Vernacular English

Standard American English	African American Vernacular English
He is nice/He's nice.	He nice.
They are mine/They're mine.	They mine.
I am going to do it/I'm gonna do it.	I gonna do it.
He is/He's nice as he says he is.	He as nice as he say he is.
John is (always) happy.	John be happy.
John is happy (now).	John happy.
She is (always) late.	She be late.
She is late (now).	She late.

Source: Labov (1969).

Some slang words are used as euphemisms, which are words that are milder, vaguer, or more indirect and are used in place of other more taboo words. For example, rather than saying that someone died, one might say that the person *passed away* or *is no longer with us*. Rather than saying *I got fired*, one might say *I was let go*, *downsized*, or *given a pink slip*. Slang words for body parts (i.e., *caboose* or *drum sticks*), bodily functions (i.e., *number one* or *number two*), and sexual acts (i.e., *hit that* or *had a go*) may start out as euphemisms and come to be used generally, even when a euphemism is not needed. Slang frequently develops for topics that are perceived to be taboo or are known to be illegal. There are numerous slang words for marijuana. Some older terms are *grass*, *pot*, *ganja*, *herb*, *weed*, *reefer*, and *Mary Jane*. More recent terms include *skunk*, *boom*, *Aunt Mary*, *gangster*, and *kif*.

▲ **Photo 6.4** Girl is using fingers to make eyeglasses. How old are children when they start engaging in this type of creative behavior?

Literal and Nonliteral Language

How Do Children Learn to Comprehend Indirect Meaning?

Children's understanding of the subtleties in the meaning of utterances develops relatively late. Most preschool children will not understand adult utterances that involve **indirect meaning,** such as statements in which the speaker's intention is not directly reflected in the grammatical structure of the statement (Ervin-Tripp, 1977). For example, you might stop by to visit a friend, and a child answers the door. You ask, "Is your mommy home?" The child will say, "Yes," but then just stand there. They may even be happy to talk to you about other topics. The child fails to appreciate that you would like to talk with Mommy. In order to get the child to understand, you likely need to state directly, "May I speak with your mommy?" Another example has likely occurred quite often with parents concluding that their child is lazy rather than concluding that their child is having trouble interpreting indirect meaning. A parent might say to a child, "Boy, you've made a big mess in your room, haven't you?" The parent may intend for the child to interpret this statement as a hint that the room should be cleaned. Young children might hear such an utterance and proclaim, "Yes, I have" but not realize that it would be wise to start straightening up the room. By the age of 6, most children are able to interpret utterances involving indirect meaning (Cherry-Wilkinson & Dollaghan, 1979). Six-year-olds can also interpret nonverbal behaviors expressing frustration, such as a loud sigh when an undesirable behavior is observed.

Children's production of utterances involving indirect meaning begins around the age of 3 and increases over time (Garvey, 1975). Children's productions contain more direct statements and requests than indirect ones. Typical children may make frequent errors until they are around 7 years of age. Some children may have trouble after age 7 (Clark & Chase, 1972).

Before the age of 4, children interpret the words and sentences that they hear literally. When they are between the ages of 4 and 6, they begin to discover figurative uses of language (Eson & Shapiro, 1982). Figurative language involves the creative use of words and phrases to express meanings that are intended to be especially descriptive or humorous. Examples of figurative language include **metaphors**, **similes,** and idioms. Table 6.9 provides examples of each. Spector (1996) found that children typically acquire idioms in elementary school; however, the ability to explain idioms to others may continue to develop throughout the teenage years (Nippold & Rudzinski, 1993).

Table 6.9 Examples of Figurative Language

Metaphor	My teacher is a mind reader. My nurse is an angel.
Simile	School is like a prison. My job is like Disneyland.
Idiom	My friend threw me under the bus. I hope you turn over a new leaf.

Source: Katz (1998).

How Do Children Develop a Sense of Humor?

Children's mastery of the social rules of language also includes their ability to appreciate humor produced by others and also their ability to produce it themselves. The study of humor in children has a long history. In 1926, Jones published the results of observations of children between the ages of 16 and 36 months. She found that children frequently produced laughter for a range of daily activities, such as when being tickled and when dressing, bathing, teasing, and playing with other children. Most important, she found that an activity in which laughter occurred during one episode might produce crying during another episode with the same child.

Some researchers have suggested that children's ability to use and to appreciate humor is positively related to their overall social competence (McGhee, 1989). In Table 6.10, the four stages of humor development are described. As you learned in Chapter 3, infants begin to smile around the age of 6 weeks and laugh around the age of 16 weeks. During the second year of life, infants find it amusing when one uses objects in atypical ways, such as when a cup is used as a hat (Snow, 1989).

Since the 1950s, researchers have recognized that there is a relationship between children's mastery of their language and their ability to use language for humorous purposes (Grotjahn, 1957; Wolfenstein, 1954). Helmers (1965) pointed out that children as young as

Table 6.10 Stages of Humor Development

Stage 1	Incongruous actions with objects	18 to 24 months
Children use an object in a way that is atypical of its usual use and show awareness of the incongruity. For example, a child might use a bowl as a hat or a hairbrush as a microphone.		
Stage 2	Incongruous labeling of objects and events	24 to 36 months
Children will make a game out of referring to objects using the wrong words. For example, a child may invent words or refer to objects intentionally using the wrong names.		
Stage 3	Conceptual incongruity	3 to 7 years
Children begin to appreciate more abstract examples of humans. They begin to learn to tell jokes or riddles (Shultz, 1974). They enjoy knock-knock jokes, slapstick humor, and jokes about bodily functions (also called toilet humor).		
Stage 4	Multiple meanings	7 to 11 years
Children become aware that some words have multiple meanings. They are able to engage in teasing of others and jokes that are considered antisocial (e.g., laughing at others when mistakes are made). For example, a child would enjoy the following joke: "Order! Order in the court!" "Ham and cheese on rye, your honor!" (Shultz & Horibe, 1974).		

Source: McGhee (1979).

▲ Photo 6.5 A boy is puting on a magic show, using props. At what age do children begin to become interested in performing in this way?

2 years old play with the sounds of words; however, they typically do not do so to amuse themselves or others until they are 3 to 4 years of age. Three-year-olds will make mistakes while speaking and will laugh if they become aware of the error. In contrast, 4-year-olds will purposely play around with words to produce humorous combinations, and by the age of 5, most children are able to understand others' jokes. Bever (1968) observed a difference in the jokes that younger and older children found humorous. Five-year-olds were amused by ambiguities involving grammatical structure of sentences, as in the following joke: Why can't you starve in the desert? Because of the *sand-witches* there. Ten-year-olds were amused by ambiguities involving the meaning of a sentence, as in the following joke: Why can you jump higher than the Empire State Building? Because it can't jump at all.

Researchers disagree about the origins of humor. One view is called the incongruity hypothesis, which states that humor arises because there is a discrepancy between what one expects and what actually occurs. Using adults as participants, the incongruity hypothesis was tested in a study in which participants were asked to lift weights. Research has found that participants laughed and smiled more often when the weight differed from prior weights (Deckers & Kizer, 1975; Gerber & Routh, 1975; Nerhardt, 1970).

How Do Children Learn to Understand and Tell Stories?

Although storytelling may be viewed as a small part of a child's language development, researchers who have studied storytelling in children have concluded that it is a critical part of the child's language development (Hedberg & Westby, 1993). Furthermore, researchers have found that young children's narrative abilities are related to their general cognitive development (Applebee, 1978) and may predict future language ability (Botting, 2002) and success in school (Bishop & Edmundson, 1987).

Hedberg and Westby (1993) identified at least three types of narratives: (1) scripts, which are descriptions of recurring events as in *you take a bath and get ready for bed;* (2) events, which are descriptions of a past, current, or future event; and (3) fictional stories, which are descriptions of events that are known not to be real. By the age of 2, most children can talk about past events (Hedberg & Stoel-Gammon, 1986), but their descriptions are composed of a series of statements that are only loosely related. They tend to use the present tense. By the age of 3, children's stories start to center upon one character, event, or setting. Between the ages of 3 and 4, primitive narratives emerge; these narratives contain structural elements that link together the series of statements, which focus on a single character, event, or setting. They usually lack temporal links between statements. Between the ages of 4 and 5, children produce a series of statements that appear to link logically, but the statements are not focused on a single setting, character, or event. By the age of 6, children begin to be able to produce true narratives that have a setting, one or more characters, and a goal (Shapiro & Hudson, 1991). Young children's narratives include more descriptions of physical states than of mental states (Kemper & Edwards, 1986). As children get older, their narratives contain more and more references to mental states (Kemper & Edwards, 1986).

Recent research suggests that adults can play an important role in how children tell stories (Melzi, Schick, & Kennedy, 2011). Adults ask questions and make helping statements as children tell their stories. Consequently, adults are guiding children through the storytelling process. Research by Wiley, Rose, Burger, and Miller (1998) compared adult–child interactions in two European American communities, one middle-class and one working-class. They found that working-class adults participated more in children's narratives and provided more supportive statements about past experiences than did middle-class parents. In the working-class group, adults emphasized that children's narratives include factual retellings. In contrast, in the middle-class group, children spoke less but were given the freedom to express opinions. Children's accuracy did not appear to be an issue.

Cross-cultural comparisons have shown that adult behaviors with children storytellers differ. Minami and McCabe (1995) compared how European American and Japanese mothers interact with their children during their children's storytelling. Japanese mothers requested less description from children as they told stories than did European American mothers. Japanese mothers also provided fewer evaluation statements of children's utterances than did European American mothers. However, Japanese mothers were more interactive with the children, paying closer attention to the child and keeping children's turns relatively short. Others have noted that the social norm in Japanese society is to refrain from excessive talk and to limit the discussion of past experiences (Han, Leichtman, & Wang, 1998).

Summary and Theoretical Implications

From the beginning of life, parents and caregivers teach infants conversational turn-taking, a core aspect of human communication. The conversations that infants have in their first 2 years of life differ a great deal from those of adults. Infants will talk to themselves when no one else is around and, when speaking with others, will produce long monologues that are not directly related to the utterances of their conversational partners. Older children are able to perceive when their utterances are not understood and are sensitive to differences that exist in conversational partners, such as Mom versus Dad. Between the ages of 15 and 48 months, children become increasingly aware of the minds of others and that others may possess knowledge that is different from their own. For individuals with autism and other developmental disorders, understanding the minds of others may be a lifelong challenge. Through the elementary school years and young adulthood, the social norms of language use continue to be acquired, specifically norms related to various social groups. While children are figuring out how to navigate the complex world of different social settings, they are also learning to understand and to use figurative forms of language. Spontaneous displays of humor and storytelling are observed as early as 2 years of age.

In terms of the four theoretical approaches to language development, the research on social aspects of language provides the strongest support that we have seen so far for the behaviorist and social-interactionist views of language development. As children learn how to use language appropriately in different social contexts, their behaviors may be shaped by classical and operant conditioning. It is also clear that parents and peers provide important input to children as they learn how to interact with others both one-on-one and also in groups. Nevertheless, the fact that some biologically based disorders are associated with social impairments suggests that our ability to learn the norms of social interactions may be, at least in part, biologically based. Autism, Asperger's syndrome, and ACC are conditions that involve deficits in social functioning. Future research is needed to determine to what extent the social aspects of language development are biologically based versus learned through experience. Last, the statistical learning approach to language development may be able to provide explanations for how social skills develop and how they are impaired in different disorders. So far, advocates of the statistical learning approach to language development have not attempted to address the social aspects of language.

KEY TERMS

African American Vernacular English (AAVE)

agenesis of the corpus callosum (ACC)

Asperger's syndrome

collective monologues

communicative competence

conversation

curse of knowledge

false belief test

genderlect

Gricean maxims

indirect meaning

jargon

joint attention

metalinguistic awareness

metaphors

mind blindness

politeness

private speech

register

request

repair

similes

slang

Southern American English (SAE)

speech acts

standard dialect

theory of mind

turn-taking

REVIEW QUESTIONS

1. What are the major conversational milestones that typically developing children experience?
2. What is turn-taking? When do children begin learning about turn-taking in interactions with others?

3. What is a speech act? Provide five examples of different speech acts.
4. How do children first learn about turn-taking?
5. What are collective monologues? How old are children who typically produce collective monologues?
6. What is private speech? What role does private speech play in children's regulating their own behavior?
7. What are the four Gricean maxims? To what extent are they believed to be universal?
8. What is theory of mind? What role does it play in communication?
9. What tasks are used to test whether an individual has theory of mind?
10. What is the evidence that individuals with autism have deficits in theory of mind?
11. What are the social abilities of someone with Asperger's syndrome?
12. What is ACC? What are the social abilities of individuals with this disorder?
13. What is a register? How is it different from a dialect?
14. How do speakers express politeness in different languages?
15. What is genderlect? How early do children display an awareness of genderlect in others?
16. How is literal language different from nonliteral language?
17. What is indirect meaning? Provide examples of an utterance with an indirect meaning.
18. At what age do children begin to understand and to use humor?
19. When do children begin learning the narrative structure of stories? How does children's storytelling ability develop?
20. How can adults influence children's narratives? To what extent have mothers' interactions with children while they tell stories been found to vary across cultures?

RECOMMENDED READING

Apperly, I. (2010). *Mind reading: The cognitive basis for theory of mind.* New York: Psychology Press.

Baugh, J. (2002). *Beyond ebonics: Linguistic pride and racial prejudice*. Oxford, UK: Oxford University Press.

Farwell, H. F., & Nicolas, J. K. (2007). *Smokey mountain voices: A lexicon of southern Appalachian speech based on the research of Horace Kephart*. Lexington: University of Kentucky Press.

Grandin, T. (2010). *Thinking in pictures: My life with autism*. New York: Vintage.

Haddon, M. (2004). *A curious incident of the dog in the night-time*. New York: Vintage.

Peek, F. (1996). *The real Rain Man: Kim Peek*. Salt Lake City, UT: Harkness.

Tannen, D. (2001). *You just don't understand: Women and men in conversation*. New York: Quill.

RECOMMENDED FILMS

Alvarez, L., & Kolker, A. (Producers). (1998). *American tongues* [DVD]. Available from www.pbs.org.
Cran, J. (Director). (1984). *The story of English* [Motion picture]. United States: HomeVision.
Cukor, G. (Director). (1964). *My fair lady* [Motion picture]. United Stated: Warner Brothers Pictures.
Jackson, M. (Director). (2010). *Thinking in pictures* [Motion picture]. United States: Home Box Office.
Jeffcoat, J. (Director). (2006). *Outsourced* [Motion picture]. United States: ShadowCatcher.
Johnston, J. (Director). (1999). *October sky* [Motion picture]. United States: Universal Studios.
Lynn, J. (Director). (1992). *My cousin Vinny* [Motion picture]. United States: 20th Century Fox.
Mayer, M. (Director). (2009). *Adam* [Motion picture]. United States: 20th Century Fox.

SUGGESTIONS FOR CLASS PROJECTS

1. Examine the different ways in which men and women tend to talk by performing an analysis of speeches given by men and women. Political speeches can be found on the World Wide Web. Select similar types of speeches, one given by a man and one given by a woman. In a paper or class presentation, discuss the differences observed between the two speeches in topics discussed, word choices, and sentence structure. You may use Deborah Tannen's book *You Just Don't Understand* to discuss the differences that are found, specifically the extent to which the observed differences are compatible with Tannen's views on why men and women approach conversations differently.

2. Investigate how the form of a question or request (i.e., polite vs. nonpolite or direct) influences the language of respondents. Develop a target question or request, and create two versions of each, a polite form and a nonpolite form. Call 10 to 20 offices on campus or in the community. Then write down verbatim the language used by the respondents. Analyze the extent to which polite forms elicit more polite responses than nonpolite forms.

3. A wonderful resource on dialects of English exists at http://web.ku.edu/ ~ idea/. Explore audio recorded samples of dialects from different countries and from different states within the United States. There are several standard scripts that speakers have used to make the recordings. Consequently, it is possible for students to pick speakers who have read the same script and to compare the differences in the pronunciation of the different dialects. Prepare either written reports or class presentations in which you analyze the differences in the speech of different speakers. An alternative project would be to record speech samples and contribute to the database.

4. Study the lyrics of popular songs from the hip-hop genre. The lyrics can be analyzed in order to determine the extent to which they reflect features of AAVE. You may also find it interesting to compare lyrics from songwriters raised in different parts of the country (i.e., New York vs. Atlanta); focus on whether there are dialectal differences.

For additional ancillary resources, please visit the companion website at www.sagepub.com/kennison.

- Video Links
- Audio Links
- Web Resources
- Internet Activities
- Flashcards
- Web Quizzes

CHAPTER 7

LIFE WITH MORE THAN ONE LANGUAGE

In some parts of the world, such as the United States, there is the misperception that knowing only one language, or monolingualism, is the norm. In fact, it is common to know more than one language. Approximately half of the world's population use more than one language in their daily life. Even in the United States, 55% of those surveyed for the 2000 census reported being proficient in English and using a language other than English at home (U.S. Census Bureau,

▲ Photo 7.1 A bilingual infant lies between Mom and Dad. Would learning two languages in childhood be appealing to you?

2000). In the United States, it is expected that the number of **bilinguals** will continue to rise in the coming decades.

The term *bilingualism* is used to describe cases in which one knows more than one language. The term *multilingualism* is used to describe cases in which one knows more than two languages. In numerous regions of the world, residents must be **trilingual** to be successful, mastering the language of the home, which may be a tribal or regional language, and also mastering official languages of the local and national government. Table 7.1 displays a list of countries with populations over 30 million in which there is more than one official language. Many countries, including those in the table, recognize regional languages. The table includes only those languages that have been recognized at the national level. There are many more countries in the world where communities, cities, and provinces are unofficially multilingual. In this chapter, you will learn about how bilingual children differ from **monolingual** children in terms of their language development as well as how they differ in terms of other aspects of their cognitive processing. Regardless of the social or political environment where children are born, they will learn whatever languages are spoken regularly in their environments. This chapter will also review the different ways in which children and adults acquire second and third languages and what is known about how bilinguals store knowledge of multiple languages in memory and access the knowledge during language processing.

Table 7.1 Countries With Populations Over Thirty Million With More Than One Official Language

Country	Population	Official Languages
India	1.2 billion	2 (Hindi and English)
Pakistan	197 million	2 (Urdu and English)
Philippines	94 million	2 (Filipino and English)
Democratic Republic of the Congo	66 million	5 (French and four others)
South Africa	50 million	11 (Afrikaans, English, and nine others)
Spain	46 million	5 (Spanish and four others)
Tanzania	43 million	2 (Swahili and English)
Kenya	39 million	2 (Swahili and English)

Country	Population	Official Languages
Algeria	36 million	2 (Arabic and Berber)
Canada	35 million	2 (English and French)
Morocco	32 million	2 (Arabic and Amazigh)
Uganda	32 million	2 (Swahili and English)
Iraq	32 million	2 (Arabic and Kurdish)
Afghanistan	31 million	2 (Dari and Pashto)
Sudan	31 million	2 (Arabic and English)

Source: Adapted from The World Factbook (2009).

Bilingual Children

How Do Bilingual Children Differ From Monolingual Children?

A common way in which infants become bilingual is that they are born to parents who speak different languages. When both languages are used regularly in the home, the infant is usually able to acquire both languages simultaneously. The term ***simultaneous bilingualism*** (or simultaneous multilingualism) is used to describe circumstances in which multiple languages are learned from birth. In contrast, when a **second language (L2)** or third language (L3) is learned after a **first language (L1)**, then the term ***sequential bilingualism*** (or sequential multilingualism) applies.

A long-standing empirical question has been whether bilingual infants experience delays in language development as compared with infants who are learning just one language. Delays for bilingual infants might be expected because they are doing twice the work of the monolingual infant. Often, there is the concern among monolingual family members or family friends that the infant will be confused by being exposed to two languages. Some may have been so concerned about the possibility that a child might be adversely affected by learning two languages that the infant was prevented from being exposed to more than one language. The remainder of this section will review the differences between bilingual and monolingual children in terms of the reaching of language milestones, the development of perceptual abilities, and vocabulary development as well as the research on the extent to which bilingual children inappropriately mix their languages.

Language Milestones

Decades of research on childhood bilingualism indicates that there is no need for concern. Bilingual and monolingual children reach the language development milestones at roughly the same times. Research suggests that the age at which first words are produced (i.e., around 12 months) does not differ for bilingual and monolingual children (Genesee, 2003; Patterson & Pearson, 2004). A study by Petitto and colleagues (2001) compared the language development of bilingual children raised in a home in which both French and English were used and children raised in a home in which French and a signed language used in Quebec were used. They found that both groups of bilingual children began speaking/signing around the same time as is normally observed for monolingual children (i.e., 12 months). The bilingual children began using two-word utterances around the age observed for monolinguals (i.e., 18 months).

Perceptual Abilities

Between birth and 12 months, bilingual and monolingual children may differ in their ability to distinguish phonemes; however, studies have produced conflicting results on this issue. Research has shown that there are differences in the perceptual abilities of bilingual and monolingual infants. Bosch and Sebastián-Gallés (2003) compared the perceptual abilities of monolingual and bilingual infants. One group of monolingual infants was reared in Spanish-speaking homes. Another group of monolingual infants was reared in homes speaking Catalan, a Romance language that is distinct from Spanish. A third group of bilingual infants was reared in homes in which both Spanish and Catalan were spoken. They tested how well infants perceived a vowel contrast that involved two distinct vowels in Catalan but involved a single phoneme in Spanish. They found that all infants could perceive the vowel contrast at 4 months. Infants reared in homes where Catalan was spoken were able to perceive the contrast when they were retested at 8 and 12 months. They found that monolingual infants reared in Spanish-speaking homes became less and less able to perceive the contrast when they were retested at 8 and 12 months. The bilingual infants showed a U-shaped pattern of performance. Their ability to perceive the contrast declined at 8 months as compared with their original performance, but when tested at 12 months, their performance increased back to the level observed at 4 months. The results suggested that the perceptual development of bilingual and monolingual infants may proceed somewhat differently.

More recent studies have found similar perceptual development for bilingual and monolingual infants (Burns, Yoshida, Hill, & Werker, 2007; Sundara, Polka, & Molnar, 2008). They compared monolingual English, monolingual French, and bilingual (English/French) infants' ability to perceive a consonant contrast (i.e., the French /d/, which differs in terms of place of articulation from the English /d/). The phonemes are not perceived as distinct by either English or French speakers but are perceived as distinct by speakers of other languages. When they tested infants who were between 6 and 8 months old, both the monolingual and bilingual infants were able to distinguish the sounds. When they were retested at 10 to 12 months, the English monolingual infants and the bilingual infants were able to distinguish the sounds, but French monolingual infants could not. The authors pointed out that French-speaking adults do not reliably distinguish the two sounds (Sundara & Polka, 2008).

In a similar study, Burns and colleagues (2007) compared the perceptual abilities of monolingual infants reared in English-speaking homes with those of bilingual infants reared in homes where both English and French were spoken. The contrast involved a pair of consonants that differed in voice-onset time. As you learned in Chapter 3, some consonants vary only in terms of the duration of silence between the beginning of the sound and when voicing begins (e.g., /p/ and /b/ in English). Both groups of infants distinguished English/French when tested at 6 to 8 months. When retested at 10 to 12 months and also at 14 to 20 months, bilingual infants performed as well as they had when tested at 6 to 8 months, but monolingual infants were able to perceive only the consonant that was a phoneme in English.

Werker, Weikum, and Yoshida (2006) suggested that bilingual infants may vary in terms of whether the two languages spoken in the home are equally dominant or whether one language is used more often or viewed as more important than the other. In their study, they tested infants reared in homes where both French and English were spoken. They tested their ability to perceive sounds found only in French and sounds found only in English. When they tested infants between 14 and 17 months, they found that some of the bilingual infants could discriminate both types of sounds, while other bilingual infants could discriminate just the sounds found in English or could discriminate just the sounds found in French.

Vocabulary Size

Studies focusing on vocabulary size for bilingual and monolingual children have observed differences (Pearson & Fernández, 1994; Pearson, Fernández, & Oller, 1993; Rescorla & Achenbach, 2002). Bilinguals appear to have slightly smaller vocabularies than monolinguals. In an early study, Pearson and colleagues (Pearson & Fernández, 1994; Pearson et al., 1993) compared the vocabulary knowledge of Spanish–English bilingual children with that of monolingual children. They found that the vocabulary knowledge and the occurrence of the word spurts were similar for both groups of children. Rescorla and Achenbach (2002) analyzed data from a national language development survey (Rescorla, 1989) and found that bilingual children had smaller vocabularies than monolingual children. The authors pointed out that the methodology might have contributed to the different results for bilingual and monolingual children, because vocabulary size was reported by parents. The task for parents of bilingual children was likely more complex and more prone to error than for parents of monolingual children.

Critics point out methodological weaknesses of the study, such as a smaller sample size for the bilingual group than the monolingual group (Patterson, 2004). A point that was not made by Rescorla and Achenbach (2002) or Patterson (2004) was that a comparison of a bilingual child's vocabulary in one of that child's languages with a monolingual child's vocabulary may be the wrong comparison. The bilingual child is likely to know two words for each concept, one word from each language. Assessing the bilingual child's vocabulary size in both languages and comparing it with the monolingual child's vocabulary in one language is likely to reveal that the bilingual child knows more words overall—even that child knows fewer words than the monolingual children in L2. However, recent research has found that bilingual school-age children have smaller productive vocabularies when vocabulary in both languages is taken into account (Yan & Nicoladis, 2009). Similar results have been obtained in studies in which the productive vocabularies of children under 3 years of age were compared (Junker & Stockman, 2002; Oller & Eilers, 2002; Pearson et al., 1993, 1995; Petitto & Kovelman, 2003).

▲ Photo 7.2 Two girls are talking on their way to school; one of the girls is bilingual. Do you know how many children in the United States speak a language other than English at home?

Studies of children's receptive vocabulary development (i.e., ability to comprehend words when spoken to them) have also been conducted with bilingual and monolingual children. Some studies have shown that bilingual children have smaller receptive vocabularies in each of their languages than do monolingual children (Bialystok, Barac, Blaye, & Poulin-Dubois, 2010; Bialystok, Luk, Peets, & Yang, 2010; Mahon & Crutchley, 2006). Other studies have found that bilingual and monolingual children have similar receptive vocabularies (Cromdal, 1999; Yan & Nicoladis, 2009). Recent research suggests that the differences in vocabulary size for bilinguals and monolinguals is also observed for young adults (Portocarrero, Burright, & Donovick, 2007); however, other studies have failed to observe differences (Bialystok, Craik, Klein, & Viswanathan, 2004).

When researchers investigated the composition of bilingual children's vocabularies, they found that many words were **translation equivalents** (i.e., words from the bilingual's two languages that mean the same thing, such as *apple* and *manzana*). Genesee and Nicoladis (2007) studied the vocabularies of children between 8 and 30 months reared in homes in which both Spanish and English were used. They found that 30% of the children's vocabulary involved translation equivalents (e.g., *apple–manzana*). Petitto and colleagues (2001) observed similar results in a study of children who spoke French and used a signed language. Between 36% and 50% of children's vocabularies were translation equivalents.

There are likely many factors, in addition to whether a child is a bilingual or monolingual, that can influence vocabulary size. Among those are being raised in a low-income home (Hart & Risley, 1995) and speaking a minority language (i.e., not the **dominant language** of a country or region) (August & Shanahan, 2006). A recent study by Vagh, Pan, and Mancilla-Martinez (2009) investigated the vocabularies of children between the ages of 24 and 36 months who were being raised in low-income homes. They relied on both parent and teacher reports of children's vocabulary knowledge. They found that vocabulary size increased more slowly for Spanish–English bilingual children than for monolingual children. Among bilingual children, those for whom English was the dominant language had larger vocabularies than those for whom Spanish was the dominant language.

Differences in bilingual and monolingual children's vocabulary acquisition may be related to their using different strategies in word learning. As you learned in Chapter 3, research has shown that children exhibit the mutual exclusivity principle when learning new words. Children tend to assume that if an object is called something then it is unlikely to be called something else. Bilinguals generally know two words for each object—a word in each language. Houston-Price, Caloghiris, and Raviglione (2010) compared how monolingual and bilingual infants between the ages of 17 and 22 months interacted with novel objects in a task using the preferential looking paradigm. Infants were shown two objects—a novel object and one whose name was familiar. They were then asked to *look at the dax*. The results showed that monolingual infants looked more often at the novel object, demonstrating the mutual exclusivity principle. In contrast, bilingual infants did not. The authors concluded that infants raised with multiple languages do not develop the mutual exclusivity strategy. Older bilingual children appear to use the strategy. In a study with children between the ages of 3 and 6, Davidson, Jergovic, Imami, and Theodos (1997) found that bilingual children relied on the mutual exclusivity principle but to a lesser extent than did monolingual children.

Language Mixing

Another common concern among parents raising bilingual children is that the children will be confused learning two languages at the same time and will mix up the languages. Again, the fear appears to be that initially mixing up the languages and later figuring out that the two languages are separate entities could cause bilingual children to lag behind their monolingual peers in the long term. Meisel (1989) has referred to this possibility as the **fusion hypothesis.** Early in life, the bilingual children's two languages are fused. They do not differentiate their languages early in life but only do so at some point later in childhood. Observations of language mixing have been reported by parents as well as researchers in children between the ages of 2 and 3 (Köppe, 1996). The fear that language mixing may indicate that the child is experiencing a problematic form of language confusion appears to be unfounded. After the age of 3, the amount of language mixing observed in these children's speech decreases dramatically. Only about 2% of bilingual preschoolers produced utterances in which languages were mixed (Lindholm & Padilla, 1978).

Bilingual children's production of sentences containing words from both languages may be not be due to their language systems being fused early in life. Critics of the fusion hypothesis suggest that language mixing could occur because children are not pragmatically competent (De Houwer, 2005; Genesee, 1989; Köppe, 1996). They may not yet understand that the language used in an utterance must be tailored to the language(s) known by listener(s). Another possibility is that children are purposely mixing the words of their languages together as a form of play or linguistic creativity. As we learned in Chapter 3, children create idiosyncratic words or idiomorphs when learning new words. It is possible that they enjoy playing with sentences as well, creating novel combinations of words. Some children spontaneously create their own language games (Cumming, 2007). For example, every so often, my 3-year-old grandson Griffin finds it very amusing to produce a long string of nonsense syllables. His father told the group that he had been doing that for a while, and when someone asks him if that is his own language, he laughs and says, "Yes. Griffin's language!"

Code-Switching

The view that language mixing by bilingual children may reflect pragmatic development is supported by the observation that language mixing occurs in the speech of adult bilinguals.

As you learned in Chapter 4, adult bilinguals often will produce utterances containing words from both of the languages that they know when speaking with other bilinguals. This form of language use has been called **code-switching.** Poplack (1980) titled her article about code-switching with an example of code-switching: "Sometimes I'll start a sentence in Spanish y termino en Espanol." Examples of code-switching by a Spanish–English bilingual from Poplack (1980) are displayed in 1. The view of Poplack (1980) and other language researchers is that code-switching is not a form of unusual or deviant language behavior; rather, it is part of the normal language repertoire of a bilingual (Gumperz, 1971, 1976).

1. a. Me iban a lay off. They were going to lay me off'.
   b. Leo un magazine. I read a magazine.

There are likely to be many reasons that adult bilinguals engage in code-switching. Heredia and Altarriba (2001) suggested that code-switching may occur when the speaker lacks proficiency in one language. When speaking in L2, bilinguals may find that they do not know or cannot remember at that moment the particular L2 word needed. An L1 word can be substituted. Heredia and Altarriba (2001) pointed out that some code-switching may be the result of word-finding difficulty. Monolingual and bilingual speakers alike sometimes experience difficulty coming up with a known word. These occurrences have been called tip-of-the-tongue states, or TOT states (James & Burke, 2000). The bilingual speaker may experience a TOT state in one language but be able to retrieve from memory the word in the other language.

Analyses of sentences containing code-switching have shown that speakers appear sensitive to syntactic structure when planning code-switched utterances. For example, speakers do not produce multimorphemic words in which some morphemes are in one language and the other morphemes are in another language. Poplack (1981) referred to this as the **free morpheme constraint.** The view that code-switching adheres to structural principles known to the speakers is supported by research showing that bilinguals show agreement about which code-switched sentences are grammatically acceptable and which are grammatically unacceptable (Aguirre, 1980; Gingras, 1974; Gumperz, 1976; Timm, 1975). For example, in Spanish, adjectives are placed after the nouns that they modify (e.g., *quiero un tomate verde*, which means I want a green tomato). Lederberg and Morales (1985) pointed out that the grammaticality of code-switched Spanish–English sentences containing adjectives depends on the rules of the language of the adjective. If the adjective appears in English, it is only grammatical when it precedes the noun as in 2a versus 2b. If the adjective appears in Spanish, it is only grammatical when it follows the noun, as in 2c versus 2d. (The asterisks indicate that the sentences are ungrammatical.)

2. a. Quiero un green tomate.
   *b. Quiero un tomate green.
   c. I want a tomato verde.
   *d. I want a verde tomato.

Code-switching is also more likely to be used in some situations than in others (Gumperz, 1976). Some of the factors that may be related to the frequency of code-switching include the topic being discussed, the social status of the individuals taking part in the conversation, the language proficiencies of those individuals, where the conversation is occurring (i.e., home, work, school, church), and the level of emotion involved in the conversation. There appears to be a particularly strong relationship between the emotionality of the topic being discussed and the likelihood that code-switching will be used.

Research has shown that when talking about stressful or traumatic past experiences, bilinguals tend to prefer to use their L2 rather than their L1. Since the 1980s, therapists working with bilingual clients have reported that bilinguals often show a preference of discussing embarrassing or upsetting topics in their L2 (Bond & Lai, 1986). Santiago-Rivera and Altarriba (2002) proposed that bilinguals use the L2 to discuss emotionally charged information because it serves to distance them from the upsetting content. Their explanation for this phenomenon is that bilinguals "represent emotional words differently in their two languages and typically associate these words with a broader range of emotions in their first language" (Santiago-Rivera & Altarriba, 2002, p. 33). They also suggested that memories for emotional events contain information about which language was used at the time of the event. When bilinguals recount memories, the language that was being used in the episode may have a privileged role in remembering and talking about the event (e.g., Javier, 1996; Rubin, Schrauf, Gulgoz, & Naka, 2007). In a recent study, Santiago-Rivera, Altarriba, Poll, Gonzalez-Miller, & Cargun (2009) interviewed nine bilingual therapists about the situations that led them to use L1 or L2 in sessions with clients and situations that led clients to use L1 or L2. They found that when clients wanted to discuss an emotionally charged topic (e.g., one involving anger), they usually used L1. At times in which they appeared to want to distance themselves from the event, they would switch to L2. In contrast, therapists reported switching from L1 to L2 as a strategy to redirect the conversation with the client. One therapist was quoted as saying, "I may challenge some beliefs, and if I feel I am not being successful in one language, I try to present it in another language" (Santiago-Rivera et al., 2009, p. 439).

The relationship between the language mixing produced by bilingual children and the code-switching behavior produced by adults is not well understood. Future research is needed to determine whether children's utterances containing language mixing appear to show the same syntactic characteristics as adult utterances involving code-switching. Furthermore, language development researchers would be interested in knowing whether the amount of language mixing observed in bilingual children is related to the amount of adult code-switching that children hear in their environments. As we learned in Chapter 1, theories of language development differ with regard to the role that language experience plays in children's acquisition, particularly in the acquisition of morphological and syntactic structures. The study of children's language mixing (or code-switching) appears to be well suited for the testing of these theories.

There are situations in which an entire community may participate in language mixing, and, over time, the language mixing leads to the formation of a new language. Text Box 7.1 describes how this can happen when a pidgin leads to the formation of a creole.

Text Box 7.1 Diversity of Human Languages: Pidgins and Creoles

When speakers of different languages and cultures work and live together on a daily basis, they often devise ways of communicating, using a small number of words drawn from one or more of the represented languages. Over time, the speakers come to understand one another. The term *lingua franca* is used to refer to communication systems that develop to enable speakers of different languages to communicate. In some cases, a lingua franca may become a pidgin, when individuals who speak different languages work together over long periods. Around the world, pidgins have developed in busy seaports, where there is an abundance of international trade (Todd, 1990). For example, in the early 20th century, Hawaii was home to immigrants from Japan, Korea, and the Philippines who worked on sugar cane plantations for English-speaking landowners. Out of this unique situation developed Hawaiian Creole English (Bickerton, 1981, 1984). The communicative power of a pidgin is in its vocabulary rather than its grammar. Most pidgins have a highly simplified grammar and most lack complex clause structures. However, an interesting thing happens when children are exposed to a pidgin and acquire it as their L1; the language that the children end up speaking is grammatically more complex. The word *creole* is used to refer to a language that was once a pidgin but has subsequently become a native language for the next generation. Bickerton (1983) pointed out that creoles are grammatically more similar to one another, despite being located in distant parts of the world, than they are to the languages from which the pidgin predecessor languages were derived. He identified 12 characteristics that creoles around the world share, including an SVO word order, the verb-tense system, and the formation of questions and negative sentences. Bickerton proposed the grammars of creoles are similar because children who learn pidgins use aspects of innate universal grammar (UG) as they learn pidgins. In his language bioprogram hypothesis, he claimed that using their innate knowledge, children who acquire a pidgin as an L1 can add in grammar where it is missing. An alternative view to explain why creoles are so syntactically similar is that some of the grammatical consistencies observed in creoles may be due to similarities found in the languages that make up the pidgins. Most creoles in the world have come from pidgins that have at least one language from the Indo-European language family. Here are examples from pidgin and Hawaiian Creole English from Bickerton (1991).

Pidgin	Hawaiian Creole English
Now days, ah, house, ah, inside, washi . . .	Those days bin get [there were] . . .
. . . clothes machines get, no? Before time . . . ,	. . . no more washing machine . . . ,
. . . ah, no more, see? And then pipe no more . . . ,	. . . no more pipe water like get [there . . .
. . . water pipe no more.	. . . is] inside house nowadays, ah?

Are There Cognitive Benefits When Children Are Bilingual?

Parents and caregivers of bilingual children are likely to be surprised to learn that there is ample evidence that there are concrete, measurable cognitive benefits when a child learns more than one language. The first of the studies suggesting the benefits of childhood bilingualism was reported in 1962 by Peal and Lambert. They showed that 10-year-old French–English bilingual children in Montreal performed better on verbal and nonverbal tasks than 10-year-old monolinguals from the same school. The authors concluded that the bilingual children were cognitively more flexible than monolingual children. Ricciardelli (1992) reported similar results in a study with a group of Italian–English bilingual children. Bilingual children who were proficient in both languages outperformed monolingual children on tests of creativity.

In a study with Hebrew–English bilingual children, Ben-Zeev (1977) found that they performed better than monolingual English-speaking children and monolingual Hebrew-speaking children on two cognitive tasks. The three groups of children were between the ages of 5 and 8. One task required children to study a display of nine cylinders of varying sizes arranged into three columns and three rows. They were then asked to describe the pattern of the cylinders and transpose it. In a second task, children played a language game with the interviewer. The interviewer told the child that an *airplane* was called a *turtle*. The interviewer then asked the children questions, such as *Can a turtle fly?* and *How does a turtle fly?* The author concluded that bilingual children are better than monolingual children at "seeking out rules and for determining which are required by the circumstances" (Ben-Zeev, 1977, pp. 1017–1018).

Hakuta and Diaz (1985) studied a large group of Spanish–English bilingual children in Puerto Rico. All children were from low socioeconomic backgrounds and were more proficient in Spanish than in English. Children who showed the highest degree of bilingualism performed better than other children on tests of nonverbal intelligence and metalinguistic awareness. The study was conducted longitudinally, enabling the researchers to test children multiple times. They found that those children who became more proficient in English over time also showed increased nonverbal intelligence and metalinguistic awareness.

Attentional Processing

Perhaps, the most convincing evidence of all has been obtained by Bialystok and colleagues (Bialystok, 1999, 2005; Bialystok & Majumder, 1998; Bialystok & Martin, 2004). They compared the attentional processing of 30 Chinese–English bilingual children and 30 monolingual English-speaking children. Children were between 3 and 7 years old. In one task, children were shown an object and a card on which the object's name was written. Children were then distracted by two toy bunnies. The bunnies caused the card to be moved to a position under a different object. After the distraction and while the card was near the wrong object, children were asked what the card had said. The interviewer then pointed out the fact that the bunnies had disrupted things. The card was moved back to the original object, and children were asked again to report what the card said. In a second task, children were given a set of cards and asked to sort the cards into two piles. They were instructed to sort according to a characteristic displayed on the cards (e.g., color or shape).

When they completed the task, they were asked to sort the cards again using a different characteristic. Both tasks required children to exert attentional control during the tasks.

Carlson and Meltzoff (2008) found that bilingual children's advantage in attentional processing over monolingual children occurs in tasks in which there are conflicting attentional demands but not in tasks in which participants must control impulses—as when a response should be given after a short delay. Their study involved 6-year-old English–Spanish bilinguals. The bilingual advantage over monolinguals in attentional processing has been observed with infants as young as 7 months old (Kovács & Mehler, 2009) as well as in adults (Bialystok, Craik, & Luk, 2008; Costa, Hernández, & Sebastián-Gallés, 2008). The bilingual advantage over monolinguals that has been observed in healthy children and adults has also been observed among older adults with dementia. Bialystok, Craik, and Freedman (2007) compared the progression of symptoms in 93 bilinguals and 91 monolinguals. The results showed that bilinguals were diagnosed with dementia an average of 4 years later than monolinguals, suggesting that the bilingualism served as a protective factor.

Language Skills

Bilingual children also appear to be better judges of grammaticality than monolingual children. Galambos and colleagues (Galambos & Goldin-Meadow, 1990; Galambos & Hakuta, 1988) compared the ability to judge and to correct syntactically ungrammatical sentences in a group of Puerto-Rican Spanish–English bilinguals and English-speaking monolinguals. Both groups were from low-income families. Bialystok and colleagues (Bialystok, 1986, 1988; Bialystok & Majumder, 1998) pointed out that grammaticality judgments require the participant to simultaneously consider the structure and the meaning of the sentence. When the two aspects of the sentence are in conflict, as occurs when one is ungrammatical, the participants must resolve the conflict by directing attention to the appropriate aspect of the sentence. Following this reasoning, they view the task of one involving attentional processing. They compared the ability of bilingual and monolingual children between the ages of 5 and 9 to judge sentences that were anomalous in terms of either grammar or meaning. They found that bilingual and monolingual children performed similarly judging grammatically correct sentences with sensible meaning (i.e., they correctly labeled sentences such as *Apples growed on trees* as ungrammatical); however, bilinguals were better than monolinguals at judging grammatically correct sentences that did not have sensible meaning (i.e., accepting sentences such as *Apples grow on noses* as grammatically correct). The results have been replicated with Italian–English bilinguals (Ricciardelli, 1992) and Swedish–English bilinguals (Cromdal, 1999).

Studies have also found that children who have learned another language show gains in their knowledge of English language structure and vocabulary (Curtain & Dahlberg, 2004; Dumas, 1999). Other studies found that bilingual children performed better than monolingual children in English as well as other subjects, such as social studies and math (Andrade, Kretschmer, & Kretschmer, 1989; Armstrong & Rogers, 1997; Masciantonio, 1977; Kretschmer & Kretschmer, 1989; Rafferty, 1986; Saunders, 1998). Armstrong and Rogers (1997) found benefits in math performance for children after they had studied an L2 for only 90 minutes per week for just one semester.

What Brain Changes Are Associated With Bilingualism?

There has been a great deal of interest among neuroimaging researchers regarding the organization and functioning of the bilingual brain. Several studies have suggested that early learned languages and late learned languages are handled somewhat differently by the brain; later learned languages involve more frontal and more bilateral activation than early learned languages (Dehaene et al., 1997; Hahne & Friederici, 2001; Hernandez, Martinez, & Kohnert, 2000; Kim, Relkin, Lee, & Hirsch, 1997; Marian, Spivey, & Hirsch, 2003; Weber-Fox & Neville, 1999, 2001). Certainly, because this area of research is so new, there are many more questions than there are answers at this point. The excitement comes in knowing that in the coming decades, there are likely to be important discoveries made that have the potential to change the way that we think about and also engage in L2 learning.

So far, multiple research studies have shown that there are brain-related differences related to proficiency. Some studies, but not all, have found brain-related differences related to the age at which the bilingual began using the L2. A study by Wartenburger and colleagues (2003) found that there were differences in bilingual brain organization related to both language proficiency and the age at which one was first exposed to the L2. Chee, Soon, Lee, and Pallier (2004) found brain differences related to language proficiency, which were unrelated to the **age of acquisition**. Kim and colleagues (1997) compared brain activations during language processing for bilinguals who learned their L2 early in childhood or later in life. They had participants use each of their languages while their brain activation was recorded. They found that there was an area of activation in the left inferior frontal gyrus that was observed during the use of each language. For early bilinguals, the areas of activation during use of L1 and L2 overlapped. For late bilinguals, the areas of activation during the use of L1 and L2 did not overlap. For both groups, there was a second region of activation that did overlap during the use of L1 and L2; the region was located in the superior temporal gyrus. Some have questioned these results because language proficiency was not assessed and the participant groups may have differed in ways other than when they learned L2 (Abutalebi, Cappa, & Perani, 2005).

One of the most intriguing studies is that by Mechelli and colleagues (2004) who compared the densities of gray and white matter in the brains of monolingual adults and two groups of bilingual adults. One group of bilinguals had acquired an L2 before the age of 5; the other group had acquired an L2 between the ages of 10 and 15. Both groups of bilinguals reported using their L2 regularly. They found that in the left parietal cortex, the density of the gray matter was greater for bilinguals than for monolinguals. The density of gray matter in this area was greater for bilinguals who had learned the L2 early in childhood versus later in childhood. Last, in correlational analyses, the density of gray matter increased as the participants' reported proficiency increased and increased as the participants' age of acquisition decreased. Past research has shown that the particular area in the left inferior parietal cortex is involved in verbal fluency (Poline, Vandenberghe, Holmes, Friston, & Frackowiak, 1996; Warburton et al., 1996). These results suggest that the differences observed between bilingual and monolingual speakers may one day be related to specific brain-related changes that occur when one begins to learn an L2.

▲ Photo 7.3 High school students are studying in a typical L2 class. Did you take an L2 in high school?

A number of recent studies have found that specific brain regions are activated when a bilingual switches languages (Hernandez et al., 2000; Price et al., 1999; Quaresima, Ferrari, van der Sluijs, Menssen, & Colier, 2002; Rodriguez-Fornells, Rotte, Heinze, Nösselt, & Münte, 2002; Rodgriguez-Fornells et al., 2005). The regions include the dorsolateral prefrontal cortex, inferior frontal cortex, anterior cingulate, and supramarginal gyrus. In these studies, bilinguals carried out two language tasks (e.g., a word-reading task and a translation task). The language in which the words were presented was either kept the same or varied so that every other word appeared in the other language. The results suggest that the differences in attentional processing that have been observed between bilingual and monolingual children may also be related to specific brain-related changes that occur when one begins to learn an L2.

Learning Second Languages

What Teaching Methods Are Used in Language Classes?

There are multiple ways for one to become bilingual. For many, learning an L2 involves some classroom instruction. Over the last 150 years, numerous methods of L2 teaching have been developed (Krause, 1916; Richards & Rodgers, 2001). Some methods have become brand names, sold to those who would like to learn a language at home (e.g., Berlitz and Rosetta Stone). Methods typically differ in the extent to which the L1 is used during L2 instruction and the extent to which grammar is explicitly taught. The **grammar-translation method** (also called the traditional method) utilizes the L1 and also explicitly teaches students the grammatical rules of L2. At the other extreme, there is the **direct method,** which involves students learning the L2 without the use of the L1. The grammar of L2 is learned inductively, through exposure to L2 sentences.

In most L2 classrooms in the United States, instructors use the grammar-translation method. Lessons are focused on having students memorize grammatical rules and vocabulary. They are also asked to translate from L1 into L2 and vice versa. Exercises involve students' practicing writing or speaking sentences containing verbs in their different tenses (e.g., present tense, past tense, future tense). In languages in which nouns must be marked with suffixes indicating gender and/or number, exercises will focus on students using the appropriate adjective–noun or subject noun–verb agreement.

There are numerous methods that place an emphasis on developing speaking and listening skills but involve varying degrees of grammar instruction. For example, the audio-lingual method (Richards & Rodgers, 1987) emphasizes the use of spoken language but also includes grammar exercises. It is based on the principles of behaviorism. Students are instructed to avoid errors at all costs. The instructor, who is a native speaker, is viewed as the model to emulate. The communicative approach emphasizes the idea that language is primarily for communication and the expression of meaning. Students are encouraged to use language with others. The long-term goal for students is to achieve communicative competence. The immersion method involves having students use only the L2 to learn the vocabulary, grammar, and pragmatics of the L2. They aim to acquire the language through interactions with speakers of the language. Classes may be taught with complete immersion (i.e., 100% of instruction is conducted in L2) or partial immersion (i.e., 50% of instruction is conducted in L2).

There are few empirical studies comparing the effectiveness of the various teaching methods. The research that has been done has been conducted in Canada, where English–French bilingualism has been a priority since the 1960s and the immersion method became popular. Numerous studies comparing students' performance in classrooms using the direct method versus the immersion method have been conducted in Canada (Baker, 1993). In the 1960s, Canada implemented immersion language programs in schools to address perceived inequities between French and English speakers in Quebec (Genesee, 1987). Programs were classified as early, middle, or late immersion. Early immersion classes began when children were 5 or 6 years old. Middle **immersion programs** began when children were 9 or 10 years old. Late immersion programs began when children were 11 or 12 years old. Studies showed that students in early immersion programs performed less well on literacy skills, such as spelling and punctuation in L2 compared with students in classrooms in which the L2 is taught using the direct method. Differences diminished over time. The evidence suggests that children can learn an L2 comparably well in either type of classroom. Following the ruling, there have been a number of changes to the law, most of them allowing schools flexibility in meeting the needs of their students.

Do All Countries Have Bilingual Education?

The extent to which L2s are used in educational settings is often determined by the law of the land. In the United States, the Bilingual Education Act was established in 1968 (Stewner-Manzanares, 1988). Originally introduced as a bill by Texas senator Ralph Yarborough, its aim was to fund programs for students entering school without adequate English language skills. The concern in Texas was schools servicing Spanish-speaking students. The Bilingual Education Act of 1968 focused generally on the needs of schools educating bilingual students. Financially, the act was a tremendous help to school districts, as it provided $7.5 million for programs, teacher training, and curriculum development. Those programs that could demonstrate success were funded for 5 years.

In 1974, the Supreme Court ruling in *Lau v. Nichol* mandated **bilingual education.** The case was a class action suit that began because an attorney had heard about a client's son who was performing poorly in school because he could not speak English. The attorney filed the suit against the San Francisco Unified School District, claiming that the Chinese students in the district (approximately 1,800) were not being provided with equal education because of the lack of bilingual education in the school. The case was lost at the lower court level but was taken all the way to the U.S. Supreme Court, where the attorney prevailed.

Canada's Official Languages Act became law in 1969 (Office of the Commissioner of Official Language, n.d.). The law established English and French as official languages for government and education and affirmed the equality of the two languages in all venues. It also affirmed the right of the individual to receive services in either English or French. Canada has supported immersion programs for students learning French. Students without prior experience with French usually can begin immersion classes in kindergarten or first grade and complete their secondary education completely in French. The availability of the program depends on the province. Some regions provide French courses starting in the fifth grade. Other regions offer French-immersion from kindergarten through ninth grade. In regions in which French is the dominant language (e.g., Quebec), students may enter similar English immersion programs. The indigenous languages of Canada have also been part of Canada's bilingual education efforts. Programs have been developed for Blackfoot, Cree, Mohawk, Ojibwe, Mi'kmaq, Inuinnaqtun, Inuktitut, and the Pacific Coast Salish languages. The development of bilingual programs for the indigenous languages was particularly important because of historical oppression of indigenous Canada. In the 20th century, many were forced into residential schools in which they were treated poorly and forbidden from using their native language. Similar residential schools were operated in the United States for Native Americans (Child, 2000).

Around the world, children's access to bilingual education varies widely. In Europe, children typically have the opportunity to learn an L2 earlier than children in the United States. For example, in Belgium, there are three official languages: (1) Dutch, (2) German, and (3) French. By law, children have the right to be educated in one of these three languages. The language of instruction varies by geographic region. In the Flanders region, the language of instruction is Dutch. In the Waloonia region, the language of instruction is French. Throughout the country, English is frequently taught and may be required. In Belgium, as in most of Europe, children begin their study of an L2 (i.e., a language different from that in which general instruction is provided) in the elementary school grades (Eurydice, 2005).

How Does First Language Influence Second Language Acquisition?

Regardless of the method used to learn an L2, one can expect to make errors. One may find that characteristics of one's L1 influence how one learns and/or uses one's L2. The term ***language transfer*** has been used to describe circumstances in which one carries over a language rule or structure from L1 to L2. **Positive language transfer** occurs when the rules in L1 are the same as in L2. Transfer helps the speaker to generate a correct usage in L2. **Negative language transfer** occurs when the rules of L1 are different from those in L2 and typically results in speakers making errors when using L2. The most common form of negative language transfer is speaking with an accent that reflects the phonological rules of the L1. For example, native speakers of Japanese who learn English typically have great difficulty pronouncing the sounds /r/ and /l/, because in Japanese these sounds are not distinct phonemes. Similarly, native speakers of English have great difficulty pronouncing many vowels in French, which are distinct phonemes in French but are not in English.

Language transfer may also be observed in vocabulary. When a bilingual's two languages share vocabulary words, one can benefit from the positive language transfer. Translation equivalents that are similar in pronunciation and in written form are called **cognates.** Some L2 words may appear similar to L1 words in sound and written form but differ in meaning, such as the Spanish word *embarazada,* which means pregnant instead of embarrassed, and the Spanish word *balde,* which means bucket. Such words are called false friends or **false cognates.** Another example of how vocabulary can be involved in language transfer comes from an interaction with a friend of mine whose L1 was not English. He once walked up to me holding his arm. He had clearly injured it somehow. Before I could ask him, "What's happened to your arm?" he said, "I hurt my hand." I was very confused. "Hand?" I asked. "Don't you mean your arm?" He paused and then realized his error. "In my language, the word for arm and hand is the same word. I make that mistake a lot." He once also said, "Something's wrong with my finger," while he was holding his thumb. On that occasion, I was able to infer that in his native language, the word for thumb is the same word as for the other fingers.

The use of pronouns can often be influenced by language transfer. If one's L1 uses the same word to refer to *he* and *she* but one's L2 has different words to refer to *he* and *she,* one may find that one comes to rely on one of the L2 words for both pronouns. In Finnish, the word *hän* is used to refer to males and females. When Finnish native speakers learn English, they sometimes may use the pronoun *he* in English to refer to female antecedents, as in **I met Mary; he is nice* (the asterisk indicates that the sentence is ungrammatical). In Chinese, the pronouns for *he* and *she* are pronounced the same but spelled differently. For native speakers of Chinese acquiring English (or other languages in which there are different words for *he* and *she*), pronoun errors in L2 tend to occur.

Language transfer may also be observed in the syntactic and morphological rules that speakers apply. Native speakers of English frequently transfer the word order of English when producing German clauses. An example of an English sentence containing two clauses is provided in 3a. An example of a grammatically correct German sentence is provided in 3b. In 3b, the adverb *heute,* which means *today,* is ordered before the verb in German, because in German, the main verb usually appears in clause-final position. The relative clause in 3c is the grammatically incorrect German form sometimes produced by English native speakers (the asterisk indicates that the sentence is ungrammatical). The positioning of the adverb in the relative clause follows the verb rather than preceding it.

3.	a.	I know when she arrives today.	English
	b.	Ich weiß, wann er heute ankommt.	Correct German
	*c.	Ich weiß, wann er ankommt heute.	Incorrect German

L2 learners may also transfer morphological rules from L1 to L2. For example, in English, one forms a comparative adjective by adding *-er* if the adjective is one syllable or ends in *-y,* as in *taller* and *happier.* For longer words, the word *more* is placed in front of the adjective, as in *more important* and *more intelligent*. Superlative adjectives in English are formed similarly, by adding *-est* to an adjective or preceding the adjective by the word *most*, as in *tallest*, *happiest*, *most important*, and *most intelligent*. Some languages, such as Spanish, form comparative and superlative adjectives using one rule. The word *mas* is placed before an adjective to form the comparative and the words *le mas* (or *la mas*) are placed before the adjective to form the superlative. When Spanish native speakers learn English, they often overuse the words *more* and *most* when forming comparative and superlative adjectives in English. In recent research, Kennison and Bowers (2011) found that the overuse of *more* and *most* in comparative and superlative adjectives in English by Spanish native speakers was negatively correlated with how long participants had resided in the United States, suggesting that exposure to the L2 in everyday life was related to fewer errors in L2 resulting from language transfer.

The prior research on language transfer explains why language learners face greater difficulty in mastering an L2 when the rules of the language differ a great deal from the rule of their native language. Conversely, L2 acquisition may be easier when one's L1 and L2 share many of the same grammatical distinctions and rules. For example, learning Italian when one's L1 is Spanish may be easier than learning Chinese when one's L1 is Spanish. As you learned in Chapter 1, languages belonging to the same language family are more similar in grammatical rules than languages belonging to different language families.

While many students in the United States struggle to learn an L2 in high school and/or in college, there are some individuals who appear to have a knack for languages. Those who have mastered three, four, or five languages usually are regarded has being particularly gifted in learning languages. The term ***polyglot*** is used to refer to individuals who speak many languages. They are able to learn multiple languages rapidly and with apparent ease. It is difficult to estimate the percentage of the population who possess above average language-learning abilities. Text Box 7.2 describes a particularly prolific polyglot: the German Emil Krebs.

Life With Two or More Languages

Those who use more than one language in daily life may be described as **balanced bilinguals**, which means that they are equally proficient in both of their languages. Most often, bilinguals report having a dominant language or a language that they feel most proficient in and use more often. The dominant language is not always L1. Some individuals may reach a high level of proficiency in two languages and end up living in a setting in which L2 is used predominantly. For example, someone raised in Canada becomes proficient in both English and French. After college, the person gets a job in France and works there for several decades, using mostly French in daily life. English is used only to speak to family or friends, which occurs only several times a month.

What Factors Predict Proficiency?

Bilinguals certainly differ from one another in terms of proficiency. Some bilinguals become proficient in speaking, reading, listening, and writing L2; however, others may not master all aspects of L2 usage. In research with bilinguals, researchers mostly rely on

Text Box 7.2 Extraordinary Individuals: Emil Krebs: An Extraordinary Polyglot

Emil Krebs (1867–1930) was born in Germany, the son of a carpenter. By the end of his life, he had become proficient in speaking and writing 60 languages (Matzat, 2000). He had studied 120 others. He is aptly described as a polyglot, which is a term used to refer to someone who knows many languages. In one written account, Krebs was described as learning languages rapidly. For example, when he studied Armenian, he mastered it in about 9 weeks (von Hentig, 1962). He spent 2 weeks learning the grammar. He studied Old Armenian for 3 weeks and focused on the spoken version of the language for the remaining 4 weeks. Between the ages of 13 and 17, he attended gymnasium (roughly equivalent to a college preparatory high school) and studied Latin, Hebrew, French, and classical Greek. By the time he completed his gymnasium studies in 1887, he spoke 12 languages, including Modern Greek, English, Italian, Spanish, Russian, Polish, Arabic, and Turkish. In that year, he entered the Berlin School of Law, where he continued studying languages. In 1890, he passed a translator's exam for Chinese. In 1913, he went to China, where he worked as a translator for the government until 1917, when diplomatic relations between Germany and China dissolved. One can find Krebs's personal library of approximately 3,500 books and his writings in over 120 languages at the National Archives in Washington, D.C. Among the other languages that he knew were Egyptian, Ainu, Albanian, Armenian, Burmese, Georgian, Hebrew, Japanese, Javanese, Korean, Manchurian, Mongolian, Nivkh, Persian, Sanskrit, Syrian, Tibetan, and Urdu. In his life, he achieved notoriety for his success with languages. As a result, following his death, his brain became part of the collection of the C. and O. Vogt Institute for Brain Research at the Heinrich Heine University in Düsseldorf. In a recent research study, Krebs's brain was compared with 11 others obtained from individuals without exceptional abilities (Amunts, Schleicher, & Zilles, 2004). The researchers found that Krebs's brain was architecturally different from the others in the areas of the left hemisphere associated with language processing.

▲ Photo 7.4 Emil Krebs lived between 1867 and 1930 and is known for being an amazing polyglot. Which language would you start learning today if you were assured that you could master it?

self-report measures of proficiency (Marian, Blumenfeld, & Kaushanskaya, 2007). Studies have found that self-reported measures of proficiency are significantly correlated with bilinguals' performance on standardized language tests. For example, in a study of Spanish–English bilinguals, Delgado, Guerrero, Goggin, and Ellis (1999) found that proficiency was significantly correlated with participants' scores on the Woodcock–Muñoz Language Survey (Woodcock & Muñoz-Sandoval, 1993). Correlations were stronger for L1 than L2. Proficiency in L2 speaking and understanding were not correlated with performance on the language survey. Other studies have investigated the relationship between self-reported proficiency and performance on language processing tasks. Bahrick, Hall, Goggin, Bahrick, and Berger (1994) observed stronger correlations between proficiency and vocabulary tasks than between proficiency and oral comprehension tasks.

Numerous studies have investigated factors that are related to L2 proficiency. One of the strongest predictors is age of acquisition (Flege, MacKay, & Piske, 2002; Hyltenstam & Abrahamsson, 2003; Johnson & Newport, 1989). As you learned in Chapter 1, since the 1960s, language researchers have recognized that childhood is the prime time for learning language (Lenneberg, 1964, 1967). If language is learned after puberty, proficiency may not be achieved. For bilinguals, those learning L2 early in life are likely to attain a higher level of proficiency than those who begin learning L2 later in life (Johnson & Newport, 1989; Kovelman, Baker, & Petitto, 2008).

Other studies have considered the length of time that L2 speakers have lived in the geographic region where their L2 is used predominantly (Birdsong, 2005; Espenshade & Fu, 1997; Flege, Yeni–Komshian, & Liu, 1999; Genesee, 1985; Stevens, 1999). For example, Flege and colleagues (1999) investigated the relationship between how long bilinguals had resided in the United States and their performance on grammatical tasks in English. They found that performance was better for those who had been in the United States for the longest period of time. For children who begin learning an L2 in school because the L2 is the language of instruction, research suggests that most are able to become proficient in the language in 5 years (Hakuta, Butler, & Witt, 2000). Conger (2009) conducted a longitudinal study of schoolchildren in New York City for whom English was an L2. He found that by the end of the first year of school, between a quarter and a third of students had become proficient in English. After 3 years, half of the students had become proficient. The results showed that children who entered school at younger ages reached proficiency faster than children who had entered school at older ages. Proficiency was influenced by demographic variables, such as sex, socioeconomic status (SES), and ethnicity.

Research by Hakuta, Bialystok, and Wiley (2003) suggests that aging may be a factor in L2 proficiency. In their study, they used 1990 census data in the United States to investigate the factors that were related to respondents' self-reported L2 ability. The census data contained responses from 2.3 million immigrants who spoke English as an L2 and Spanish or Chinese as an L1. The results showed that several factors were related to respondents' proficiency, including age of immigration, socioeconomic level, and amount of formal education. They found that in both groups of speakers (i.e., L1 speakers of Spanish and L1 speakers of Chinese), as the years since their immigration to the United States increased, proficiency in English declined. Stevens (2004) suggested that different results might be observed using different sampling techniques and statistical procedures. Wiley, Bialystok, and Hakuta (2005) reported the results of additional analyses, following the recommendations of Stevens (2004). They observed results similar to those reported by Hakuta et al. (2003). They concluded that the most plausible explanation for the steady decline in L2 proficiency was aging, but other factors may also play a role (e.g., amount and quality of daily experience with the language).

Over the past three decades, an impressive amount of research has investigated language processing in bilingual individuals. In the past decade, researchers have begun to report studies in which processing differences for bilinguals and monolinguals have been observed. For example, in picture-naming tasks, research has shown that bilinguals are generally slower than monolinguals, even when bilinguals named pictures using their L1 (Ivanova & Costa, 2008). Roberts, Garcia, Desrochers, and Hernandez (2002) found that bilinguals made more errors in picture-naming than did monolinguals. Rosselli and colleagues (2000) found that on tasks of fluency in which one must produce as many words as possible within a minute that belong to a specific semantic category (e.g., fruits or animals) or begin with a particular letter (i.e., *s, a,* or *f*), Spanish–English bilinguals did not perform as well as monolinguals who were the same age. Rogers, Lister, Febo, Besing, and Abrams (2006) observed that bilinguals' ability to perceive words embedded in white noise was poorer than monolinguals'. Bilinguals also experience TOT states more often than monolinguals (Gollan & Acenas, 2004; Gollan & Silverberg, 2001).

A topic that has received relatively little attention is how bilinguals use both languages in school and work. Sometimes, one may acquire information in one language but have to use that information in an L2. There is preliminary research suggesting that the knowledge that can be gained in one situation and applied in another is not dependent on the language that is used to gain the information. This research is described in Text Box 7.3.

How Are Multiple Languages Stored in Memory?

Among researchers interested in bilingualism, there is still a debate regarding whether the languages that a bilingual knows are stored in memory in a single, shared memory system (Francis, 1999; Klein, Milner, Zatorre, Zhao, & Nikelski, 1999) or whether there are separate memory systems, one for each language (Dehaene et al., 1997; Durgunoğlu & Roediger, 1987).

Numerous studies involving adult bilinguals have revealed how the knowledge about words from multiple languages is organized in memory. Of particular interest is whether words in the language being used are selectively accessed or whether all the words that the bilingual knows are activated, regardless of the language being used. Numerous studies have shown that there are differences in how quickly bilinguals can retrieve L2 and L1 words during language processing tasks. Some studies have observed differences using the bilingual-translation task. Bilinguals view a series of words in one language and are instructed to pronounce translation equivalents as rapidly as possible. Kroll and Stewart (1994) have observed that participants can translate an L2 word into L1 faster than they can translate an L1 word into L2. They proposed that memory links between L2 and L1 words are stronger than links between L1 and L2 words because of the processes involved in L2 learning. One more frequently associates an L2 word with an L1 word than vice versa.

Text Box 7.3 Research Discovery: Bilingual Problem Solving

In everyday life, bilinguals often acquire knowledge from one language, either via reading or listening, and then apply the knowledge to a task in which another language is used (García, 2008). Very little is known about how knowledge transfer of this type occurs. Researchers have begun to investigate knowledge transfer by bilinguals in studies involving problem solving in which there is the opportunity for participants to apply a previously encountered solution to a subsequent problem. The research utilizes the analogical transfer paradigm (Gick & Holyoak, 1980, 1983; Holyoak & Koh, 1987; Holyoak & Thagard, 1989; Spellman & Holyoak, 1992). When participants successfully apply a solution to a subsequent problem, they are viewed as having drawn an analogy between the two problems and having transferred solution of the first problem to the subsequent problem. Thus far, there have been only three bilingual studies using the analogical transfer paradigm (Bernardo, 1998; Francis, 1999; Fukumine & Kennison, 2011). The results of these studies indicate that bilinguals can acquire knowledge from a source problem in one language and transfer it to solve a problem in another language.

In a large study, Francis (1999) investigated how Spanish–English bilinguals solved problems that were written either in the same language or in different languages. She tested four groups of participants. For two groups, the source problem and the target problem were in the same language, either L1 or L2. For the other two groups, the source problem and the target problem were in different languages. One group received the source problem in L1, and the target problem in L2. The other group received the source problem in L2 and the target problem in L1. The results showed that analogical transfer occurred comparably often when the problems were presented in the same language as when they were presented in different languages. Across conditions, analogical transfer occurred most of the time (i.e., over 70%). The most surprising result in the study was that participants' language proficiencies were not strongly related to problem-solving performance. Consequently, Francis concluded that the knowledge that is transferred during problem solving across languages is stored in memory in a manner that is language-free or language-neutral.

Francis's (1999) results and those similar to them (Fukumine & Kennison, 2011) suggest that in settings in which bilinguals are asked to acquire knowledge rapidly, there is no reason to limit them to materials in one language. Educators and policymakers are likely to believe that using materials prepared in different languages might create barriers to learning. The research suggests the opposite. Once a bilingual acquires knowledge from a text or speech, it appears to be stored in memory in a representation that is not linked to any language. When the goal is acquiring knowledge, bilinguals' performance is likely to be facilitated by allowing them to access any useful materials, regardless of the language in which they are prepared.

Kroll and colleagues (Kroll & Stewart, 1994; Kroll, Van Hell, Tokowicz, & Green, 2010) proposed the revised hierarchical model to describe the organization of bilinguals' memory for L1 and L2 words. Figure 7.1 displays the model, showing that conceptual representations are connected to representations for L1 words and to representations for L2 words. The representations for L1 words are connected to the representations for L2 words and vice versa. The lines that connect the boxes indicate these connections. The darker the line in the figure, the stronger the link is believed to be. Research by Bowers and Kennison (2011) compared bilingual translation for words learned early in childhood and words learned after the age of 8. They found translation from L2 to L1 was faster than L1 to L2 only for L1 words that were learned early in childhood—perhaps because memory links are stronger for words learned early in childhood than for words learned later in childhood.

Critics of the revised hierarchical model (Brysbaert & Duyck, 2010) believe that the Bilingual Interactive Activation Plus (BIA+) model (Dijkstra & Van Heuven, 2002) emphasizes the interconnectedness of the bilinguals' language knowledge, particularly when recognizing words. Numerous studies have shown that when bilinguals recognize words, they retrieve and use information about the words from both of their languages (Lam & Dijkstra, 2010). For example, studies have observed evidence that Dutch–English bilinguals who view the letters *work* briefly activate not only similar English words (e.g., *word* and *cork*): they also activate similar Dutch words (e.g., *werk*, *wolk*, and *worp*). The studies require participants to hit a key on a keyboard as quickly as possible when deciding whether a group of letters (e.g., *work*) is an actual word or is a nonword (e.g., *wark*). Research with monolinguals has shown that the time that participants take is related to the number of words that the person knows that look similar to the target

Figure 7.1 Kroll and Stewart's revised hierarchical model proposes that bilingual memory involves separate representations for L1 and L2 words. Both types of representations can activate the meaning or conceptual representation. The solid arrows indicate stronger memory links than the dotted arrows

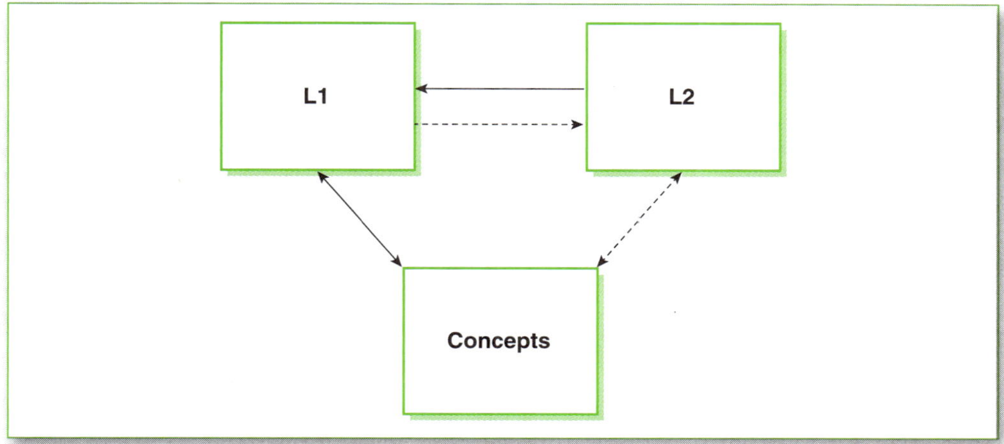

Source: Kroll and Stewart (1994).

Table 7.2 Examples of Translation Equivalents (Noncognates), Cognates, and False Cognates for German and English

	German Word	English Translation
Noncognates	Zorn	Anger
	Brust	Chest
	Hund	Dog
	Saft	Juice
	Leute	People
	Frage	Question
Cognates	Tiger	Tiger
	Steak	Steak
	Garten	Garden
	Freund	Friend
	Gitarre	Guitar
	Papier	Paper
False Cognates	Gang	Hallway
	Mantal	Coat
	Tag	Day
	Kind	Child
	Bad	Bath
	Teller	Plate

Source: Friel and Kennison (2001).

word. When the target word is similar to many other known words of the same length, participants respond more quickly than when the target word is not similar to other known words of the same length (Andrews, 1989; Grainger, O'Regan, Jacobs, & Segui, 1989; Grainger & Segui, 1990). In studies with bilinguals, participants generally respond more quickly when a target word is similar to a word in the other language than when it is not.

The jury is still out regarding which of the two theories of bilingual memory best describes bilingual language processing. Proponents of both theories have found that there are some types of words that are particularly useful in testing the models. The types of words vary in terms of how similar they are to L1 and L2 words in terms of sound and memory. Most L2–L1 translation equivalents are completely dissimilar in terms of pronunciation and written form, as in the English word *apple* and the Spanish equivalent *manzana*. In contrast, some L2 and L1 translation equivalents are highly similar, as in the English word *map* and the Spanish equivalent *mapa*. Such cases are referred to as cognates. Bilinguals—particularly less proficient bilinguals—have been found to translate cognates more quickly than noncognates (de Groot, 1992). Research by Lotto and de Groot (1998) found that in an L2 learning task, participants learned cognates faster than noncognates. A second type of word that has been particularly useful to researchers has been called false cognates or false friends. Beauvillain and Grainger (1987) showed that when bilinguals process a false cognate, they retrieve multiple meanings for the word from L1 and L2. For example, when French–English bilinguals view *coin*, they activate both the translation equivalent in English *corner* and the English word *coin*, which is related to money. In order to better understand how words from different languages can be similar in sound and meaning, refer to Table 7.2, which displays samples of translation equivalents (noncognates), cognates, and false cognates for German and English.

Can One Forget One's First Language?

Among middle-aged and older bilinguals, there are some who may feel that they are not entirely proficient in any language. They may have acquired English and worked in an

English-speaking environment for many years. As a result, they may have grown unaccustomed to speaking their native language if it is not used daily in their home. Without the opportunity to speak the native language regularly, one may become rusty at speaking the native language. One example of this was reported by Isurin (2000); a native speaker of Russian was adopted by English-speaking parents at the age of 9. After that point, the child did not speak Russian. After 1 year, the child's vocabulary knowledge decreased by 20%. It is common for children who are adopted before the age of 9 and do not use their native language in their new homes to have little or no memory of their L1 (e.g., Pallier et al., 2003).

Several studies have found that bilinguals can experience some loss of their L1 over time (de Bot, 1999; Levy, McVeigh, Marful, & Anderson, 2007; Linck, Kroll, & Sunderman, 2009; Seliger & Vago, 1991). Linck and colleagues (2009) investigated the language performance of Spanish–English bilinguals in Spain. One group of bilinguals was learning Spanish in an immersion experience. The other group of bilinguals was learning Spanish in a classroom setting. Participants' comprehension and production abilities were measured in L2. Those participants learning L2 through immersion outperformed those learning in a classroom. The results also showed that L1 performance was worse for participants learning through immersion than for those learning in a classroom. The authors concluded that the learning of L2 inhibited L1.

A recent study by Levy and colleagues (2007) reached similar conclusions in a study with Spanish–English bilinguals. They utilized a procedure known as retrieval-induced forgetting (RIF) (Anderson, Bjork, & Bjork, 1994). In the typical RIF experiment, participants were instructed to study category-exemplar pairs (e.g., fruits-apple, fruits-pear, drinks-whiskey). After the pairs were learned, they were then instructed to practice remembering half of the items (i.e., every other pair of items in the previous list). Last, they were instructed to recall all of the words in the word pairs. Before performing the RIF task, they were asked to name pictures either in Spanish (L2) or in English (L1). The number of picture-naming trials was varied. The results showed that performance in the memory task in English was worse when participants had previously named pictures in Spanish. Participants who struggled the most with Spanish pronunciation showed the most interference in English on the memory task. Their third experiment showed that the sounds of L2 words, rather than the meaning of the L2 words, interfered with L1 words.

Summary and Theoretical Implications

Children who are reared in homes in which multiple languages are spoken develop language along the same time frame as children who are reared in homes in which only one language is spoken. Research shows that the perceptual abilities of bilingual and monolingual infants may differ early in childhood; however, it is unclear whether these differences impact later language development. Some studies have shown that bilingual children may have smaller vocabularies than monolingual children. Related research shows that other factors influence vocabulary size, such as sex, SES, and whether the L2 is a culturally dominant or minority language. An increasing number of studies have shown that bilingual children outperform monolingual children in tasks requiring attentional processing.

There are many methods that have been developed to teach L2s. Few studies have compared how well learners fare using the different methods. When languages share vocabulary and grammatical rules, learning is easier than when languages do not share vocabulary and grammatical rules. Researchers debate about the extent to which the words in bilinguals' two languages are stored together or separate in memory. In language processing studies involving adult bilinguals, they tend to be slower and more error prone than monolinguals. Language studies have shown that the L1 is vulnerable to interference from the L2.

The four theoretical approaches to language development each have potential to account for some aspects of L2 acquisition and bilingualism. The behaviorist approach predicts that learning an L2 would be facilitated through the mechanisms of classical and operant conditioning. The social-interactionist approach predicts that learning an L2 would be facilitated when the learning occurs in a supportive social environment, which is characteristic of a total immersion language learning environment. The statistical learning approach is likely to have compelling explanations for language transfer effects. Last, because of the generative approach's claim that aspects of language learning are innate, the approach is compatible with the evidence that there is a critical period for language acquisition. With the current state of knowledge on the topic of L2 acquisition, we are not able to judge any view superior to any other. As research continues to be conducted, it may be possible to determine whether any of these approaches can completely account for how L2s are learned and processed.

KEY TERMS

age of acquisition
balanced bilinguals
bilingual education
bilinguals
code-switching
cognates
direct method
dominant language
false cognates
first language (L1)
free morpheme constraint
fusion hypothesis
grammar-translation method
immersion programs
language transfer
monolingual
negative language transfer
polyglot
positive language transfer
second language (L2)
sequential bilingualism
simultaneous bilingualism
translation equivalents
trilingual

REVIEW QUESTIONS

1. What is the difference between simultaneous and sequential bilingualism?
2. In terms of perceptual abilities, how do bilingual and monolingual infants differ?
3. In terms of vocabulary size, how do bilingual and monolingual children differ?
4. What factors are related to vocabulary size in bilingual children?

5. What is code-switching? What evidence is there that code-switching involves grammatical rules?
6. What are the cognitive benefits for children who know more than one language?
7. What evidence is there that attentional processing differs for bilingual and monolingual children?
8. What differences have been observed in brain-imaging studies between bilingual and monolingual brains?
9. What area of the brain has been identified as involved when a bilingual switches from one language to another?
10. What is the evidence that one's L2 interferes with one's L1?
11. What is language transfer? What is the difference between positive language transfer and negative language transfer?
12. How does the performance of adult bilinguals and monolinguals on language processing tasks differ?
13. What are the most common ways in which people learn L2s and L3s? Has research shown that one or more methods are more effective than others?
14. What has research shown about the effectiveness of immersion programs as compared with other programs?
15. How did bilingual education come about in the United States in the 1960s and 1970s?
16. What is a polyglot? How did the brain of one famous polyglot, Emil Krebs, differ from the brains of non-polyglots?
17. What is a cognate? What has research shown about how bilinguals process this type of word?
18. What is a false cognate? What has research shown about how bilinguals process this type of word?
19. What evidence is there that bilinguals activate words from both languages when they are recognizing words?
20. What are the two competing models of bilingual memory? How do they differ?

RECOMMENDED READING

Altarriba, J., & Heredia, R. R. (2008). *Introduction to bilingualism.* New York: Psychology Press.

Bialystok, E., & Hakuta, K. (1995). *In other words: The science and psychology of second language acquisition.* New York: Basic Books.

Gardner-Chloros, P. (2007). *Codeswitching.* Cambridge, UK: Cambridge University Press.

Grosjean, F. (2010). *Bilingual: Life and reality.* Cambridge, MA: Harvard University Press.

Hakuta, K. (1997). *The mirror of language: The debate on bilingualism.* New York: Basic Books.

Kroll, J., & de Groot, A. (2009). *Handbook of bilingualism.* Oxford, UK: Oxford University Press.

Myers-Scotten, C. (2005). *Multiple voices: An introduction to bilingualism.* New York: John Wiley.

RECOMMENDED FILMS

Jeffcoat, J. (Director). (2006). *Outsourced* [Motion picture]. United States: ShadowCatcher.
Nava, G. (Director). (2008). *El norte* [Motion picture]. United States: Criterion Collection.
Pakula, A. (Director). (1982). *Sophie's choice* [Motion picture]. United States: Universal Pictures.
Schrieber, L. (Director). (2006). *Everything's illuminated* [Motion picture]. United States: Warner Home Video.
Wang, W. (Director). (1982). *Chan is missing* [Motion picture]. United States: Koch Lorber Films.

SUGGESTIONS FOR CLASS PROJECTS

1. Investigate public opinion about bilingual education. Work alone or in groups to develop a brief questionnaire. Questions may relate to the public's understanding of the federal law that mandates bilingual education, the ways in which local school districts implement bilingual education, and attitudes about the advantages and disadvantages of bilingual education. Determine whether respondents' opinions are changed after they learn about the recent research demonstrating that children who acquire more than one language experience cognitive benefits.

2. Survey your peers in other classes about their views on whether an infant should be raised to be bilingual from birth or whether parents of an infant who speak different languages should pick a single language of the household to use with the infant. The questionnaire can be designed to assess others' opinions about which approach they would use if placed in that situation and also to assess why they believe that option to be the best. The reasons that respondents give may reveal beliefs about bilingualism that may or may not be supported by empirical research. Discuss either in a class presentation or in a paper what commonly held views are or are not supported by empirical research.

3. Survey up to five peers about their experiences trying to learn an L2. Be sure to ask respondents their age when they began trying to learn the L2; whether they learned in a formal classroom setting or in a naturalistic, immersion setting; and the level of proficiency that they achieved in the L2. Share your findings with the entire class. The class can then analyze the group results, speculating about the role of age of acquisition and method of acquisition in success in learning an L2.

For additional ancillary resources, please visit the companion website at www.sagepub.com/kennison.

- Video Links
- Audio Links
- Web Resources
- Internet Activities
- Flashcards
- Web Quizzes

CHAPTER 8

LANGUAGE AND THOUGHT

The relationship between language and thought has intrigued scholars for hundreds, if not thousands, of years. Many have wondered to what extent the language that one speaks determines *what* one can think and/or *how* one thinks. We know that the question was recognized by the ancient Greeks. The Greek historian Herodotus attributed the difference in the thinking

▲ Photo 8.1 Members of a Masai tribe gather beneath a tree at sunset in Kenya.

of Greeks and Egyptians to the fact that the languages were written in different directions. Greek is written left to right, and Egyptian is written from right to left (Fishman, 1980; see also Hunt & Agnoli, 1991). More than a millennium later, in the 19th century, the German philosopher Wilhelm von Humboldt (1767–1835) suggested that a nation's language and spirit are integrally linked (von Humboldt, 1841–1852/1988; see also Miller, 1968). About a century later, in the early 20th century, the anthropologist Edward Sapir (1884–1939) and his student Benjamin Lee Whorf (1897–1941) proposed that language has a profound effect on human thinking, determining entirely how one conceptualizes the world. The view became known as the **Sapir–Whorf hypothesis** (also called the **Whorfian hypothesis**). In this chapter, you will learn about the origins of the Sapir–Whorf hypothesis and how researchers have tested its predictions in empirical studies. You will learn about the characteristics of the languages that inspired Sapir and Whorf to develop their views. You will also learn about the numerous studies that have attempted to determine how language influences cognitive processes. Last, you will learn about how children's acquisition of concepts is influenced by their lexical development.

The Sapir–Whorf Hyphothesis

What Were the Origins of the Sapir-Whorf Hypothesis?

Edward Sapir was an anthropologist and linguist who was born in Germany in 1884 but who immigrated to New York City with his family in 1888 (Koerner & Koerner, 1985). He received his bachelor's and master's degrees from Columbia, as well as his PhD in anthropology in 1909. For his PhD, he studied under Franz Boas, who was at the time a well-known anthropologist. While at Columbia, Sapir specialized in the study of the indigenous languages of North America. After receiving his PhD, Sapir's first job was as a researcher in California, documenting endangered, indigenous languages. Later, he served as the director of the Geological Survey of Canada from 1910 to 1925. He went on to teach anthropology at the University of Chicago and Yale University. He became the head of the Anthropology Department and worked until his death in 1939. In the 1930s, a graduate student who shared his interest in the indigenous languages of North America entered his department at Yale. This student was Benjamin Lee Whorf.

Benjamin Lee Whorf was a native of Massachusetts. In 1918, he received his bachelor's degree in chemical engineering from the Massachusetts Institute of Technology (MIT). After receiving his degree, he worked for the Hartford Fire Insurance Company. In his spare time, he was able to pursue his interest in linguistics and anthropology. His passion for the topic is evident in the many works that he published, which include books on Mayan hieroglyphics;

the grammar of the Hopi language; and dialects of Nahuatl, a Uto-Aztecan language spoken in central Mexico. In 1931, he began studying at Yale University, where he met and studied under Edward Sapir. In the few years that they worked together, the core elements of the Sapir–Whorf hypothesis came together. In 1941, Whorf died of cancer at the young age of 44, leaving one to wonder how much more he might have contributed to the field had he lived longer.

Their view about the relationship between language and thought has been described in multiple ways. Whorf is quoted as saying, "We dissect nature along lines laid down by our native languages" (Carroll, 1956, p. 213). One version of the view is called **linguistic relativity**—one's language shapes one's thinking, thus cognitive processes may vary for speakers of different languages. Contemporary researchers refer to this version of the view as the **weak version of the Sapir–Whorf hypothesis.** A second version of the view is called **linguistic determinism**—language determines all thinking, specifically what one can think about and how the cognitive processes occur. Contemporary researchers refer to this as the **strong version of the Sapir–Whorf hypothesis.** Scholars have noted that Sapir's view may have reflected a weaker version of the hypothesis than the view of Whorf (Rollins, 1980). However, others point out that Whorf's own writings suggest that his view may have fluctuated between the weak and strong versions (Carroll, 1956).

The Role of Vocabulary

One of the observations that Whorf used as evidence that language influences thought was that languages varied in their vocabulary—specifically regarding whether the language had a word to refer to a particular concept or how many words the language used to refer to concepts. Whorf picked up on an observation made by Boas (1911) that languages spoken by people living in cold climates, such as the Yup'ik people who live in Alaska and Southeastern Russia, seem to have many more words for *snow* than people living in warmer climates around the world. He once wrote the following:

> We have the same word for falling snow, snow on the ground, snow packed hard like ice, slushy snow, wind-driven flying snow—whatever the situation may be. To an Eskimo, this all-inclusive word would be almost unthinkable; he would say that falling snow, slushy snow, and so on, are sensuously and operationally different, different things to contend with; he uses different words for them and for other kinds of snow. (Carroll, 1956, p. 216)

Whorf was correct that languages differ in terms of how many words there are to refer to specific concepts. The term **lexical differentiation** is often used to describe this characteristic of languages. A language with more words to refer to a concept would be described as having greater lexical differentiation than a language with fewer words for a concept. However, it turns out that Whorf was incorrect about the number of words that languages spoken in the Arctic regions have to refer to snow (Martin, 1986; Pullum, 1991). In a thorough investigation, Martin (1986) found only about a dozen different words for snow in languages spoken in the Arctic (e.g., Yup'ik). She also pointed out that there were also quite a lot of snow-related words in English (e.g., *snow, sleet, slush, blizzard, flakes, avalanche, dusting, frost*, and *powder*). Martin's analysis also noted that the Yup'ik language is a synthetic language that has many suffixes and prefixes; thus, for the nonnative speaker of the language, it is difficult to know what should be called a word or part of a word.

The Case of Hopi

A second famous example used by Whorf involved the Hopi language, an indigenous language spoken in Arizona and New Mexico. Whorf observed that Hopi had no word for *time* and no way to indicate time in the grammar of the language. Following his view about the impact of language on thought, he suggested was that speakers of Hopi perceived time (i.e., the past, present, and future) differently than speakers of English (Whorf, 1956, p. 57). Although the way in which the grammar of Hopi specifies verb tense (i.e., past, present, and future) is different from how verb tense is specified in languages of the Indo-European language family, such as English, it does enable to speakers to express different tenses (Deutscher, 2010).

Some have criticized Whorf for making broad claims about languages with which he had minimal experience (Deutscher, 2010; Pinker, 1994). His knowledge of languages was primarily through textual analysis (e.g., written dictionaries and grammars) rather than through direct contact with speakers of the language. Carroll (1956), a friend of Whorf's, has stated that Whorf had visited a Hopi village and spent time with speakers of Hopi in 1938. It is likely that the time that he spent with the Hopi may not have been enough for him to formulate a complete understanding of the culture and the cultural norms involving the conceptualization of time. In the field of linguistics, there are many scholars who spend years doing fieldwork, documenting and studying one or more languages in a region. They live among the speakers of the language so that they can come to understand not only the languages' vocabularies and grammars but also their cultures (Newman & Ratliff, 2001).

Whorf's observations about languages other than English and the speakers of those languages now appear to be made from a European-centric perspective in which the European languages, such as English and German, appear to be used as a standard by which other languages are measured. Other languages are viewed as the exotic *other* language. Although much was made of the numerous words for *snow* in languages of the Arctic, no mention was made of the many words that English has for some concepts for which other languages have relatively few words. Table 8.1 displays the words in English used to refer to the concept *to see.* Also, there are some concepts for which there are no words in English. For example, in English, there is only one word to refer to the color blue, but in Russian there are two words to refer to blue: *sinii*, which is dark blue, and *goluboi*, which is light blue. There are two words to refer to red in Hungarian: *piros*, which is light red, and *vörös*, which is dark red. Table 8.2 displays other examples. Despite the fact that one can find examples in which the thinking of speakers of English might have been impoverished when compared with the thinking of speakers of other languages, such comparisons were not emphasized by Whorf or Sapir.

Those who are employed in the literary professions, such as writing and editing, may embrace the Sapir–Whorf hypothesis, regardless of the language or languages that they speak. Such individuals understand how the choice of words for a speech or a written document (e.g., a poem, essay, story, or novel) can have a substantial influence on how one interprets the language and also thinks about the subject being described. Text Box 8.1 addresses this issue with the example of euphemisms, which are used routinely in politics as well as in daily life to influence how listeners and readers experience linguistic messages.

Table 8.1 Fifty Words in English Referring to the Concept *to See*

To see	To look	To gawk	To scan	To watch
To glimpse	To gape	To case	To glance	To gaze
To spot	To take in	To glare	To peer	To espy
To scry	To peek	To peep	To discern	To absorb
To observe	To ogle	To scout	To leer	To examine
To inspect	To perceive	To eye	To glean	To notice
To identify	To peruse	To view	To witness	To recognize
To study	To search	To spy	To sight	To detect
To behold	To experience	To scrutinize	To find	To stare
To survey	To regard	To reckon	To visualize	To rubberneck

Table 8.2 Examples of Words Not Found in English

Word	Language	Meaning
Schadenfreude	German	To take pleasure from another's misery
Schlimazel	Yiddish	Someone with extremely bad luck
Toska	Russian	Deep, spiritual anguish, usually with no singular cause
Litost	Czech	Agony caused by the sudden sight of one's own misery
Iktsuarpok	Inuit	To go outside to check to see if someone is coming
Jayas	Indonesian	A joke told so badly that it is funny
Saudade	Portuguese	A feeling of longing for something that is lost
Ya'aburnee	Arabic	Something that is simultaneously morbid and beautiful
Cafune	Portuguese	To run fingers through another's hair affectionately
Tingo	Pascuense	To borrow objects from your neighbor until there is nothing left
Bakku-shan	Japanese	A woman who appears attractive from behind but unattractive from the front
Hygge	Danish	Complete absence of unpleasant things and presence of only a pleasurable feeling
Uitwaaien	Dutch	To take a trip to the countryside to clear the mind
Zalatwic	Polish	To use charm, bribes, friends, and connections to get things accomplished
Age-otori	Japanese	To look worse after a haircut

Sources: Cafemom.com (n.d.); Wordsmith.org (n.d.).

Text Box 8.1 Diversity of Human Languages: Are Euphemisms Evidence for Linguistic Relativity?

As you learned in Chapter 6, a euphemism is a word that is milder, vaguer, or more indirect and is used in place of another more objectionable word. Does using a milder word to describe an event or object change one's perception of the event or object? One might assume that the answer is no. However, those who understand the power of propaganda can provide examples of how changing the words used to refer to events or objects can influence how others think of them. In George Orwell's (1948) novel *1984,* the totalitarian state created a form of English called Newspeak, which was a greatly simplified form of the former version of English referred to as Oldspeak. In the following passage, Orwell explained some of the vocabulary particulars of Newspeak:

> As we have already seen in the case of the word free, words which had once borne a heretical meaning were sometimes retained for the sake of convenience, but only with the undesirable meanings purged out of them. Countless other words such as honour, justice, morality, internationalism, democracy, science, and religion had simply ceased to exist. A few blanket words covered them, and, in covering them, abolished them. All words grouping themselves round the concepts of liberty and equality, for instance, were contained in the single word crimethink, while all words grouping themselves round the concepts of objectivity and rationalism were contained in the single word oldthink. (Orwell, 1948, p. 379)

Real-world government leaders take advantage of the power of words in influencing how people think. One of the most cited examples is that of Adolf Hitler and his cunning manipulation of the German language to describe heinous projects. The plan to murder millions of the Jewish faith in gas chambers was referred to as *die Endlösung,* or the final solution (Brustein, 1996). An earlier plan to euthanize German children with mental and physical handicaps referred to such individuals as *Lebensunwertes Leben,* or life unworthy of life. There were many expressions used to refer indirectly to the murder of Jewish people, including *Entsprechend* (treated appropriately), *Sonderaktionen* (special actions), and *Sauberung* (cleansing). The expression *Auswertung* (utilization) was used to refer to the process of removing gold teeth and hair from the bodies of the dead so that these items could be sold for profit.

In modern politics, the careful crafting of language still occurs. Some refer to the manipulation of language for political purposes as *spin.* In order to soften the meaning of deaths experienced during military action, the terms *collateral damage* and *friendly fire* have been created. Wars are referred to as *humanitarian operations* or *preemptive strikes.* When politicians choose to advance favored positions, they may refer to those positions with words chosen to influence the listeners' perceptions and cognitions. On the topic of abortion, some describe themselves as *pro-life;* others describe themselves as *pro-choice.* Some refer to social programs as *entitlements.* Others refer to tax cuts as *economic stimulus.*

In our personal lives, we may also take advantage of the power of words in shaping how listeners perceive events. A mistake that we make with friends or family could be acknowledged as an *uncharacteristic lapse in judgment* or a *youthful indiscretion*, rather than a *deliberate act of rebellion*.

How Did Researchers Test the Sapir–Whorf Hypothesis?

Color Words

The first empirical tests of the Sapir–Whorf hypothesis did not occur until the 1950s. Brown and Lenneberg (1954) investigated how English speakers processed colors. They noted that while languages vary in terms of the number of color words they have, no language has a word for every possible color. In the study, they showed participants a chart with 24 colors, which were selected at random. In the first study, participants viewed the chart for 5 minutes. The chart was taken away, and participants viewed individual presentations of a color. They were instructed to name the color as they would in daily life. Participants' response times were recorded. In the second study, participants were given a recognition memory test. They were shown 4 of the 24 colors used in the previous experiment. The chart of four colors was removed, and participants were asked to point out on a large chart of 120 colors the colors they had just seen. The results showed that participants were better able to recognize hues that were easily named than hues that were not easily named.

▲ Photo 8.2 Color array are commonly used by painters. The colors include both focal and nonfocal examples of the primary colors.

Brown and Lenneberg's (1954) study was the first in a long line of color perception studies investigating the relationship between lexical differentiation of color in languages and the processing of color by speakers of different languages. In 1969, Berlin and Kay published a comprehensive analysis of how color words vary across languages. They found that the color words found in various languages followed a regular pattern. They identified seven stages of color differentiation across languages. Their seven stages are described in Table 8.3. Because they found a pattern across languages that appeared to be universal, they were critical of the Sapir–Whorf hypothesis. They conducted a study in which they presented participants with chips of varying colors. They asked participants to point to the color that best represented each **color term** in their language. They also asked participants to point to the other hues that could be referred to by the same word. They found that individuals who spoke different languages pointed out the same core or **focal color** for 11 color words. Even when participants spoke languages that did not have all 11 color words, their judgments of focal colors were the same as the judgments of those who spoke languages having all 11 color words. Participants did vary in how they categorized colors that represented the boundaries between the focal colors (e.g., the region of the color spectrum between blue and green). Berlin and Kay's (1969) work has received criticism (Saunders, 2000; Wierzbicka, 2008); however, at the time that it was published, the work played a big role in inspiring researchers to investigate color processing across cultures.

In 1972, Eleanor Heider Rosch compared the color processing abilities of speakers of English and speakers of Dani, a language spoken in New Guinea (Heider, 1972). Dani has only two color words: (1) *mili* to refer to black and (2) *mola* to refer to white. For both groups of participants, she presented them with a painted color chip and asked them to name the color. She also tested participants' memory for color. Participants were shown one chip for 5 seconds and then shown an array of 40 colors. They had to point to the color that they had seen. She found that both groups of participants showed better memory for colors that represented the best example of the category (also called focal colors) versus other examples of that category. In a separate study with Dani speakers, she found that they could learn names for colors. She concluded that the perception of color and color processing was not influenced by the language spoken by the perceiver. Lucy and Schweder (1979) criticized the study because there might have been differences in how easily participants could visually discriminate focal and **nonfocal colors.** They carried out a similar study controlling for this variable and found no differences between participants' memory for focal and nonfocal colors in short-term recognition; however, they did observe differences in long-term recognition.

Table 8.3 Seven Stages of Color Differentiation Across Languages

Stage I	A language has two color words, one to refer to dark-cool entities and one to refer to light-warm entities.
Stage II	A language will have three color words, Stage I words, and also a word referring to *red*.
Stage III	A language will have Stage I and Stage II words and also a word referring to either *green* or *yellow*.
Stage IV	A language will have Stage I, II, and III words and also a word referring to *green* and a word referring to *yellow*.
Stage V	A language will have Stage I–IV words and also a word referring to *blue*.
Stage VI	A language will have Stage I–V words and also a word referring to *brown*.
Stage VII	A language will have Stage I–VI words and also words referring to *purple, pink, orange,* or *gray*.

Source: Berlin and Kay (1969).

After the publication of Rosch's results (Heider, 1972), many viewed the Sapir–Whorf hypothesis disproven once and for all. The results of Rosch and other similar studies may have shown that the strong version of the Sapir–Whorf hypothesis is incorrect; however, researchers are still debating the extent to which the weak version of the Sapir–Whorf hypothesis may be true. In the past several decades, there have been a number of studies whose results appear to support the weak version of the Sapir–Whorf hypothesis. For example, Kay and Kempton (1984) compared the processing of color by speakers of English and speakers of Tarahumara, a Mexican Indian language in which there is one word to refer to blue and green. The language does not have other words to refer to blue and green separately. In the study, they showed participants three colors, an example of blue, an example of green, and an example of a third color that was in between blue and green. Participants were asked to decide whether the third color was more similar to blue or to green. English-speaking participants' responses showed that they reliably categorized the colors as either green or blue. In contrast, speakers of Tarahumara did not. In a follow-up study, they instructed English-speaking participants to refer to the in-between color as *both blue and green*. The results showed that their subsequent performance categorizing in-between colors was similar to that of those who spoke Tarahumara. The authors concluded that the categorization of colors is influenced by the words that are used to refer to them.

Processing Color

Other researchers have reported studies in which the memory for color appears to be directly related to the color words that speakers have in their languages. For example, Roberson, Davies, and Davidoff (2000) compared color memory for speakers of British English and speakers of Berinmo, a language spoken in Papua New Guinea that has only five basic color terms. They used a research design similar to that used by Rosch (Heider, 1972). In Berinmo, the word for blue and green is the same; however, when speakers of Berinmo were tested, the way in which they distinguished the colors along the blue–green continuum differed from speakers of English. Roberson and colleagues found a similar pattern of results when comparing speakers of British English and speakers of Himba, a language spoken in Namibia (Roberson, Davidoff, Davies, & Shapiro, 2005) that also has fewer color words than English.

Research has also shown that the processing of color may be carried out differently by the left and right hemispheres of the brain. Gilbert, Regier, Kay, and Ivry (2006) tested the hypothesis that language's effect on color processing may be stronger when the color is processed by the left hemisphere, which is dominant for language, than when it is processed by the right hemisphere. In their studies, participants were shown a display of 12 squares, arranged in a circle. One of the squares was presented in a different hue than the others, which had the same hue. The position of the differently colored square was varied from trial to trial. Participants were asked to indicate whether the distinct square appeared in the left or right visual field. Reaction times varied significantly only when targets appeared in the right visual field (i.e., left hemisphere). Participants were faster when the hues of target square and other squares belonged to different lexical categories (e.g., green among blues) than when the hues belonged to the same lexical category (e.g., two different hues of blue). Gilbert, Regier, Kay, and Ivry (2008) showed that the same pattern of results was obtained when the stimuli were black-on-gray silhouettes of animals (e.g., dogs and cats) rather than colors. Participants were faster when the target and distracter pictures belonged to different lexical categories (e.g., dog and cat) than when the pictures belonged to the same lexical category (e.g., two different pictures of cats). Roberson, Pak, and Hanley (2008) have obtained similar results in tasks involving color perception with speakers of Korean.

Converging evidence has also been obtained in studies in which researchers measured event-related brain potential (ERP) as participants view colors. As you learned in Chapter 2, in an ERP study, electrical activity is measured from a participant's scalp as processing occurs. Holmes, Franklin, Clifford, and Davies (2009) tested English-speaking participants and found differences in brain activity when participants were shown colors belonging to the same category versus a different category (i.e., hues of blue vs. hues of blue and green). In conditions in which hues represented different categories, brain responses occurred more rapidly than in conditions in which hues were from the same category. Thierry, Athanasopoulos, Wiggett, Dering, and Kuipers (2009) tested speakers of Greek, which has different words to refer to dark blue and light blue. Differences were observed in brain activity approximately 200 milliseconds after light and dark hues of blue were processed. When English speakers were tested in the same task, no difference in brain activity was observed, presumably because the hues of blue corresponded to just one color word in English.

Counterfactuals

Other research testing the Sapir–Whorf hypothesis focused on how speakers of different languages could or could not easily interpret specific types of statements. Starting in the early 1980s, a series of studies investigated the role of language in influencing how easy or difficult it is for individuals to comprehend statements that are not true (i.e., counter-to-fact). Such statements are referred to as **counterfactual** statements. Sample statements are provided in 1. In English, in such conditional statements, the subjective mood is used to describe a state of affairs known to be false. Bloom (1981) pointed out that Chinese lacks a way for speakers to grammatically distinguish counterfactual statements from factual ones. He also observed that because of this, the Sapir–Whorf hypothesis would predict that speakers of Chinese would have difficulty processing texts containing counterfactual statements.

1. a. If Mary had come earlier, they would have arrived at the party on time.
   b. If I were president, I would create more national holidays.

Bloom (1981) reported a series of studies in which passages containing counterfactual content were presented to either Chinese or American students. Students read the passages in their native language. They then answered comprehension questions in order to ascertain whether they had interpreted the content of the passage factually or counterfactually. In one study, Bloom found that 98% of the American students interpreted the passage counterfactually but only 6% of the Chinese students did. In a second study, Bloom found that 96% of the American students interpreted the passage counterfactually but only 59% of the Chinese students did. Bloom interpreted the results as supporting the Sapir–Whorf hypothesis.

▲ Photo 8.3 Traditional huts from the New Guinea Highlands are commonly used even today. What types of vocabulary words would you likely use if you lived in this village?

Bloom's methods and conclusions have been criticized by Au (1983, 1984) and Liu (1985). The criticisms suggest that the passages that were prepared in Chinese by a native speaker of Chinese and used by Bloom may have been particularly difficult for Chinese readers to interpret (i.e., more difficult than the English passages that American students read). Au (1983, 1984) carried out replications of the research design used by Bloom, using a revised set of materials, and found that Chinese students performed much better on the comprehension questions. Au concluded that the results and those by Bloom (1981) do not support the Sapir–Whorf hypothesis. Bloom (1984) followed up with criticisms of his own, regarding Au's (1983, 1984) studies. The Chinese participants tested by Au had received about 12 years of training in English. They were essentially Chinese–English bilinguals. Bloom concluded that their experience with English could explain why they performed so well on comprehending counterfactual statements. Unfortunately, the question of whether there are differences in how speakers of Chinese and speakers of English process counterfactual statements is still open.

What Are the Methodological Challenges in Testing the Sapir–Whorf Hypothesis?

Conducting cross-cultural research is particularly challenging, regardless of the particular focus of the research. Cross-cultural studies of language and thinking may be among the most challenging to conduct. Researchers face challenges in at least three areas. The first challenge is identifying the specific focus of inquiry (i.e., color words, perception of time), as one must be knowledgeable of at least two languages and the cultures in which the languages are used. If researchers investigate a topic for which there is an existing body of literature and previously tested materials, they may be able to use those prior materials and not be challenged with the task of creating new materials. The challenges are greatest when researchers aim to investigate an aspect of thinking that has not previously been investigated, which also could be particularly valuable to the field. However, if one uses the materials from prior studies and the materials have flaws, one may replicate prior results even when the results do not reflect something true about how language is processed in general. Researchers who conduct research with language should routinely attempt to replicate their results and the results of others using different sets of materials as well as different sets of participants (Clark, 1973).

Assuming that one has developed a reasonable hypothesis to test, the second challenge is to develop the materials that will be used. Creating an ideal set of materials requires that one be able to see the materials from the perspective of individuals in both of the cultures being tested. Prior to conducting the study, researchers may pilot test the materials with one or more native speakers of each language group. Problems with materials can involve words that have more than one meaning in one or both languages. Typically, researchers would want to exclude materials that involve ambiguity, as it is important that participants do not experience difficulty above and beyond that anticipated and/or manipulated by the researcher.

The biggest challenge is gaining access to the participants in two or more language and cultural groups. Ideally, researchers would want to compare the performance of native speakers of different languages and also to ensure that the two language groups have comparable experience with second or third languages. Proponents of the Sapir–Whorf hypothesis would predict that the thinking of bilinguals would be expected to be influenced by both of the languages that are known and would be different from the thinking of those who spoke only one of the languages known by the bilingual. For researchers, the most convenient populations to access are those living in the country where the researchers live and/or work. Individuals from convenient populations are more likely to have experiences with multiple languages; thus, they are less likely to be truly monolingual, native speakers.

Despite the challenges of cross-cultural research, there has been a steady stream of studies over the past three decades. Some of these studies advance our understanding of the relationship between language and thought. As with the past studies that have tested the Sapir–Whorf hypothesis, there are often conflicting results. As you learn about some of these studies in the next section, keep in mind the methodological challenges, as these may be responsible, at least in part, for the different pattern of results observed across studies.

Recent Evidence for Language's Effect on Thinking

How Does Language Influence Thinking About Time and Space?

Direction of Writing

Recent research has shown that the Greek historian Herodotus might have been on to something after all when he suggested that the direction in which a language was written had an impact on how speakers of that language thought and lived. Studies have shown there are measurable differences in thinking for speakers whose languages are written in different directions. Tversky, Kugelmass, and Winter (1991) investigated how speakers represented the events of the day and found that speakers of English, which is written left to right, generally created horizontal timelines. Earlier events were ordered to the left of later events. When they tested speakers of Arabic, which is written right to left, they found that earlier events were ordered to the right of later events. Maas and Russo (2003) compared how Italian and Arabic speakers represented events described in sentences. Participants were asked to draw the events. Speakers of Italian, a language that is written left to right, drew the sequence with the agent of the action appearing to the left of the patient of the action. In contrast, speakers of Arabic drew the sequence in the opposite direction, with the subject of the action appearing to the right of the patient of the action.

If such results occur because of the direction in which the languages are written, then one would not expect to observe any particular direction bias for those who do not yet write, such as young children, or for those who are illiterate. Dobel, Diesendruck, and Bölte (2007) tested preliterate children and illiterate adults who spoke either German, which is written left to right, or Hebrew, which is written right to left. The results showed that neither group of preliterate children showed a general spatial bias. However, illiterate adults showed the spatial bias of literate adults in the two languages. The results suggested that the spatial bias is learned either through the writing direction of the language or through the directional bias present throughout the culture.

The direction in which a language is written has also been found to be related to how speakers of the language conceptualized the **mental number line** (Dehaene, Bossini, & Giraux, 1993). Speakers of languages that are written from left to right envision the number line with small numbers at the left and increasingly larger numbers at the right. In contrast, those who speak languages that are written right to left envision the reverse (Zebian, 2005). Other studies have shown speakers of English and French respond more quickly to small numbers using a left response (versus right response) and to large numbers using a right response (versus left response) (Bächtold, Baumüller, & Brugger, 1998; Dehaene et al., 1993). Some research suggests that the spatial bias of the mental number line occurs only after 2 years of formal schooling (Berch, Foley, Hill, & Ryan, 1999). Wood, Willmes, Nuerk, and Fischer (2008) conducted a meta-analysis of over 100 studies that investigated spatial bias of the mental number line and found that the observed effect size appears to be larger in studies with older participants. However, it is not clear whether the spatial bias of the mental number line increases as a person ages.

Fuhrman and Boroditsky (2010) found additional evidence for the effect of language on speakers' conceptualization of time. They compared how speakers of English and Hebrew spatially represented time. Their results showed that speakers of English arranged temporal sequences with earlier events positioned to the left of later events, and speakers of Hebrew showed the reverse pattern. In subsequent experiments, participants were shown two events from a sequence and were asked to judge which event occurred first. The results showed that English speakers responded more quickly to earlier events in the sequence with a left key press than a right key press. Speakers of Hebrew responded more quickly to earlier events in the sequence with a right key press than a left key press.

Conceptualizing Time

Other studies have also suggested that speakers of different languages may conceptualize time differently. Boroditsky (2001) suggested that speakers of English and Chinese differ in their conceptualization of time. In English, time is often described in a horizontal manner. Past events are described as behind us. Future events are described as in front of us or ahead of us. When describing appointments, we often say that we must move appointments forward or move deadlines back. In contrast, speakers of Chinese use a vertical metaphor for time in addition to the horizontal metaphor that English speakers use (Scott, 1989). Events occurring earlier in time are described as *shàng* or *up*.

Events occurring later in time are described as *xià* or *down*. In a series of experiments, Boroditsky compared the processing of native speakers of English and Chinese and found differences in processing. Chinese native speakers were Chinese–English bilinguals. Bilinguals' performance when asked to use the vertical metaphor for time was related to their age when they began learning English. English speakers could be taught to use the vertical metaphor for time.

Later attempts to replicate the results of Boroditsky (2001) have not been successful. Chen (2007) investigated how often Chinese speakers used horizontal and vertical metaphors for time. She found that Chinese speakers used horizontal metaphors most frequently. She also conducted four attempts to replicate the results of Boroditsky (2001) but was unable to obtain similar results. She concluded that the way in which Chinese speakers think about time does not differ from the way in which English speakers think about time. Other researchers have also attempted to replicate the results of Boroditsky (2001) using similar methods and have also failed to replicate the results (January & Kako, 2007; Tse & Altarriba, 2008). Because of the conflicting results, we must be cautious and realize that the extent to which speakers of English and Chinese think differently about time is not yet known.

Other studies have shown that speakers of Chinese, when asked to create a temporal sequence of pictures or events of a story, will use a vertical representation of time. For example, Chan and Bergen (2005) asked Chinese speakers to arrange a sequence of pictures. Thirty percent (between 18% and 39%) of participants used a vertical arrangement. In contrast, English speakers performing the same task never used a vertical display. Boroditsky (2008) asked speakers of Chinese and English to locate the events within the three-dimensional space around the body. The results showed that Chinese speakers arranged the events vertically 42% of the time and English speakers did so only 5% of the time. Boroditsky, Fuhrman, and McCormick (2011) found similar results in a recent study using an improved methodology. Miles, Tan, Noble, Lumsden, and Macrae (2011) found that Chinese–English bilinguals used both ways to represent time.

Comparisons have also been made between speakers of English and languages other than Chinese. Boroditsky and Gaby (2010) described the Australian Aboriginal community of Pormpuraaw, which is located in Western Australia. In that community, time is represented spatially, relying on directional terms (e.g., *north*, *south*, *east*, and *west*). For example, one might say, "Move your cup over to the north-northwest a little bit" or "the boy standing to the south of Mary is my brother" (Boroditsky & Gaby, 2010, p. 1635). Unlike speakers of English from the United States, when the Pormpuraaw faced south, they generally spatially represented time from left to right. When they faced north, they generally spatially represented time from right to left. When they faced east, they generally spatially represented time as moving toward them. When they faced west, they generally spatially represented time as moving away from them. The participants' performance occurred even when they were not explicitly told which direction they were facing.

In some languages, the ways in which speakers communicate about space and time may seem quite exotic to speakers of English. Text Box 8.2 describes the case of Tzeltal. Researchers conducted a simple, elegant study illustrating how speakers of Tzeltal refer to spatial direction and make decisions about spatial direction.

Text Box 8.2 Research Discovery: Spatial Direction in Tzeltal

An important way in which languages differ is in terms of how speakers use words to refer to objects in three-dimensional space. Many languages, such as English and Dutch, use terms that allow the speaker to describe an object's relative position (e.g., front/back, left/right, up/down). Some languages do not rely on this relative positioning strategy; rather, they use an absolute reference system similar to when one distinguishes north from south in English. These languages contain words for north and south but lack words to express relative positioning.

One such language is Tzeltal, which is a Mayan language spoken by over 300,000 people in the Chiapas region of Mexico. In Tzeltal, objects that are north are described as being *downhill,* and objects that are south are described as being *uphill* (Levinson, 1996; Li & Gleitman, 2002; Polian & Bohnemeyer, 2011). In 1996, Levinson investigated the extent to which speakers of different types of languages perform differently in tasks involving spatial relations. She compared the performance for speakers of Tzeltal and speakers of Dutch. In the task, participants were seated at a table on which rested an arrow that was pointed either right, which was in the direction of north, or left, which was in the direction of south. They were then asked to rotate 180 degrees and sit at a second table on which rested two arrows—one pointing left (north) and one pointing right (south). They were asked to choose the arrow that was like the one on the first table. When the first table displayed an arrow pointing right (north), speakers of Dutch tended to choose the arrow on the second table that was pointing right (south), and when the first table displayed an arrow pointing left (south), they chose the arrow on the second table that was pointing left (north).

In contrast, speakers of Tzeltal chose the arrow on the second table that matched the absolute relation of the arrow on the first table. When the first table displayed an arrow pointing right (north), they chose the arrow on the second table that was pointing north, even though it was pointing to the left, and when the first table displayed an arrow pointing left (south), they chose the arrow on the second table that was pointing south even though it was pointing right. The results suggest that how one's language handles spatial referencing can have a strong influence on how one thinks about space.

How Does Language Influence Thinking About Gender?

As you learned in Chapter 4, some languages have grammatical markings for gender. English does not, but some of the Romance languages do. In Spanish, nouns occur with different suffixes and different articles to indicate the gender of the noun. For nouns referring to humans, the same root word occurs with masculine or feminine gender, as shown in 2. For inanimate nouns, the noun is classified as either masculine, as *el carro* (*car*), or feminine, as in *la luna* (*moon*). In other languages, such as German and Russian, all nouns are assigned a particular gender category (i.e., masculine, feminine, neuter). The classification of the noun is not predictable from the noun's meaning, as shown in 3. Across languages that classify nouns in terms of gender, there is a great deal of variability. For example, the word *moon* is feminine in Spanish but masculine in German. The word *sun* is masculine in Spanish, feminine in German, and neuter in Russian (Boroditsky, Schmidt, & Philips, 2003).

2. a. el carpentiero — Male carpenter
   b. la carpentiera — Female carpenter
   c. el enfermiero — Male nurse
   d. la enfermiera — Female nurse

3. a. der Hund — The dog
   b. die Katze — The cat
   c. das Vögelein — The bird

If one's language affects how one thinks, as claimed by the weak version of the Sapir-Whorf hypothesis, then one would expect to find that differences in how languages use gender in the grammatical system would be related to processing differences in speakers of those languages. In several studies, researchers have investigated this possibility. Boroditsky and colleagues (2003) investigated how Spanish and German speakers are influenced by the gender classification of the nouns in their languages. They identified 24 objects for which the words in Spanish and German had the opposite gender classifications. They presented the list of objects to two groups of participants. One group was composed of native speakers of Spanish. The other group was composed of native speakers of German. Both groups of participants were fluent in English. Each participant was asked to write down in English the first three adjectives that came to mind to describe each object. The adjectives that were generated by participants were later rated by native speakers of English on a continuum of masculinity and femininity. For both groups of participants, the adjectives generated for nouns classified with masculine gender were rated as more masculine than adjectives generated for nouns classified with feminine gender. Consequently, it was the way that the language classified the noun as masculine or feminine rather than what the object was that influenced whether masculine or feminine adjectives were used to describe the object.

Sera, Berge, and del Castillo Pintado (1994) tested Spanish-speaking and English-speaking adults and children. They showed participants pictures of animals and everyday objects. They said that in the future the objects may be used in an animated film, and they wanted to know whether the object should be used with a male voice or a female voice. The results showed that responses varied across participant groups. For Spanish-speaking adults, responses reflected the grammatical gender of the object. For Spanish-speaking children, responses did not reflect the grammatical gender of the objects as closely. The responses of English-speaking adults and children showed no consistent pattern. In a study with 3- to 4-year-old children, Martinez and Shatz (1996) asked children to sort pictures of objects into groups "that go together." Spanish-speaking children tended to sort items into groups according to grammatical gender. A later study found similar results for adult speakers of French but not for adult speakers of German (Sera, Elieff, Burch, Forbes, & Rodríguez, 2002).

Kurinski and Sera (2011) investigated how the acquisition of a second language that marked grammatical gender influenced participants' categorization of inanimate objects in a laboratory task. They tested native speakers of English four times throughout an academic year as they learned Spanish. Participants were asked to assign male or female voices to objects. Among the list of items were control items, such as doctor and nurse, which would be unambiguous. The results showed participants had acquired the grammatical gender information after 10 weeks of instruction. Their classification of objects as masculine or feminine became increasingly similar to the grammatical gender of the noun in Spanish after 20 weeks of instruction.

How Does Language Influence Thinking About Numbers?

In the past decade, a number of studies have investigated the possibility that the language one speaks influences one's ability to carry out tasks involving numbers and varying quantities. Several studies have focused on the Pirahã people, a hunting-gathering tribe living in Brazil in the rainforests of the Amazon River. The language of the Pirahã has few words to refer to different quantities (Everett, 2005; Gordon, 2004). In Pirahã, there is one word *hói* (falling tone) that refers to *one*, and one word *hoi* (rising tone) that refers to *two*. There is a third word *baagi* or *aibai* that refers to *many*. Gordon (2004) investigated how speakers of Pirahã performed matching tasks that required them to reproduce quantities of objects. The interviewer would arrange an array of object (e.g., AA batteries) and would ask the participant to create the same array. The results showed that participants did relatively well when reproducing arrays with relatively small numbers of objects but performed increasingly inaccurately as the number of objects in the array increased. In one of the matching tasks, participants were allowed to view the target array for a brief period of time and then asked to re-create the array from memory. Gordon concluded that speakers did not represent numbers as exact amounts; rather, they used a strategy in which they approximated quantities. Frank, Everett, Fedorenko, and Gibson (2008) also investigated how speakers of Pirahã carry out numerical tasks. They found that participants were able to perform the matching tasks, even for large quantities; however, when participants had to rely on memory to perform the matching tasks, they performed very poorly. The results were consistent with Gordon's (2004) previous results.

Research has also been conducted with speakers of Mundurukú, which is also a language spoken in the Amazon rainforest and has few number words. Speakers used the term *pũg/pũg ma,* which refers to *one*; the term *xep xep*, which refers to *two*; and the term *ebadidip*, which refers to *around four*. In a study by Pica, Lemer, Izard, and Dehaene (2004), speakers of Mundurukú were shown arrays with varying numbers of dots (i.e., between 1 and 15). The results showed that speakers used the word *ebadidip* to refer to arrays of four dots but also to arrays of 1, 2, 3, 5, and so on all the way up to 15 dots (Pica et al., 2004). The authors concluded that speakers of Mundurukú use a strategy of approximation for quantities rather than using exact numerical calculations. As with the Pirahã, the Mundurukú are good at approximating quantities (with a nonrandom error margin) but not at exact calculation (see also Izard et al., 2008).

A third example of the role of language on mathematical processing is Butterworth, Reeve, Reynolds, and Lloyd (2008). They compared the mathematical abilities of English-speaking children and children who speak Walpiri and Anindilyakwa, which are indigenous languages spoken in Australia. As in Pirahã, these languages have relatively few words to refer to numbers. They use separate words for *one, two*, and *many*. The results showed that the children's performance did not differ in the matching and recall tasks but did differ on tasks involving addition and other tasks requiring them to use exact quantities when reasoning. The strategies used by children who spoke Walpiri or Anindilyakwa appeared to differ from those used by the English-speaking children. For example, they used spatial cues more often that did English-speaking children. In a more recent study, Butterworth, Reeve, and Reynolds (2011) investigated how children who speak Walpiri and Anindilyakwa carried out addition problems. The results showed that the children performed well by relying on spatial strategies rather than on the enumeration strategy used by English-speaking children.

The topic of how languages vary in their handling of number words and how those differences can influence processing would not be complete without a discussion of studies showing that the language one speaks is related to one's verbal working memory. Working memory refers to the type of memory that one uses when holding information in memory for a short amount of time and also when attempting to solve a problem in a short time frame (Baddeley, 1990). Since the work of Miller (1956), researchers have recognized that the capacity of working memory is about seven items with most people having a capacity between five and nine units. Hoosain (1982) showed that working memory capacity as measured by how many numbers one can retain in memory is directly related to how quickly one can pronounce the digits. For example, a series of monosyllabic digits could be pronounced more quickly than a series of multisyllabic digits.

Later work showed that the number of digits that could be retained in working memory varied for speakers of different languages, because the length of the number words varies across languages. Working memory capacity for digits has been found to be larger for languages in which the digit words are phonologically shorter (Hoosain, 1982). Table 8.4 displays the number words for the numbers 1 through 10 for eight languages. Speakers of different languages also speak at different rates, which can also lead to differences in working memory capacity (Naveh-Benjamin & Ayres, 1986). For example, speakers of Arabic generally speak more slowly than speakers of English. Naveh-Benjamin and Ayres (1986) found that the working memory capacity for speakers of Arabic was smaller (i.e., 5.77 units) than for speakers of English (i.e., 7.21 units). Similar conclusions were reached in studies comparing the working memory capacity for digits of bilinguals. Multiple studies have shown that working memory capacity for digits was larger in the language that was spoken faster by the participant (Chincotta & Hoosain, 1995; Chincotta & Underwood, 1996).

Conceptual and Lexical Development

Understanding the relationship between language and thought has also been investigated from a developmental perspective. Researchers have conducted studies in which the focus is children's acquisitions of concepts and the words that are used to refer to them. If language influences thinking, then children's thinking is likely to vary with age because age is related to language development. Older children typically have larger vocabularies than

Table 8.4 The Digits One to Ten From Eight Languages

English	Spanish	Russian	Greek	Persian	Arabic	Hindi	Aymaraa
One	Uno	Odin	Ena	Yek	Wahid	Ek	Maya
Two	Dos	Dva	Dio	Dou	Ethnan m	Do	Phaya
Three	Tres	Tri	Tria	Seh	Thalatha	Tin	Kimsa
Four	Cuatro	Cheterie	Tesera	Chahaar	Arrbaa	Char	Pusi
Five	Cinco	Piat	Pente	Pang	Khamsa	Panch	Pheschka
Six	Seis	Shest	Exi	Shash	Sitta	Chha	Sojjta
Seven	Siete	Siem	Epta	Haft	Sabaa	Sat	Pakallko
Eight	Ocho	Vosiem	Octo	Hasht	Thamania	Ath	Kimsakallko
Nine	Nueve	Dievit	Enea	Nouh	Tisaa	Nau	Llatunca
Ten	Diez	Diesit	Deka	Dah	Ashra	Das	Tunca

Source: Ardila and Rosselli (2002).

younger children. If language influences thought, then evidence can likely be found in studies investigating children's conceptual development. Children's processing of the world before they have acquired the words for specific concepts is likely to differ from their processing of the world after they have acquired the words for those concepts.

How Do Children Learn About Color?

Prior research has shown that children experience a surprising amount of difficulty learning color words (Bornstein, 1985). Infants as young as 4 months of age are able to perceive different colors (Bornstein, Kessen, & Weiskopf, 1976); however, children do not usually master the appropriate use of color words until between the ages of 4 and 7 years. In an early study of vocabulary development, Wolfe (1890) found that among the first 300 to 500 words of a sample of 2-year-olds, there were no color words. Subsequent research confirmed that children's use of color words before the age of 2 years was rare (e.g., Gesell, 1940). A study by Cook (1931) tested children between the ages of 2 and 6 and found that children were able to match colors better than they were able to name colors. The difference in performance between matching and naming was greatest for the youngest children. A recent study by Fernald, Thorpe, and Marchman (2010) indicated that children's ability to use color words appropriately developed gradually over the third year of life. Other research has shown that children continue to make some errors with color terms until 6 years of age (Roberson et al., 2005; Sandhofer & Smith, 1999).

Studies in which researchers have attempted to teach children color words have yielded conflicting results. Rice (1980) found that children learned the color words *red, green,* and *yellow* only after a very large number of learning trials (i.e., 1,080). Other researchers have found that children could learn a new color word on a single learning trial if the new word were contrasted with a known color word (Carey & Bartlett, 1978; Heibeck & Markman, 1987). Au and Markman (1987) found that when 3- and 4-year-old children were tested in a word learning task, they acquired new terms to refer to textures of objects more readily than they acquired new terms to refer to color. In the color condition, an interviewer said to children, "Pass me the mauve square. See this. It is not green and it is not red. It is mauve." In the texture condition, the interviewer said, "Pass me the rattan square. See this. It is not cotton and it is not wood. It is rattan."

A variety of theories have been proposed to explain why children have such difficulty acquiring color words. Bornstein (1985) proposed that the cause was related to maturational processes related to brain mechanisms. Sandhofer and Smith (1999) proposed that children must have acquired multiple color terms before they could conceptually represent color as a feature of objects, because learning the color words may draw children's attention to color as a property and lead to increased conceptual understanding of color. In contrast, Kowalski and Zimiles (2006) proposed that children's ability to learn color terms may be influenced by their ability to form an abstract conceptual representation of color. A third view, proposed by Soja (1994), claimed that children understand color conceptually before they learn color words, but children are biased in their attentional processing to attend more to object shape than to object color because shape is less variable (more predictable) than color when one categorizes objects in terms of function.

Kowalski and Zimiles (2006) tested children who were between the ages of 20 and 39 months on their ability to conceptually understand color and their knowledge of color words. The results showed that that there was a strong relationship between knowledge of color words and children's ability to draw inferences involving color. Children who comprehended color terms performed better on a conceptual task involving color than children who did not comprehend color terms.

Franklin and colleagues (2008) compared infant and adult performance on a visual search task much like that of Gilbert and colleagues (2006). The researchers measured eye movements during the task and calculated the time that participants took to move their eyes from a central fixation point to the target, which was a dot of one color embedded in a different colored background. The color of the dot was either from the same or different color category as the color of the background. The results from adults showed faster eye movements when the target and background colors were from different categories than when they were from the same categories only when the target was in the right visual field (i.e., left hemisphere). There were no significant differences when the target was presented in the left visual field (i.e., right hemisphere). In contrast, the results from infants showed that when the target and background colors were from different categories than when they were from the same categories when the target was in the left visual field (i.e., right hemisphere), there were no significant differences when the target was presented in the right visual field (i.e., left hemisphere).

In a follow-up study, Franklin, Catherwood, Alvarez, and Axelsson (2010) used the same eye movement monitoring procedure as Franklin and colleagues (2008) and compared the performance of two groups of toddlers whose ages ranged from 2 to 5 years. One group of toddlers knew the color words *blue* and *green*. The other group of toddlers did not. The performance for the toddlers who already knew the color words was similar to the performance of adults in the Franklin and colleagues' (2008) study; they showed faster eye movements when the target and background colors were from different categories than when they were from the same categories only when the target was in the right visual field (i.e., left hemisphere). There were no significant differences when the target was presented in the left visual field (i.e., right hemisphere). The performance for the toddlers who did not know the color words was similar to the performance of infants in the Franklin and colleagues' (2008) study; they were faster when the target and background colors were from different categories than when they were from the same categories when the target was in the left visual field (i.e., right hemisphere). There were no significant differences when the target was presented in the right visual field (i.e., left hemisphere). The results suggested that children's acquisition of color words affects their processing of colors.

▲ Photo 8.4 A boy is counting apples with his mom. Do you remember what age you were when you started learning to count?

Roberson and colleagues (2005) examined how children from different cultures acquired color terms over a 3-year period. One group of children involved 3- and 4-year-olds from England. The other group of children included 3- and 4-year-olds in Northern Namibia, who belonged to the Himba tribe. The Himba people have lived in a seminomadic lifestyle for the past century and have existed in cultural isolation. Unlike the English children, the Himba children did not received formal schooling. Their results showed that both groups of children exhibited a gradual acquisition of color categories. Early acquisition was influenced by perceptual aspects of colors. As children aged, their conceptual knowledge of color and color terms was influenced by language-specific distinctions, such as the number of color terms. They also found that when children did not receive "intensive adult input, color category acquisition [was] universally slow and effortful" (Roberson et al., 2005).

How Do Children Learn About Numbers?

As you learned in Chapter 3, infants show an understanding of number and appear able to perform rudimentary addition and subtraction (Wynn, 1992). However, children's ability to count using number words emerges much later, between the ages of 3 and 4 years. Children's understanding of amount and its relation to numbers appears to develop when they are about 4.5 (Le Corre & Carey, 2007). Gelman and Gallistel (1978) proposed that children master five principles as they learn to count. These principles

Table 8.5 Five Concepts Children Learn About Counting

Principle 1	Each object is assigned only one number word.
Principle 2	Numbers are always assigned in the same order.
Principle 3	The total number of objects is equal to the last number assigned in the counting sequence.
Principle 4	Objects can be counted in any order.
Principle 5	Principles 1 to 4 can be applied to any set of objects.

Source: Gelman and Gallistel (1978).

are described in Table 8.5. Toddlers between the ages of 2 and 3 engage in learning numbers and practice counting. Gelman and Gallistel (1978) suggested that by the age of 5, children understand the five principles of counting displayed in Table 8.5. A number of more recent studies have suggested that children do well with counting before they have truly grasped these counting principles (Bermejo, 1996; Frye, Braisby, Lowe, Maroudas, & Nicholls, 1989).

However, children can have difficulty in using the numbers 1 to 10 to refer to sets of objects (Miller, Smith, Zhu, & Zhang, 1995). The appropriate use of numbers as descriptions of sets of objects is likely to involve symbolic representations. Mix (1999) investigated how 3-year-olds matched cards containing varying numbers of dots to an array of disks, shells, or a random set of objects or a set of shells. They performed best when matching the dots to the array of disks. In contrast, 4-year olds performed better matching the dots to the array of disks and shells than to the random set of objects. The author concluded that children's performance was strongly influenced by how perceptually similar the dots were to the objects (i.e., disks and shells were more similar to dots than were random objects).

Lipton and Spelke (2006) investigated the number knowledge of 5-year-olds. They found that the children understood large numbers (i.e., amounts) that were larger than they could count. Barth, Starr, and Sullivan (2009) compared the number knowledge of three groups of children: (1) those who were unable to count to 35, (2) those who could count higher than 35 but not higher than 60, and (3) those who could count higher than 60. All children were asked to estimate the number of items displayed on a card. The results showed that children's accuracy was related to counting ability. The best counters were most accurate. They found that number estimates were correct for large quantities—even for the least skilled counters.

Prior research has shown that specific areas of the brain appear to be involved in processing numerical information. Relatively recent research demonstrates that how the brain processes numerical information may vary across cultures. Text Box 8.3 describes brain-imaging research comparing the areas of activation in the brains of speakers of English and speakers of Chinese as numerical tasks are performed.

Text Box 8.3 Research Discovery: Language, Numbers, and the Brain

Tang and colleagues (2006) investigated the extent to which the brain processing of tasks involving numbers and arithmetic was influenced by participants' language experience. They compared the brain processing of a group of native speakers of English and a group of native speakers of Chinese as they performed arithmetic. The authors hypothesized that the brain processing occurring as individuals carried out math tasks would be affected by culture and educational experiences. The results showed that different brain areas were involved in addition and numerical comparison. For both groups of participants, numerical comparison resulted in activation in the inferior parietal cortex. However, addition appeared to involve more language-related processing, as there was greater activation observed in the language areas of the left hemisphere. The results also showed that the brain areas involved in number processing differed for the two participant groups.

How Do Children Learn Words About Space?

Choi and Bowerman (1991) compared the words used by English-speaking and Korean-speaking children to refer to spatial relations. They found that the Korean-speaking children used words referring to vertical motion later than English-speaking children. Children reared in English-speaking environments usually produce the words *up* and *down* by the age of 16 months, while children reared in Korean-speaking environments express vertical motion using the verb *ollita*, which means to cause to go up, after the age of 24 months.

McDonough, Choi, and Mandler (2003) investigated preverbal infants from English-speaking and Korean-speaking homes and found that infants as young as 9 months of age developed spatial concepts, which were not represented in the language being acquired. The particular concept investigated was related to containment, specifically whether containment was achieved through a tight fit or a loose fit. Korean has a linguistic form to distinguish tight fit from loose fit, but English does not. Similar results were obtained by Hespos and Spelke (2004), who tested infants as young as 5 months old. The infants in their study were able to distinguish the two types of containment. Choi, McDonough, Bowerman, and Mandler (1999) found that despite the fact that English-speaking toddlers distinguished the tight fit and loose fit spatial distinction, they did not use it in a semantic categorization task, which can be viewed as involving linguistic expression. Casasola, Wilbourn, and Yang (2006) showed that English-speaking toddlers around the age of 21 to 22 months were able to learn to associate a novel word with the tight-fit relation.

Choi (2006) further showed that as infants reared in English-speaking environments aged and acquired more and more English, their sensitivity to the tight fit and loose fit distinction declined by the age of 29 months. They tested infants at 18 months, 24 months, 29 months, and 36 months. They also found that infants reared in Korean-speaking homes maintained the distinction when tested at 29 months and 36 months of age.

How Do Children Learn Words About Time?

Researchers who have studied children's vocabulary development have noted that they generally produce words referring to time later than they produce words referring to space (Clark, 1973). For example, children will use the preposition *in* to refer to a location (e.g., *in the bed*) earlier than they use the preposition to refer to time (e.g., *in a second*), despite the fact that the temporal use of *in* is more frequent in adult speech and writing than the spatial use (Clark, 1973). Similarly, they use the words *here* and *there,* which are spatial words, earlier than they use the words *now* and *then,* which are temporal words.

Between the ages of 18 months and 2 years, children begin to talk about events that occurred in the past or events that are expected to occur in the future (Ames, 1946; Antinucci & Miller, 1976). Harner (1975, 1980, 1982) showed that children's comprehension of the words *yesterday* and *tomorrow* developed between 2 and 4 years of age. Recent research has shown that most 4- and 5-year-olds could correctly identify events as having occurred *yesterday* or occurring *tomorrow* (Busby & Suddendorf, 2005). The words *before* and *after* are usually mastered between the ages of 2 and 4 years (Clark, 1973; Harner, 1980).

When children are around the age of 4, they start to be able to judge whether an event happened the week before or several weeks before (Friedman, 1991). However, in another study, when 4-year-olds were tested one week before Valentine's Day, they were generally unable to say whether Valentine's Day would occur before Christmas. Children's ability to judge when past events occur appears to develop relatively late, between the ages of 8 and 10 years (Friedman, 2000). One study showed that the ability to judge the recency of events that occurred more than 60 days in the past did not emerge until around the age of 9 (Friedman, Gardner, & Zubin, 1995).

Children's ability to estimate the duration of events lasting up to 30 seconds usually develops by the age of 5 years (Fraisse, 1982). As children age, they appear to use counting as the primary strategy for keeping track of time intervals. This strategy can result in inaccurate estimates if the rate of counting fluctuates (Levin & Wilkening, 1989). Children's judgments of time occurring over long periods are more variable. When children enter school, they learn about the cultural norms regarding temporal events, such as weeks, months, years, decades, and centuries. In the United States, most children are taught to tell time using an analog clock in either the first or second grade. One of my own stressful memories from second grade was being asked to go to the front of the classroom to adjust the hands on a clock to the particular time that the teacher requested. I am not sure why, but I found the various ways to refer to time exceedingly confusing. Was it 10:15 or a quarter after 10:00? Why could you say 5 of 10:00 but not 5 of 10:30?

Casasanto, Fotakopoulou, and Boroditsky (2010) tested children who were native speakers of Greek. Two groups of children were tested: a group of 4- to 6-year-olds and a group of 9- to 10-year-olds. They watched movies showing two animals (e.g., snails) moving in parallel. The distance that the animals moved was varied. Children were asked to judge the distance moved or time it took for the animals to move. The results showed that children's judgments about duration were influenced by distance information to a greater degree than judgments about distance were influenced by duration judgments.

Summary and Theoretical Implications

The view that language influences thought has a long history, going back to ancient Greece. In the 20th century, the Sapir–Whorf hypothesis inspired numerous empirical studies that have investigated the extent to which the characteristics of one's native language influence perception, comprehension, and memory. These studies have investigated how speakers of different languages perceive and remember color, comprehend texts containing counterfactuals, carry out numerical tasks, and make judgments about objects' apparent masculinity or femininity. The results of studies in the 1970s and 1980s convinced many people that the strong version of the Sapir–Whorf hypothesis was incorrect. A growing body of research suggests that the weak version of the Sapir–Whorf hypothesis may, in fact, be true. This research is particularly challenging to conduct, because the researcher must be knowledgeable of multiple languages and cultures to be able to construct materials in multiple languages and to test groups of participants who are native speakers of the languages being studied. Recent research showing that the thinking of bilingual participants may differ from the thinking of monolingual participants suggests that researchers should be careful to make comparisons between monolingual groups or bilingual groups rather than comparing English monolinguals to native speakers of another language who are also fluent in English.

In terms of the four theoretical approaches to language development, the research that has been conducted exploring the effect of language on thought does not help us in determining which theory is most accurate, because none of the theories has directly addressed this topic. In the future, there is the possibility that any or all of the theories could offer a satisfactory account of the empirical evidence. In order for this to occur, not only is more research needed; more theory development is needed as well. Advocates of each approach will need to extend their theories to make specific predictions about how language may affect perception, memory, and performance in decision making.

KEY TERMS

color term
counterfactual
focal color
lexical differentiation
linguistic determinism
linguistic relativity
mental number line
nonfocal colors
Sapir–Whorf hypothesis
strong version of the Sapir–Whorf hypothesis
weak version of the Sapir–Whorf hypothesis
Whorfian hypothesis

REVIEW QUESTIONS

1. What is the Sapir–Whorf hypothesis? According to the hypothesis, how does language influence thought?
2. What is the distinction that is made between the strong and weak versions of the Sapir–Whorf hypothesis?
3. What was the relationship between Sapir and Whorf? What prior experience with language did each person have?
4. How does the number of color words vary across human languages?
5. What has research shown about how readers of Chinese comprehend passages containing counterfactual statements?
6. What has research shown about the role of language in the processing of color?
7. What has research shown about how the two hemispheres of the brain are involved in the processing of color?
8. What methodological challenges do researchers who test the Sapir–Whorf hypothesis routinely encounter?
9. What has research shown about how the direction in which a language is written might influence how speakers of the language think of the mental number line?
10. What evidence is there that the way in which a language marks gender in its grammatical system influences how speakers of the language think about everyday objects?
11. What has research shown about how the direction in which a language is written might influence how speakers of the language think about time?
12. How do the world's languages vary in terms of the number of words that they contain?
13. What has research with the Amazonian Pirahã tribe suggested about how language influences numerical processing?

14. What is known about how children acquire color words?
15. What is known about how children acquire number words?
16. What factors has research shown to be related to how working memory capacity for digits varies across speakers of different languages?
17. What is known about how children acquire words for time and space?
18. How do English and Korean differ in their vocabulary for spatial relations?

RECOMMENDED READING

Berlin, B., & Kay, P. (1969). *Basic color terms: Their universality and evolution*. Berkeley: University of California Press.

Hunt, E., & Agnoli, F. (1991). The Whorfian hypothesis: A cognitive psychology perspective. *Psychological Review, 98*, 377–389.

Lakoff, G. (1990). *Women, fire, and dangerous things*. Chicago: University of Chicago Press.

Lakoff, G. (2003). *Metaphors we live by*. Chicago: University of Chicago Press.

MacLaury, R. E., Paramei, G. V., & Dedrick, D. (2008). *Anthropology of color*. Philadelphia: John Benjamins Publishing Company.

Pullum, G. K. (1991). The great Eskimo vocabulary hoax and other irreverent essays on the study of language. Chicago: University of Chicago Press.

Whorf, B., & Carroll, J. B. (2011). *Language, thought, and reality: Selected writings of Benjamin Lee Whorf*. Eastford, CT: Martino Fine Books.

RECOMMENDED FILMS

Coppola, S. (Director). (2003). *Lost in translation* [Motion picture]. United States: Universal Studios.

Edwards, B. (Director). (2009). *The Yup'ik way* [Motion picture.]. United States: Beth Edwards.

Shapiro, D., & Shapiro, L. G. (Directors). (2000). *Keep the river on the right: A modern cannibal tale* [Motion picture]. United States: IFC Films.

SUGGESTIONS FOR CLASS PROJECTS

1. Survey friends and family members about their views on the relationship between language and thought. In the survey, ask respondents whether they believe that the language one speaks influences thinking. If the respondent believes that it does, he or she should be asked to elaborate on why he or she believes this. Also ask respondents whether they have heard the claim that there

are lots of words for *snow* in languages used by people who live in cold climates or the claim that speakers of some Native American languages do not have the same understanding of time as English speakers due to their language. Discuss the results in terms of the long-term impact of the Sapir-Whorf hypothesis on popular culture.

2. Interview a bilingual for the purposes of determining how similar the conceptual meanings are for the words in the bilingual's two languages. Select a list of words that will be used for the interview. In order to increase the chances that there will be conceptual differences for the translation equivalents in the bilingual's languages, include words that refer to abstract concepts (e.g., loneliness, friendliness) as well as words that refer to concrete concepts (e.g., table, chair). Words that occur infrequently and appear to be specific to certain situations may be good candidates for having slight conceptual differences in the bilingual's two languages. For example, when a German–English bilingual considers the translation equivalents (e.g., *erringen* and *achievement*), there may be subtle differences in when one would use the words in context. The German word *erringen* is used often to refer to situations in which there is a score, and the English word *achievement* can be used more generally.

For additional ancillary resources, please visit the companion website at www.sagepub.com/kennison.

- Video Links
- Audio Links
- Web Resources
- Internet Activities
- Flashcards
- Web Quizzes

CHAPTER 9

PLANNING SPEECH: FROM THOUGHT TO ARTICULATION

When we are speaking in formal settings—to elected officials, to bosses, to older members of our communities, or to other authority figures—we may choose our words carefully, taking our time to ensure that each utterance is exactly what we mean to say. We also may be extra mindful that our utterances do not contain undesirable connotations or ambiguities. At other times, when we are speaking to peers, we may speak quickly and without much conscious effort. It may seem to take no time at all for us to go from an idea of what we would like to say to actually saying it. Because speech planning occurs so rapidly, understanding the cognitive processes that are involved in speaking is challenging. Researchers must find ways to tap into processes that are carried out in only several hundred milliseconds or even less. In this chapter, you will learn that research on the topic of speech production began with the study of the errors that are made during everyday language use. You will also learn about how the speech production research has evolved over the past 40 years. Research has yielded theoretical models to explain how language is planned from thought to articulation. These models have been tested in studies conducted both in the field and in the laboratory. Last, you will learn about aspects of physical and psychological development that sometimes affect speech production and how problems involving speech production are treated by physicians, educators, and speech therapists.

▲ Photo 9.1 A girl whispers something to her sister. Do you remember when you learned about keeping secrets?

Spoonerisms and Other Speech Errors

A **slip of the tongue** is an error made during speaking. Every speaker has made such errors. They are part of normal language processing. In popular culture, one of the most familiar types of speech errors is the **Freudian slip.** Sigmund Freud (1901/1971) proposed that some speech errors held clues to the speaker's mind, particularly the unconscious mind. Among the most frequent examples are slips related to sex and/or one's possible sexual attraction to another. Imagine someone who is showing another to a seat in an office. Instead of saying, "Please, make yourself comfortable in my office" the person says, "Please, make yourself comfortable in my bed." One might aim to compliment another about having cute dimples but instead say *nipples.* Slips may also involve other socially inappropriate topics. For example, one may be interacting with another who has a prominent nose. Instead of saying, "Would you mind passing me the notebook?" the person says, "Would you mind passing me the nose?" While a small number of speech errors may be of the type described by Sigmund Freud, most are not—for example, saying a *luck troad* instead of a *truck load* or *is there car in the gas* instead of *is there gas in the* *car*. Research has shown that most speech errors are produced because of predictable processes occurring during the planning of speech rather than because of mysterious processing occurring in our subconscious, where our repressed urges are kept under wraps.

Some individuals may be more prone to make speech errors than others. Some people, in fact, become famous for their misspeaking. One such individual is Reverend William Archibald Spooner (1844–1930). In the early 1900s, he was head of one of the colleges at Oxford University. He became well known for his frequent slips of the tongue. Some examples of Spooner's most famous speech errors are displayed in Table 9.1. To this day, the term ***Spoonerism*** is used to refer to slips of the tongue as well as speech in which the sounds of words are deliberately shuffled around for the purposes of humor. By all accounts, Spooner was a beloved member of the university community, so his errors became one of his endearing eccentricities (Hayter, 1977).

Table 9.1 Examples of Speech Errors Attributed to Reverend Spooner

The Lord is a shoving leopard.

You have hissed all my mystery lectures. You have tasted a whole worm. Please leave Oxford on the next town drain.

Three cheers for our queer old dean!

It is kisstomary to cuss the bride.

Fighting a liar in the quadrangle.

Noble tons of soil.

When our boys come home from France, we will have the hags flung out.

I believe you're occupewing my pie. May I sew you to another sheet?

Source: Hayter (1977).

What Are the Different Types of Speech Errors?

Fromkin (1971) analyzed a collection of speech errors that she recorded over a period of 3 years, and she analyzed the errors collected by other researchers. She was able to identify a relatively small number of different types of speech errors. The eight types that she discussed have also been observed by other researchers in subsequent studies (Garrett, 1975; Jaeger, 2005; Shattuck-Hufnagel, 1979). These are (1) **substitutions,** (2) **exchanges,** (3) **shifts,** (4) **additions,** (5) **deletions,** (6) **anticipations,** (7) **perseverations,** and (8) **blends.**

Substitutions and Malapropisms

Substitutions refer to errors in which an unintended word or morpheme appears in the utterance—essentially, something shows up in the utterance that should not be there. Consider the examples in 1. In 1a, the adjective *cold* appears in the utterance instead of the adjective *hot*. In 1b, the word *Friday* appears in the utterance instead of *Monday*. The element does not have to be an entire word but could be a morpheme, as in the suffix *-ful,* in 1c, or a phoneme, as in the /k/ in 1d.

1. a. The coffee isn't *cold* enough. (hot)
   b. The exam has been moved to next *Friday*. (Monday)
   c. She was very *productful*. (productive)
   d. What *kime* is it? (time)

A substitution may be hard to distinguish from a **malapropism,** which occurs when a speaker has confused words that sound similar. Consider the examples in 2. If the error is a substitution, then the speaker intended to produce the correct word, but the similar-sounding incorrect word ends up being spoken. If the error is a malapropism, then the error originated with the speaker; there was an intention to include the incorrect word in the utterance from the moment that the utterance was planned. From the listener's perspective, it may be impossible to determine which type of error was made.

2. a. We cannot let terrorists and rogue nations hold this nation *hostile* or hold our allies *hostile*. (*hostage*, George W. Bush, former president of the United States)
   b. He was a man of great *statue*. (*stature*, Thomas Menino, former mayor of Boston)
   c. Republicans understand the importance of *bondage* between a mother and child. (*bonding*, Dan Quayle, former vice president of the United States)
   d. Texas has a lot of *electrical* votes. (*electoral*, Yogi Berra, baseball player and manager)

Exchange Errors

Of all speech errors, listeners are most likely to notice exchanges, which involve either a phoneme or morpheme appearing to have switched places in the utterance. Consider the examples in 3. In 3a, the nouns *book* and *shelf* have ended up in the wrong locations in the sentence. In 3b, the first phonemes in the word *shake* and *finger* have exchanged places. Phonemes have also switched locations in 3c. In 3d, two suffixes appear in each other's locations.

3. a. Put the *shelf* on the *book*. (book, shelf)
   b. Don't *fake* your *shinger* at me! (shake, finger)
   c. He dreamed of a *minx* in the *sphoonlight*. (sphinx, moonlight)
   d. She was always *conscientful* and *helpous*. (conscientious, helpful)

Shifts, Additions, and Deletions

Some errors involve only a single phoneme or morpheme appearing in the incorrect location in an utterance. Shifts, as the name implies, occur when a phoneme or morpheme shows up in the wrong location in an utterance. Consider the examples in 4. In each of the examples, a bound morpheme shows up in the wrong place in the utterance. In contrast, additions are speech errors that contain an unintended phoneme or morpheme somewhere in the utterance. Consider the examples in 5.

4. a. She *point outed* that. (pointed out)
   b. He had *forgot aboutten* it. (forgotten about)
   c. That *add ups* to. (adds up)
   d. Sometimes he *try* to *wins* it. (tries, win)

5. a. He cleaned the gun *clarefully*. (carefully)
   b. He assured that there were no strings *attrached*. (attached)
   c. She is *sometrimes* afraid of the dark. (sometimes)
   d. He forgot to unload the *weaplon*. (weapon)

Some errors result in intended phonemes or morphemes being absent from the utterance. These errors are called deletions. Examples are provided in 6.

6. a. It happened in a *slit* second. (split)
   b. The grass is always *geener*. (greener)
   c. The instructions weren't *pacific*. (specific)
   d. The neighbor seemed *capable* of murder. (incapable)

Anticipations, Perseverations, and Blends

Among the most complex types of errors are anticipations, perseverations, and blends. Anticipations refer to errors in which phonemes or morphemes appear earlier in an utterance than they should, as in 7. The anticipation may include a whole word, as in 7a, or a single phoneme as in 7b, 7c, and 7d. In contrast, perseverations refer to errors in which phonemes or morphemes from earlier in an utterance appear in place of another phoneme or morpheme later in the utterance. Examples are displayed in 8. The perseverating element may be a morpheme, as in 8a; the suffix *-ful* appears on the word *helpful* but then appears again attached to the *friend*. In 8b, 8c, and 8d, the persevering element is a phoneme. Last, blends refer to errors in which two or more morphemes get merged into one in an utterance. Consider the examples in 9. Each of the examples contains a made-up word or neologism that reflects a merging of two semantically related words. Blends are rare, but when they are produced, they are usually noticed by both the speaker and the listener.

7. a. He bought chocolate *dough* cookie dough. (chip)
   b. She passed out a *leading* list. (reading)
   c. It happened on the *mirst* of May. (first)
   d. I told him to *bake* my bike. (take)

8. a. He was helpful and friendful. (friendly)
   b. She *gave* the *goy* the book. (boy)
   c. It is a *rule* of *rum*. (thumb)
   d. There were several *black bloxes*. (boxes)

9. a. My *stummy* hurts (stomach, tummy)
   b. They *misunderestimated* me. (misunderstood, underestimated)
   c. I wasn't bothered in the *sleast*. (slightest, least)
   d. There was a noticeable *dreeze*. (draft, breeze)

How Are Dysfluencies Different From Speech Errors?

Speech errors are just one example of how speech can go awry. Besides slips of the tongue, there are dysfluencies, which include silences; filled pauses, such as *er, um, uh;* inappropriate silence; and self-corrections (also called repairs). Researchers have estimated that typical speech involves a dysfluency about six times per 100 words (Bortfeld, Leon, Bloom, Schober, & Brennan, 2001; Fox Tree, 1995). The frequency of dysfluencies has been found to vary across different types of speech (Duez, 1982). Formal speeches contain fewer impromptu utterances. While dysfluencies are virtually nonexistent in scripted television programs and films, they can be observed during live recordings and reality programs.

Silences

The last type of dysfluency is an inappropriate silence during speaking. Only recently have researchers treated silences as a distinct type of behavior; early research on dysfluencies typically categorized them with filled pauses (Bortfeld et al., 2001). Goldman-Eisler (1968) pointed out that when one closely examines naturally occurring speech, silences can make up about half of the total duration. This author found that silences range from between a quarter of a second and 2.5 seconds. Research suggests that some silences during speech may be related to processes related to planning the utterance and retrieving upcoming words (Goldman-Eisler, 1958a, 1958b; Kircher, Brammer, Levelt, Bartels, & McGuire, 2004).

There are definitely individuals who experience speech errors and dysfluencies more than others, such as Reverend Spooner; however, there are also individuals who are exceptionally good speakers. Their impressive speaking ability certainly involves a great deal of practice but may also involve being born with an exceptionally developed vocal tract. Text Box 9.1 describes people who are fast talkers.

Filled Pauses

A second type of dysfluency is a filled pause, such as *um* and *uh* for English speakers, as in *Barbara phoned to say that she...um...is going to be a little late.* Interestingly, speakers of other languages use different sounds for filled pauses. Speakers of British English use *er* or *erm*. Speakers of German use *ah*. Speakers of French use *euh*. Some researchers have proposed that speakers' production of a filled pause is automatic, occurring soon after some glitch in speech planning (Levelt, 1989). Others have proposed that the filled pauses are used purposely by the speaker to signal listeners that there is more to come and that the speaker is not ready to yield the floor (Maclay & Osgood, 1959). Clark and Fox Tree (2002) investigated the factors influencing the use of *uh* and *um* in everyday speech. They

Text Box 9.1 Extraordinary Individuals: Fast Talkers

Some individuals are extraordinarily fast talkers. In the 1990s, John Moschitta Jr. was listed as the World's Fastest Talker in the Guinness Book of World Records. He was able to produce 596 words per minute. The speaking rate for the average speaker is 150 to 160 words per minute. Moschitta's title brought him a great deal of fame. He has appeared in over 750 television commercials and was a frequent guest on talk shows. One of the most famous of these commercials was the 1981 Federal Express advertisement. One can find video clips of his fantastic speaking ability on the Internet. Since then, his record has been broken by other fast talkers. Currently, the fastest female talker in the world is Fran Capo, who can produce 603 words per minute. The fastest male talker in the world is Steve Woodmore, who can produce 637 words per minute. The term ***tachylalia*** is used to refer to extremely rapid speech. Although it may occur in some speech disorders, it is not considered a speech disorder. Fast talking is a skill. Individuals who want to become fast talkers must devote themselves to long-term practice. Auctioneers are well known for fast talking. Their speaking rate is about 250 words per minute.

analyzed a large corpus of speech recordings and examined instances in which speakers used a filled pause. The results suggested that speakers tend to use *uh* to signal a minor delay in the fluency of speech, and they use *um* to signal a longer delay. The authors concluded that *um* and *uh* function as words during speech planning and are selected for use by speakers in the same way that other words are selected. Other research has found that fillers tend to be produced before low frequency and unpredictable words (Beattie & Butterworth, 1979; Levelt, 1983; Schnadt & Corley, 2006) as well as in cases where one's utterance involves semantic or syntactic ambiguity (Schachter, Christenfeld, Ravina, & Bilous, 1991).

Repairs

A prominent type of dysfluency is a repair, which occurs when speakers realize an error has been made and offer a correction, such as *The committee decided to move the motion, I mean meeting, to next Tuesday.* Nooteboom (1980) investigated the repairs that speakers make and do not make during speaking. He found that 64% of the 648 speech errors in his sample were repaired. Usually, the repair occurred immediately following the error. The existence of repairs suggests that an important part of speech production is self-monitoring on the part of the speaker. Levelt (1983) described repairs as involving three stages. In the first stage, speakers detect the error during self-monitoring and then interrupt themselves. Nooteboom's (1980) study revealed that speakers rarely interrupt themselves in the middle of a word; they produce the repair following the word containing the error. In the second stage, speakers produce a filled pause or filler phrase, such as *uh, sorry, I mean*, or some other expression to indicate that an error has been detected. Last, the utterance is repaired when the speaker restates the entire utterance or repeats the word or phrase that contained the error.

Do Researchers Study Naturally Occurring Speech Errors?

Research has shown that although speech errors occur infrequently, they do occur regularly (Fromkin, 1971). History reveals that the study of speech errors is not a recent endeavor. Fromkin discovered that the earliest study of speech errors using **naturalistic observation** occurred in the 8th century; the Arab linguist al-Kisaa'i published the book *Errors of Populace* in which speech errors we discussed (cited by Fromkin, 1988). Fromkin also noted that researchers in the late 1800s were also intrigued by speech errors; Meringer and Mayer also used naturalistic observation to collect over 8,000 slips of the tongue (Meringer, 1908; Meringer & Mayer, 1895/1978).

Fromkin pointed out that the primary method for studying speech errors involved collecting them over time through naturalistic observation, which refers to researchers observing events as they occur in the environment without intervention on the part of the researcher. Naturalistic observation is used frequently by researchers studying the behavior of animals in the wild. The same technique was well suited for observing speech errors produced by people during the course of their typical conversations and interactions. Because speech errors occur relatively infrequently, researchers must carry out naturalistic observations over long periods of time, carefully recording each observation. This procedure is referred to as the **diary method.**

Observations of Children's Speech Errors

An important question is whether everyone is susceptible to making speech errors. There was a time not so long ago in which many people assumed that children did not produce speech errors. Jaeger (2005) reported a personal communication in which a fellow researcher expressed this belief. As you learned in Chapter 4, most children do make mistakes when speaking. For example, they make overregularization errors (e.g., **braked* or **holded*; the asterisks indicate ungrammatical words). Jaeger (2005) wondered whether children experience the same types of slips of the tongue that adults experience. She carried out a naturalistic observation of her own children over a period of 10 years. Her collection of errors totaled 1,383. The types of errors observed in her corpus were highly similar to the types of errors identified by Fromkin (1971). One of the most memorable examples from Jaeger's corpus was produced by her daughter Anna when she was 4 years old. Anna was performing a play of *The Three Little Pigs* with her cousins for her parents, grandparents, and uncles and aunts. After the Big Bad Wolf demanded, "Little Pig, Little Pig, let me come in!" Anna, the Little Pig, replied, "No! Not by the chair of my hinny hin hin!"

Observations of Signing Errors

One might also wonder whether speech errors are limited to only speaking. Do users of signed languages experience the same types of errors as speakers of oral languages?

Research suggests that the answer is yes. Users of signed languages make errors during sign production that are similar to the errors that speakers make. Newkirk, Klima, Pedersen, and Bellugi (1980) reported a study of **slips of the hand** for which they collected 131 sign language errors.

Methodological Issues

The collection of speech errors generally involves writing down, as precisely as possible, what the speaker produces and also what the intended utterance was. In some cases, this may be straightforward, such as when a speaker notices the error and follows up with a repair. For example, one might say, "The papers are due on Friday… I mean, Monday. The papers are due on Monday." Often, speakers will not realize that they have made an error. In these cases, the listener may also not realize there has been an error and will take away an unintended message. Occasionally, the listener may be able to use contextual information or prior knowledge to infer that the speaker has misspoken. Anytime that one must rely on speculation, there is the opportunity for researcher bias to influence the results.

Another challenging aspect of naturalistic observation for the study of speech errors is that the researcher may experience **a slip of the ear**, which involves the mishearing of what another person says. In environments with background noise and poor acoustics, hearing others clearly can be challenging and slips of the ear may occur more often than in settings in which the noise level can be kept at acceptably low levels and be tightly controlled. In order to reduce the extent to which slips of the ear influence one's conclusions about speech errors, it is useful to make notes in the diary entry for each error recorded about the situational context in which the error occurred. A researcher might record whether the error occurred indoors or outdoors, how far away the speaker was when the error occurred, what the noise level was in the area, and whether there were any other distractions occurring at the same time.

Models of Speech Planning

What Stages of Processing Occur During Speech Planning?

Understanding how speakers plan utterances is challenging, because it occurs so rapidly with little conscious effort. One might envision the process of planning an utterance as involving at least four phases. First, one must choose the meaning of the utterance. As you learned in Chapter 3, the meaning of an utterance is referred to as its illocutionary intent (Austin, 1962). Second, one must choose the structural characteristics of the utterance, such as whether the utterance will be in the form of a question or a statement and whether the utterance will be formed to be overly polite or more direct given the norms of the speaker's language. Third, one must choose the specific words and phrases that one wants to include in the sentence structure. Last, one must physically articulate the utterance. The production of the utterance is referred to as the locutionary act.

Table 9.2 Fromkin's Six Stages of Speech Production

Stage 1:	Selection of Meaning for the Utterance
Stage 2:	Selection of Syntactic Structure
Stage 3:	Selection of Intonational and Stress Contour of Utterance
Stage 4:	Selection of Content words
Stage 5:	Selection of Function Words, Suffixes, and Prefixes
Stage 6:	Selection of Phonetic Units

Source: Fromkin (1971).

Fromkin's Model of Speech Errors

In 1971, Victoria Fromkin published a groundbreaking article on speech errors. She developed a classification system for different types of speech errors. She also published a model of speech planning using her own sample of naturally occurring speech errors. Her theory explained that different types of errors occur during different stages of speech planning. Fromkin proposed that there are at least six stages involved in speech production. These stages are described in Table 9.2. Imagine that you wanted to say, "Do you plan on going to the free movie at the student union tonight?" First, you would need to select the meaning of the utterance and then the sentence structure. You would need to determine the amount to stress the words in the sentence and the overall intonational pattern of the sentence. The model stipulates that you would select the content words that you would use in the sentence before you would select the function words, suffixes, and prefixes. Last, you would determine the phonetic units that you would set out to articulate.

Fromkin (1971) explained that in many cases, speech errors may be linked to one of the six stages. For example, a speech error such as *put the car in the gas* is an exchange error involving the two nouns *car* and *gas*. Such an error would likely occur during Stage 4, when speakers select content words. Consider a second type of speech error, saying *singing sewer machine* instead of *Singer sewing machine.* This is an exchange error involving the two suffixes *-er* and *-ing*. Such an error would likely occur during Stage 5, when speakers select function words, suffixes, and prefixes for their utterance. If one were to say *put down you weaplon,* the error would likely occur during Stage 6, when speakers are selecting the individual phonemes of each word.

Garrett's Model of Speech Errors

Garrett (1975) analyzed different types of exchange errors. He found that most of the exchanges occurred within the clause. Some involved the exchange of whole words and others involved morphemes or phonemes. His analysis revealed that the locations within the utterance that were involved in the morpheme and phoneme exchanges were one word apart or less. In contrast, word exchanges can involve locations that are relatively far apart in the utterance. Word exchanges also usually involved two words of the same part of speech (e.g., noun vs. verb). Garrett's (1975) conclusion was that word exchanges are likely occurring earlier than Stage 4, probably as early as Stage 2, when speakers select the syntactic structure of the utterance.

Garrett (1980) noted that for some speech errors, the element that appears to have moved in the utterance undergoes a phonetic change. Consider the examples in 10. The morpheme that has wound up in the wrong location in utterance 10a is pronounced as an /s/ in *outs*. Had the sound been produced in the word *runs*, it would have been a /z/ as in *zoo*. The term ***accommodation*** is used to refer to cases in which the moved element takes on the phonetic characteristics of the location to which it moved. In 10b, the phoneme /l/ is moved, leaving its original word to begin with a vowel. The phonological change of the *a* to *an* must have occurred following the move. Had the phonetic segments of the utterance been selected before the move, the error would have been produced as "a anguage lacquisition." In 10c, the verbs *know* and *heard* are exchanged. The phonological changes associated with past tense morphology must have been made after the exchange occurred, because the verbs are produced in the tense that is consistent with the location where moved versus where the verbs originated in the utterance.

10. a. It certainly *run outs* fast. (runs out)
    b. An *anguage lacquisiton* (a language acquisition)
    c. I don't *know if I'd hear one if I knew it*. (know one if I heard it)

Dell's Model of Speech Errors

While both Fromkin's and Garrett's models of speech production envision speech production as occurring in a series of nonoverlapping stages, other researchers have claimed that speech production may be carried out via multiple stages of processing operating in parallel. Dell (1986) proposed that speech is planned using four interactive levels of processing: (1) semantic, (2) syntactic, (3) morphological, and (4) phonological. When information is activated at one level, activation can spread within the same level and also to other levels. Dell's approach can be viewed as similar in nature to the statistical learning approach to language, as processing is not assumed to be carried out in a modular fashion.

Some individuals become exceptionally skilled in planning utterances. They are able to change the sound of their voice so that the sound appears to be emanating from locations away from their bodies. Such individuals are called ventriloquists. Text Box 9.2 provides information about ventriloquism and its unusual history.

What Have Laboratory Studies of Speech Production Revealed?

Over the past four decades, researchers have moved the study of speech production into the laboratory. Unlike studies carried out in the natural environment, laboratory studies enable researchers to exert some control over the setting, such as making sure that the room is quiet and utterances are recorded. Researchers may also manipulate what participants experience. In speech production experiments, participants are often asked to produce carefully prescribed utterances, while the researcher waits patiently for speech errors to occur. One technique that has been used is the **phonological bias technique.** Participants

Text Box 9.2 Extraordinary Individuals: Ventriloquists

Ventriloquism is a form of entertainment in which performers work with one or more puppets or dummies for which they speak by *throwing* their voice. Ventriloquists became a staple of Vaudeville acts in the 19th century (Warren, 2011). The term *ventriloquism* comes from the Latin word *venter*, which means *belly*, and the word *loqui*, which means *speak*. The most skilled ventriloquists can produce the speaking voice for their dummy without appearing to move themselves. Mastering the art of ventriloquism requires that one not only learn how to vary the acoustic properties of one's voice but also learn how to divert the attention of audience members during the performance. The performer must convince the audience that the dummy is alive and talking (Mason, 2010). Varying the acoustics of the voice involves using the muscles of the diaphragm and the placement of the tongue in the mouth differently than during typical speaking. With practice, the skilled ventriloquist can project speech to appear to be originating from any location in the nearby environment. The most talented performers can produce a wide variety of speech sounds for the dummy without others being able to detect the movement in their lips, face, and throat.

▲ Photo 9.2 A ventriloquist performs on stage. Have you ever tried to *throw* your voice as a ventriloquist does?

In the ancient world, ventriloquism was part of religious practices. The Greeks believed that the noises produced by the stomach were the voices of the dead (Warren, 2011). This practice was called *gastromancy*. The person who produced the noises interpreted the voices for others and also made predictions about the future. One such individual was the priestess of the temple of Apollo in Delphi. She channeled the wisdom of the Oracle, using this type of ventriloquism. By the Middle Ages in Europe, the practice was categorized among the practices included in witchcraft. However, similar practices have been observed in other cultures around the world, including the Maori in Australia, the Zulus in Africa, and cultures of the Arctic Circle.

Famous ventriloquists of the 20th century include Edgar Bergen and his dummy Charlie McCarthy and Señor Wences, who created the character Johnny, which was a childlike face drawn on Wences's hand. One of the most familiar ventriloquists working today is Jeff Dunham.

are presented with word pairs that they are instructed to pronounce quickly. Consider the following list from Baars (1980):

ball doze

bash door

bean deck

bell deck

darn bore

The first four pairs of words share the same initial phonemes. The last word pair has those phonemes switched. When participants process the last word pair, they are in the practice of producing a b-word followed by a d-word; researchers would say that they are *primed* to produce the phonetic pattern. Baars (1980) found that on 30% of the trials, participants made a slip of the tongue, saying *barn door* instead of *darn bore.*

Lexical Bias Effect

Studies using this technique showed that some types of errors occur more frequently than others. Baars, Motley, and MacKay (1975) found that errors that result in real words, as *barn door,* are more common than errors that result in nonwords. Consider the following list of word pairs. Participants produced an error on the last word pair only 10% of the time, producing *bart doard.* The term ***lexical bias effect*** refers to the fact that speech errors are more likely to result in an actual word versus a nonword.

big Dutch

big doll

bill deal

bark dog

dart board

Preventing Errors

Other studies utilizing the phonological bias technique have shown that participants' speech errors can be influenced by the situation. For example, Motley (1980) showed that male participants made more sexually related errors (e.g., when asked to pronounce the nonsense word pair *bine fody,* they sometimes said *fine*) when the research assistant who administered the task was an attractively dressed female (vs. a male). In similar study, Motley, Camden, and Baars (1981, 1982) presented participants with word pairs that primed them to produce a vulgar word. The results showed that participants were significantly less likely to produce an error involving a vulgar word than an error involving a neutral word (e.g., reading *tool kits* as *cool tits* occurred less often than neutral errors). The results suggested that speakers must have an ability to inhibit, at the last moment before production, an utterance containing a vulgarity. In order to do so, participants must be engaging in active self-monitoring.

Verb Agreement Errors

Bock and Cutting (1992) investigated how speakers produced sentences such as those in 11. A common error that they observed involved the agreement between the verb and the subject noun phrase. Singular nouns are used with the singular form of the past tense version of the verb *to be*: *was.* Plural nouns are used with the plural form of the past tense version: *were.* They found that more subject–verb agreement errors were made when the subject noun and the verb were separated by an intervening clause, as in 11a, than when there was no intervening clause, as in 11b.

11. a. *The report that they controlled the fires were printed in the paper.
    b. The report of the destructive fires were accurate.

What Are the Units of Speech Planning?

The Role of Clauses in Speech Planning

Researchers also disagree about how the lexical aspects of speech are planned. Some have argued that the clause is the unit of planning (Martin, Crowther, Knight, Tamborello, & Yang, 2010; Martin & Freedman, 2001; Martin, Miller, & Vu, 2004; Smith & Wheeldon, 1999). Others believe that lexical planning occurs word by word (Griffin, 2003; Griffin & Bock, 2000; Meyer, Sleiderink, & Levelt, 1998). Support for the view that lexical planning occurs word by word was obtained in experiments in which speakers' eye movements were recorded during a laboratory session in which visual displays were presented (Griffin & Bock, 2000; Meyer et al., 1998). Of particular interest was how quickly participants fixated on an object in the display prior to saying the word. The authors reasoned that if speakers planned more than one word of an utterance at the same time, then they might spend more time looking at objects that are named earlier in the utterance than objects that are named later in the utterance. Griffin and Bock (2000) found that participants looked at objects for a similar amount of time regardless of whether they were produced earlier or later in an utterance. The time between looking at the object and speaking the word also did not differ as a function of whether the word occurred earlier or later in the utterance. Participants' fixations and time to start speaking were influenced by the words' frequency of occurrence. They concluded that speakers plan the phonological sequence for the first object in the utterance before they plan the phonological sequence for the second object in the utterance. Similar conclusions were drawn by Spieler and Griffin (2006), who recorded eye movements as speakers produced sentences having the form *the A and the B are above the C.* The time that participants spent looking at the object and the time between the fixation and the pronunciation of the words in the A position of the utterance were influenced by word frequency of the name for the object but were not influenced by characteristics related to the object named in the B position.

The view that utterances are planned in clausal units suggests that speakers activate the lexical representations of all content words in the clause before they start speaking (Allum & Wheeldon, 2007; Martin et al., 2004; Smith & Wheeldon, 1999). A number of studies have shown that the speed at which participants start utterances is influenced by how similar the words in the clause are. For example, Damian and Dumay (2007) discovered that English speakers started their utterances faster when the final word in a determiner–adjective–noun sequence (e.g., the blue bell) began with the same phoneme as a distractor word. Miozzo and Caramazza (1999) observed similar results in Italian and Alario and Caramazza (2002) in French.

The Role of Syllables in Speech Planning

A growing body of research suggests that when speakers plan the phonological aspects of speech, the unit of planning is the syllable rather than the individual phoneme (Cholin, Levelt, & Schiller, 2006; Levelt, Roelofs, & Meyer, 1999; Levelt & Wheeldon, 1994). Levelt and Wheeldon (1994) reasoned that when frequency effects are observed, such as in word-naming tasks or word perception tasks, they occur only for units of information stored in memory. They investigated the possibility that syllables, like words, are stored in memory. If so, then during speech production, syllable frequency should influence processing. They compared the time that participants took to name two-syllable words that contained either high or low frequency syllable while controlling for overall word frequency. The results showed that syllables frequency strongly influenced naming time. Similar results were also observed in English by Cholin, Dell, and Levelt (2011). O'Seaghdha, Chen, and Chen (2010) failed to find similar effects for Chinese; they proposed that the planning units can vary across languages.

How Are Gestures Related to Speech?

When people speak, they often will simultaneously move their bodies and their hands. Sometimes gestures are described as nonverbal communication or body language (Fromkin, Rodman, & Hyams, 2013). The term *gesture* is used to describe bodily movements that are used in communication. For example, one might wave with one's hand to express *goodbye* or give a nod of the head to signal *yes*. Most of the gestures that people use have meanings that are defined within the culture. Text Box 9.3 describes a number of gestures that while offensive in some cultures are not offensive in others.

Text Box 9.3 Diversity of Human Languages: Gestures Around the World

A small number of gestures appear to mean the same thing across cultures. For example, when individuals point their index finger, they are likely signaling that others should turn their attention toward that direction. However, the meanings of most gestures vary across cultures. The most familiar examples of gestures whose meanings vary across cultures are those that have a positive or neutral meaning in one culture and a negative or taboo meaning in another culture. For example, in the United States, one would signal *okay* by touching the index finger and the thumb to form a circle, while the other

(Continued)

(Continued)

fingers are held straight. An American might use the gesture to say that everything is good. However, in other parts of the world (e.g., South America and Europe) the gesture is offensive. The gesture is interpreted as meaning that the person to whom the gesture is made is a zero or a nothing. When one makes the gesture to another, it is extremely insulting. A second example is the peace sign that is used in the United States; one holds up the hand facing outward away from the gesturer with the index and middle fingers forming a V. The same sign was used during World War II to indicate *victory*. Today in England, when this sign is made with the palm of the hand held facing the gesturer, the gesture is offensive, meaning something similar to the middle finger in the United States. In many Asian nations, the same sign is often used in photographs when the person being photographed would like to appear cute. In the United States, the thumbs-up gesture is sometimes used by people needing a ride (i.e., hitchhikers) or to signal that everything is good. In some countries, such as Afghanistan, Iran, Nigeria, and parts of Greece, the gesture is similar in meaning to the middle finger in the United States. In the United States, people sometimes gesture for another to come closer by raising the index finger and moving it toward the gesturer several times. The gesture is viewed as generally neutral and something that you might use with anyone, even a stranger. However, in the Philippines, the gesture is considered extremely rude and demeaning, as though the gesturer were treating the recipient as lowly as a dog.

▲ **Photo 9.3** A young girl gestures to the photographer. Which gestures do you use on a regular basis?

▲ **Photo 9.4** A young girl makes the *peace* sign. Have you ever made the peace sign when interacting with your friends?

▲ **Photo 9.5** A boy makes the *okay* sign. How common is it for people your age to use the okay sign?

McNeill's Approach to Gestures

The role of gestures in language production is not completely understood. However, researchers such as David McNeill (2005) have made a career out of trying to figure out the extent to which the production of speech and gestures are interrelated. McNeill's views are strongly influenced by the work of Lev Vygotsky (1934/1986), whose work emphasized the importance of the social context in learning. You may recall reading about Vygotsky in Chapter 1. McNeill proposes that the planning of speech and gestures involves the same conceptual processes. He views gesture and speech as intimately connected to one another and also tied directly to the social context. Research by others has shown that the production of gesture occurs almost exclusively during speaking Stephens (1983) analyzed hundreds of hours of videotaped interactions and found only one gesture that was produced by a listener, rather than a speaker.

Since the publication of a McNeill's (1985) article in *Psychological Review*, there have been numerous studies of gesture production during speaking. He described the performance of participants who were asked to describe an event presented to them in a cartoon. He found that speakers' nonverbal behavior was closely linked to their verbal behavior. Examples are shown in 12.

12. a. He tries going up, inside the drainpipe.
    [hand rises and points upward]
    b. He tries climbing up the drain spout of the building.
    [hand rises and starts to point upward]
    c. He goes up through the drainpipe this time.
    [hand rises quickly and fingers open to form a basket]

McNeill (1985) reported that 90% of all the gestures that his participants produced occurred when they were speaking. The remaining 10% of the gestures were made during a period of silence that was followed by additional speech. Of the gestures made, McNeill identified two distinct types. The first type was "simple and rapid hand movement of a type that usually accompanies words whose importance depends on multi-sentence text relations" (McNeill, 1985, p. 354). He referred to these as beats; they functioned as an indication of a break in the utterance. The second type of gesture "symbolize[d] concepts about language and communication." He referred to these as conduit gestures. They were produced during times in which it appeared that speakers were unable to verbalize. In such cases, gestures maintained the flow of the interaction. McNeill concluded that because gestures are so closely tied to speaking and because they function in important ways in communicating meaning, they must be viewed as a core part of the language production process, rather than a peripheral, unimportant part of the process.

Since 1985, McNeill and his students have conducted numerous studies of gestures using the careful analysis of videotapes of individuals gesturing while speaking (McNeill, 2005). By analyzing long stretches of interactions, McNeill has been able to dissect the utterances in terms of their conceptual planning. The results have shown that observers can identify points in time that correspond to the start of an utterance. McNeill (2003) referred to this as the growth point. These studies have also shown that speakers sometimes produce the same gesture multiple times over the course of an interaction. McNeill referred to this as catchment. The repetition of gestures in an interaction appears to maintain the cohesion in the discourse. The conclusion of such studies strongly supports the view that the careful coordination of speech with gestures enables speakers to achieve effective communication.

Despite the mounting evidence in favor of the view that gestures play a vital role in human communication, there are researchers who believe that they do not. For example, Krauss and colleagues (Krauss, 1998; Krauss, Chen, & Chawla, 1996; Krauss, Chen, & Gottesman, 2000) claim that gestures contribute little to verbal interactions under most circumstances. Hostetter (2011) conducted a meta-analysis of 63 studies in which gesturing during speaking was investigated. She concluded that gestures play a role in communication. She found that gestures vary systematically as topics vary. For example, gestures that represent actions facilitate communication more than gestures representing more abstract topics. Second, when gestures do not overlap with the accompanying verbal information, communication is facilitated more than when gestures are redundant with the accompanying verbal information. Last, children appear to benefit more from a speaker's gestures than do adults.

Evidence That Gesturing Facilitates Communication

The view that gestures facilitate communication was supported by a particularly clever experiment conducted by Rauscher, Krauss, and Chen (1996). They investigated how preventing gesturing during verbal interaction might affect one's speech. They had participants describe an event presented to them in a cartoon, as McNeill (1985) had done previously. Half of the time, they allowed participants to gesture freely during their descriptions. For the other half of the time, participants were instructed to keep their hands touching an electrode, which would measure activity. The researchers also varied the lexical content that participants should use in their descriptions. A third of the time, they were asked to use normal speech. A third of the time, they were asked to use obscure words to describe the event. A third of the time, they were asked to avoid using words with a particular phoneme in their descriptions. The results showed that participants' speech was more fluent, involving fewer pauses, when they could gesture than when they could not gesture but only when participants described spatial aspects of events. The authors concluded that gesturing during speaking facilitated participants' lexical processing during speech planning—specifically access to the mental lexicon. Other research has shown that speakers produce more gestures and also larger gestures when listeners can see the speaker (Alibali, Heath, & Myers, 2001; Cohen & Harrison, 1973). Speakers also gesture more frequently when they perceive that the topic of conversation is unfamiliar to listeners (Jacobs & Garnham, 2007). Speakers also use larger gestures when the content of the communication is unknown to the audience (Holler & Stevens, 2007).

Recent research suggests that the view that gesturing during speaking is associated with processing difficulties may be incorrect. Hostetter and Alibali (2008) proposed that gestures are brought about during speech because of increased activation in the mind's mental imagery system. They referred to their theory as gesture-as-simulated action (GSA). Sassenberg and van der Meer (2010) conducted an experiment designed to test the contrasting predictions of the two views of gestures. In the experiment, participants described a mapped route to listeners. The complexity of the route was varied. They found that gestures were produced when the spoken information was easier versus more difficult, such as when information was repeated from previous utterances. The authors concluded that their data were most consistent with Hostetter and Alibali's (2008) GSA view.

Intriguing research by Goldin-Meadow and colleagues demonstrated that the gestures of children can convey information to adults in the environment. Goldin-Meadow, Goodrich, Sauer, and Iverson (2007) investigated how 10 mothers responded to the gestures of their infants over a 24-month period. Observations occurred from the time the infants were producing one-word utterances until the infants produced two-word utterances (i.e., between 10 and 23 months of age). The results showed that mothers interpreted infants' gestures into words and acted upon their inferred meaning. Previous research by Goldin-Meadow and Sandhofer (1999) showed that adults could reliably infer meaning from children's gestures that was not present in the children's speech. The children were between the ages of 5 and 8. In the study, adults viewed videos of children performing a series of conservation tasks in which speech and gestures occurred. The results showed that adults engaged in accurate "gesture-reading," suggesting that even in children, gestures can provide an important source of information during communication.

Problems Affecting Speech

Problems with speech can result from a variety of causes. Most often, the cause is a problem with physical aspects of the parts of the body involved in producing speech. The vocal cords play a vital role in the production of speech. Sound is produced when air is forced through the vocal folds. Phonation refers to the production of sound by the opening and closing of the vocal folds. The vocal folds of men are generally larger than the vocal folds of women. The larynx in men is also larger than the larynx in women. These size differences result in men's voices being lower than women's voices. The term *pitch* is used to refer to the lowness or highness of the voice, which is determined by the number of sound waves produced by the vibration of the vocal folds in an interval. High-pitched sounds involve a greater number of sound waves per interval than low-pitched sounds.

The physical act of producing speech is the result of a highly orchestrated set of muscle movements involving three different systems. Problems with speech production may involve one or more of these systems. The supralaryngeal system involves the vocal tract above the larynx, which includes the jaw, tongue, teeth, and lips. The laryngeal system includes the vocal cords, which are two folds of muscle that vibrate to create voicing. In Chapter 3, you learned about the difference between voiced and voiceless phonemes. Whispered speech is also produced without the vocal cords in motion. The respiratory system includes the lungs and the muscles of the torso that are used to bring air into the lungs and to force air out of the lungs. The air that moves in and out of the lungs passes through the laryngeal and supralaryngeal systems. As you learned in Chapter 3, the position of the larynx in the vocal tract enables speakers to produce rapid consonant and vowel changes typical of adult speech. The position of the larynx is low in the throat near the opening of the lungs, which makes choking a hazard.

What Are the Most Common Problems Affecting Speech?

Approximately six to eight million people in the United States have some type of language impairment (National Institute on Deafness and Other Communication Disorders, n.d.). As you learned in Chapter 2, adults who experience a brain injury may experience difficulty when speaking. The term *aphasia* refers to any language deficit resulting from brain damage. Broca's aphasia involves extreme difficulty in articulating speech, as well as problems producing function words. Individuals who experience some form of aphasia affecting speech may have to relearn how to speak and spend long hours in speech-related rehabilitation therapy. Unlike aphasia, which is an acquired disorder, many speech production disorders are observed in young children and are described as developmental disorders. Other speech production disorders affect individuals who have no trouble producing sound; however, their speech is deficient in some way. For example, some individuals have an articulation disorder, which is characterized by difficulty producing speech sounds accurately. Speech sounds may be omitted from utterances, such as the ends of words (e.g., /fi/ for *feet*). Such errors are called **omissions** or deletions. Errors may involve also substitutions, as when one sound appears instead of another (e.g., *Tandace* instead of *Candace*). Sometimes, extra sounds may be produced in a word, such as *flish* instead of *fish*. These errors are referred to as additions.

Speaking Problems Affecting Children

Between the time that the child begins speaking and the time the child enters the early grades of elementary school, problems with speech may be detected. Speech therapy is a fairly common experience for elementary school children; most children need only a few sessions to correct their particular speaking problems. However, there are some speech problems that do not resolve themselves during childhood but may affect individuals into adulthood. Children may routinely mispronounce certain speech sounds or repeatedly substitute one sound for another. They may be able to perceive the difference between two confused speech sounds in the speech of others but be unable to perceive the distinction in their own speech. This has been referred to as the **fis phenomenon.** The term stems from Berko and Brown's (1960) description of a child who referred to a toy fish as a *fis.* When others asked him, "Is this your fis?" he would say no. When others asked him, "Is this your fish?" he said, "Yes, my fis." This case and others like it indicate that one should not assume that a child's errors in producing speech sounds always correspond to the child's perceptions of speech sounds.

Some difficulties with phoneme production persist into adulthood. For example, some individuals speak with a **lisp**, which is a speech impediment in which one pronounces /s/ as the *th* in *thin* and pronounces /z/ as the *th* in *that.* There are four types of lisps. Some individuals have an interdental lisp, which is produced with the tongue extended in between the front teeth. A dentalized lisp occurs when the speaker pushes the tongue against the front teeth. A lateral lisp occurs when one produces air around the tongue during speaking, which results in an excessively wet sound. Last, a palatal lisp occurs when the speaker's tongue frequently touches the roof of the mouth. Some individuals have difficulty producing /r/, such as Barbara Walters and the beloved cartoon character Elmer

Fudd. The term ***rhotacism*** is used to describe this condition. Often, individuals replace the /r/ sound with /w/, as in *I am going to get you, you wascally wabbit!*

The expression **tongue-tied** is sometimes used to describe individuals with lisps or other speech impediments; however, the expression stems from the tongue being abnormally connected to the floor of the mouth (Williams, 2012). The connective tissue that connects the tongue to the mouth is called the **frenulum.** The condition is known as ankyloglossia and is known to affect speech in children, particularly the phonemes /s/, /z/, /l/, and /r/ (Messner & Lalakea, 2002). Typically, these children do not experience delays in language development. A surgical procedure to cut the frenulum to allow greater movement of the tongue can lead to improvement in articulation. However, some individuals experience difficulties in articulation long after surgery.

Most children will show a behavior known as tongue thrust, which can also negatively affect speech (Williams, 2012). It occurs when the resting tongue extends through the incisor teeth during speech and also during swallowing. The behavior typically disappears in most children by the age of 6; however, in some children, it persists due to an orofacial muscular disturbance. Children with enlarged tongues, large tonsils, or allergies that cause chronic nasal congestion may be affected more often than others. Children who suck their thumbs or who are bottle-fed are also affected more often than other children. Over time, children with tongue thrust may develop problems with their bite due to the tongue applying pressure on the teeth, which can cause teeth to become misaligned.

A common cause for articulation problems is hearing loss. If children are unable to hear clearly the sounds of their language, they often are unable to reproduce those sounds. In some cases, children may have multiple ear infections, which result in a buildup of fluid in the middle ear. This condition is referred to as **otitis media.** Research has shown that otitis media is associated not only with problems in language development, such as articulation problems, but also with the development of social skills (Roberts, 1997).

Stuttering

One of the most familiar speech production disorders is **stuttering.** Stuttering occurs in about 5% of the population (Mansson, 2000). Some individuals who stutter may have tremendous difficulty communicating at all with others. Other cases of stuttering may be mild, enabling speakers to communicate reasonably well. The causes of stuttering remain elusive. Four types of stuttering are generally recognized. The most common type of stuttering is developmental stuttering, which affects children who are acquiring language. Mechanisms involved in speech planning may be developing and cause dysfluencies. This type of stuttering resolves itself as the child matures. Approximately 65% of children who stutter will recover within 2 years (Yairi, 1993); about 74% will recover by the time that they are teenagers. A second type of stuttering is genetic in origin. Stuttering has long been known to occur within families. Recent breakthroughs in genetic research have identified three genes associated with stuttering. Stuttering that has a genetic origin does not resolve itself with maturation but may be a lifelong issue. The third type of stuttering is neurogenic stuttering, which occurs following brain damage, such as a stroke or trauma to the head. The areas of the brain affected by the brain injury are those involved in the planning and/or articulation of speech. The fourth type of stuttering is psychogenic stuttering, which occurs following extreme emotional trauma or because of a thought disorder. It is interesting to note that at one time in history, all stuttering was believed to be rooted in psychological, rather than physical processes.

The speech of those who stutter may include three types of dysfluencies (Williams, 2012). Part-word repetitions involve the repeating of a sound or syllable during articulation, such as *buh buh buh buh bucket* and *muh muh muh muh muh mouse.* Prolongations occur when one continues making a sound for much longer than needed, perhaps because there is difficulty transitioning to making the other sounds contained in the word, such as *mmmmmouse* or *ssssssssssnake.* Last, **hesitations** involve filled or unfilled pauses during speech, such as *My name is um um um......Billy* or *My ..... name ......is ..... Billy.* Those who stutter may also exhibit nonverbal behaviors, which have been developed over time in response to the physical and social difficulties experienced during speaking. These might include specific patterns of eye blinking, foot-tapping, and arm moving.

Those who stutter may face problems dealing with the stigma associated with stuttering. Peers may not be understanding and isolate the stutterer. Negative social experiences may lead to an increased desire to conform to the will of others (Heaven, 2001) and also social anxiety (Blood, Blood, Tellis, & Gabel, 2001). Some studies have found that most stutterers report having been bullied during their school years (Hugh-Jones & Smith, 1999). Many stutters develop negative attitudes toward school and fail to thrive academically (Blood et al., 2001). A possible contributor to academic problems is low self-esteem, as stutterers tend to view themselves as incompetent. Van Riper (1982) suggested that stutterers are chronic underperformers because of their excessive fear of failure. Because stutters expend so much effort in producing speech and attempting to control their stuttering, they may develop problems concentrating on other tasks. Over time, the problems concentrating may contribute to problems learning (Daniels, 2007).

Stutterers are found among the world's most famous individuals. Winston Churchill was a stutterer as was King Edward IV, whose experiences with speech therapy were the inspiration for the 2010 film *The King's Speech.* James Earl Jones, the voice of Darth Vader in the *Star Wars* films, as well as Jack Welch, business tycoon, were both stutterers.

Apraxia and Dysarthria

Some problems with speech production arise due to motor disorders. For example, **apraxia** of speech is characterized by problems producing sounds, syllables, and words. The disorder stems from the brain's inability to plan speech properly. Speech plans involve sending signals to the different parts of the vocal tract (i.e., jaw, tongue, and lips). Individuals with apraxia experience a great deal of frustration, because they know what they would like to say, but they are unable to coordinate the physical movements required to produce the utterance. Speech production may also be impaired when one's muscles are paralyzed or weakened in one or more of the systems involved in speech (e.g., respiratory system, face, mouth, or throat). **Dysarthria** is the term for this type of speech disorder. Individuals with dysarthria may exhibit one or more of the following symptoms: slurred speech, slowed speech, mumbled speech, whispered speech, an odd rhythm in speech, and unusual vocal characteristics (e.g., hoarseness or breathiness).

Cleft Palate

Speech-language pathologists treat numerous rare conditions that impact speech. For example, clefts of the palate and/or lip generally lead to a child receiving help with speech. Clefts are caused by a variety of factors, including genetic abnormalities, maternal nutrition during pregnancy, and toxins ingested by the mother during pregnancy (e.g., medications). **Cleft palate** involves a hole in the roof of the mouth, which is also the floor of the nose; the hole involves missing or misplaced bone and muscle (Peterson-Falzone, Hardin-Jones, & Karnell, 2001). The prevalence of cleft palate is between 1 in 2,500 to 1 in 3,000 live births. In contrast, cleft lip is more common, affecting between 1 in 250 and 1 in 1,000 births. Females are more likely to be born with cleft palate, but males are more likely to be born with cleft lip. There are differences in prevalence of cleft palate and cleft lip across ethnicities. Cleft lip occurs most frequently among people with Asian and Native American ancestry. Cleft palate occurs comparably often among people with Asian, Caucasian, and Native American ancestry; it is least common among people with African ancestry. There are more than 350 developmental disorders that involve cleft palate and cleft lip. For example, cleft palate is associated with most cases of Pierre Robin sequence. Others syndromes that may involve cleft palate include Apert syndrome, Crouzon syndrome, Stickler syndrome, Treacher Collins syndrome, Van der Woude syndrome, and velocardiofacial syndrome.

In the developed world, surgery to close the cleft is recommended around the age of 9 months (Peterson-Falzone et al., 2001). In undeveloped countries, surgery may occur when the child is older if at all. If the cleft is not closed, language development can be adversely affected. If there is a cleft of both the lip and the palate, the child may require multiple surgeries. Before the cleft is closed, children may produce hypernasal speech, which is caused by air being allowed into the nasal cavity through the opening of the cleft. After the cleft is closed by surgery, there may be some hypernasality, because the soft palate may be unable to open and close sufficiently during the production nasal sounds, such as /m/, /n/, and /ng/, and nonnasal sounds. Consonant sounds are often articulated with a burst of air being emitting from the nostrils. Children with cleft palates may also have difficulty producing the /k/ and /g/ sounds, because these are normally produced in the back of the mouth near the opening to the throat. Children with cleft palates may produce them using the front rather than the back of the mouth. In addition, children with cleft palates may have additional speech problems stemming from other problems such as hearing loss. The incidence of hearing loss for non–cleft palate children is about 20%; among children with cleft palates, it is 90% (Wysyznski, 2002). Most children born with clefts are able to achieve normal speaking ability by 5 years of age.

Cluttering

A particularly rare speech disorder is **cluttering.** It can be severe, and its cause is unknown. Individuals with cluttering are difficult to understand because of their fast rate of speaking and erratic speech rhythms. They may also produce utterances that are ungrammatical and/or contain words that are nonsensical. Those diagnosed with cluttering may not appear to be aware that their speech is difficult to understand. They may appear to be inattentive, disorganized, and unable or unwilling to listen. Unlike stutterers who experience muscle tension and difficulty with the physical movements associated with speech, clutterers find the physical act of speaking easy to achieve. Some clutterers may have problems with reading and writing. Individuals with cluttering rarely seek speech therapy on their own; they typically do not realize that there is a problem with their speech. Therapy is usually the suggestion of a parent or friend.

What Are Some Common Treatments for Speech Disorders?

Individuals who wish to improve their speech can receive speech therapy, which can focus on improving one's ability to produce speech and/or improving one's ability to comprehend speech. Professionals with a master's degree who help children and adults improve their speaking ability are speech therapists (also called speech-language pathologists).

Many of the techniques that are used to improve speech have been built on the principles of **operant conditioning,** which was learned about in Chapter 1. Desirable speech behaviors, such as clear, correct articulations, can be increased through the application of **reinforcements,** which are applied in the attempt to increase the frequency of desirable behaviors. With very young children, colorful stickers may be awarded for each successful production of the target speech sound or word. Undesirable speech behaviors, such as hesitations and distortions in speech, can be decreased through the application of what B. F. Skinner called **punishments.** Punishments include anything that is designed to decrease the frequency of an unwanted behavior, such as the removal of something desirable, such as a prize or candy.

The use of the principles of operant conditioning for speech therapy and other types of therapies is called **behavioral modification.** A typical treatment plan for speech therapy involves presenting patients with examples or models of the ideal speech sounds that they have trouble producing. The speech pathologist constructs tasks and exercises that help the patient to progress toward being able to produce speech similar to the model. Patients may be taught how to become more aware of their speech and their breathing and how to exert better control of their speaking and breathing rate. They may be asked to practice producing speech under various conditions, such as talking fast or slow, using short or long utterances, or singing words versus simply speaking them.

Summary and Theoretical Implications

Although the study of speech production started with the study of naturally occurring speech errors, over the past 40 years, researchers have developed a variety of laboratory tasks to investigate how speakers plan their utterances. Research supports the view that utterances are planned in a series of stages. Exchanges involving whole words are likely to result from a mistake occurring earlier than exchanges involving individual phonemes. Speakers also appear capable of inhibiting the production of a word immediately prior to articulation. Researchers have developed a variety of techniques to study speech errors and speech production in the laboratory. Some of this research supports the view that speakers plan utterances a word at a time. Other results suggest that speakers plan utterances a clause at a time. There are a wide variety of disorders related to speaking and speech planning. Some speech problems are acquired following physical injury; other speech problems are developmental in nature (i.e., arising early in childhood). Those who have problems speaking can receive services from professionals specializing in speech and language disorders.

The data that have been obtained in the numerous studies investigating speech production can most easily be addressed by the generative and statistical learning approaches to language and are less easily accounted for the behaviorist and social-interactionist approaches. Researchers advocating for by the generative approach would presume that processing occurring during speech planning and speech production would involve the application of grammatical rules carried out in a modular fashion (i.e., grammatical information being used differently in time or in manner from nongrammatical information). Researchers advocating for the statistical learning approach would presume that all types of information could be used interactively throughout processing. In terms of treating speech-related problems, the behaviorist and social-interactionist approaches may each have some benefit to offer. Behaviorist learning principles are often applied when trying to improve speaking ability. It is also possible that improving speech, particularly in children, can be influenced by the extent to which the individual and the therapist have a productive social relationship.

KEY TERMS

accommodation
additions
anticipations
apraxia
behavioral modification
blends
cleft palate
cluttering
deletions
diary method
dysarthria
exchanges
fis phenomenon
frenulum
Freudian slip
hesitations
lexical bias effect
lisp
malapropism
naturalistic observation
omissions
operant conditioning
otitis media
perseverations
phonological bias technique
punishments
reinforcements
rhotacism
shifts
slip of the ear
slips of the hand
slip of the tongue
Spoonerism
stuttering
substitutions
tachylalia
tongue-tied

REVIEW QUESTIONS

1. How are Freudian slips different from other slips of the tongue that occur in daily life?
2. What are the eight types of speech errors that have been identified by speech error researchers?
3. What did Fromkin's work on speech errors suggest about how speakers plan their utterances?
4. Describe Fromkin's (1971) stages of speech production. What types of planning occur in each stage?
5. Garrett (1975, 1980) proposed that speech errors provide evidence that phonological planning occurs after some exchange errors. Provide an example of an exchange error that serves as an example, and explain what aspects of the error support his view.
6. What evidence is there that children produce speech errors?
7. What evidence is there that speech errors occur in manually produced speech (i.e., signed languages)?
8. Discuss the extent to which human gestures are similar or different in meaning across cultures.
9. What did Baars's experiments on speech errors suggest about speakers' ability to cancel an utterance relatively late into its planning?
10. What evidence is there that the context in which speech is produced can influence the likelihood that speech errors will occur?
11. What evidence is there that utterances are planned a word at a time?
12. What evidence is there that utterances are planned a clause at a time?
13. What are David McNeill's views about the role of gestures in speaking?
14. What did Hostetter's (2011) meta-analysis suggest about the role of gestures in human communication?
15. What is known about the causes of stuttering?
16. What is behavior modification? How is it used to treat speech problems?
17. What are the symptoms of cluttering? How do they differ from stuttering?
18. What is a cleft palate? How does it influence speech production?

RECOMMENDED READING

Crimmins, C. (2001). *Where is the mango princess: A journal back from brain injury.* New York: Vintage.

Duncan, S. D., Cassell, J., & Levy, E. T. (Eds.). (2007). *Gesture and the dynamic dimension of language.* Philadelphia: John Benjamins Publishing Company.

Fromkin, V. (Ed.). (1980). *Errors in linguistic performance: Slips of the tongue, ear, pen, and hand*. San Francisco: Academic Press.

Jaeger, J. (2005). *Kids' slips: What young children's slips of the tongue reveal about language development*. Sussex, UK: Psychology Press.

Levelt, J. M. W. (1993). *Speaking: From intention to articulation*. New York: A Bradford Book.

Lucas, S. (2003). *The art of public speaking*. New York: McGraw-Hill.

RECOMMENDED FILMS

Apted, M. (Director). (1994). *Nell* [Motion picture]. United States: Fox Home Entertainment.

Gosse, B. (Director). (1997). *Niagara, Niagara* [Motion picture]. United States: Shooting Gallery.

Hooper, T. (Director). (2010). *The king's speech* [Motion picture]. United States: The Weinstein Company.

Pritikin, G. (Director). (2003). *Dummy* [Motion picture]. United States: Lions Gate.

Winick, G. (Director). (1999). *The tic code* [Motion picture]. United States: Gun for Hire Films.

SUGGESTIONS FOR CLASS PROJECTS

1. Categorize the following list of speech errors collected by Fromkin (1971) as one of the eight different types of speech errors: (1) substitution, (2) exchange, (3) shift, (4) addition, (5) deletion, (6) anticipation, (7) perseveration, and (8) blend.

    1. She did all her homework on the commuter.
    2. At the end of today's lection, everything should be resolved.
    3. He wanted a kice cream cone.
    4. The words he missed were both prulal.
    5. The thought it hears different after the repair.
    6. She took a walk on the beak.
    7. It's too hot, I mean, cold in here.
    8. She was studying the heft lemisphere.
    9. When she got upset, she starting shelling at everyone.
    10. It was maistly her doing.
    11. The teacher tried to plan the seed of curiosity in the students.
    12. Try to fop on one foot.
    13. They were having a meating arathon.
    14. You have to square your problems facely.
    15. Paul put the pot on the pable.
    16. The money is likely better spelt elsewhere.
    17. He felt like he was having a planic attack.
    18. Every so often, he has insomia.
    19. She forgot to change the car in the oil.
    20. They are Turking talkish.

2. Keep a diary of your own errors for a period of a week, month, or semester. Then analyze the types of errors that occurred. Write a report in which you compare your errors with those observed in prior research. Set out to observe speech errors produced by others, peers, or individuals on television or radio programs. For these observations, you will not always be certain what the intended utterance was. Nevertheless, the exercise provides you with hands-on experience collecting errors in the field.

3. Keep a diary of your tip-of-the-tongue states, or TOT states. Because TOT states are rare, perhaps more rare than slips of the tongue, you may need to collect observations for 1 or 2 months. At the end of the collection period, the instructor can combine the data from the entire class for a class discussion and/or for you to analyze in a written assignment.

For additional ancillary resources, please visit the companion website at www.sagepub.com/kennison.

- Video Links
- Audio Links
- Web Resources
- Internet Activities
- Flashcards
- Web Quizzes

CHAPTER 10

LANGUAGE COMPREHENSION

The study of language comprehension has a relatively long history. Researchers were studying the topic using scientific methods as far back as the late 1800s (Huey, 1908/1968). However, it was not until after the cognitive revolution around 1960 that most of the major advances in understanding how language is processed were made. With the invention of the computer and the popularity of the computer as a metaphor for how the human mind functions, research into human cognition has flourished. Over the past 50 years, a great deal has been learned about how natural language is processed as text and as speech. Researchers generally agree that comprehension occurs incrementally—word by word or

▲ Photo 10.1 A student reads to his classmates. When you read or listen to stories, do you find the process of comprehension effortful or effortless?

morpheme by morpheme—over time. As you are reading this paragraph, you move your eyes from word to word and are able to interpret each word in terms of what it means and in terms of how it fits with what you have read before. As you read along, you are storing in memory not only information about what individual words you have seen but also information about what the words mean when taken together as a phrase or a clause or a whole passage. In this chapter, you will learn about how researchers have investigated the processing that occurs during the interpretation of single sentences and longer discourses. and about the role of memory in comprehension. You will also learn about the research comparing adults' and children's memory for sentences and discourses and research investigating whether there are differences in how adults and children process sentences and discourses.

Syntactic Parsing

What Is Syntactic Parsing?

Since the 1970s, researchers have envisioned the process of interpreting a sentence as involving incremental processing (Frazier & Fodor, 1978; Kimball, 1973). During listening, the incremental processing is influenced by the rate of the speaker. When one is listening to a fast talker, one experiences the words of the sentence relatively close together in time and must rely on sensory memory and short-term memory to store the speech until further processing can be carried out. In contrast, during reading, one normally has more control over how quickly or slowly the processing occurs. As you learned in Chapter 5, the length of time that readers spend looking at individual words during reading is directly related to characteristics of those words (i.e., word length, word frequency). Research has also shown that across readers, there is great variability in normal reading time. One cannot make the generalization that all highly skilled readers read quickly (Rayner, Pollatsek, Ashby, & Clifton, 2012). Some highly skilled readers process texts at relatively slow rates, while other highly skilled readers read quite quickly. Also, there are those who are labeled poor readers (i.e., individuals who demonstrate relatively low levels of comprehension during reading) who read fast and others who read exceedingly slowly.

For all comprehenders, the first step in comprehension is the identification of the first word in the stream of speech or the line of text. In Chapter 5, you learned about how words are recognized in processes described as lexical access. The identification of an incoming word results in the comprehender knowing quite a lot about the word (e.g., what the word means, its part of speech, its pronunciation). Using the information gained from lexical access, the comprehender can begin to integrate it with the material that has come before. This integration is occurring in the comprehender's mind, specifically within working memory. The incoming word is integrated into the syntactic structure that is being constructed for the sentence. The word is also eventually integrated into the meaning of the sentence in which it occurs and also into the meaning of the overall discourse. Theories of language comprehension differ in how they envision syntactic structure and semantic interpretation occurring. Some theories of sentence processing, such as the garden path model, envision the syntactic integration and semantic integration of a word with a preceding context as occurring in nonoverlapping (i.e., serial) stages (Frazier & Rayner, 1982). This approach grew out of the generative approach to language, assuming that key aspects of language are innate and that processing is carried out in a modular fashion. Other theories of sentence processing, such as the constraint satisfaction model, envision syntactic integration and semantic integration of a word with a preceding context as occurring simultaneously and having the ability to influence the other types of processing (MacDonald, Pearlmutter, & Seidenberg, 1994). This approach is an example of a statistical learning approach to language, as individuals are believed to retain in memory detailed information about the frequency with which words co-occur in everyday language use and to use this information to process language as it unfolds in real time.

When researchers refer to the processing involved in the construction of a sentence's syntactic structure, they use the term ***syntactic parsing.*** During syntactic parsing, comprehenders assign words (or morphemes) of a sentence to linguistic categories (e.g., subject, verb, object, as well as others). The process occurs incredibly rapidly within working memory. Comprehenders do not have an awareness that it is happening. Researchers use tree diagrams and phrase structure rules to illustrate what they believe is happening during comprehension. As you learned in Chapter 4, the phrase structure rules of English are as shown in 1. Researchers assume that during syntactic parsing, readers and listeners use their knowledge of phrase structure rules to assign each incoming word to a potential role in the sentence structure. Figure 10.1 displays how a comprehender is likely to syntactically parse a simple sentence in English. The term ***phrase marker*** is generally used to refer to the syntactic representation that the comprehender constructs. Phrase markers are typically represented using tree diagrams.

1. a. S → NP VP
   b. VP → (ADV) V (NP) (PP) (S)
   c. NP →Det (ADJ) N (PP) (S)
   d. PP → P NP

Researchers have debated how syntactic parsing decisions are made. One of the debates is whether comprehenders always analyze incoming words immediately in terms of their syntactic relationship with the preceding sentence context. Researchers disagree about whether parsing decisions are made immediately or whether there are circumstances when a comprehender might delay a decision. An example of this is when a word can be used in more than one part of speech, such as the word *silver,* which could be a noun referring to an object, as in *She found the silver,* or an adjective referring to a characteristic of an object,

Figure 10.1 Word-by-word parsing of a simple English sentence

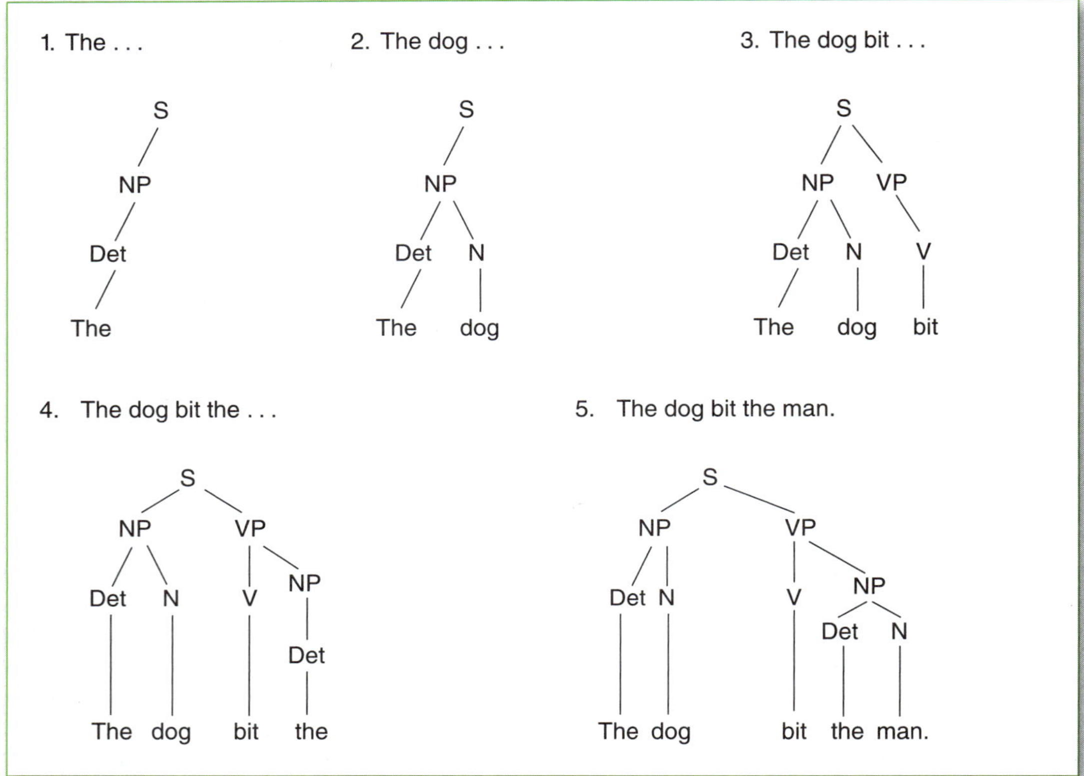

as in *She found the silver amulet*. In 1980, Just and Carpenter proposed the immediacy principle, which claims that all words are interpreted and parsed as soon as they are heard or viewed. A strict interpretation of their view is that there are no circumstances in which comprehenders postpone the syntactic parsing of an incoming word. Proponents of language comprehension models that envision language comprehension as a highly interactive process with all types of information influencing all types of decisions, such as the constraint satisfaction approach (MacDonald et al., 1994; Trueswell & Tanenhaus, 1994), have claimed that delays in syntactic parsing would not likely occur, because comprehenders would be able to make some determination of an incoming word's likely role in the sentence using the information that is available in the context (Sedivy, Tanenhaus, Chambers, & Carlson, 1999).

Advocates of the garden path model have proposed that in some circumstances the comprehender may delay assigning an incoming word to a particular category in the phrase marker. For example, Frazier and Rayner (1987) proposed that words that can be used as more than one syntactic category are not immediately attached into the phrase marker.

Consider the examples in 2. The word *desert* is ambiguous, as it can be a noun and serve as the subject of the subordinate clause, as in 2a and 2c, or an adjective and serve as the modifier of the following noun, as in 2b and 2d. The word *trains* is also ambiguous; it can be a verb, as in 2a and 2c, or a noun, as in 2b and 2d. In the experiment, Frazier and Rayner (1987) measured reading time on sentences that did and did not disambiguate the syntactic category ambiguity of *desert*. When the word *desert* is preceded by the word *the,* the syntactic category of *desert* is ambiguous, as in 2a and 2b. When the word *desert* is preceded by the word *this* or *these,* the syntactic category of *desert* is unambiguous. The results showed that reading time on the disambiguating words following *trains* was longer in ambiguous conditions than in unambiguous conditions.

2. a. I know that the desert trains young people to be especially tough.
   b. I know that the desert trains are especially tough on young people.
   c. I know that this desert trains young people to be especially tough.
   d. I know that these desert trains are especially tough on young people.

Other research by Frazier and colleagues (Frazier & Rayner, 1990) has shown that readers sometimes delay interpreting words that have more than one meaning if the word is preceded by a context that does not strongly support any one particular meaning (see also Frazier, Pacht, & Rayner, 1999, for additional cases of immediate semantic commitment). Consider the examples in 3. The sentences in 3a and 3b contain biasing information before the ambiguous word *pitcher.* In 3a, the context supports the meaning of *pitcher* as a container for liquid. In 3b, the context supports the meaning of *pitcher* as one who throws balls in a game. In 3c and 3d, the biasing material follows the ambiguous word. The results showed that readers spent less time reading the ambiguous word when the disambiguating information followed it (i.e., 3c and 3d) than when the disambiguating information preceded it (i.e., 3a and 3b). Frazier and Rayner (1990) concluded that comprehenders do not always make semantic commitments about how a word will be interpreted as soon as a word is processed.

3. a. Being so elegantly designed, the pitcher pleased Mary.
   b. Throwing so many curve balls, the pitcher pleased Mary.
   c. Of course the pitcher pleased Mary, being so elegantly designed.
   d. Of course the pitcher pleased Mary, throwing so many curve balls.

Other researchers have also provided empirical support for the view that there are some circumstances in which syntactic parsing decisions are delayed. Kennison (2005) investigated the possibility that readers of English parsed plural nouns preceded by an adjective more quickly than they processed singular nouns preceded by an adjective, because plural nouns are less likely to be followed by another noun than are singular nouns. She reasoned that readers could not always be sure that when a singular noun was encountered, the word was the head noun of the phrase, as it could be followed by other nouns that could potentially serve as the end of the phrase (e.g., the *purple bus . . .* vs. *the* *purple bus driver*). In contrast, because of the grammatical rules of English, when a plural noun was encountered, the word was far more likely to be the head of the noun phrase, as plural nouns cannot be compounded (e.g., *the purple buses* . . . vs. **the purple buses driver*; the asterisk indicates an ungrammatical phrase). She measured reading time on sentences similar to those in 4. The results showed that when the adjective–noun combination was implausible, readers slowed down during the processing of plural nouns, as in 4c and 4d. In conditions containing singular nouns, the slowing down due to implausibility was observed on the word following the singular noun, as in 4a and 4b. The results suggested that the time course of semantic processing occurs earlier during processing for plural nouns than for singular nouns.

4. a. John said that the ancient castle . . . Plausible combination
   b. John said that the careful castle . . . Implausible combination
   c. John said that the ancient castles . . . Plausible combination
   d. John said that the careful castles . . . Implausible combination

Despite the strong evidence that there are some circumstances in which syntactic parsing decisions are delayed, there is also convincing research supporting the view that comprehension occurs rapidly without delays. Sedivy et al. (1999) conducted an experiment in which participants' eye movements were recorded as they viewed a visual scene and listened to speech presented over headphones. The speech was another person describing the scene. This technique has been referred to as the **visual world paradigm** (Cooper, 1974; Tanenhaus, Spivey-Knowlton, Eberhard, & Sedivy, 1995). In this experiment, every scene contained four objects. Among the sentences that participants heard were short imperative statements (e.g., *Touch the tall glass*). For example, in one experiment, the scene contained the object mentioned in the imperative sentence (i.e., a tall glass), a contrasting version of the same object (i.e., short glass), a different object having a similar characteristic as the target (i.e., tall pitcher), and an object that is altogether unrelated to the object described in the target sentence (i.e., key). Participants' eye movement during listening indicated that they were interpreting the adjective before they heard the noun. Analysis of listeners' eye movement behavior indicated that prior to the presentation of the noun, participants began interpreting the adjective in relation to the visual display.

One question that you might have had as you started reading this chapter is whether comprehension processes occur the same when one is listening and when one is reading. Researchers generally assume that most of the processes that are involved in comprehension are the same for listening and reading. During listening, there are sources of information that are not available during reading, such as the intonation and stress pattern of the sentence, the speaker's voice, and the speaker's nonverbal elements of communication. Because experiments involving text are easier to set up and to control than experiments involving speech, most of the research on language comprehension has focused on reading. However, now that technology is making experiments involving speech stimuli easier to conduct, more and more researchers are turning to the study of speech. One of the most fascinating phenomena in language comprehension involves speech perception. Since the

1970s, researchers have known that the act of listening involves not just hearing but also seeing. Text Box 10.1 describes the experiments demonstrating what has come to be known as the **McGurk effect.**

Text Box 10.1 Research Discovery: Do We Listen With Our Eyes and Our Ears?

In 1976, McGurk and MacDonald published a paper titled "Hearing Lips and Seeing Voices." In their lab, they accidentally discovered the effect that has come to be referred to as the McGurk effect. MacDonald, then a research assistant, had a lab technician splice together a video of a face pronouncing one phoneme and an audio recording of the person saying a different phoneme. For example, the video showed someone saying *bah bah bah*, but the audio was someone saying *gah gah gah*. MacDonald observed that when the video and audio were experienced together, the listener heard neither /ba/ nor /ga/ but, instead, a third syllable /da/. The syllable that was perceived was predictable; it appeared to reflect an averaging of the syllable viewed and the syllable heard. For example, /ba/ is a syllable that is articulated at the lips (i.e., the airflow is stopped by closing the lips), and /ga/ is a syllable that is articulated at the back of the throat at the velum. The syllable that is perceived (i.e., /da/) is one that typically is articulated at the alveolar ridge (i.e., just behind the upper teeth), which is about halfway between the lips and the velum. Studies have also shown that when listeners see a video with a face producing /pa/ and hear an audio recording of /ka/, the syllable that is perceived is /ta/. All of these sounds have the same place of articulation as /ba/, /ga/, and /da/, respectively. They differ only in voicing: /b/, /g/, and /d/ are voiced consonants and /p/, /k/, and /t/ are voiceless consonants. The effect has proven to be very robust, being observed again and again in research as well as in informal demonstrations. The effect is an auditory illusion that demonstrates the fact that auditory perception can be influenced by visual processing. In order to experience the illusion, one must attend to both streams of input (i.e., the video and the audio). Massaro and Cohen (2000) showed that listeners may be more likely to experience the illusion if the video information is high quality, but the auditory information is degraded or poor quality.

How Is Syntactic Ambiguity Resolved?

Most of the research conducted on the topic of syntactic parsing has investigated how comprehenders resolve **syntactic ambiguity,** which is a word or phrase that can be integrated into the syntactic structure of the sentence in more than one way (Clifton & Duffy, 2001). An example of a sentence containing a syntactic ambiguity is *The spy saw the man with the binoculars.* Typically, the phrase *with the binoculars* is interpreted as a modifier of the verb, specifying how the spy carried out the seeing. The phrase could also be interpreted as modifying the noun *the man,* with the man being the one in possession of the binoculars. Figure 10.2 displays the phrase marker for this sentence. Much of the time,

Figure 10.2 Parsing a syntactic ambiguity

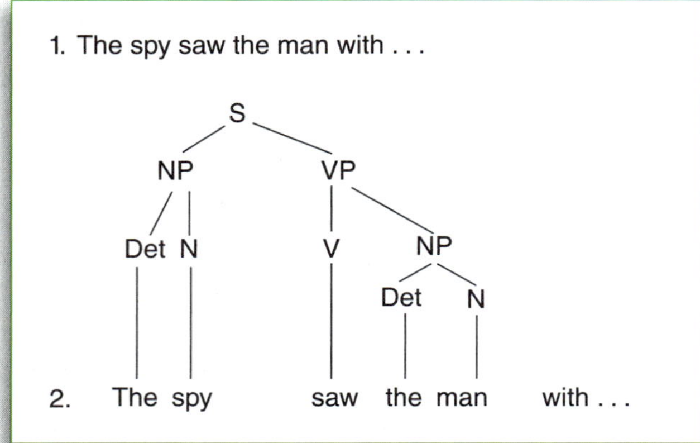

comprehenders will process sentences containing a syntactic ambiguity and not even become aware that there was an ambiguity. They arrive at an interpretation of the ambiguity that makes sense and is not in conflict with the following information.

There are other times during sentence comprehension when a comprehender may abruptly realize that the interpretation that she was constructing is not the one intended by the author of the text or the speaker. The interpretation of the sentence appears to be going fine until a word or phrase is encountered that does not fit in with the syntactic analysis that has been constructed. Consider the sentence *The student accepted the trophy for most valuable player of the year had been broken during shipment and would have to be reordered.* As you were reading the previous sentence, did you experience any type of confusion when you got to the phrase *had been broken*? If you did, you are not alone. Most readers would find those words to be unexpected, because earlier the phrase *the trophy* . . . was interpreted as the direct object of the verb *accepted.* When one encounters the words *had been broken,* one realizes that the noun phrase *the trophy* actually serves as the subject of a subordinate clause.

Researchers have identified a number of sentence structures that involve syntactic ambiguities. Most of these have been identified in English; however, as the number of researchers working in other languages increases, more and more syntactic ambiguities are being identified. The examples in 5 illustrate one of the most well-researched syntactic ambiguities—the direct object/clause ambiguity. When comprehenders encounter the phrase *the answer,* they could construct a direct object noun phrase or they could construct a new clause. In 5b, the sentence continues with a context that supports the direct object usage. In 5c, the sentence continues with a context that supports the clause usage.

5. a. Alicia knew the answer . . .
   b. Alicia knew the answer by heart. Direct object continuation
   c. Alicia knew the answer was correct. Clause continuation

Researchers disagree about how syntactic ambiguity is resolved during comprehension. Advocates of the garden path model (Frazier & Clifton, 1996; Frazier & Rayner, 1982; Rayner, Carlson, & Frazier, 1983) claim that syntactic parsing occurs in a serial fashion with syntactic analysis preceding semantic interpretation. They believe that comprehenders use information about the incoming word's syntactic category to assign the incoming word a role in the phrase marker. The parsing decision is carried out using only syntactic information. The model assumes that when faced with a syntactic ambiguity, comprehenders apply one of two syntactic parsing strategies—(1) **minimal attachment** or (2) **late closure.** Minimal attachment applies when an ambiguity involves a choice between two syntactic attachments that involve different amounts of complexity in terms of how much structure must be added to the phrase marker. The garden path model's minimal attachment principle predicts that in such cases, comprehenders will always choose to resolve the ambiguity as the least complex structure. For example, in the ambiguity involved in the examples in 5, the direct object analysis would involve adding less syntactic structure (i.e., in terms of syntactic nodes and syntactic branches that must be added to the phrase marker) than the clause analysis. Consequently, minimal attachment predicts that whenever comprehenders are faced with a phrase that could be attached in these two ways, comprehenders will always interpret the phrase as a direct object.

In contrast, late closure applies when an ambiguity involves a choice between syntactic structures of comparable complexity (i.e., there is no difference between the two analyses in terms of how much new syntactic structure is added to the phrase marker); comprehenders are predicted to choose the attachment involving the most recent part of the phrase marker. An example of a late closure ambiguity is *John admired the son of the general, who was killed recently*. Who was killed recently? Was it the general or the son? The clause *who was killed recently* could be attached into the phrase marker as a modifier of the noun phrase *the son* or a modifier of the noun phrase *the general.* Both of the possible attachments involve relatively little new syntactic structure. Late closure predicts that when faced with this type of ambiguity, comprehenders will attach the phrase to the most recent part of the phrase marker, which would result in most readers interpreting the phrase as indicating that it was the general who died recently.

Other researchers, such as advocates of the constraint satisfaction model (MacDonald et al., 1994; Sedivy et al., 1999; Tanenhaus et al., 1995), envision the process of interpreting a sentence as being highly interactive with all sources of information being able to influence all stages of processing. They argued that when comprehenders resolve syntactic ambiguity, they use a variety of information sources and generally interpret the ambiguity in the way that is most strongly predicted by all the sources of information taken together. In their view, syntactic information does not have a special status in comprehenders' decision-making process. They suggest that if a comprehender has information that the syntactically more complex analysis of an ambiguity is more plausible in the context, it would not make sense for the comprehender to ignore that information.

Relatively recently, a third type of model represents a mixture of the features of the other two—the unrestricted race approach (Traxler, Pickering, & Clifton, 1998; van Gompel, Pickering, Pearson, & Liversedge, 2005; van Gompel, Pickering, & Traxler, 2001). This view assumes that multiple analyses are constructed simultaneously with the analysis constructed most quickly influencing processing time and that initial analyses can be influenced by all sources of information. Also, the multiple syntactic analyses constructed in parallel do not compete. Consequently, the approach makes similar predictions as the garden path approach for the processing of the syntactically ambiguous material but makes similar predictions as the constraint-satisfaction approach for the processing of the disambiguating material in a sentence.

Syntactic Reanalysis

The garden path theory claims that comprehenders resolve syntactic ambiguity by selecting one particular analysis. If the following sentence context proves to be inconsistent with that analysis, comprehenders must carry out a **syntactic reanalysis** (i.e., give up the initial analysis and construct the alternative analysis). The process of syntactic reanalysis has been traditionally viewed as occurring when the initial analysis of the ambiguity is found to be incorrect and must be abandoned. The syntactic analysis of the ambiguity must be recomputed (Ferreira & Henderson, 1990; Frazier & Rayner, 1987). Research in which eye movements were recorded during reading has shown that when readers process the word in a sentence that disambiguates an earlier syntactic ambiguity, they will look back to earlier parts of the sentence, suggesting that they are carrying out a recomputation of the syntactic analysis (Frazier & Rayner, 1982).

Some research has suggested that the processing difficulty that is experienced during syntactic reanalysis increases as the length of the ambiguous material increases (Ferreira & Henderson, 1991; Frazier & Rayner, 1982; Kennedy & Murray, 1984; Warner & Glass, 1987). Consider the examples in 6. Frazier and Rayner (1982) compared the processing of sentences containing a short ambiguous phrase, as in 6a, versus a long ambiguous phrase as in 6b. They found that when readers encountered the disambiguating clausal verb *was,* the time they took to resolve the ambiguity was greater when the ambiguity had been long than when it had been short.

6. a. Sally found out the answer was in the book.
   b. Sally found out the answer to the difficult physics problem was in the book.

These results are consistent with the view that syntactic reanalysis will take longer when one has spent more time pursuing the initial, incorrect analysis, because it takes longer to abandon the analysis (Tabor & Hutchins, 2004). Presumably, the initial, incorrect analysis has been constructed within working memory and the representations that have been formed must be erased. The exact nature of the processing that occurs during syntactic reanalysis remains unclear. In at least three reading studies, length effects in syntactic reanalysis have been investigated and not observed (Kennison, 2001, 2004, 2009).

The Role of Verb Information

Since the 1970s, there have been numerous studies conducted on the topic of syntactic ambiguity resolution. Typically, these studies have focused on determining the extent to which comprehenders rely solely on syntactic information to resolve a syntactically ambiguous phrase or whether comprehenders' initial interpretation of syntactic ambiguity can be influenced by some nonsyntactic source of information. For example, many studies have investigated how lexical information associated with a verb might influence the analysis of following syntactic ambiguity. Consider the examples in 7. In both 7a and 7b, the noun phrase *the answer* is temporarily ambiguous.

7. a. The student revealed the answer . . . Biased for direct object
   b. The student believed the answer . . . Biased for clause

In 7a, the verb *revealed* is a verb that can be used with both a direct object and a following clause but is used most frequently with a direct object. In contrast, in 7b, the verb *believed* is a verb that can be used with both a direct object and a following clause but is used most frequently with a following clause. Advocates of the constraint satisfaction model propose that as soon as the comprehender processes the verb in the sentence and achieves lexical access for the verb, he or she retrieves from memory information about all the verb's possible usages as well as their frequencies of usage (MacDonald et al., 1994). The most frequent usage for the verb is to build the syntactic structure that the comprehender is building and drives the interpretation of following information. Because verb information is lexical information, advocates of the garden path model assert that it would not be used initially to resolve a syntactically ambiguous phrase, but it could be used later in processing (Frazier & Rayner, 1982). Of the many studies that have investigated the role of verb information in syntactic ambiguity resolution, some have suggested that comprehenders use verb information to guide their interpretation of a following noun phrase (Garnsey, Pearlmutter, Myers, & Lotocky, 1997; Trueswell, Tanenhaus, & Kello, 1993). Other studies have found that verb information does not influence syntactic ambiguity resolution when other factors, such as semantic plausibility, are controlled (Kennison, 2001).

Are There Individual Differences in Sentence Processing?

An important question is whether all comprehenders resolve ambiguity the same way. Since the early days of language processing research, there has been an acknowledgment that there is a distinction between one's knowledge of language and one's use of language, because the latter involves biological processes that might vary somewhat from individual to individual. Chomsky (1985) distinguished **language competence**, which is language users' knowledge of their language, from **language performance,** which is language users' production and comprehension of their language. Chomsky saw knowledge of language as one's knowledge of grammar and vocabulary in its pristine state, essentially perfect or error-free. Language errors could occur through language performance, when one's knowledge of language is accessed and manipulated by the biological systems, such as the sensory systems, memory, and the systems involved in movement planning.

Working Memory Capacity

The possibility that there are individual differences in syntactic ambiguity resolution has intrigued researchers. As comprehenders syntactically parse sentences and interpret their meanings, they are using working memory. As you learned in Chapter 5, working memory is just one of the many types of human memory. Working memory is the short-term memory buffer that people use for problem solving and comprehension. Some researchers have described it as *the mental blackboard* (Ashcraft & Radvansky, 2009). Since the 1970s, the role of working memory in syntactic parsing has been acknowledged (Frazier, 1979). Since that time, it has also been recognized that individual working memory capacity is limited (Miller, 1956). Research in the 1980s revealed that there is a relationship between working memory capacity and measures of language performance, such as verbal SAT scores and performance on nonstandardized comprehension questions (Daneman & Carpenter, 1980). A relationship between working memory capacity and comprehension has also been observed when standardized comprehension questions have been used (e.g., Baddeley, Logie, Nimmo-Smith, & Brereton, 1985; Masson & Miller, 1983).

Research by Just and colleagues (Just & Carpenter, 2002; MacDonald, Just, & Carpenter, 1992) has suggested that how readers resolve syntactic ambiguity may depend on the their working memory capacity. Those with relatively large working memory capacities may be able to pursue multiple analyses of an ambiguity, while those with relatively small working memory capacities may not. Consequently, when disambiguating material is encountered, syntactic reanalysis would be expected to occur more often for readers with small versus large working memory capacities. Their results suggested that prior research showing different results in how comprehenders resolved syntactic ambiguity may have occurred because there were differences in the working memory capacities of participants across the studies. Other researchers have argued that working memory capacity influences how comprehenders use information about semantic plausibility during syntactic ambiguity resolution (Traxler, Williams, Blozis, & Morris, 2005). Gibson and colleagues (Gibson, 1998, 2000; Warren & Gibson, 2002) have claimed that linking referents locally (i.e., within a clause) imposes less of a burden on working memory than linking referents globally in a discourse (i.e., across clause boundaries). Other research has shown that the perceptual processes during reading do not differ for readers who are high versus low in working memory capacity (Kennison & Clifton, 1995).

Semantic Interpretation

How Is Meaning Interpreted?

The least well-understood aspect of language comprehension is semantic processing. As you might recall from Chapter 1, one of Chomsky's early observations was that the syntactic structure of a sentence is distinct from its meaning. He created the example *Green ideas sleep furiously* (Chomksy, 1957) to point out that a sentence can be grammatically well formed but be meaningless. Researchers have developed terminology to describe the semantic information contained within sentences. Dowty (1990) identified a set of **thematic roles,** which categorized noun phrases in sentences in terms of the relation that they serve. Consider the examples in 8. An agent is a phrase that serves as the performer of an action. In 8a and 8b, the agent is *the bear.* The patient is a phrase that is changed by the action. In 8a and 8b, the patient is *the honey.* The **instrument** is a phrase that is used to carry out an action. In 8c, the instrument is *its paw.* The location phrase specifies where an action occurs. In 8a and 8b, *under the tree* is the location. As these examples show, the thematic roles of a sentence are not restricted to a particular position in the sentence. It is true that in English most of agents occur in subject position, but they may also occur in a prepositional phrase, as in 8b. Most patients occur in direct object position, but they may also occur in subject position. Most instruments occur in prepositional phrases, but they may also occur in subject position, as in *The spatula broke the egg*.

8. a. The bear ate the honey under the tree.
   b. The honey was eaten by the bear under the tree.
   c. The bear retrieved the honey with its paw.

There is research suggesting that in addition to the word-by-word interpretation that is occurring during processing, there is likely to be semantic processing occurring much later in processing, such as the end of a sentence or end of a clause. Since the 1970s, it has been known that readers take longer to process words that come at the end of a clause or sentence than the same words if they do not correspond to the end of a clause or sentence (Jarvella, 1971; Just & Carpenter, 1980; Kennison, 2004; Kennison, Sieck, & Briesch, 2003). The pattern of processing has been called the **end-of-clause wrap-up effect.** One interpretation of these results is that comprehenders spend time at the ends of clauses and sentences integrating the meaning of the words within the clause or sentence (Just & Carpenter, 1980; Rayner, Sereno, Morris, Schmauder, & Clifton, 1989).

Kennison et al. (2003) investigated the time course of phonological processing in silent reading. They measured reading time on sentences that either did or did not contain repeated phonemes. Participants were instructed to read silently. Sentences were similar to those in 9. They were interested in whether phonological processing would result in slower reading time as each new word was processed in the sentence. The results showed that the slowing in reading time due to repeated phonemes did not arise incrementally; rather, reading time following the end of the clause in which the repeated phonemes occurred, as in 9a, was longer than when phonemes were not repeated in the clause, as in 9b. They argued that although phonological processing occurs during initial processing of individual words, there is also some phonological processing that occurs at the ends of clauses.

9. a. Tom and Tanya took Timmy's truck on Tuesday afternoon after he said it was okay.
   b. Bob and Wanda got Randy's boat on Tuesday afternoon after he said it was okay.

Despite the decades of research that have been carried out on the topic of language comprehension, there is relatively little known about how semantic interpretation occurs. The assumption is that semantic interpretation occurs relatively rapidly either soon after syntactic analysis or at the same time as syntactic analysis. Relatively little is also known about how comprehenders resolve figurative language (i.e., groups of words whose meaning when taken together does not correspond to the meaning of each word considered with each of the other words). When the meaning of a phrase does correspond to the meanings of the words in the phrase, it is described as having literal meaning or being an example of literal language use. Literal language contrasts with figurative meaning or figurative language use, in which the meaning of a phrase is not derived from the meanings of the individual words within the phrase. Consider the examples in 10.

10. a. The boulder went through the roof. Literal
 b. The foreman went through the roof. Figurative
 c. The bulldozer brought down the house. Literal
 d. The rock band brought down the house. Figurative

There are numerous types of figurative language use. Table 10.1 provides some examples. In the comprehension literature, many studies have focused on the processing of metaphors. Those who study metaphors use the terms *tenor, vehicle,* and *ground* to refer to different parts of a metaphor. In the metaphor *the teacher was a drill sergeant,* the tenor is the topic of the metaphor, which is *teacher.* The vehicle refers to *drill sergeant,* which is the predicate of the expression. The similarity between the tenor and the vehicle is the ground.

A traditional view suggests that interpreting figurative language occurs only after the comprehender attempts to construct a literal interpretation but determines that this interpretation is unlikely to be correct (Fogelin, 1991; Grice, 1989; Searle, 1979). Searle (1979) proposed that there are at least three steps. First, one determines the literal meaning. Second, one determines whether the literal meaning is compatible with the context. In the event that the meaning is not compatible with the context, one must search for a figurative meaning that does make sense in the context. This view predicts that examples of figurative language should always take longer to process than similar examples of literal language. Gibbs (1979) investigated the processing of indirect requests, such as *Can you pass the salt?* Participants read short discourses a line at a time. He compared how quickly participants paraphrased the last line of the discourse. He found that reading time on indirect requests was not longer than reading time on either direct requests or literal sentences.

Table 10.1 Examples of Figurative Language

Type	Examples
Idiom	The foreman went through the roof. His plans went up in smoke. She was feeling under the weather.
Proverb	Birds of a feather flock together. A stitch in time saves nine. Absence makes the heart grow fonder.
Simile	The homework assignment was like a maze. The movie was like a roller coaster. The friends were like the Three Musketeers.
Metaphor	The car was a rocket. The chocolate was medicine. The quarterback was a cheetah.
Metonymy	The United States won the gold medal in archery. Hollywood has embraced the starlet. The White House had no comment.

How Are Pronouns Comprehended?

As sentences are viewed as being comprehended incrementally a word at a time, discourses are viewed as being comprehended incrementally a sentence at a time. The processes that enable a comprehender to form a coherent mental representation of long narratives, which have been heard or read, are still not well understood. There has also been a great deal of research focused on understanding how incoming words trigger processing that results in the comprehenders linking in memory information from distant parts of the discourse. An example of this is referential processing, which occurs when an incoming word is used to refer to a discourse entity that has been previously mentioned. Among the most common referential words are **pronouns** such as *he, she, it,* and *they.* Pronouns can refer to either previously mentioned discourse entities or following ones. Consider the examples in 11. In 11a, the pronoun *she* refers back to the antecedent Jane. A pronoun that refers to previously mentioned antecedents is called an **anaphoric pronoun.** In contrast, in 11b, the pronoun *she* refers to a following name. A pronoun that refers forward in a discourse is called a **cataphoric pronoun** (Kennison, Fernandez, & Bowers, 2009).

11. a. On Saturday, Jane arrived late. She apologized profusely to the group.
    b. On Saturday, she arrived late. Jane apologized profusely to the group.

Research has also shown that when comprehenders are resolving pronouns, they use information about gender relatively late in the processing. If gender information is used early during the resolution of a pronoun, then a pronoun would only be linked with a discourse entity of the same gender. Consequently, in a sentence such as 12a, the pronoun *he* would be interpreted as referring to Terrell, but in sentences such as 12b, the pronoun *she* would not be interpreted as referring to Terrell. Several studies have shown that reading time on pronouns in sentences such as 12b is longer than pronouns in sentences such as 12a (Kennison et al., 2009; van Gompel & Liversedge, 2003). Van Gompel and Liversedge (2003) showed that the longer reading time in sentences such as 12b was not due to processing ramifications of introducing a new entity into the discourse. The results were interpreted as indicating that pronouns are linked with preceding **antecedents,** regardless of whether they are compatible in gender. After the link is formed, then the genders of the pronoun and antecedent are compared. When the genders are incompatible, the comprehender must abandon the link in favor of an interpretation in which the pronoun refers to a new discourse entity.

12. a. After Terrell arrived, he asked if there was coffee.
    b. After Terrell arrived, she asked if there was coffee.

Pronouns, such as *he, she,* and *they,* are used to maintain cohesion in the discourse. Numerous studies have been conducted to understand how comprehenders interpret pronouns, specifically linking the pronoun with the previously processed discourse. Consider the examples in 13. In 13a, the pronoun *she* refers to the same person as the proper name *Chelsea.* In 13b, the pronoun *he* refers to the same person as the proper name *Trevor.*

13. a. Chelsea checked into the hotel at six.
    She was exhausted and went straight to bed.
    b. Trevor arrived at the meeting late.
    He had woken up late and rushed into the office.

Information About Gender

As you learned in Chapter 5, our mental lexicons store information about individual words. During processing, when the word is accessed, all the information that we associate with that word is activated in memory and may influence processing. One type of information that plays an important role in the comprehension of pronouns is gender information. In English, there are a small number of words where gender is unambiguous. For example, the word *bachelor* refers to an unmarried male who is typically young. In contrast, the word *spinster* refers to an unmarried female who is typically quite old. Some other words that refer specifically to one gender or the other include *man, woman, girl, boy, grandmother, grandfather, king, emperor, empress, priestess,* and *goddess.* The vast majority of words in English that are used to refer to people can refer to either males or females.

Research has shown that information about gender is activated and used during comprehension. In 1996, Banaji and Hardin demonstrated that gender information is activated following the processing of words that are strongly associated with one gender (e.g., *doctor, nurse*). They presented participants with pairs of stimuli, such as a word followed by the pronoun *he* or *she.* Participants were asked to judge whether the word *he* or *she* referred to a male or female. The word that preceded the pronoun was either strongly stereotyped to refer to mostly males (e.g., *doctor, judge, sheriff*) or strongly stereotyped to refer to mostly females (e.g., *nurse, typist, florist*). The results showed that participants' judgments were faster when the gender of the pronoun matched the stereotype of the preceding word (e.g., *doctor-he, nurse-she*) than when the gender mismatched (e.g., *doctor-she, nurse-he*). Table 10.2 displays some common English words that are strongly stereotyped to refer to males and females and words that are neutral (i.e., not strongly stereotyped for either gender).

Table 10.2 Common English Words That Are Strongly Stereotyped to Refer to Males or Females and That Are Gender Neutral

Male Stereotyped	Female Stereotyped	Gender Neutral
Judge	Secretary	Artist
Sheriff	Receptionist	Counselor
Carpenter	Manicurist	Newscaster
Hunter	Babysitter	Student
Boxer	Florist	Client
Doctor	Nurse	Patient
Farmer	Dancer	Teenager
Lawyer	Paralegal	Juror
Accountant	Bookkeeper	Customer
Deputy	Nanny	Passenger

Source: Kennison and Trofe (2003).

▲ Photo 10.2 A female firefighter stands in a firefighting uniform.

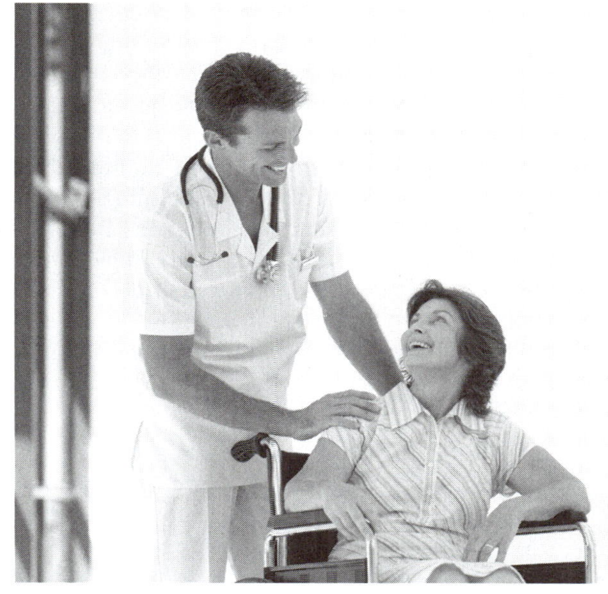

▲ Photo 10.3 A male nurse pushes a patient in a wheelchair. Are you considering a career that is predominantly male or female?

When readers comprehend sentences containing pronouns, they also use gender information. Kennison and Trofe (2003) showed that comprehenders took longer to read the pronouns *he* and *she* if the gender stereotype of the antecedent for the pronoun did not match the gender of the pronoun than when the genders of the antecedent and pronoun matched. Consider the examples in 14. Readers took longer to read the pronoun *she* in sentences similar to 14b than 14c and took longer to read the pronoun *he* in sentences similar to 14d than 14a.

14. a. The executive distributed an urgent memo. He made it clear . . .
    b. The executive distributed an urgent memo. She made it clear . . .
    c. The secretary distributed an urgent memo. She made it clear . . .
    d. The secretary distributed an urgent memo. He made it clear . . .

Carreiras, Garnham, Oakhill, and Cain (1996) observed similar results in English and in Spanish and argued that when a comprehender encounters a noun that is associated with a strong gender stereotype, the character is introduced into the comprehender's mental model as having the sex of the stereotype. If the comprehender encounters a following pronoun with a different gender, the comprehender must revise the mental model to reflect the gender indicated by the pronoun. The revision would be associated with increased processing time. When a comprehender encounters a noun that is not strongly stereotyped, the processing of a following pronoun is easy because no updating of the mental model is required.

In 2004, Duffy and Keir reported two eye tracking experiments investigating how reflexive pronoun anaphors were resolved with gender ambiguous antecedents. Participants read sentences containing a reflexive pronoun anaphor, such as *himself* or *herself*, preceded by a strongly stereotyped antecedent (e.g., *electrician* or *babysitter*). Examples are displayed in 15. They found that readers took longer to resolve an anaphor when its gender mismatched the gender stereotype of the antecedent than when the genders matched. They also found that the processing difficulty related to the gender mismatch did not occur if the gender of the antecedent was disambiguated earlier in the discourse. Their results were compatible with the Carreiras et al. (1996) mental model approach, suggesting that the pronouns and anaphors are processed similarly when resolved with gender stereotyped antecedents.

15. a. The firefighter burned himself/herself while rescuing victims. . . .
    b. The babysitter found herself/himself humming while walking up . . .

Over the past several decades, there have been advances in the comprehension of speech by computers. The term ***artificial intelligence*** is used to refer to the branch of computer science that aims to create machines that demonstrate humanlike intelligence. When you call your credit card company or an airline, your speech may be interpreted by a computer. In Text Box 10.2, you will learn about the computer named Watson, whose abilities to understand and to produce human language are nothing short of remarkable.

Text Box 10.2 Extraordinary Individuals: The Amazing Watson

In 2011, viewers of the television game show Jeopardy were treated to a most unusual matchup. Two former champions on Jeopardy competed against a computer named Watson. The players were Bill Rutter, who had won the most games in Jeopardy history, and Ken Jennings, who was the Jeopardy player with the longest winning streak. The computer that they played against was developed in the DeepQA Project under the direction of David Ferrucci at IBM. Watson is the namesake of IBM's first president, Thomas J. Watson. Watson received the input as electronically submitted text at the same time as the human competitors heard Alex Trebek read the question. As in the regular version of the game show, the first contestant pressed a key and then responded using spoken language. Watson spoke with an electronic voice. With each answer that Watson attempted to generate, his programming also generated a confidence level. With each question, the audience was allowed to see the three answers deemed most probable by Watson's programming. Watson's data banks involved around 200 million pages of information, including the content of Wikipedia, other encyclopedias, and dictionaries. Ultimately, Watson beat both Rutter and Jennings. The winnings of Rutter and Jennings went to charities of their choosing. Watson's winnings were split between the charities chosen by Rutter and Jennings. IBM also developed the chess-playing computer Deep Blue, which in 1997 played the human world champion chess player Garry Kasparov. Deep Blue won the six-game match with two wins and three draws. Kasparov won just one game.

How Do We Remember What We Comprehend?

The process of understanding a discourse ultimately results in one forming a long-term memory for the information contained in the discourse. Long-term memory, as the name implies, refers to the permanent memory store. Research has not identified any limits on the size or capacity of long-term memory (Ashcraft & Radvansky, 2009). There have been numerous studies demonstrating the unreliability of long-term memory. For example, research by Elizabeth Loftus (2005) has shown that memory for events can be incorrect. Researchers have also shown that when answering questions about their memory for events, participants may perform less accurately when the questions contain incorrect details. These results have been referred to as the **misinformation effect**.

As you learned in Chapter 5, memory for events or episodic memory is one type of long-term memory. Another type is semantic memory, which refers to our general knowledge of word meanings and facts. Our memory for information that we acquire through the processing of discourses, either spoken or written, is one part episodic (i.e., those details of the time and place when we heard or read the discourse) and one part semantic (i.e., those details reflecting acquired knowledge). There are some discourses that we remember extremely well; we remember every word. This verbatim recall of discourses is rare. Americans may remember the Pledge of Allegiance well because of having to recite it so often during the elementary school years. Followers of Christianity may be able to recite by heart the Lord's Prayer or the 23rd Psalm. Remembering discourses word for word is rare and generally requires effortful practice. I vividly remember studying long hours trying to commit the Gettysburg Address to memory for an elementary school class assignment.

Sir Frederic Charles Bartlett (1886–1969) carried out a classic study about how people remember information from narratives. He constructed narratives for the study; the most famous was titled *The War of the Ghosts* (Bartlett, 1920). Table 10.3 displays this story in its entirety. The stories involved a plot that followed a seemingly logical sequence; however, in subtle ways, the events were only loosely related to one another. He asked participants to read a story and then to recall as much of the story as they could. The results showed that participants made many errors in retelling the story. Most found the task quite difficult. On close inspection, he found that the pattern of remembering and forgetting of specific details related to how well the details fit with the theme of the story.

The limitations of human memory have been well researched over the past 40 years, particularly in relation to what we see. Eyewitness testimony has been shown to be unreliable (Loftus, 1996). Our memory for what we read is also far less than perfect.

Bartlett (1920) studied how people's memory for stories changed over retelling. He found that with each successive telling, stories became shorter. Storytellers often changed the order of events to increase the coherence of the story. Furthermore, fewer of the inconsistent story elements were included, which resulted in the story seeming more coherent. For example, over retellings, the ghosts were often omitted. Details in the story were frequently changed to be more consistent with the storytellers' stereotypes about the context of the story. While most people experience routine failures of memory, a few individuals have exceptional memory, recalling virtually everything that they experience. Text Box 10.3 describes one such individual, described in a case study by the neuropsychologist Alexander Luria (1968).

Table 10.3 War of the Ghosts

One night two young men from Egulac went down to the river to hunt seals and while they were there it became foggy and calm. Then they heard war-cries, and they thought: "Maybe this is a war-party." They escaped to the shore, and hid behind a log. Now canoes came up, and they heard the noise of paddles, and saw one canoe coming up to them. There were five men in the canoe, and they said:

"What do you think? We wish to take you along. We are going up the river to make war on the people."

One of the young men said, "I have no arrows."

"Arrows are in the canoe," they said.

"I will not go along. I might be killed. My relatives do not know where I have gone. But you," he said, turning to the other, "may go with them."

So one of the young men went, but the other returned home.

And the warriors went on up the river to a town on the other side of Kalama. The people came down to the water and they began to fight, and many were killed. But presently the young man heard one of the warriors say, "Quick, let us go home: that Indian has been hit." Now he thought: "Oh, they are ghosts." He did not feel sick, but they said he had been shot.

So the canoes went back to Egulac and the young man went ashore to his house and made a fire. And he told everybody and said: "Behold I accompanied the ghosts, and we went to fight. Many of our fellows were killed, and many of those who attacked us were killed. They said I was hit, and I did not feel sick."

He told it all, and then he became quiet. When the sun rose he fell down. Something black came out of his mouth. His face became contorted. The people jumped up and cried.

He was dead.

Source: Bartlett (1920).

Development of Comprehension

As you learned in Chapter 4, children typically master most of the syntactic rules of their native language(s) by 4 to 5 years of age. You also learned in Chapter 3 that young children's understanding of language or their receptive language ability generally precedes their production of language or their **expressive language**. Most researchers agree that children's parsing involves the same stages of processing as adult parsing. The **continuity hypothesis** (Clahsen & Felser, 2006; Pinker, 1984) refers to the view that the mechanism that is used to parse sentences is essentially the same for children and for adults. Nevertheless, the knowledge that children activate and use during parsing is likely to be different from that used by adults. Furthermore, because parsing relies on the use of working memory and children's working memory capacity differs from adults' working memory capacity, differences between children's and adults' parsing are expected.

Text Box 10.3 Extraordinary Individuals: A Man With an Incredible Memory

In 1968, Alexander Luria published a case study describing a man with an astonishing memory. The individual was Solomon Veniaminovich Shereshevsky (1886–1958). He was born in Russia and worked as a journalist. Because of his excellent memory, he had no need to take notes for his work. The meeting of Luria and Shereshevsky occurred after Shereshevsky was reprimanded at work for not taking notes. He was able to recall verbatim the speech that had been spoken. He regularly demonstrated his extraordinary powers of memory in public settings. He was able to memorize tables of random numbers, poems in foreign languages, and mathematical formulas. He demonstrated near-perfect recall for a table of 100 random numbers after 16 years. He reported using the method of loci as his primary technique for remembering vast amounts of information. This method involves selecting a familiar route. Each item that is to be remembered is mentally placed at some point along the route. The technique is effective because of its reliance on mental imagery. His powers of memory were also likely influenced by the fact that he had synesthesia, a sensory disorder in which one experiences inappropriate sensations during information processing. Some people with synesthesia may see images when hearing sounds or hear sounds when viewing visual stimuli. Shereshevsky had fivefold synesthesia; each sensation that he experienced triggered sensations in each of his other sensory systems. A sound could trigger the visual experience of a color, a smell, a taste, and a sensation along the skin. His unusual sensory system likely contributed to his ability to develop effective memory strategies, but it also may have been responsible for odd feelings that sometimes caused him difficulty in life. Reading was sometimes problematic, because a word would trigger unpleasant sensations. The words that he heard could also lead to unpleasant associations. Consider the following example reported by Luria (1968):

> One time I went to buy some ice cream. . . . I walked over to the vendor and asked her what kind of ice cream she had. "Fruit ice cream," she said. But she answered in such a tone that a whole pile of coals, of black cinders, came bursting out of her mouth, and I couldn't bring myself to buy any ice cream after she had answered in that way. (p. 82)

How Do Children Process Passive Sentences?

The processing of passive sentences by children has been the focus of research since the 1970s. Horgan (1978) investigated how children comprehended sentences containing two types of passive constructions: (1) reversible and (2) nonreversible passives. Passive constructions contrast with active constructions. Consider the examples in 16. In active constructions, the subject of the sentence is the agent of the action, as in 16a. In passive constructions, the subject of the sentence is the patient of the action, as in

16b, and the agent of the action is either expressed in a preposition phrase or is omitted, as in 16c. For **reversible passives,** the noun phrases that serve as the agent and patient can be exchanged in the sentence, and the sentence still makes good sense. Consider the examples in 17a and 17b. For **irreversible passives,** the sentence does not make good sense if the positions of agent and patient noun phrases are switched, as in 17c and 17d.

16. a. The cat chased the rat. — Active
    b. The rat was chased by the cat. — Passive
    c. The rat was chased. — Passive

17. a. The cat was chased by the dog. — Reversible passive
    b. The dog was chased by the cat. — Reversible passive
    c. The cookie was eaten by the toddler. — Irreversible passive
    d. The toddler was eaten by the cookie. — Irreversible passive

Recent research has found that one can observe passive constructions in children's speech as early as the age of 3 years (Budwig, 1990). By the age of 4, children are producing full passives. Horgan (1978) found that children under 4 years of age used more reversible passives than irreversible passives. Furthermore, the use of both reversible and irreversible passives appeared to be related to age; the children who used both types of passives were older than 11. Still 50% of 11- to 13-year-olds in the study used only one type of passive or the other. Other studies have found that children experience little trouble comprehending adjectival passives, such as *The man was embarrassed* (Fox & Grodzinsky, 1998).

Children's processing of passives has also been investigated using a sentence-picture matching task (Stromswold, 2004). Stromswold compared performance by adults and children who were between 4 and 7 years old. Participants saw a pair of pictures and then heard either a passive sentence (e.g., *The girl was pushed by the boy*) or an active sentence (e.g., *The girl was pushing the boy*). Participants then judged which picture matched the meaning of the sentence. A matching picture was always present. As participants carried out the task, their eye movements were recorded. Analysis of eye movements showed that adults had interpreted passives at the word *pushed.* In contrast, children took longer interpreting passives; they did not look to the matching picture until after the sentence was completely processed. Stromswold (2004) concluded that children and adults process passives differently. Unlike adults, children appear to interpret the first noun in a sentence as the agent of the action and do not rapidly integrate morphosyntactic information as words.

Recent research clearly shows that the processing of passive sentences differs for young versus older children. Hahne, Eckstein, and Friederici (2004) measured the evoked brain potentials of children as they listened to passive sentences. The researchers were particularly interested in knowing whether childrens is brain activity included a pattern observed in adults' brain activity when they processed passive sentences. The pattern is referred to as **early left anterior negativity (ELAN)**. The pattern involves a waveform pattern observed on the left hemisphere anterior area of the scalp occurring relatively early in processing. They found that 6-year-old children's brain waves did not contain the ELAN pattern. Nevertheless, children were able to understand passive sentences. In a follow-up study, Oberecker, Friedrich, and Friederici (2005) measured brain waves for 2-year-old children as they processed syntactically grammatical or ungrammatical sentences. The results showed that children's brain waves contained the ELAN when processing ungrammatical sentences.

Researchers disagree about how very young children's knowledge of passive constructions begins. Tomasello (1992) argued that children's use of passives stems from their acquisition of specific verbs in their passive form. His view of children's early knowledge of grammar is not one that involves the arrangement of words and morphemes using grammatical rules, such as syntax or morphology. Instead, he argued that children learn language in chunks (e.g., groups of words), which they put together with other chunks to produce utterances. His view of children's grammatical development is consistent with the statistical learning approach. In contrast, other researchers believe that children's initial grammatical knowledge that is involved in their producing and comprehending passives is abstract and not tied specifically to individual words that they have learned (Conwell & Demuth, 2007; Naigles, Bavin, & Smith, 2005). The abstract knowledge would resemble grammatical rules, which are applied during both comprehension and production. This view of processing is consistent with the generative approach.

How Do Children Use Contextual Referential Information During Comprehension?

Relatively few studies have been conducted to investigate how children carry out syntactic parsing and to what extent their parsing is similar to the parsing of adults. The studies that have been conducted provide strong evidence that children's performance during parsing differs from performance by adults. It is premature to conclude that the results are evidence against the continuity hypothesis. It is possible that while the syntactic parser is the same for children and adults, other processing mechanisms may be different (e.g., attention, memory). The most revealing research so far is that conducted by Trueswell and colleagues (Snedeker & Trueswell, 2004; Trueswell, Sekerina, Hill, & Logrip, 1999). Trueswell and colleagues (1999) have used the visual world paradigm in which the eye movements of children were recorded as they looked at a visual scene of objects and listened to imperative sentences, such as *Put the frog on the napkin in the box*. The sentence contains a syntactic ambiguity (i.e., should the child move the frog to the napkin or is

▲ Photo 10.4 Children listen to an adult reading a story. Do you remember having story time when you were in first or second grade?

there a frog that is sitting on a napkin already that the child should move to the box). They manipulated the visual scene to disambiguate the meaning. On some trials, there were two frogs, which could lead listeners to interpret the phrase *on the napkins* as a modifier of the noun rather than a modifier of the verb. On other trials, the scene contained a single frog, which could lead listeners to interpret the phrase as a modifier of the verb. The results of the study showed that children around 5 years of age, unlike older children, interpreted the sentence incorrectly, initially attaching the phrase *on the napkin* to be where the frog should be moved. They spent a considerable amount of time looking at the incorrect location (i.e., an empty napkin). Sixty percent of the time, children incorrectly moved a frog to a napkin before placing the frog in the box. Children's performance was flawless when processing sentences without a syntactic ambiguity (e.g., *Put the frog that's on the napkin in the box*)—thus, the children's trouble stemmed from their failure to use the contextual information to guide their processing of syntactic ambiguity. Further studies showed that children 8 years old or older did not experience the trouble understanding the syntactic ambiguity; their performance was comparable to that of adults.

Other studies have shown that children may rely on other types of information during parsing more than do adults. Snedeker and Trueswell (2004) compared children's and adults' use of verb bias information in resolving a syntactic ambiguity. They also used the visual world paradigm. As participants viewed a visual scene of objects and listened to imperative sentences, their eye movements were recorded. Participants experienced three different trials. Example sentences are presented in 18. In each sentence, the prepositional phrase beginning with the preposition *with* could be initially interpreted as an instrument phrase, indicating how the action would be executed, or it could be interpreted as a modifier of the noun phrase, indicating which one of multiple cows, frogs, or pigs is being described. The type of verb in the sentence was varied. Some verbs, such as *tickle*, are used frequently with instrument phrases. Other verbs, such as choose, are used frequently with modifiers.

18. a. Tickle the pig with the fan. — Instrument bias
    b. Feel the frog with the feather. — Equi bias
    c. Choose the cow with the stick. — Modifer bias

The performance of children around 5 years of age showed that they were influenced by the type of verb in the sentence. In contrast, the performance of adults was not influenced by the type of verb in the sentence. Using a similar methodology, Kidd, Stewart, and Serratrice (2011) found that 5-year-old children failed to revise an initial syntactic analysis even when disambiguating information from the sentence was presented soon after the ambiguity.

Young children also appear to ignore semantic plausibility when they encounter a syntactic ambiguity. Kidd and Bavin (2005) used the visual world paradigm and found that children who were about 4 months away from their 4th birthday interpreted the sentence *Chop the tree with the leaves* differently than adults. Adults were shown to interpret the prepositional phrase *with the leaves* as a modifier of the noun *tree*. The results showed that the children interpreted the phrase as a modifier of the verb, despite the fact that *leaves* cannot be used to chop something else.

Joseph et al. (2008) measured children's and adults' reading time on sentences containing implausible words. Sample sentences are shown in 19. In each sentence, the direct object of the sentence serves as an instrument for a following verb. Later, the reader encounters the noun phrase *horrible mouse*. In 19a, the instrument is plausible. In 19b, the instrument is implausible, and in 19c, the instrument is anomalous (i.e., nonsensical). The results showed that adults and children processed anomalous sentences similarly with slower reading times at the word *mouse* in the anomalous condition than the plausible condition. However, when processing conditions were implausible, children took longer to show an effect of plausibility than did adults.

19. a. Robert used a trap to catch the horrible mouse that was very scared.
    b. Robert used a hook to catch the horrible mouse that was very scared.
    c. Robert used a radio to play the horrible mouse that was very scared.

Summary and Theoretical Implications

Language comprehension is believed to involve multiple stages of processing. Researchers disagree about whether processing occurs with serial, nonoverlapping stages or with parallel, completely interactive processing. Studies investigating how readers resolve temporary syntactic ambiguities have found that sometimes an initial interpretation of an ambiguous phrase turns out to be incorrect. Processing difficulty occurs when readers detect that the following sentence context does not support the initial analysis. Comprehenders use information from the context to comprehend sentences.

As comprehenders process discourses, they form mental links between distant parts of the text. As comprehenders process sentences and discourses, they form mental representations of the content or mental models. Comprehenders construct a mental model over time, updating it and revising it as new information is processed. Processing difficulty can occur when one must revise the mental model. Memory for discourses is far from perfect; details are routinely omitted and erroneous details are inserted. One's knowledge of the setting described in the discourse can serve as a schema with which one organizes the details of the discourse.

Although there are far fewer studies investigating comprehension in children than investigating comprehension in adults, the studies that have been conducted have produced strong evidence that there are differences in the comprehension strategies of adults and children in the processing of passive constructions and the resolution of syntactic ambiguity. As more work is conducted in this area, it is likely that more differences will be discovered.

In terms of the theoretical approaches to language development, there is research supporting the generative perspective, which argues that aspects of language are innate and that comprehension processes are carried out in a modular fashion, and research supporting the statistical learning approach, which argues that knowledge of language is stored in memory in a manner that is similar to other types of knowledge and that during language processing, stored knowledge can be used interactively throughout processing (i.e., not in a modular fashion). In contrast, the research on language comprehension provides little or no support for the behaviorist and social-interactionist views.

KEY TERMS

anaphoric pronoun
antecedents
artificial intelligence
cataphoric pronoun
continuity hypothesis
early left anterior negativity (ELAN)
end-of-clause wrap-up effect
expressive language
instrument
irreversible passives
language competence
language performance
late closure
McGurk effect
minimal attachment
misinformation effect
phrase marker
pronouns
reversible passives
syntactic ambiguity
syntactic parsing
syntactic reanalysis
thematic roles
visual world paradigm

REVIEW QUESTIONS

1. When researchers describe language comprehension as occurring incrementally, what do they mean?
2. What are the two theoretical approaches to syntactic parsing? How do these approaches differ?
3. What do the syntactic parsing principles *minimal attachment* and *late closure* predict about how readers interpret ambiguous words and phrases?
4. What roles does working memory play in determining how comprehenders process sentences containing long-distance dependencies?
5. What is the end-of-clause wrap-up effect? What do researchers believe is occurring at the ends of clauses?
6. What is a thematic role? How are thematic roles related to the syntactic roles in sentences, such as subject and object?
7. What is known about how comprehenders process figurative versus literal language?
8. What are some ways in which speakers and authors maintain cohesion in discourses?
9. What is the evidence that comprehenders activate and use information about gender during processing?
10. What roles does working memory play in determining whether comprehenders draw an inference during discourse processing?
11. What is the continuity hypothesis? What prediction does it make about how children process sentences?
12. How does syntactic parsing develop during early childhood? How is the syntactic parsing of children similar to or different from the syntactic parsing of adults?

13. What has research revealed about how children's memory for discourses differs from adults' memory for discourses?
14. How do readers use mental models to understand discourses?
15. How does children's ability to use passive sentences change during childhood?
16. What evidence is there that children use contextual information during processing differently than adults?

RECOMMENDED READING

Carreiras, M., & Clifton, C. (2004). *The on-line study of sentence comprehension: Eyetracking, erps and beyond.* New York: Psychology Press.

Fernandez, E. (2004). *Fundamentals of psycholinguistics*. Hoboken, NJ: Wiley-Blackwell.

Garnham, A. (2001). *Mental models and the interpretation of anaphora.* New York: Psychology Press.

Perfetti, C. (1995). *Reading ability.* New York: Oxford University Press.

Rayner, K., & Pollatsek, A. (1989). *The psychology of reading.* New York: Prentice Hall.

Townsend, D. J., & Bever, T. G. (2001). *Sentence comprehension: The integration ofhabits and rules.* Cambridge, MA: MIT Press.

Traxler, M., & Gernsbacher, M. (2004). *Handbook of psycholinguistics* (2nd ed.). New York: Academic Press.

RECOMMENDED FILMS

Jayanti, V. (Director). (2003). *Game over: Kasparov and the machine* [Motion picture]. Canada: The National Film Board of Canada.

Searchinger, G., Male, M., & Wright, M. (Writers). (2005). *Human language series* [DVD]. United States: Equinox Films/Ways of Knowing Inc.

SUGGESTIONS FOR CLASS PROJECTS

1. Become more familiar with the structure of sentences by diagramming them. One method of diagramming sentences is the Reed-Kellogg System (Reed & Kellogg, 1899). Select 10 to 20 sentences varying in complexity to diagram. The following diagrams can be used for different sentence types.

a. Subject + Verb

Subject | Verb

b. Subject + Verb + Object

Subject | Verb | Object

c. Subject + Verb + Object + Adverb

Subject | Verb | Object

\Adverb

2. Identify the agent, patient, and instrument thematic roles in the following sentences. Draw a circle around the noun phrases that serve as agents, a square around the noun phrases that serve as patients, and a triangle around the noun phrases that serves as an instruments. Some sentences will not have all three thematic roles.

a. The horses trampled the flowers.
b. The flower was trampled by the horses.
c. The chef stirred the mixture with a spatula.
d. The pony kicked the steer by the barn.
e. The kitten licked the child with its tongue.
f. The burglar bludgeoned the resident with a lamp.
g. The doctor was sued by the former patient.
h. The hamburgers were eaten by the children with ketchup and mustard.
i. The tornado flattened the town.
j. The detective watched the couple using binoculars.

3. Test your classmates' and/or friends' memory for written narratives using a method similar to that used by Bartlett (1920). Select a narrative from collections of short stories from the Internet or the library. Ask someone to read the story and to try to remember all the details of the story that he or she can. Once he or she is finished reading the story, the reader can be asked to write down what was remembered of the story. Responses can be collected and analyzed for content. Note details that were provided that were actually in the story, note details that were in the story but not mentioned, and also note details that were not in the story but were provided in responses. Analyze the responses to find trends. Summarize your findings in a paper or class presentation.

For additional ancillary resources, please visit the companion website at www.sagepub.com/kennison.

- Video Links
- Audio Links
- Web Resources
- Internet Activities
- Flashcards
- Web Quizzes

CHAPTER 11

LANGUAGE DURING THE SCHOOL YEARS

By the time children begin school, they have usually become fairly skilled speakers of their native language. They may have made some progress in learning how to write the alphabet and some basic words, including their names and the names of family members. Depending on their home environment, they may have also been provided with opportunities to learn to read. There is a great deal of variability in children's experience with reading and writing as they enter kindergarten. In this chapter, you will learn about how children around the world continue to develop language skills during the elementary school years. Becoming literate is a journey; after basic reading skills are developed, students hone their comprehension skills for many years. In this chapter, you will learn how children are taught language skills in classroom settings. You will learn about the research that has investigated which methods are most effective in teaching children to read. You will also learn about some of the reasons why children have difficulty reading. The most familiar problem is dyslexia. You will learn about the different varieties of dyslexia that have been observed, the causes of dyslexia, and the most common treatments for dyslexia. Last, you will learn about other disorders that impact children's ability to succeed in school. These include **attention deficit disorder** (ADD), **dyscalculia,** and **dysgraphia,** as well as others.

▲ Photo 11.1 A group of children are reading the same book. Do you remember any group reading activities from your childhood?

Learning to Read

What Is Phonological Awareness?

Among the first language skills that children learn, either when they enter a school setting or before, is the learning of the alphabet for their language. English-speaking children learn their ABCs. Most learn how to sing the alphabet song:

A-B-C-D-E-F-G H-I-J-K L-M-N-O-P Q-R-S T-U-V W X Y and Z

Now, I've said my ABCs. Next time won't you sing with me?

The song appeared in 1835 when the Boston music publisher Charles Bradlee registered it for a copyright. The tune was the creation of Louis Le Maire, a French composer who lived in the 18th century. In other languages, there are well-known children's songs or rhymes that contain the phonemes or syllables of the languages (Le Maire, 1835).

The learning of the ABCs is the beginning of children's **phonological awareness**, which refers to their awareness of the sounds and sound structure of their language. Children's phonological awareness also involves both speaking and listening ability. Research from the 1970s showed that only about 50% of children between 4 and 5 years of age could accurately identify the number of syllables in multisyllabic words (Liberman, Shankweiler, Fischer, & Carter, 1974). The researchers asked children to tap their hands on a table's surface once for each syllable that they heard. When 6-year-olds were asked to perform the same task, 90% of the children were able to do so correctly.

Table 11.1 Examples of Onsets and Rimes for Ten Common English Words

Word	Onset	Rime
Top	/t/	/op/
Bland	/bl/	/and/
School	/sk/	/ool/
Fox	/f/	/oks/
Flock	/fl/	/ok/
Pep	/p/	/ep/
Please	/pl/	/ez/
Sixths	/s/	/iksθs/
Splints	/spl/	/ints/
Strippec	/str/	/ipt/

Source: Dictionary.com.

Awareness of Syllables

Research by Rebecca Treiman (1985) showed that children's understanding of the phonological structure of words includes information about the parts of a syllable. A syllable is composed of an **onset**, which is the initial consonant or consonant cluster, and a **rime**, which is the rest of the syllable. Table 11.1 provides examples of the onsets and rimes for some common English words. Treiman (1985) found that children who were as young as 4 years old could identify word onsets. In an activity involving a puppet, children were told that the puppet had a favorite sound: /f/. Later, the children were asked which nonword syllables (e.g., /fol/ or /zol/) the puppet would like. Children were able to identify syllables that began with the onset favored by the puppet. A later study showed that many 3-year-olds could also perform this task (Maclean, Bryant, & Bradley, 1987).

Young children are not as successful in dissecting the individual phonemes that are contained in the rime of a syllable. Liberman and colleagues (1974) found that children 5 years of age performed poorly when asked to count the number of phonemes in a word by tapping them out; only 17% of the children were successful. Four-year-olds were also tested, and none could count phonemes by tapping them out. The researchers found that among 6-year-olds, approximately 70% were successful tapping out individual phonemes. In later research, Treiman (1985) also found that young children were unable to identify individual phonemes within the rime of a syllable. She found that the 3-year-olds who demonstrated an awareness that the word *dog* could be broken down into the onset /d/ and the rime /og/ were not able to break down the rime into the phonemes /o/ and /g/.

Children's Word Games

Before children enter school, they may have practiced identifying syllables and phonemes through the recitation of nursery rhymes. Research has shown that knowledge of nursery rhymes by 3-year-olds was strongly related to their reading ability 3 years later (Bryant, Bradley, Maclean, & Crossland, 1989). Table 11.2 displays examples of English nursery rhymes that are typically familiar to preschool children. The relationship

▲ Photo 11.2 Three girls are jumping rope. Do you think that they are singing a jump rope song?

between knowing nursery rhymes at an early age and reading ability later in childhood is not well understood. Research has shown that when one factors out family income level and intelligence (as measured by IQ), there is a strong, positive relationship between nursery rhyme knowledge and knowledge of syllable structure (i.e., onsets and rimes) (MacLean et al., 1987). It is possible that knowing nursery rhymes increases phonological awareness, which aids children in learning to read. Once children begin reading, it is also likely that the more that they read, the more they will become aware of the phonological aspects of words and syllables.

Table 11.2 Examples of English Nursery Rhymes

Mary had a little lamb its fleece was white as snow; And everywhere that Mary went, the lamb was sure to go. It followed her to school one day, which was against the rule; It made the children laugh and play, to see a lamb at school. And so the teacher turned it out, but still it lingered near, And waited patiently about till Mary did appear. "Why does the lamb love Mary so?" the eager children cry; "Why, Mary loves the lamb, you know," the teacher did reply.
Georgie Porgie pudding and pie, Kissed the girls and made them cry When the boys came out to play, Georgie Porgie ran away.
Peter Peter pumpkin eater, Had a wife and couldn't keep her! He put her in a pumpkin shell, And there he kept her very well!
Old King Cole was a merry old soul, and a merry old soul was he; He called for his pipe in the middle of the night And he called for his fiddlers three. Every fiddler had a fine fiddle, and a very fine fiddle had he; Oh there's none so rare as can compare With King Cole and his fiddlers three.

Source: Opie and Opie (1951/1997).

Many children's games involve wordplay. These games can increase phonological awareness. One of the most common word games that English-speaking children play demonstrates their knowledge of basic syllable structure. The game is **pig Latin.** I vividly recall the first time that I was introduced to pig Latin. I was fascinated with first trying to understand it and later trying to speak it. The rules of pig Latin are as follows:

> First, if an English word begins with a consonant, then you form the pig Latin word by adding the syllable /ay/ to the initial consonant sound or consonant cluster and moving that new syllable to the end of the English word. For example, *galloping horses* becomes *alloping-gay orses-hay* and *hungry children* becomes *ungry-hay ildren-chay.* Second, if an English word begins with a vowel sound, one would add /ay/ to the end of the word. For example, *almost always* becomes *almost-ay always-ay.* In some regions of the country, vowel-initial pig Latin words may be formed differently than this. Last, for English compound words, one may apply the first rule to each word in the compound. For example, *playground* becomes *ay-play ound-gray.*

There are variants of pig Latin among speakers of German, Swedish, and French. For pig Latin and similar language games in other languages, children will practice identifying the individual syllables of words and identifying the onsets and rimes of syllables.

Once children enter school, they typically get a lot of practice dissecting words into syllables and phonemes. They may be asked to generate words that begin with the same sound as another word (e.g., *pony* and *puppy*). They may be asked to produce words that rhyme with other words (e.g., *air* and *hair*). As children learn the relationship between the sounds of their language and how the words are spelled (i.e., the spelling-sound correspondence), they may be asked to produce words that start with the same sound as *phone,* such as *flower, friend,* and *fox*. They may be asked to produce words that rhyme with each other despite the fact that they are spelled differently (e.g., *bare* and *fair*). Such exercises may be assigned as in-class activities or drills. They may also be given them as written homework.

How Do Children Learn to Read?

For children living in industrialized countries, learning to read is an important milestone in life. The age at which children start to read varies a great deal across the world. In the United States, children begin formal instruction in reading in the first grade. I still recall vividly the day in first grade when my teacher Mrs. Ballard passed out to each of us a Dick and Jane reader (Gray & Sharp, 1930). She then brought small groups of us over to the reading circle. She had each of us, in turn, sit on her lap. She pointed to a word on the page and pronounced the word, having us repeat the word out loud. Then she said, "Congratulations, you have read your first word." I recall being filled with awe, sensing that it was an important event. Now, I look back, considering that my academic interests have always been language, and know that this event played an important role in my life.

Chall (1996) studied how children progress from beginning to learn to read to becoming skilled readers. She identified three stages: (1) learning to read, (2) reading to learn, and (3) reading independently. Children in the first, second, and third grades are typically in the first stage of reading development. Their comprehension is better during listening than during reading. They are learning the grapheme-to-phoneme rules and learning how to comprehend words in context. Children in the fourth through ninth grades use reading as a tool for increasing knowledge. Their reading contributes to an impressive growth in their vocabularies. Their comprehension ability during reading comes to match their comprehension ability during listening. By the time children are in the 10th grade and beyond, they are typically independent readers. They read a wide variety of materials and continue to expand their vocabularies.

Learning to Read Outside of the United States

In China, learning to read proceeds differently than in English-speaking countries. Children do not receive formal training in Chinese characters until the third grade (Cheung & Ng, 2003). In the first and second grades, they are taught **pinyin,** which is a phonologically transparent writing system that allows children to read and write words based on their sound. Pinyin is written using Roman characters, which are the same characters that are used to spell English words. Pinyin is a system for representing words in terms of 21 different syllable onsets, which are consonants, and 35 syllable rimes. There are also four tones that specify how the syllable should be pronounced either with a high intonation, a rising intonation, a falling intonation, or a falling intonation quickly followed by a rising intonation. The pronunciation of words written in pinyin is completely unambiguous. In a study of 6-year-old children in Beijing, Lin and colleagues (2010) investigated the relationship between the children's ability to read and write in pinyin for words and nonwords and their ability to read using the Chinese logographic script 1 year later. They found that children's ability to read and write nonwords in pinyin predicted how well they read words in Chinese 1 year later. The results suggest that the phonological awareness that children develop while using pinyin is related to their future success in learning to read and write words using logographic symbols, which do not provide visual cues regarding the pronunciation of the word.

Outside of the United States, children learn to read at different ages. For example in England, children learn to read around the age of 5. In Sweden and Denmark, children do not start formal reading instruction until the age of 7. In Japan and France, children learn to read at the age of 6. Today, in the United States, many children enter preschool and/or kindergarten with some experience reading. Over the past several decades, the practice of teaching very young children to read has become increasingly popular. A common question that many parents have for educators and reading researchers is whether there is an ideal age for children to begin learning to read. The term *reading readiness* is used to refer to when children are ready to begin to learn how to read.

When children enter school and are provided with formal reading instruction, they might receive one of two popular teaching methods. The **phonics method** focuses on teaching children the spelling-to-sound regularities of the language. The method teaches children to be more phonologically aware. Instruction involves the use of phonologically based drills and exercises. The **whole language method** focuses on teaching children to read by emphasizing the whole reading experience and the use of reading strategies to extract meaning. Instruction involves students reading in groups, reading aloud, and encouraging students to develop a love of reading. Because curriculum decisions are made on the local level in the United States (i.e., at the level of the school district), the method used to teach reading varies widely from school to school.

Determining which of the methods used to teach children to read is most effective has proven not to be easy. In the 1990s, a panel was convened by Congress to study the issue and to report their findings (MacGuinness, 2004). They released their report in 2000. They found that thousands of studies had been conducted; however, most studies lacked the scientific rigor to yield conclusive findings. The conclusion recommended by the panel was viewed as a political compromise; they recommended that reading is best taught using multiple methods, which included training in phonics.

Table 11.3 Examples of Multiple Pronunciations for the Past Tense

-ed Pronounced as	Examples
/d/	Shunned, fried, used, creamed
/t/	Smacked, cooked, walked, kicked
/id/	Rated, tainted, friended, collected

In 1999, the American Federation of Teachers pointed out that in the United States, schools of education, which train reading teachers, are not keeping up with the scientific knowledge that has been gained about reading. Surveys of reading teachers have in the United States revealed that their knowledge of spelling in English and their ability to identify speech sounds is no better than the average native speaker of English. The explanation for these results is that reading teachers are not required to take any courses on the structure or history of the English language. Having a course that covers the grammatical properties of English would be helpful, as there is a close relationship between the phonological and morphological structure of words. The relationship impacts the spelling of words. The morphological rules apply to how prefixes and suffixes are used. With the addition of these morphemes, there are predictable changes in the part of speech for the word. For example, adding the suffix *-al* to a verb creates a noun (e.g., *approval, denial, recusal*). Table 11.3 displays the various forms of past tense verbs. The pronunciation of the suffix *-ed* varies predictably.

The American Federation of Teachers also points out that it is too often the case that beginning reading is taught by someone who does not have a fundamental understanding of the psychological processes that make up skilled reading. They do not know what constitutes typical or atypical reading development. Unlike other professional groups, such as beauticians, art therapists, and optometrists, reading teachers are not credentialed. They are not required to demonstrate proficiency prior to obtaining employment. Furthermore, most boards of education and colleges of education seem to believe that teaching reading can be done by anyone; it does not require any technical expertise or specialized training. If this misconception is not dispelled quickly, children in the United States will continue to lag behind other countries in literacy.

How Are Social Factors Related to Learning to Read?

Teachers and researchers alike will agree that regardless of the method used to teach reading in the classroom, some students end up becoming better readers than others. One of the best predictors of reading ability as a child ages is the child's vocabulary knowledge.

A study by Ouellette (2006) investigated the reading abilities of children in fourth grade and found that it was their vocabulary depth versus their vocabulary breadth that was the stronger predictor of reading comprehension. Similar results were obtained by Tannenbaum, Torgesen, and Wagner (2006) in a study with third graders.

One of the best predictors of children's reading ability as well as academic success is their family income. Researchers often refer to family income level as socioeconomic status (SES). The relationship between SES and reading ability is not completely understood. Children from low SES backgrounds are more likely to be born prematurely and more likely to have substandard nutrition and health care than children from higher SES backgrounds. They are also more likely to reside in communities and/or buildings near polluted areas (e.g., superfund sites). Exposures to toxins, such as lead, can permanently lower IQ and cause information processing deficits.

Research has found strong relationships between SES and a range of measures related to academic achievement, including IQ (Liaw & Brooks-Gunn, 1994; Smith, Brooks-Gunn, & Klebanov, 1997) and performance on standardized tests (Brooks-Gunn, Guo, & Furstenberg, 1993). One study found that children's SES explained about 20% of the variance observed in childhood IQ scores (Gottfried, Gottfried, Bathurst, Guerin, & Parramore, 2003). Studies have also found that SES is related to children's performance in tasks involving language in general and involving reading specifically. Several studies have found that children from higher SES homes have shown advantages in language development over children from lower SES homes (Hoff, 2003; Noble, Norman, & Farah, 2005; Whitehurst, 1997). Baydar, Brooks-Gunn, and Furstenberg (1993) found a strong relationship between children's SES and functional literacy (see also White, 1982). SES has also been found to be related to children's ability to read individual words and nonwords (Bowey, 1995; Hecht, Burgess, Torgeson, Wagner, & Rashotte, 2000; Raz & Bryant, 1990). Children from low SES backgrounds also are less likely than those from higher SES backgrounds to have books in the home, to be talked to by caregivers, and to be read to (Levitt & Dubner, 2005).

The Matthew Effect

In 1986, Keith Stanovich proposed that there is a **Matthew effect** for education generally and for reading ability specifically. The term *Matthew effect* had been used in sociology to refer to the proverb that "the rich get richer and the poor get poorer." The origin of this proverb is the Christian Bible:

> For to all those who have, more will be given, and they will have an abundance;
> but from those who have nothing, even what they have will be taken away.
>
> —Matthew 13:12 (see also Matthew 25:29)

Stanovich (1986) suggested that children who start out as strong readers are likely to become better and better readers over time. His view is captured well by the following statement:

> The effect of reading volume on vocabulary growth, combined with the large skill differences in reading volume, could mean that a "rich-get-richer" cumulative advantage phenomenon is almost inextricably embedded within the

> developmental course of reading progress. The very children who are reading well and who have good vocabularies will read more, learn more word meanings, and hence read even better. Children with inadequate vocabularies—who read slowly and without enjoyment—read less, and as a result have slower development of vocabulary knowledge, which inhibits further growth in reading ability. (Stanovich, 1986, p. 381)

Numerous longitudinal studies of children's reading ability have shown that young children who start out as poor readers continue to struggle with reading, and young children who are typical or above average readers continue to be typical or above average (Aarnoutse, van Leeuwe, Voeten, & Oud, 2001; Bast & Reitsma, 1998; Juel, 1988; Scarborough & Parker, 2003; Shaywitz et al., 1995). In a study by Juel (1988), the probability that a student in first grade who is a poor reader will go on to become a typical or above average reader by fourth grade was very low (i.e., 0.12). Shaywitz et al. (1995) pointed out that Stanovich's (1986) Matthew effect predicts that difference in reading performance for good and poor readers would increase over time. In a longitudinal study that followed kindergarten children for 7 years, Shaywitz found a small Matthew effect for IQ but none for reading. Kempe, Eriksson-Gustavsson, and Samuelsson (2011) studied a group of children from the time they were in first grade until the third grade. They found Matthew effects in measures of reading comprehension and vocabulary knowledge but did not find Matthew effects on measures of word recognition, spelling, or nonverbal ability.

Since 1965, low-income families with children between the ages of 3 and 5 have qualified for Head Start. Head Start is a government program that aims to help prepare low-income children for kindergarten. Numerous studies have shown that children who attend Head Start do better in reading and math than other children (Lee, 2011). Just as children's homes can differ in terms of language environment and the extent to which the environment supports learning to read, so do classrooms. Snow and Dickinson (1990) found that Head Start classrooms varied a great deal in terms of how much class time was spent on reading-related activities. The research showed that Head Start classrooms used between 0% and 18% of class time on such activities. Research by the National Institute of Child Health and Development (NICHD) (2002, 2005) found that when classrooms contained more adults per child and children received more exposure to adult speech, their performance on cognitive and language tests was better than when classrooms contained fewer adults per child. Huttenlocher, Vasilyeva, Cymerman, and Levine (2002) found that the pupils whose teachers were more likely to use complex sentence structures developed complexity in their own speech over the course of an academic year to a greater extent than pupils whose teachers were less likely to use complex sentence structures.

Without a doubt, children's language skills are one predictor of academic success (Jordan, Snow, & Porsche, 2000; Whitehurst & Lonigan, 1998). However, educators have long recognized that success in school involves more than just academic skills, such as reading, writing, and arithmetic (see Anderson et al., 2003, for review; see also Raver & Zigler, 1997). An important predictor of academic success is social competence, which refers to a child's ability to interact effectively with peers and superiors (Ladd, Kochenderfer, & Coleman, 1996; O'Neil, Welsh, Parke, Wang, & Strand, 1997). Gertner, Rice, and Hadley (1994) found that children's expressive language was significantly related to their social competence, as judged by teachers. They reasoned that children who have difficulty in expressing themselves to peers may be rejected more often by peers and fail to have social experiences because of lack of opportunities. A recent study showed that the relationship between language ability and social competence stemmed specifically from verbal aspects of social competence, such as using the names of others, than from nonverbal aspects of social competence, such as using classroom equipment safely (Longoria, Page, Hubbs-Tait, & Kennison, 2009).

Cultural Differences

There are cultural differences in children's socialization, specifically in how much they are encouraged to speak their minds in conversations with adults and with peers. For example, children reared in Western cultures (e.g., the United States, Canada, and the United Kingdom) often speak their minds, ask questions, refute the statements of others, and even engage in arguments. In contrast, children reared in Eastern cultures (e.g., China, Japan, Korea, and India) display more respect toward those with whom they converse; they are less likely to spontaneously report opinions and express disagreement with others. Within the United States, there are also cultural differences. Children reared in Southeast Asian, Mexican American, and some African American communities are less likely to initiate conversations with adults than children reared in other communities (Grant & Gomez, 2001).

Becoming a skilled reader is important if one is going to be successful in school and in a future career (McCardle & Chhabra, 2004). Children who have severe reading problems generally do not complete college (Neuman & Dickinson, 2002). Poor readers also have a higher chance of becoming incarcerated. The literacy rates around the world have been compared by the Organisation for Economic Cooperation and Development (OECD) since the mid-1990s (OECD, 2000). Their results have found that literacy rates are generally lower in English-speaking countries (i.e., United States, Canada, United Kingdom, Ireland, Australia, and New Zealand) than in European countries in which English is not spoken. Among the English-speaking countries, the lowest ranked country was the United States. Table 11.4 displays these countries in ranked order.

The low literacy levels in the United States have received a great deal of attention. Low literacy levels impact all age-groups. The American Federation of Teachers found that about 20% of elementary school children do not read well and generally cannot read on their own (Moats, 1999). When results for ethnic minority students were examined separately, they found that 60% to 70% fell within the category of *poor reader*. They found that children from the most privileged families also struggled with reading; 30% of poor readers were from families headed by a college-educated parent. Studies have also shown that about one quarter of adults in the United States do not possess the literacy skills to be successful in the workforce (Kirsch, Jungeblut, Jenkins, & Kolstad, 1993). Individuals who are illiterate appear to be overly represented in prison populations. Linacre (1996) estimated that approximately two thirds of people incarcerated in prisons in the United States are functionally illiterate.

The low literacy levels in English-speaking countries may be due in part to the fact that English is a language with many irregularly spelled words (Kennison, 2007). The irregularity

Table 11.4 Top Ranked OECD Countries in Literacy

	Type of Literacy		
Country	**Prose**	**Document**	**Quantitative**
Australia	5	6	6
Belgium	7	4	4
Canada	9	8	9
Germany	3	2	2
Ireland	8	10	11
Netherlands	2	3	3
New Zealand	4	7	10
Sweden	1	1	1
Switzerland	10/11[a]	9/11[a]	5/8[a]
United Kingdom	6	5	7
United States	12	12	12

Source: OECD (2000).

[a]Values are for the German-speaking/French-speaking regions of Switzerland.

in English spelling is not trivial. A study of 17,000 words revealed that only 50% followed regular spelling-to-sound rules; over 25% were found to follow an irregular spelling-to-sound pattern, and 10% followed no pattern (Hanna, Hodges, & Hanna, 1971). Many irregularly spelled words are also among the most frequently used words in the English language (e.g., *the, one, caught*) (Francis & Kučera, 1982). In contrast, European languages, such as German, French, Swedish, and Dutch, have transparent spelling-to-sound systems, which means that when a reader reads a new word, the pronunciation of the word is completely predictable and when one hears a word pronounced, the word's spelling is also completely predictable. Interestingly, spelling bees, which you learned about in Chapter 5, typically occur only in English-speaking countries.

The problematic spelling of English is due to the fact that the language has been written since approximately the 8th century AD (Baugh & Cable, 2002). Over time, the English language has been influenced by other languages, such as Greek, German, French, the Scandinavian languages, and the language of the Celts (Lass & Hogg, 2000). Only about 25% of words in the *Merriam-Webster Dictionary* of English have English origins (Tilque, 2000). Table 11.5 displays a list of common English words that have foreign origins.

Since the late 1800s, there have been those who have suggested that the spelling of English should be changed to reduce the amount of irregularity. Such changes would likely make learning to read and to spell English words easier (Ives, 1979). There have been changes in spellings before. In 1876, eleven new spellings were adopted by the American Philological Association (e.g., *ar, catalog, definit, gard, giv, hav, infinit, liv, tho, thru,* and *wisht*). In 1879, the Spelling Reform Association was formed, 3 years after the International Convention for the Amendment of English Orthography was held. In the decades that followed, there were many English spelling changes introduced. Attempts to make more substantial changes failed, including an attempt by President Theodore Roosevelt, who proposed changing the spelling of 300 commonly used words. His order was blocked by the U.S. Congress. Other countries have been more successful in spelling reform efforts. Many believe that it is unlikely that spelling reform efforts will prevail. Opponents of spelling reform claim that changes in the spelling system would result in widespread confusion and costs stemming from the need to change computer applications used for the processing of texts.

For example, in 1443, King Sejong the Great completely changed the spelling system of Korean (Kim-Renaud, 1997). At the time, only the wealthiest individuals typically learned to read and write. The writing system was difficult to learn, as it was logographic, using Chinese-based symbols called Hanja. King Sejong directed a team of scholars to reform the Korean writing system. They created a 28-character alphabet that could be learned relatively quickly. His long-term goal for the new alphabet was to increase literacy among all levels of Korean society. The new alphabet was called Hangul. It was not until the 19th and 20th centuries that the use of Hangul became widespread (Fischer, 2004).

Table 11.5 Common English Words With Foreign Origins

Word	Original Language
Candy	Arabic
Algebra	Arabic
Silk	Chinese
Ketchup	Chinese
Massacre	French
Sex	French
Lasso	Spanish
Patio	Spanish
Ski	Norwegian
Futon	Japanese
Tycoon	Japanese
Hooligan	Irish
Whiskey	Irish
Golf	Scottish
Blackmail	Scottish
Breeze	Portuguese
Cobra	Portuguese

Source: Oxford English Dictionary (OED) (2012).

Dyslexia

What Is Dyslexia?

In popular culture, dyslexia is viewed as persistent problems with reading that do not disappear as one ages. Among researchers, a more formal definition is used. Dyslexia is defined as follows:

> A disorder manifested by difficulty learning to read, despite conventional instruction, adequate intelligence and sociocultural opportunity. It is dependent upon fundamental cognitive disabilities which are frequently of constitutional origin. (Critchley, 1975)

This definition points out that individuals diagnosed with dyslexia must have an IQ in the normal range; otherwise, the problems with reading could be attributed to intellectual disability. Young children who go on to be diagnosed with dyslexia may exhibit a range of symptoms when they begin to learn to read and write. They may exhibit speech delays, such as being a late talker. When learning nursery rhymes, they may have trouble identifying rhyming words. They may have trouble learning new words and manipulating letters. Some children may reverse letters during writing.

In the United States, it is common practice to delay diagnosis of dyslexia until children are at least two grade levels behind their same-age peers in reading (Rayner, Pollatsek, Ashby, & Clifton, 2012). Assuming that reading instruction does not begin until first grade, most children with dyslexia will not receive a diagnosis until the third grade. By the time that a child is assessed for dyslexia, the parents and the child may be desperate to understand why the child is struggling so much with schoolwork. They may be disappointed if they are told that the child does not have dyslexia. Approximately 5% to 10% of students who have reading difficulties are labeled as *poor readers* (i.e., they are not dyslexic) (Nation & Snowling, 1997; Yuill & Oakhill, 1991). Children who are poor readers have difficulty with language comprehension (Catts, Adolf, & Weismer, 2006; Nation, Clarke, Marshall, & Durand, 2004) but often perform well on tasks involving single word reading. Some researchers have claimed that poor readers struggle mostly with oral language comprehension (Catts et al., 2006). Other researchers have documented that poor readers have trouble with multiple aspects of sentence-level comprehension (e.g., Conti-Ramsden & Botting, 1999). Poor readers typically exhibit weaker semantic skills (Nation & Snowling, 1998) and syntactic skills (Nation & Snowling, 2000) than their same-age peers who read well.

What Are the Different Types of Dyslexia?

Researchers make a distinction between **developmental dyslexia,** which occurs in children who have no known physical problems, and **acquired dyslexia,** which occurs in individuals who have problems reading after they have sustained a brain injury. Several different types of developmental dyslexias have been identified (Heim et al., 2008). Early research with individuals who acquired dyslexia relatively late in life found that most cases could be categorized into two types: (1) **surface dyslexia,** which is characterized by the ability to read words that follow the regular grapheme-to-phoneme rules but difficulty reading irregular words or (2) **phonological dyslexia,** which is characterized by the ability to read familiar words but difficulty reading function words and words with inflectional morphemes. Unfamiliar words cannot be sounded out or read.

Castles and Coltheart (1993) reviewed the types of acquired and developmental dyslexias and argued that there were striking similarities between the types of acquired dyslexia and the types of dyslexia observed in children. The phonological dyslexia observed in some who experience brain injury is similar to a developmental variety of dyslexia known as **dysphonetic dyslexia.** It is characterized by the ability to read familiar words but difficulty reading function words and words with inflectional morphology. Surface dyslexia, also an acquired dyslexia, is similar to the developmental dyslexia known as **dyseidetic dyslexia.** It is characterized by the ability to read words that follow the regular grapheme-to-phoneme rules but difficulty reading irregular words. A third common type of developmental dyslexia, **deep dyslexia,** appeared to be least similar to acquired dyslexia. It is characterized by semantic errors involving individual words in reading, such as seeing the word *ape,* but experiencing the word as *monkey.* Those with deep dyslexia may also have problems reading function words, abstract words, and words that are previously unknown.

Castles and Coltheart (1993) noted that the phonological/dysphonetic dyslexias and the surface/dyseidetic dyslexias appear to be compatible with Coltheart's (1978) **dual-route model** of reading (see also Morton & Patterson, 1980). He proposed that word reading occurs via two parallel processes. One process involves the recognition of whole words. In this whole word reading route, a reader matches to memory the whole visual form of the word. A second process involves the use of grapheme-to-phoneme correspondences. In this spelling-to-sound reading route, a reader computes the sound of the word and matches to memory the phonological representation of the word. In nondyslexic readers, the two processes are carried out simultaneously. Sometimes, word recognition may result from one route being completed first, and other times, it may result from the other route being completed first. In phonological and dysphonetic dyslexias, the spelling-to-sound route appears not to be functioning; thus, word reading is carried out only through the use of the whole-word reading route. In surface and dyseidetic dyslexias, the whole word reading route appears not to be functioning; thus, word reading is carried out only through the use of the spelling-to-sound route. Castles and Coltheart (1993) argued that some individuals with dyslexia have problems with decoding orthography specifically; the problem is not related to phonological processing.

Prevalence of Developmental Dyslexia

Research suggests that in the United States between 5% and 17% of people have dyslexia (Shaywitz & Shaywitz, 2004). Girls and boys are equally likely to be dyslexic. Dyslexia can be inherited. If one child has a biological family member with dyslexia, there is a greater likelihood that the child will receive a diagnosis of dyslexia than a child who has no biological relatives with dyslexia. In order for a child to be diagnosed with dyslexia, his IQ must be in the normal range. The average IQ score is 100 and has a standard deviation of 15 (Schachter, Gilbert, & Wagner, 2011). Ninety-five percent of the population has an IQ score between 70 and 130. Recent research has investigated the extent to which dyslexia is related to IQ (Ferrer, Shaywitz, Holahan, Marchione, & Shaywitz, 2010). This research is consistent with the observation that some of the most successful and creative individuals in society are dyslexic. Text Box 11.1 provides examples of those who have become famous despite having dyslexia.

Text Box 11.1 Extraordinary Individuals: Many Successful People Have Dyslexia

Receiving a diagnosis of dyslexia may leave one feeling different from others and alone. Over the past century as the stigma associated with being diagnosed with dyslexia has decreased, many individuals with dyslexia have spoken publically about their personal struggles. Many of these individuals are among the most successful individuals in society. The English businessman Richard Branson is CEO of Virgin Group, which owns over 400 companies. He has spoken openly about being diagnosed with dyslexia and having a great deal of difficulty succeeding in school (Branson, 2011). Some of the world's most creative individuals have struggled with dyslexia. Henry Winkler, the American actor, now speaks frequently about how dyslexia affected his life (Murfitt, 2008). Others include the actors Orlando Bloom and Tom Cruise, the businessman Charles Schwab (Peterson, 2008), and the bilingual actress Salma Hayek (Potter, 2012).

In the book *An Autobiography of a Dyslexic,* Abraham Schmitt (1994) described his life in a farming community in western Canada. He struggled through school but persevered. He was able to receive his PhD by using a variety of techniques to avoid having to read text. He could understand material if it was read aloud to him. He was able to compose written documents by dictating them to another person and having all of his documents carefully proofread by someone else—his wife. His stories (as well as the stories of others who have gone on to be successful despite dyslexia) demonstrate that with hard work and perseverance, students with dyslexia may be able to fulfill their goals.

Causes of Developmental Dyslexia

The causes of dyslexia are not completely understood. Our understanding of dyslexia has evolved a great deal over the last century. An early view of the cause of dyslexia was that it stemmed from a visual impairment (Orton, 1925). Studies investigating the eye movements of dyslexics as they read have found that the patterns of eye movements during reading do differ for dyslexics and nondyslexics; however, the difference appears to be the result of readers hesitating as they read rather than a cause of reading problems (Ashby & Rayner, 2007).

A growing body of research suggests that there may be a biological basis for some varieties of developmental dyslexia. There is an impressive number of studies showing that dyslexia can run in families. An abundance of research has demonstrated that it is highly inheritable (Barr & Couto, 2007; Svensson et al., 2011). Grigorenko (2001) reviewed prior studies of families in which dyslexia appeared to have been heritable. In all, there were 516 families across the eight studies. Grigorenko found that approximately 37% of the parents of children diagnosed with reading problems also were found to have reading problems when tested. Wadsworth, Corley, Hewitt, Plomin, and DeFries (2002) showed that there was no reliable correlation between child and parent reading problems for children who had been adopted.

Since the 1980s, researchers have searched for specific genes associated with dyslexia and/or poor reading ability. The research has yielded nine different genetic markers. Smith, Kimberling, Pennington, and Lubs (1983) identified a location on chromosome 15 as related to dyslexia. Studies conducted by other researchers later found dyslexia to be associated with regions on chromosomes 1, 2, 3, 4, 6, 11, 17, and 18 (see Svensson et al., 2011). Researchers have also identified individual genes associated with dyslexia (C2ORF3, Anthoni et al., 2007; DCDC2, Menga et al., 2005; DYX1C1, Taipale et al., 2003; KIAA0319, Cope et al., 2005; MRPL19, Anthoni et al., 2007; ROBO1, Hannula-Jouppi et al., 2005). Of these genes, all but ROBO1 has been observed to be related to dyslexia in more than one study (Paracchini et al., 2011). DYX1C1 is the gene found to be associated with dyslexia most often (Paracchini et al., 2011), and it is a good candidate as the "dyslexia gene" because it is known to be involved in determining how neurons migrate in the brain during development (Galaburda, Sherman, Rosen, Aboitiz, & Geschwind, 1985; Wang et al., 2006). However, there have been a number of studies in which DYX1C1 was not found to be related to dyslexia. One study by Marino and colleagues (2005) was conducted in Italy. Another study was conducted in the United States by Menga and colleagues (2005). The studies that have observed the relationship were conducted in Finland (Taipale et al., 2003) and Canada (Wigg et al., 2004). With the increasing amount of research being conducted using genetic analysis, researchers may solve the mystery of dyslexia within our lifetimes.

In general, dyslexia is most likely determined by a number of genes with small to moderate effects where the genes contribute both to a general and to an explicit phenotype (Smith, 2007). However, other genes have been found in other studies (Fagerheim et al., 1999). Quite a few investigations have failed to replicate earlier genetic findings (e.g., Brkanac et al., 2007; Cope et al., 2005; de Kovel et al., 2008; Field & Kaplan, 1998; Petryshen, Kaplan, Liu, & Field, 2000; Schumacher et al., 2006; Svensson, 2003), and reasons for failures have been discussed (Barr & Couto, 2007; Fisher & DeFries, 2002; Grigorenko, 2005) focusing on three main issues: (1) the ascertainment procedure, (2) the choice of analytical method, and (3) the characteristics of the phenotype (Fisher & DeFries, 2002; Grigorenko, 2005; Grigorenko, Ngorosho, Jukes, & Bundy, 2006; Grigorenkgo et al., 2007).

Galaburda and colleagues (1985) conducted autopsies on the brains of four men who had lived with dyslexia from childhood. When the brains of these men were compared with the brains of individuals without any history of developmental dyslexia, the researchers found a difference in the planum temporal, which is located in the posterior area of the superior temporal lobe. In nondyslexic individuals, this region is asymmetric, larger on the left than on the right in 65% of the population. In 11% of the population, the right side is larger than the left. In the brains of the dyslexic men, there was no asymmetry. Researchers disagree about why the brains of dyslexics may lack asymmetry in the planum temporal (Grigorenko, 2001). In subsequent research, Galaburda, Schrott, Sherman, Rosen, and Denenberg (1996) found that there were size abnormalities in the parts of the thalamus in the brains of dyslexics. The thalamic structures in the brains of dyslexics were smaller than those areas in the brains of nondyslexics. Booth and Burman (2001) compared the brains of dyslexics and nondyslexics and found that there was less gray matter in an area of the brain associated with phonological processing (i.e., the left parietotemporal area) in the brains of dyslexics than the brains of nondyslexics. In other research, Deutsch and colleagues (2005) found that reading skill was strongly correlated with having more white matter in the brain; white matter is the communication system of the brain.

Shaywitz and colleagues (2002) used brain imaging to investigate the brain processing of children who did and did not have problems reading. They found that there was greater activation in brain regions related to letter and sound processing for children who were skilled readers than for children who had reading problems. One of the important regions in which there were observed differences was the posterior area of the left hemisphere at the edge of the occipital lobe.

Tarnopol and Tarnopol (1981) surveyed 26 countries to learn what percentage of schoolchildren were dyslexic. Their survey revealed that the prevalence of dyslexia varied widely across countries. Japan and China reported an extremely low percentage of children with dyslexia (i.e., 1%). English-speaking countries reported an average of 20%. Venezuela reported a relatively high rate of 33%. Glezerman (1983, cited in Grigorenko, 2001) found relatively low rates of dyslexia in Germany (i.e., 5%) and in Scandinavian countries (i.e., 10%).

Reading researchers often categorize languages in terms of their orthographies, specifically in how easily the pronunciation of words can be determined from the spelling of the words. In languages in which the pronunciation of words can be determined unambiguously from their spelling, the orthography is described as shallow. Languages with shallow orthographies include German, Spanish, Italian, Greek, Norwegian, and Finnish. Languages in which the pronunciation of words cannot be easily determined from their orthography are described as having a deep orthography. English, French, Danish, and Chinese have deep orthographies. Some languages have orthographies that fall in between the shallow and the deep; these include Swedish, Icelandic, Dutch, and Portuguese. Across the world's languages, there are a variety of scripts used; these scripts differ in the shallowness of the orthography. Text Box 11.2 provides more information about the variety of writing systems that exist around the world.

During elementary school, children are usually not identified as having trouble reading until they fall behind their peers in reading in terms of grade level. Most children will not be tested for dyslexia until their reading level is two grades behind that of their same-age peers. Consequently, children are usually in the third grade or higher before they are diagnosed as having dyslexia. Estimates suggest that between 5% and 17% of children in the United States are dyslexic (Rayner et al., 2012). Dyslexia is more common in boys than in girls.

Is There a Cure for Dyslexia?

Unfortunately, there is no cure for dyslexia. There has been a great deal of research on dyslexia and much has been discovered. Researchers have developed a number of techniques to treat individuals with reading problems. This research has shown that

Text Box 11.2 Diversity of Human Languages: Writing Systems Across Languages

The earliest evidence of human writing found, thus far, dates back to 6600 BC; the artifacts were found in 1999 in Henan, China (Li, Harbottle, Zhang, & Changsui, 2003). Historians speculate that writing may have been developed to keep track of commercial transactions. Despite the prevalence of writing around the globe, there are some languages that have never been written. Of the languages that are written, there is a variety of writing systems. As you learned in Chapter 5, the Chinese writing system involves ideographs. In contrast, languages such as English, Italian, and German use roman characters. Other examples of writing scripts include Cyrillic, which is used for Russian, Serbian, and Yupik, and Devanagari, which is used for Hindi, Marathi, and Nepali. Writing systems can be classified into three types: (1) logographic, (2) syllabic, or (3) alphabetic. **Logographic writing systems** represent whole words with an individual symbol. **Syllabic writing systems** represent individual syllables with an individual symbol. In **alphabetic writing systems,** individual symbols represent phonemes. Table 11.6 provides examples of each type of writing system.

▲ Photo 11.3 Eye chart with letters printed in Cyrillic script.

Table 11.6 Examples of Logographic, Syllabic, and Alphabetic Writing Systems

Language	Type of Writing System	Direction of Writing
Egyptian Hieroglyphics	Logographic	Either direction
Chinese	Logographic	Top to bottom or left to right
Japanese	Logographic and Syllabic	Top to bottom Columns ordered right to left
Mycenaean Greek	Syllabic	Left to Right
Cherokee	Syllabic	Left to Right
Cree	Syllabic	Left to Right
English	Alphabetic	Left to Right
Arabic	Alphabetic	Right to Left
Hebrew	Alphabetic	Right to Left

Source: Rogers (2005).

treatment for dyslexia must be highly individualized, targeting specifically the reading-related skills that the individual needs to improve. Federal law entitles a child who is 3 years of age or older and receives a diagnosis of dyslexia to have an individualized educational plan (IEP). The IEP is a treatment plan that is developed by the teachers, school counselors, and special education teachers in consultation with the parents. A child's IEP should be reevaluated every year. IEPs for dyslexia typically include tutoring for the child, extra practice reading aloud, increasing phonological awareness, and spelling.

Despite how much has been learned about dyslexia over the past four decades, the information is not likely to have reached all areas of our society. Many school districts may still not have up-to-date information about the latest views on dyslexia and the currently recommended treatments. As I was preparing this chapter, I found that most college textbooks do not include detailed discussions of dyslexia. Those textbooks that discuss dyslexia devote only a few pages to the topic. Consequently, the training that teachers and school officials have received about dyslexia is most likely to have come after their college training through professional development experiences. The extent to which school districts in the United States provide professional development training for their faculty varies widely. Those who are most prepared to diagnose and treat dyslexia in schoolchildren are **school psychologists.** The most highly trained school psychologists have a PhD, which usually requires 5 years of training after college. School psychologists are also usually licensed, which requires additional on-the-job training and completion of a licensing exam. The specific requirements for licensure vary by state in the United States. Obtaining a license usually requires that one has obtained a degree from a graduate training program accredited by the American Psychological Association (APA). There are master's level school psychologists who hold the MEd degree and also those who are called specialists who hold the EdS degree. Depending on their educational background and training, they may or may not have up-to-date knowledge about the types of dyslexia, their causes, and the treatments that have been demonstrated to be effective in scientifically rigorous studies.

Over the past few decades, there have been some experimental treatments for dyslexia. One such treatment was developed and promoted by Wynford Dore, a Welsh businessman. In his book *The Miracle Cure* (Dore, 2006), he claimed that dyslexia may be caused because of a cerebellar developmental delay. He developed physical exercises that can be done in short 10-minute intervals several times a day. In 2002, the program was featured on British television. Since then, Dore has established dozens of Dore Centers in Britain, Australia, and the United States. The program costs $4,000, which Dore compares to braces for a child's teeth or the price of a family vacation. Dore cited a research study that found evidence that the treatment resulted in substantial improvement in reading and writing. Independent research studies using autopsy and brain imaging have found cerebellar abnormalities in individuals with developmental dyslexia (Bishop, 2002). However, to date, there has been no convincing, scientific evidence that the Dore program is effective (Barth et al., 2010; Bishop, 2007).

Other Learning-Related Disorders

Dyslexia is just one disorder that can affect children's success in school. During the elementary, middle, and high school years, children may experience problems learning. Some may be diagnosed as having a learning disability. Between 4% and 7% of schoolchildren are told that they have a learning disability (Geary, 2006; Hasselhorn & Schuchardt, 2006; Mercer & Pullen, 2005). A useful definition for learning disability was developed by the National Joint Committee on Learning Disabilities (NJCLD) (1990):

> [A learning disability is] a heterogeneous group of disorders manifested by significant difficulties in the acquisition and use of listening, speaking, reading, writing, reasoning or mathematical abilities. These disorders are intrinsic to the individual and presumed to be due to Central Nervous System Dysfunction. Even though a learning disability may occur concomitantly with other handicapping conditions (e.g. sensory impairment, mental retardation, social and emotional disturbance) or environmental influences (e.g. cultural differences, insufficient/inappropriate instruction, psychogenic factors) it is not the direct result of those conditions or influences.

The term ***comorbidity*** is used to describe the circumstance when one is diagnosed with more than one disorder. The disorders are described as being comorbid. Among the disorders that most commonly affect schoolchildren are ADD, dysgraphia, dyscalculia, and Asperger's syndrome. When schoolchildren struggle in school and are unable to keep up with the expected level of performance, they may experience low self-esteem and anxiety. These responses may also lead to conduct problems (Alexander-Passe, 2006). When children experience problems with schoolwork, they may believe that if they could only read faster their problems would be solved. Speed-reading courses have been around for about 100 years. Text Box 11.3 describes research that has beenconducted to investigate how well people comprehend what they read when they are speed-reading.

What Is Attention Deficit Disorder?

ADD is characterized by difficulty in maintaining focused attention. Some individuals may have **attention deficit/hyperactivity disorder** (ADHD) and may also experience problems with impulsivity. The symptoms that are typically observed in individuals who have difficulty maintaining attention include being easily distracted, being forgetful, missing details that others do not, becoming easily bored, switching frequently between activities, losing track of possessions frequently, appearing not to listen to others, appearing in a daze, and showing problems when asked to follow instructions. The symptoms that are typically observed in individuals who experience hyperactivity include being fidgety when sitting, having trouble sitting still during meals, talking excessively, seeming to prefer to always be moving, and having an inability to work on tasks quietly. The prevalence of ADD/ADHD in children without intellectual delay is between 3% and 7% (American Psychiatric

Text Box 11.3 Research Discovery: Speed-Reading: Fact or Myth?

▲ Photo 11.4 A young boy reads a book on his own. At what age did you start reading books by yourself?

Speed-reading courses often promise customers that after the course they can go from reading around 200 to 400 words per minute to over 10,000 words per minute or more. Courses typically teach readers to move their eyes more quickly over the text and to follow a moving pencil or finger. Often, courses teach readers to inhibit completely their inner speech and subvocalization. Reading researchers have investigated the quality of the comprehension of speed-readers. For example, Just, Carpenter, and Masson (1982) measured the eye movements of speed-readers during reading. They also measured eye movements for a group of non-speed-readers when they read at a normal pace and when they skimmed the text. They observed reading rates between 600 and 700 words per minute for the speed-readers and for the typical readers when they were skimming the text. The typical readers achieved a reading rate of about 250 words per minute when they were reading normally.

The comprehension performance for the speed-readers was similar to the comprehension level of the typical readers when they were skimming. Speed-readers were generally able to answer comprehension questions covering content on which they had fixated during reading (i.e., the eye looked directly at the information). Speed-readers were generally not able to answer questions about information that had been skipped (i.e., not fixated) during processing. Speed-readers did fairly well comprehending the gist of the text. However, those who purchase speed-reading courses are likely interested in recalling more than just the gist of a page of text. Most reading researchers generally view speed-reading courses as promising more than they can deliver. One who would like to read faster is likely better off spending more time reading. The greater familiarity that one has with the topic matter and the vocabulary used in that genre, the faster one will be able to read (Rayner et al., 2012).

Association, 2000). ADD/ADHD is between two and four times more common in boys than in girls (Sciutto, Nolfi, & Bluhm, 2004).

Individuals with ADHD may also be affected by psychiatric disorders. Barkley (2006) reported that 25% to 55% of children who are referred to clinics for ADHD also have one or more psychiatric disorders. The most common comorbid psychiatric disorders are major depression, anxiety disorders, and conduct disorder, which involves persistent acting out in which social norms for behavior are violated and the rights of others may be disregarded.

Causes of Attention Deficit Disorder and Attention Deficit/Hyperactivity Disorder

The causes of ADD/ADHD are not fully known. Studies suggest that the disorder can run in families; thus, there is likely to be a genetic link. Researchers suggest that genetics play a role in about 75% of cases (Acosta, Arcos-Burgos, & Muenke, 2004; Volkow et al., 2007). A number of genes have been identified as possibly involved in causing ADD/ADHD symptoms. Many of these genes play a role in transporting dopamine in the brain. A brain-imaging study identified four locations within the frontostriatal region as playing important roles in ADHD (Bush, Valera, & Seidman, 2005). These locations were the lateral prefrontal cortex, dorsal anterior cingulate cortex, caudate, and putamen. Environmental factors have been found to be related to ADD/ADHD. Expectant mothers who use alcohol and tobacco during pregnancy are more likely to have a child with ADD/ADHD than those who do not use these substances (Braun, Kahn, Froehlich, Auinger, & Lanphear, 2006). The link between ADD/ADHD and smoking during pregnancy may exist because fetuses exposed to nicotine may experience low oxygen levels (Pomerleau et al., 2003). Viral and bacterial infections contracted during pregnancy may increase the probability that a child goes on to develop ADD/ADHD (Millichap, 2008).

Treatments for Attention Deficit Disorder and Attention Deficit/Hyperactivity Disorder

The treatments for ADD/ADHD include medication and psychological therapies (Elia, Ambrosini, & Rapoport, 1999). The medications that are most commonly prescribed to treat ADD/ADHD are stimulants such as Ritalin, Metadate, Concerta, Dexedrine, and Adderall. These medications are believed to increase the levels of dopamine in the brain. The medications are not recommended for preschool children. Those who take them should do so under medical supervision. There are unpleasant side effects; some side effects are life threatening. The psychological therapies for ADD/ADHD include school-based interventions, family therapy, social skills training, behavior therapy, and cognitive-behavioral therapy.

What Is Dysgraphia?

Writing disability, or dysgraphia, is a learning disability involving handwriting, the organization of ideas during composition, and sometimes spelling (Berninger & Wolf, 2009; Nicolson & Fawcett, 2011). The term *dysgraphia* is composed of the root *graphia*, which means *to write,* and the prefix *dys-,* which means *difficulty.* Dysgraphia usually involves difficulties with handwriting in particular. Researchers have viewed the problem as one directly related to how the brain controls movements of the hands and fingers (Hamstra-Bletz & Blote, 1993; Rosenblum, Werner, Dekel, Gurevitz, & Heinik, 2010; Smits-Engelsman & Van Galen, 1997). Research studies investigating this aspect of dysgraphia generally assess one's writing speed and the legibility of the handwriting, which is influenced by the size of letters, how well letters are drawn, and the spacing of letters and words (Feder & Majnemer, 2007; Tseng & Chow, 2000).

With the availability of technology that can measure the force applied to a digital tablet during writing, researchers are able to measure a writer's grip force and force changes throughout the process of writing. In a study using this technology, Longstaff and Heath (1997) found that dysgraphic writers showed greater variability in forces throughout the course of writing than did nondysgraphic writers. Rosenblum, Weiss, and Parush (2004) found that dysgraphic writers took longer to write overall than nondysgraphic writers both when they were asked to write on paper and when they were asked to write in the air.

Problems with handwriting can be observed in individuals with other developmental disorders. For example, individuals with ADHD usually exhibit handwriting problems (Racine, Majnemer, Shevell, & Snider, 2008), such as writing that is larger than normal (Frings et al., 2010) and slower than normal due to having the pen or pencil off the paper frequently (Rosenblum, Epsztein, & Josman, 2008). The link between dysgraphia and ADHD is not well understood. Some researchers have suggested that the disorders may be caused by the same underlying genetic origin (Stevenson et al., 1993).

What Is Dyscalculia?

Math disability or dyscalculia is a learning disability involving mathematics. The disorder has not received much attention in the news media; thus, many physicians, teachers, and parents may not have realized that a child who demonstrates excessive difficulty learning math may, in fact, be suffering from a learning disability. Current estimates indicate that dyscalculia affects 3% to 6% of the population. Generally, to be diagnosed with dyscalculia, a child must score below the 20th or 25th percentile on the math portion of a standardized test and have an IQ in the normal range (Geary, Hamson, & Hoard, 2000; Gross-Tsur, Manor, & Shalev, 1996). Those with dyscalculia tend to also have problems with spatial reasoning and working with measurements and have trouble keeping track of time and telling time. As with dyslexia, dyscalculia is generally believed to be unrelated to IQ, as it occurs in individuals in the low, medium, and high IQ ranges.

Research in the 1970s revealed how children typically become proficient in math. Gelman and Gallistel (1978) identified five principles that children learn to use. The first principle that children learn is that there is a one-to-one correspondence between an object that is being counted and each number that is counted out (e.g., *one, two, three*). The second principle is that when one counts, there is a stable order for the counting that is the same regardless of what objects are being counted. The third principle is that the number of the final object that is counted is the total number of items in the set. This principle is referred to as **cardinality.** The fourth principle, which is referred to as abstraction, refers to the fact that any set of objects can be counted; the objects need not be related in any way. The fifth principle is that the order in which a set of objects is counted is irrelevant; the objects may be counted in any order with the same outcome. Gelman and Meck (1983) showed that these five principles serve as the foundation for how children learn to count.

Children's knowledge of counting principles has been investigated in studies in which children observe a puppet counting objects. Sometimes the puppet counts correctly. Sometimes the puppet makes mistakes while counting. The child is asked to judge whether the puppet's counting is *okay* or *not okay*. Geary, Bow-Thomas, and Yao (1992) used the puppet technique to study the counting knowledge of children with dyscalculia. They found that the children were able to detect errors in the puppet's counting for four of the five counting principles identified by Gelman and Gallistel (1978). Children's performance indicated that they assumed that objects close to each other had to be counted in a particular sequence. They did not understand that the order in which a set of objects were counted was irrelevant. Geary and colleagues (1992) also observed that children viewed trials as correct when the first item being counted was counted twice; however, when the last object in a set was counted twice, children were able to label it as an error. The results suggested that the children may have problems retaining information in working memory. Geary (1993) investigated the possibility that the working memory abilities for individuals with dyscalculia differed from those without dyscalculia. He concluded that working memory problems are observed in a broad range of learning disabilities, including dyscalculia. It is not possible to conclude that working memory deficits are the sole cause of dyscalculia.

Some individuals can acquire dyscalculia following an injury to the brain (Levy, Reis, & Grafman, 1999; Mayer et al., 1999). These individuals have provided clear evidence that there are specific brain locations involved in the processing of number information and in mathematical reasoning. The regions of the brain that have been identified as involved in math processing may develop abnormally in individuals who experience problems in math in early childhood.

What Is Conduct Disorder?

Children who frequently misbehave in school and display a disregard for the rights of others may be diagnosed with conduct disorder. Conduct disorder is often comorbid with ADD/ADHD but may also occur in individuals with dyslexia, dysgraphia, and dyscalculia. Studies of individuals with conduct disorder indicate that between 20% and 25% also have a learning disability (Frick, Lahey, Christ, Loeber, & Green, 1991). Other studies have found that approximately 25% to 30% of boys with conduct disorder and 50% to 55% of girls with conduct disorder also have ADHD (Waschbusch, 2002). The age of onset is an important diagnostic consideration. Those exhibiting symptoms after 10 years of age may be diagnosed with the adolescent type of conduct disorder. Those exhibiting symptoms before 10 years of age may be diagnosed with the childhood type (Berkout, Young, & Gross, 2011).

Those diagnosed with conduct disorder may have been truant from school, gotten into fights, threatened others, physically injured others with their hands and/or with weapons, and displayed cruelty toward people and/or animals. In severe cases, they may have destroyed the property of others, set fires, broken into another's home or car, carried out or attempted a robbery, or committed a sexual assault. Conduct disorder has a negative impact on communities. It has been estimated that a single adolescent with conduct disorder might cost a community $70,000 over 7 years (Foster & Jones, 2005). Many consider conduct disorder a major public health problem (Garland et al., 2001).

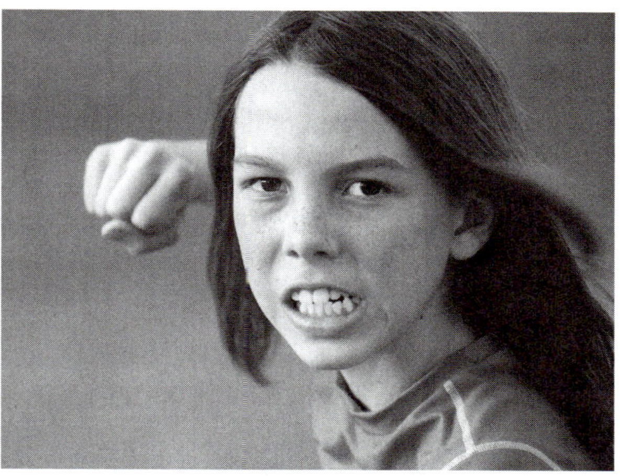

▲ **Photo 11.5** A young boy is ready to fight. Why do you think some children have a harder time controlling their anger than others?

Recent research suggests that children with conduct disorder show problems with social aspects of language (Oliver, Barker, Mandy, Skuse, & Maughan, 2011). Two measures of social function predicted conduct problems. These are the 12-item Social and Communication Disorders Checklist (Bishop, 1998) and the 70-item Children's Communication Checklist (Skuse, Mandy, & Scourfield, 2005). The Social and Communication Disorders Checklist assessed a range of nonverbal abilities, such as being aware of the feelings of others, detecting anger in others, being aware of how one's behavior affects family members, and being able to interpret body language. The Children's Communication Checklist assessed the ability to build rapport with others through body language (e.g., smiling appropriately when talking to others), using context to interpret the speech of others (e.g., misinterprets what others say), expressing one's ideas clearly (e.g., sometimes says things that others do not understand), and following social norms when starting conversations.

Summary and Theoretical Implications

During the school years, children are faced with the challenge of acquiring increasingly complex knowledge at a steady pace. They learn to read, write, and perform mathematical computations. Not every child flourishes in school. Those who struggle are sometimes diagnosed with dyslexia and/or other learning disorders. As we learn more about the functioning of the brain and how genes are involved in brain development, we are learning that dyslexia and other learning disorders have a basis in biology. Other disorders than can affect children in school include ADD, dysgraphia, and dyscalculia. The disorders tend to be comorbid with a child receiving two or more diagnoses.

In terms of the four theoretical approaches to language development, the research related to language in the school years does not provide definitive clues regarding which of the four approaches is accurate. The fact that research is increasingly supporting the view that reading problems, such as dyslexia, are biologically based appears to be most consistent with the generative approach to language development. Nevertheless, the statistical learning approach to language development may be able to provide a reasonable account; some researchers have attempted to demonstrate that dyslexic-like performance can be produced in computer models that do not assume specialized language processing modules. Treating dyslexia, ADHD, and behavioral problems during the school years is likely to involve techniques inspired by behaviorist principles, such as classical and operant conditioning. Last, individuals with these problems during the school years may be benefited by close, positive relationships with parents and peers.

KEY TERMS

acquired dyslexia
alphabetic writing systems
attention deficit disorder (ADD)
attention deficit/hyperactivity disorder (ADHD)
cardinality
comorbidity
deep dyslexia
developmental dyslexia
dual-route model
dyscalculia
dyseidetic dyslexia
dysgraphia
dysphonetic dyslexia
logographic writing systems
math disability
Matthew effect
onset
phonics method
phonological awareness
phonological dyslexia
pig Latin
pinyin
rime
school psychologists
surface dyslexia
syllabic writing systems
whole language method
writing disability

REVIEW QUESTIONS

1. At what age do children typically learn to read?
2. What is pinyin? What role does it play in Chinese children's reading development?
3. What has research revealed about which techniques for teaching reading are most effective?
4. In terms of literacy levels, where do English-speaking countries rank?
5. What is the Matthew effect? What evidence, if any, supports the existence of the Matthew effect in education?
6. What is the distinction between developmental dyslexia and acquired dyslexia?
7. What are the different types of acquired dyslexia?

8. What are the different types of developmental dyslexia?
9. How are developmental types of dyslexia similar to the types of acquired dyslexia that have been observed?
10. What is the dual route theory of word reading? What are the two routes in word recognition according to the theory?
11. How does the prevalence of dyslexia vary across countries?
12. How is SES related to IQ and reading ability?
13. What is an IEP?
14. What are the types of treatments that are used with dyslexic children?
15. What are the symptoms of ADD? How is ADHD believed to differ from ADD?
16. What interventions are used with children with ADD in school settings?
17. What are the symptoms of someone who has dysgraphia?
18. What is dyscalculia? With which disorder is dyscalculia often comorbid?
19. What is conduct disorder? What research demonstrates that children with conduct disorder may have difficulty with the social aspects of language comprehension?

RECOMMENDED READING

Butterworth, B. (2004). *Dyscalculia guidance: Helping pupils with specific learning difficulties in maths.* New York: David Fulton Publications.

Daheane, S. (2009). *Reading in the brain.* New York: Viking.

Davis, R. D., & Braun, E. M. (2003). *The gift of dyslexia, revised and expanded: Why some of the world's smartest people can't read . . . and how they can learn.* Bel Air, CA: Perigree Books.

Eide, B. L., & Eide, F. F. (2011). *The dyslexic advantage: Unlocking the hidden advantage of the dyslexic brain.* New York: Penguin.

Harris, M., & Hatano, G. (Eds.). (1999). *Learning to read and write: A cross-linguistic perspective.* Cambridge, UK: Cambridge University Press.

Leong, C. K., & Joshi, R. M. (Eds.). (1997). *Cross language studies of learning to read and spell.* Dordrecht, The Netherlands: Kluwer Academic Publishers.

Schmitt, A. (1992). *Brilliant idiot: An autobiography of a dyslexic.* Intercourse, PA: Good Books Publishing.

RECOMMENDED FILMS

Khan, A. (Director). (2007). *Like stars on earth* [Motion picture]. India: Aamir Khan Productions.

Ritt, M. (Director). (1989). *Stanley and Iris* [Motion picture]. United States: MGM Home Entertainment.

SUGGESTIONS FOR CLASS PROJECTS

1. Investigate public opinion about dyslexia. Do people know what it is? Do some people think that it is not real but rather is just an excuse for why some students do not do well in school? Do people believe that a person can be cured of it or "grow out of it"? Do people know anyone personally who has been diagnosed with dyslexia? Are they aware of any famous individuals who have or have had dyslexia? Work alone or in groups to develop a brief questionnaire, and present your results in either a class presentation or a written report.

2. Investigate public opinion about ADD/ADHD. Do people know what it is? Do some people think that it is not real but rather is just an excuse for why some students do not do well in school? Do people believe that a person can be cured of it or "grow out of it"? Do people know anyone personally who has been diagnosed with ADD or ADHD? Are they aware of any famous individuals who have it? Work alone or in groups to develop a brief questionnaire, and present your results in either a class presentation or a written report.

3. Select three to five pages of text of prose, such as from a novel or short story. Then, choose one type of dyslexia, and circle all the words that a person with that form of dyslexia would have difficulty reading.

For additional ancillary resources, please visit the companion website at www.sagepub.com/kennison.

- Video Links
- Audio Links
- Web Resources
- Internet Activities
- Flashcards
- Web Quizzes

CHAPTER 12

LANGUAGE ACROSS THE LIFE SPAN

Over the past 100 years, the average life span in industrialized countries has dramatically increased. The National Institute on Aging (NIA) (n.d.) estimates that by 2030 there will be 1 billion adults who are 65 years old or older in the world. Currently, the number is about 500 million. The number of people who are over 100 years old is increasing

▲ Photo 12.1 An older couple kayaks in view of icy mountain peaks. Do you ever imagine how your life will be when you are much older?

about 5.5% each year. There were 455,000 people over 100 years of age in 2009, but it will likely be approximately 4.1 million in 2050 (United Nations, 2002). The worldwide changes in the median age of the population will have far-reaching effects on society. Many of us will be able to live to be 80, 90, or 100 years old or older. Some of us may even approach the maximum life span (i.e., approximately 120 years) (Santrock, 2007) or may live long enough to develop an illness or health condition associated with aging. As we age, we are likely to experience many physical and cognitive changes, many of which will affect how we understand and produce language. In this chapter, you will learn about the physical, cognitive, and language-related changes that occur during normal aging. Because there are many misconceptions about normal aging, some of the changes that you might think are certain to happen may not be and other changes that are likely to occur may be unfamiliar to you. The reason that there are many misconceptions about aging is because the vast majority of us hold negative attitudes toward aging. The term that is used to refer to this form of prejudice is **ageism.** Ageism has profound effects on how we live our lives and how we treat older adults. At the end of the chapter, you will learn about some of the illnesses that are most stereotypically associated with older adults and how language processing is usually affected. These illnesses include **dementia, Alzheimer's disease, Parkinson's disease, multiple sclerosis,** and **Huntington's disease.**

Normal Aging

What Is Gerontology?

Over the past century, the number of professionals who specialize in the study of aging and the treatment of age-related illnesses has grown dramatically. The term ***gerontology*** is used to refer to the study of the processes involved in aging, including biological, psychological, and social processes. A **gerontologist** is a professional who specializes in the study of aging. The term *gerontology* was introduced by Ilya Ilyich Mechnikov around 1903. The word comes from the combination of the Greek suffix *-logy*, which means the study of, and the word *geron*, which means old man. Currently, the field of gerontology includes professionals from a wide range of fields whose work focuses specifically on older adults. For example, medical doctors and nurses may specialize in treating older adults. Social workers and physical therapists may choose to work predominantly with older adults. Researchers in psychology, sociology, public health, and economics may devote their careers to answering questions focusing primarily on aging, older adults, and/or the impact of aging on society (Hooyman & Kiyak, 2011) In the coming decades, as the number of older adults in the world increases, it is likely that the number of professionals who specialize in gerontology will also continue to increase.

Gerontologists have pointed out that there are at least four aspects of aging. These are (1) chronological aging, (2) biological aging, (3) psychological aging, and (4) social aging. As the term implies, chronological aging refers to the aging that is simply how long one has lived. Biological aging refers to the age-related changes that occur in the body. Psychological aging refers to the age-related changes that occur in our processing abilities, including cognitive and language processing. Last, social aging includes the age-related changes that occur in one's relationships with family and friends as well as with organizations. An individual may have similar or different levels of aging for each of these categories. For example, a person may be very advanced in years but have the physical condition of a much younger person. A person may be relatively young (i.e., in her 50s or 60s) but go about her social relationships in a manner similar to someone who is much older (i.e., those in their 80s or 90s), relying more on children and younger family members than is typical for a person of that chronological age.

Gerontologists are particularly interested in identifying those factors that are related to aging well (i.e., having a physical age that is lower than one's chronological age). One of the most important factors appears to be education, with well-educated people aging better than less educated people (Scarmeas & Stern, 2003). A second important factor is exercise. An increasing body of research shows that there are benefits of exercise for older adults. Older adults who are healthy have been shown to experience increased cardiovascular functioning after implementing regular exercise (Colcombe et al., 2004; Kramer, Larish, & Strayer, 1999). Other studies have shown that exercise in older adults was associated with those adults being better able to perform the activities of daily living (Gu et al., 2009). Brain studies have found that exercise is related to a reduction in the loss of brain volume due to aging (Colcombe et al., 2003; Erickson et al., 2009). As people age, they tend to spend less time engaging in physical activity (Lindwall, Rennemark, & Berggren, 2008). Some older adults may become less active because they have a misconception that exercise may harm them, particularly sports-related activities (Irwin et al., 2004). More commonly, older adults become less and less active over time due to changes in their social life. They may spend less time interacting with friends and family and have fewer opportunities to be active. Other factors include staying mentally engaged with activities that are meaningful, maintaining friendships and ties to a community, and diet.

Centenarians

One approach in the study of successful aging has been to conduct in-depth studies of individuals who live a very long time. Table 12.1 displays five categories of older adults that are often used in studies investigating processes related to aging. **Centenarians** (those over 100 years of age) have received a great deal of attention because they are the rarest group and because researchers think that they may hold the secret to aging well. The other categories are **sexagenerarians, septuagenerians, octogenerians,** and **nonagenerians.** By studying the daily habits of those who live for a very long time, it may be possible to identify factors that may be useful in extending the chronological ages of others. Such characteristics might involve personality traits as well as lifestyle choices. If such commonalities exist, then there is the hope that younger adults could make changes that could lead to a longer life span. Givens and colleagues (2009) investigated the personality traits of centenarian siblings. The study focused on the Big 5 personality traits: (1) conscientiousness, (2) openness, (3) neuroticism, (4) extraversion, and (5) agreeableness. They found that siblings' openness and conscientiousness were similar to levels for noncentenarians; however, they had lower levels of neuroticism and higher levels of extraversion than noncentenarians. They found that female siblings had higher than average levels of agreeableness.

Research with centenarians pointed out that there are places in the world where living to a very old age occurs more frequently than in most other places. For example, the island of Okinawa, Japan, has a population of 1.3 million people. In 2007, it was home to 792 centenarians, which is the highest per capita number of centenarians in Japan (Willcox et al., 2008). Willcox and colleagues (2008) analyzed data from the longitudinal study of centenarians in Okinawa and found that their average life span was 110 to 112 years. Considering their advanced age, centenarians were remarkably healthy. None had cancer or diabetes. Only 8% had cardiovascular disease. Only 8% had experienced a stroke. Thirty-three percent had experienced a fracture. Only 42% had cataracts. Few individuals required residential care (i.e., nonindependent living arrangements) before the age of 105 years. The study did show that by the age of 105, most had begun to show signs of decline. Table 12.2 displays the average life spans for countries around the world.

So far, the research has not identified a secret to long life; however, studies of centenarians around the world have yielded some interesting results about the characteristics of the oldest among us. In a study of 602 centenarians in Italy, Motta, Bennati, Ferlito, Malaguarnera, and Motta (2005) found that centenarians fell into three groups. Twenty percent had good health. They were generally free of chronic illness, able to live independently, and with good cognitive abilities. A little over 33% had medium health. About 47% had poor health. Many reports about older adults on television, in newspapers, and in magazines may emphasize particularly surprising lifestyle choices (e.g., not going to doctors, smoking, using alcohol). It would be a mistake to draw the conclusion that those lifestyle choices are compatible with good health generally and longevity specifically. Such claims are viewed as **anecdotal**

Table 12.1 Categories of Older Adults

Category	Definition
Sexagenarians	Individuals in their 60s (i.e., between 60 and 69)
Septuagenarians	Individuals in their 70s (i.e., between 70 and 79)
Octogenarians	Individuals in their 80s (i.e., between 80 and 89)
Nonagenarians	Individuals in their 90s (i.e., between 90 and 99)
Centenarians	Individuals who live to be 100 years or older

Source: Hooyman and Kiyak (2011).

Table 12.2 Top Ten Countries With the Longest Life Spans for Men and Women

Men		Women	
Country	**Life Span**	**Country**	**Life Span**
Japan	79	Japan	87
Iceland	79	Hong Kong	84
Switzerland	79	Switzerland	84
Australia	79	Italy	84
Hong Kong	79	Spain	84
Sweden	78	Australia	84
Italy	78	Iceland	83
Singapore	78	Sweden	83
Israel	78	Israel	83
Canada	78	Canada	83

Source: United Nations (2011).

evidence by researchers. **Anecdotal evidence** is notoriously unreliable, as it may be vulnerable to exaggeration and subjectivity. Research suggests that one's genes play an important role in how long we live. Hjelmborg and colleagues (2006) studied 20,502 identical and fraternal twins in Denmark, Finland, and Sweden. **Identical twins** share the same genetic profile. **Fraternal twins** share 50% of their genes, as do biological siblings who result from different pregnancies. They found strong evidence that genes are determining life span.

While none of us is responsible for our genes nor can we change the genes that we have, we can improve our odds of living to a ripe old age by following a healthy lifestyle in terms of our diet and our daily activities, such as exercise. Some also believe that avoiding high levels of stress and having a supportive social network may be helpful in maintaining good health. Certainly, it is easy to see how the lifestyle choices that we make when we are young could have a profound influence on how long we live. As more people recognize the real possibility of living a very long time, more people may choose to modify their lifestyles to maximize their chances of doing so.

What Occurs During Normal Aging?

Normal aging involves many predictable changes. For example, the skin generally becomes less elastic, causing lines to form. These lines or wrinkles become deeper and more prevalent over time. The skin becomes drier and may be more sensitive to the sun and wind. Hair loses its color, becoming gray or white. It can also become much thinner over time. Fingernails grow more slowly and may become more brittle. Our metabolism also slows as we age. The body uses less energy to carry out the same processes. The body tends to lose muscle mass and replace it with fat. If one does not reduce one's caloric intake or increase one's physical activity, one can easily gain weight. In fact, it may be quite a challenge to maintain one's weight if one does not make adjustments in diet and in exercise. In order to counter the natural loss of muscle mass due to aging, older adults are encouraged to engage in activities involving strength, such as strength training exercises or working out with weights. Our bones do become weaker as we age. Some individuals develop the condition referred to as **osteoporosis,** which is characterized by reduced bone density. Individuals with osteoporosis are at a higher risk of bone fractures than others. Bone loss also results in a loss of height in most people. Typical 80-year-olds may be as much as 2 inches shorter than they were at their tallest. Osteoporosis affects women more often than men. Taking supplements containing vitamin D and calcium can help prevent bone loss. There are also prescription medications that can be taken.

Sensory Processing

Aging-related changes in our sensory processing can affect our cognitive processing and our ability to understand and to produce language. The most noticeable changes occur in our seeing and hearing. Most people require glasses in their 40s, because the eyes' lenses are less flexible. Lin, Thorpe, Gordon-Salant, and Ferrucci (2011) estimates that two thirds of adults over the age of 70 in the United States experience some form of hearing loss. Hearing loss for older adults tends to affect high frequency sounds more than low frequency sounds. The high frequencies are important for speech perception (Morrell, Gordon-Salant, Pearson, Brant, & Fozard, 1996). Some hearing changes can cause speech to sound garbled even with the use of hearing aids (Jerger, Jerger, Oliver, & Pirozzolo, 1989; Schneider & Pichora-Fuller, 2000). Older adults who experience hearing loss are more likely to become socially isolated and have decreased functioning in daily life than other adults (Strawbridge, Wallhagen, Shema, & Kaplan, 2000). Viljanen et al. (2009) found that older adults with hearing problems were more likely to experience falls than those with better hearing.

As we age, we tend to carry out physical and cognitive tasks more slowly (Salthouse, 1996). Older adults may move more slowly, speak more slowly, and carry out mental computations more slowly than younger adults. Cognitive psychologists refer to one's ability to think logically, reason, and solve problems as **fluid intelligence.** As we age, we can expect to experience measurable declines in fluid intelligence. On the other hand, **crystallized intelligence,** which refers to one's acquired knowledge, increases throughout our life span. Examples of crystallized intelligence include factual knowledge, skills, and abilities derived from experience. Research with older adults has shown that crystallized intelligence increases with age. Brain studies have shown that as one ages, the brain becomes slightly smaller and also becomes less able to repair itself following injury (Peters, 2006).

Changes in our physical and cognitive function as we age can affect all levels of language processing: phonological, lexical, syntactic, semantic, and pragmatic. In terms of phonological processing, the changes in sensation and perception may result in difficulty identifying words (e.g., Humes, 1996). There is also a particular difficulty distinguishing voiceless consonants from voiced consonants, such as /p/ and /b/ or /k/ and /g/ (Tremblay, Piskosz, & Souza, 2002). As you learned in Chapter 3, voiceless consonants have a period of silence (i.e., between 20 and 40 milliseconds) at the beginning before the voicing begins; voiced consonants have no silence at the beginning, as there is voicing throughout the production of the consonant. Older adults may mishear words containing voiceless consonants, such as perceiving *the great legs* instead of *the Great Lakes*. Older adults are less able than younger adults to adapt to the speaking voices of others. There is a tremendous amount of variability in the acoustic characteristics of speech. Pisoni (1993) has referred to the process of adjusting to others' speech patterns as **perceptual normalization.** Older adults appear to be less flexible when listening to many speakers than are younger adults (Yonan & Sommers, 2000). Numerous studies have revealed that older adults may experience more difficulty remembering speech and interpreting what they hear or read than do younger adults (Gordon-Salant & Fitzgibbons, 1997; Wingfield, 1996; Wingfield, Tun, Koh, & Rosen, 1999).

▲ Photo 12.2 An older couple is talking. How important are conversations with friends?

Language Processing

As you learned in Chapter 5, speakers sometimes experience difficulty finding the word that they have in mind. These experiences have been called tip-of-the-tongue states, or TOT states. Research has shown that the frequency of TOT states increases with age. Age differences in TOT states have been found in studies conducted in the laboratory (e.g., James & Burke, 2000; Maylor, 1990; White & Abrams, 2002) as well as in diary studies in which participants record the TOT states that they experience in daily life (e.g., Burke, MacKay, Worthley, & Wade, 1991; Heine, Ober, & Shenaut, 1999). Researchers do not fully understand why older adults are more susceptible to TOT states than young adults. One view is that our ability to inhibit unimportant information decreases as we age; the activated information interferes with our ability to retrieve from memory the word that we would like to produce (Hasher & Zacks, 1988; Zacks & Hasher, 1997). A second view claims that as we age, the connections in memory weaken; activating the phonological representation of a word when we are planning speech is more difficult (Burke et al., 1991; James & Burke, 2000; MacKay, Abrams, & Pedroza, 1999; MacKay & James, 2004). A growing number of studies have tested the contrasting predictions of these views and have generally supported the view that connections in memory weaken as we age, and retrieving words may be easier for younger adults than for older adults (Burke et al., 1991; Cross & Burke, 2004; James & Burke, 2000; White & Abrams, 2002, 2004).

Older adults are more likely to experience a TOT state on particular types of words. For example, proper names (e.g., *Larry* and *Cordelia*) are more likely to be involved in a TOT state than content nouns, such as *moose* and *flower* (e.g., Burke, Locantore, Austin, & Chae, 2004; James, 2004). Interestingly, James (2004) showed that older adults did not differ from young adults in remembering another person's occupation, but they experienced more difficulty remembering proper names than did younger adults. James's (2004) study was quite clever in that participants were asked to remember occupations that could also be used as proper names (e.g., *farmer* and *Mr. Farmer*). Difficulty retrieving names may also take the form of producing an incorrect name, such as intending to say George Clooney but instead saying Brad Pitt. I have been able to observe this phenomenon in my own mother, who now at the age of 70 routinely calls me by my sister's name and sometimes even by my brother's name. Over the past few years, I have found myself experiencing a noticeable slowness in retrieving proper names.

Change in semantic processing has also been observed in older adults (Light, 1992; Madden, 1988). When one hears or reads a word, there is a chain reaction of processing that is set off within one's semantic network. As you learned in Chapter 9, this chain reaction of processing is called spreading activation. You also learned that the processing of a word can be facilitated (or primed) when a related word is processed immediately before. The amount of facilitation (or semantic priming) observed for older adults has been found to be greater than that observed for younger adults (Laver & Burke, 1993; Myerson, Ferraro, Hale, & Lima, 1992). The term ***hyperpriming*** has been used to describe the greater priming for older adults than younger adults.

Researchers have speculated about the possible causes of hyperpriming. Stanovich and West (1983) pointed out that older adults generally have longer response times than younger adults in all conditions; thus, following the processing of the prime word there is more time for activation to spread throughout the semantic network for older adults than for younger adults (see also Myerson, Hale, Chen, & Lawrence, 1997). Another proposal is that hyperpriming occurs because older adults have larger vocabularies than young adults and have larger semantic networks (Laver & Burke, 1993). They reason that the size of the semantic priming effect increases with the number of indirect conditions between nodes in the network. A third possible explanation is that the decision criteria for responding changes as we age; older adults use more conservative criteria than younger adults (Ratcliff, Thapar, Gomez, & McKoon, 2004). It should be noted that hyperpriming for older adults is consistently a relatively small effect; thus, it is not clear what practical implications it has (Laver & Burke, 1993; Myerson et al., 1992). There have been a number of studies that have investigated semantic priming in younger and older adults and have failed to find significant differences (Burke, White, & Diaz, 1987; Quaireau, 1995; White & Abrams, 2004). Such studies suggest that hyperpriming may not be experienced by all older adults. More research is needed to determine which categories of older adults may be more likely to experience it.

Working Memory

Research investigating the working memory differences between younger and older adults has revealed a relatively large decline in the working memory capacity for older adults (e.g., Craik, 1968; Craik & Byrd, 1982). As you recall from Chapter 9, working memory refers to the relatively short-term memory buffer that is used for solving problems and briefly retaining information, such as a telephone number or street address. Researchers have developed a variety of techniques to measure working memory capacity. One of the oldest methods is the **digit span task** (Baddeley & Hitch, 1974). Participants hear or see a list of digits (e.g., 2, 7, 8) and are asked to re-create the list, preserving the order. They are first given short lists of two or three items. If they are able to respond correctly, they are given longer lists. The participants' digit span corresponds to the number of items in the last list that they could re-create correctly. Digit span has been found to decrease as people age—with sharper declines occurring after 70 (Gregoire & Van der Linden, 1997).

Another method of measuring working memory capacity is Daneman and Carpenter's (1980) **reading span test.** This test was developed to assess individual differences in working memory that is used in comprehension. In the reading span test, participants are asked to read sets of sentences aloud to an experimenter. The sets of sentences vary from sets containing two to six sentences. Participants receive five sets of each quantity of sentences. After each set of sentences is read, the participant is asked to recall the last word from each of the sentences from that set. Participants are allowed to keep going until they miss three or more sets in one of the groupings of two to six sentences. The reading span score corresponds to the number of sentences in the set that the person could correctly respond to three or more sets in a row. If one goes on to answer two sets from the next higher grouping correctly, then an additional half credit is awarded. Since the reading span test was first introduced, a number of different varieties have been developed (LaPointe & Engle, 1990).

Research with older adults has shown that reductions in working memory may result in difficulties in both formulating speech—particularly complex sentences—and also in understanding language (Kemper, 1992; Wingfield & Stine-Morrow, 2000). There are numerous investigations showing that older adults perform similarly to young adults in a wide range of tasks involving language comprehension (De Beni, Borella, & Carretti, 2007). Recent research by De Beni et al. (2007) suggests that declines in comprehension that are observed in older adults are related to declines in working memory and other aspects of processing rather than chronological age. Consequently, it is possible that someone over 100 years of age could have a working memory capacity of someone much younger and be less likely to experience working memory related difficulties when speaking and when comprehending speech or text.

▲ Photo 12.3 A grandmother blows bubbles with her granddaughter. Did you spend time with older adults when you were a child?

What Are Some Methodological Issues Involved in Aging Research?

Over the past 30 years, there have been numerous studies comparing the language processing of older adults with the processing of young adults. When one reviews these studies, one can find conflicting results. There is an important methodological reason that conflicting results are to be expected. Studies in which different types of people are compared are more likely to lead to erroneous results than studies in which participants can be randomly assigned to the different conditions in the study. The latter type of study is referred to as a **true experiment** (Goodwin, 2009). Experimenters can manipulate a variable that they believe can produce a change in some measurable outcome, such as language processing or memory performance. When experimenters randomly assign participants to conditions, they can ensure that there is no systematic way in which characteristics of participants have an opportunity to influence performance in any one condition. When experimenters carry out true experiments with random assignment and sufficiently large sample sizes, they can conclude, with confidence, that the manipulated variable caused a change in the outcome.

When researchers compare differences in processing for older and younger adults, the variable that is being compared cannot be manipulated. Age is a **subject variable** that cannot be randomly assigned; a person is the age that they are and cannot be assigned to some other age for a research study. Other subject variables include sex (i.e., male vs. female), smoking status (e.g., smoker vs. nonsmoker), or religion. A research design comparing groups of people who differ on one or more subject variables is called a **quasi-experiment**. The results of quasi-experiments cannot be used to conclude that the compared variable is the sole cause of an observed difference. Because the different types of people being compared in the study may vary in ways in addition to the variable of interest, the researcher cannot conclude that it is the subject variable, such as age, that is responsible for the outcome. There are likely many variables that differ systematically with age, which can contribute to a research finding. For example, studies have shown that older adults are more conscientious and more agreeable than middle-aged and younger adults (Allemand, Zimprich, & Hendriks, 2008).

A design in which groups of individuals differ in age is called a **cross-sectional design**. Another type of research design that is often used to investigate age-related changes in cognitive processing is the **longitudinal design,** which involves investigating the same group of individuals followed over time. Studies that use longitudinal designs have the advantage in that the effects of age can be established in the same individuals. The disadvantage of longitudinal designs is that they must be conducted over long periods of time. Consequently, there is the risk that participants may start the study but fail to complete it. In order to ensure that the sample size of the study is acceptable by the study's conclusion, researchers must enroll a relatively large sample at the beginning. The term that is used to describe the problem of participants not continuing with a research study is **attrition.** In studies in which there is a high level of attrition, researchers must be cautious in interpreting the results. It is possible that the results obtained, while reflecting what is true for the sample of participants at the conclusion of the study, may not reflect what is likely to be true for everyone.

A second disadvantage of longitudinal designs is that the participants contribute data at regular intervals, such as each year or every several years or even more frequently. When the data are derived from tasks, participants have the opportunity to improve their performance with each experience with the task. Researchers refer to this issue as a **testing confound.** Performance on a task can be influenced by prior experiences with the same task or similar tasks. When researchers analyze data collected over time, it can be difficult to conclude that changes in performance for an individual are the sole result of the aging of the individual.

Diseases Affecting Older Adults

As we age, our likelihood of developing certain illnesses increases. When young adults think of the illnesses most strongly associated with aging, they are likely to think of Alzheimer's disease. To the surprise of many young adults, most older adults will not go on to develop Alzheimer's disease. The diseases that commonly affect older adults and the causes for death in older adults differ from those that commonly affect young adults. Table 12.3 displays the leading causes of death in older adults. As the table shows, the top three are heart disease, cancer, and stroke. Accidents are responsible for a relatively small percentage of deaths in those over the age of 65. In contrast, accidents account for a far greater percentage of deaths in young adults. Understanding the changing vulnerabilities

Table 12.3 Leading Causes of Death in Percentages for Persons Ages Sixty-Five Years and Older by Age

	Age-Group			
Cause of Death	**65 and Older**	**65–74**	**75–84**	**85 and Older**
Heart disease	31.8	26.6	30.2	36.7
Cancer	21.6	34.2	23.6	11.6
Stroke	7.9	5.2	7.8	9.8
Chronic lower respiratory diseases	6.0	7.0	7.0	4.3
Influenza and pneumonia	3.2	1.6	2.8	4.7
Diabetes	3.0	4.0	3.3	2.2
Nephritis, nephritic syndrome, and nephrosis	1.9	1.7	2.0	1.9
Accidents	1.9	1.9	1.8	1.9
Septicemia	1.5	1.5	1.6	1.4

Source: Anderson and Smith (2005).

as we age is important because we can make better lifestyle choices, which can contribute to our living longer. It is also important to understand the aging-related disorders that affect a relatively small percentage of older adults. Even if we are not diagnosed with a disorder ourselves, our friends or family members may be.

What Is Dementia?

The term *dementia* refers to the loss of general cognitive ability that can arise from a variety of circumstances. The word has its origins in Latin. The root word *mens,* which means mind, and the prefix *de-*, which means without. Although individuals diagnosed with dementia are statistically more likely to be over the age of 65, dementia can occur in those younger than 65. One with dementia is someone without a mind or without the mind that they once had. A more old-fashioned term for dementia is **senility.** Dementia can occur following an accident involving a brain injury or can be the result of a progressive brain disease. Individuals with dementia display problems with attention, memory, problem solving, and language. Problems must persist for at least 6 months to warrant a diagnosis of dementia. The term ***delirium*** is used to refer to such cognitive problems that manifest and resolve themselves over a shorter period. There are several types of dementia. The most familiar is Alzheimer's disease. Others include frontotemporal dementia, vascular dementia, and semantic dementia. It is difficult to obtain accurate estimates of the number of older adults with dementia, because many older adults may not seek a diagnosis. Doctors estimate that between 5% and 8% of people 65 years of age or older develop dementia (Cleveland Clinic, n.d.). As people age, the chances of developing dementia double every 5 years after the age of 65.

Individuals with dementia typically experience some problems with language (Appell, Kertesz, & Fisman, 1982). Bayles (1979) pointed out that measures of language ability were better indicators of dementia than were other cognitive measures, such as memory or learning. Some researchers have suggested that problems using language may actually appear earlier than general memory problems (Wechsler, 1977). Some aspects of language are affected more than others. Studies have shown that phonological and morphological processing were intact (Irigaray, 1973) as well as syntactic processing (Bayles, 1979; Schwartz, Marin, & Saffran, 1979). Word order in the speech of dementia patients is typically correct (Kemper, 1984). In Kemper's (1984) study of 400 utterances produced by eight dementia patients, only 12 errors involving syntax or morphology were observed. Despite producing syntactically well-formed sentences, individuals with dementia generally are often not successful in producing sensible speech.

Hier, Hagenlocker, and Shindler (1985) described the speech of dementia as being devoid of meaning, because of errors made in word choice or lexical selection. Numerous studies have documented that individuals with dementia experience word-finding difficulty or tip of the tongue (Sabat, 1994). Appell and colleagues (1982) tested individuals with Alzheimer's disease and found that they would frequently substitute the word *thing* for a word that they could not produce. They would also produce errors, such as *firebug* for *match*. Kemper (1984) asked patients to name objects that were presented as line drawings. Patients made many errors. Most errors were semantic in nature, such as saying *fishing* instead of *rowboat* or saying *moon* instead of *sun*. Kemper observed that the incorrect word usually was the same part of speech as the correct word. The results suggested that patients were able to activate in memory a set of potential candidates but were not able to select the correct word (see also Irigaray, 1973).

What Is Alzheimer's Disease?

Alzheimer's disease is a type of dementia that was first described in 1906 by the German psychiatrist and neuropathologist Alois Alzheimer. Current estimates suggest that approximately 5 million people in the United States have Alzheimer's disease. Today, most people who are diagnosed with Alzheimer's disease are 65 years of age or older. However, a small percentage of cases develop in much younger adults. The numbers of individuals with Alzheimer's worldwide are staggering. In 2006, it was estimated that around the world, there were 26.6 million with the disease. It is predicted that by 2050, about 1 in 85 people will be living with the disease. Alzheimer's disease can run in families, but often there is no family history. Inherited forms of the disease make up approximately 5% of the cases (Serretti, Olgiati, & De Ronchi, 2007). Contrary to popular belief, there is no definitive test to diagnose Alzheimer's disease. There is no brain scan that can be conducted or battery of cognitive tests whose results can say with certainty that one has or does not have Alzheimer's disease.

Problems With Cognition, Memory, and Language

Individuals with Alzheimer's disease may show problems with cognition, memory, and language well before they or family members seek a diagnosis. Problems with episodic memory (i.e., memory for life events) are common (Bäckman, Small, & Fratiglioni, 2001; Small, Herlitz, Fratiglioni, Almkvist, & Bäckman, 1997). These individuals also take increasingly longer to carry out perceptual tasks (Albert, Moss, Tanzi, & Jones, 2001; Fabrigoule et al., 1998; Fox, Warrington, Seiffer, Agnew, & Rossor, 1998), tasks involving attention (Nielsen, Lolk, Andersen, Andersen, & Kragh-Sørenson, 1999; Rubin et al., 1998; Tierney et al., 1996), and tasks involving visuospatial processing (Albert et al., 2001; Fowler, Saling, Conway, Semple, & Louis, 2002). Problems with language appear quite early in the development of Alzheimer's disease and become progressively more severe (Convit et al., 2000; Grober & Kawas, 1997; Small et al., 1997).

The conversations of individuals with Alzheimer's disease may contain speech errors and roundabout ways of expressing meaning, which is called **circumlocution.** Rather than expressing meaning in a simple, direct manner, one uses many more words than others would use. For example, instead of saying *comb,* one might say *the thing that people use to smooth out their hair and to part their hair.* Instead of saying *eat,* one might say *the thing that people do when they feel their stomach feeling funny because so much time has passed since breakfast.* When speaking, an individual with Alzheimer's may experience difficulty finding and producing words (Nebes, 1989). The difficulty is likely due to the loss of semantic memory due to the deterioration of brain tissue. There is also an overuse of pronouns in place of full noun phrases (Kempler & Zelinski, 1994; Nebes, 1989; Ripich & Terrell, 1988), which is likely because speakers have difficulty finding and producing the nouns but less difficulty finding and producing pronouns.

As Alzheimer's disease progresses, patients' speech may become unintelligible due to problems with phonology and articulation (Murdoch, Chenery, Wilks, & Boyle, 1987). Patients' writing ability gradually deteriorates with spelling errors increasing, which reflect problems with phonological processing (Forbes, Shanks, & Venneri, 2004; Glosser, Grugan, & Friedman, 1999). Research suggests that individuals with Alzheimer's disease do particularly poorly when asked to comprehend figurative language, such as metaphors (Winner & Gardner, 1977). As you learned in Chapter 6, a metaphor is an expression that compares two different entities in a manner to bring about a novel meaning, such as *college is torture*. Winner and Gardner (1977) asked patients with Alzheimer's disease to match metaphors with pictures depicting the correct relationship. In the task, patients viewed four pictures, three incorrect and one correct. They also asked patients to explain the meanings of metaphors in their own words. Results from seven patients showed that they were generally unable to match metaphors to pictures and unable to explain the meanings of metaphors in their own words. Other studies have found that patients with Alzheimer's disease also have difficulties understanding proverbs, idioms, irony, and sarcasm (Rapp & Wild, 2011). There is a gradual loss of ability to comprehend spoken language (Groves-Wright, Neils-Strunjas, Burnett, & O'Neill, 2004) that ultimately leads to the complete loss of the ability to interact verbally through either speaking or listening.

Compelling research has shown that the language use of individuals during the teenage years may hold clues about who will and will not go on to develop Alzheimer's disease later in life. Text Box 12.1 describes this research.

Text Box 12.1 Research Discovery: Teenage Language Use Predicts Alzheimer's Later in Life

▲ Photo 12.4 Nuns enter a building in a line. Are you familiar with the culture of convents?

In 2001, Kemper, Greiner, Marquis, Prenovost, and Mitzner reported research suggesting that one's language use at 18 can predict whether one goes on to develop Alzheimer's disease much later in life. They analyzed autobiographical writings from individuals participating in the Nun Study, which is a longitudinal study investigating the risk factors associated with Alzheimer's disease (Snowdon, 1997; Snowdon et al., 1997). The study included 678 members of the School Sisters of Notre Dame. Each year, participants were tested for cognitive functioning and health. They also provided information about health and lifestyle. Language samples were provided by 180 participants. Language samples were compared for the participants who went on to develop dementia and those who did not. Differences were observed between the writings of the two groups when they were young adults with the grammatical complexity being higher for those who did not go on to develop dementia. It is not clear why language use during late adolescence and early adulthood would be related to the likelihood of developing dementia late in life. Biological factors may be related to both language use early in life and brain changes later in life. It is also possible that life experiences that are correlated with developing language skills early in life are also correlated with factors contributing to dementia later in life.

What Is Parkinson's Disease?

Parkinson's disease is a disorder of the central nervous system, caused by the degeneration of cells in the part of the brain called the *substantia nigra,* which is located in the midbrain. The cells produce the neurotransmitter dopamine, which is involved in producing movements. Early symptoms of Parkinson's disease include tremors, involuntary shaking, and difficulty controlling movements. As more and more cells of the substantia nigra are destroyed, symptoms worsen. Studies have shown that individuals with Parkinson's disease experience cognitive problems, specifically related to executive functioning and memory (Elgh et al., 2009). Surprisingly, by the time an individual with Parkinson's disease begins to notice symptoms, approximately 70% or more of the cells in the substantia nigra have already been lost.

The prevalence of Parkinson's disease varies across countries. In the United States, approximately a half million people likely live with the disease (National Institute of Neurological Disorders and Stroke [NINDS], 2012). Fifty thousand new cases or more are diagnosed each year. Men are more likely to be affected than women. Most cases of Parkinson's disease are diagnosed around the age of 60. Although it is rare for individuals younger than 40 to develop Parkinson's disease, it can occur. Michael J. Fox, the actor, is an example of someone who received the diagnosis in his late 30s. Text Box 12.2 provides information about his career before and after being diagnosed with the disease.

Problems With Language

Research also has shown that language can be affected (Williams-Gray et al., 2009). Recent research has shown that problems with language processing manifest following the onset of Parkinson's disease. Language problems appear to be the result of reduced working memory capacity brought about because of deterioration of pathways connecting the frontal lobe to other parts of the brain (Gilbert, Belleville, Bherer, & Chouinard, 2005; Monetta & Pell, 2007). Verreyt, Nys, Santens, and Vingerhoets (2011) found that language performance differs for individuals whose symptoms are more severe on the right side of the body than on the left side of the body. As you learned in Chapter 2, movements of the body are controlled by the opposite hemisphere; thus, the left hemisphere controls the right side of the body, and the right hemisphere controls the left side of the body.

Text Box 12.2 Extraordinary Individuals: Michael J. Fox

Michael J. Fox is the Canadian-born actor who starred in popular American television shows and films. He was raised in a military family, moving around a lot as a child. When his father retired from the Canadian military, they settled in Vancouver. He began acting in Canada at the age of 15. By the time that he was 21, he had landed the lead role of Alex on the American television show *Family Ties.* For the role, he won three Emmys and a Golden Globe. He also started in the popular *Back to the Future* trilogy and the horror comedy *Teen Wolf.* In 1991, at the age of 30, he was diagnosed with Parkinson's disease. Five years later, he starred on another hit American television show, *Spin City.* His diagnosis was not publicly known until 1998. In 2000, he retired from acting and established a foundation that raises money to fund scientific research on Parkinson's disease. The foundation is currently the largest funder of Parkinson's research, having raised over $179 million. Current research is focused on identifying biomarkers of Parkinson's disease. Fox is an active fundraiser and speaks frequently about Parkinson's disease. He also came out of retirement and is still involved in acting. He's had recurring roles on the television shows *Boston Legal* and *The Good Wife.* He has made guest appearances on *Rescue Me, Scrubs,* and *Curb Your Enthusiasm.* He has authored three bestselling books: *Lucky Man* (2002); *Always Looking Up: The Adventures of an Incurable Optimist* (2009); and *Funny Thing Happened on the Way to the Future, a Compendium of Wisdom for Graduates* (2010).

Most Parkinson's patients develop difficulties producing speech. Approximately 90% develop dysarthria, which is characterized by weakened or paralyzed muscles in the speech tract (Ho, Iansek, Marigliani, Bradshaw, & Gates, 1998; Logemann, Fisher, Boshes, & Blonsky, 1978). Problems with the voice are also common. Many individuals with Parkinson's disease also struggle with language comprehension, particularly sentences with complex syntactic structures (e.g., Lieberman et al., 1992; Natsopoulos et al., 1991, 1993). For example, sentences similar to those that follow are somewhat difficult for most speakers of English.

a. The boy that the bully hit dropped the book bag.
b. The cat that the dog chased scrambled up the tree.

A number of studies with Parkinson's patients have found that they experienced a great deal of difficulty processing such sentences (e.g., Grossman, Kalmanson, Bernhardt, Stern, & Hurtig, 2000; Kemmerer, 1999; Natsopoulos et al., 1991). Recent research also suggests that patients with Parkinson's disease have difficulty comprehending pragmatic aspects of language (Holtgraves, McNamara, Durso, & Cappaert, 2010). In the end stage of the disease, an individual becomes less and less able to understand the speech of others. Ultimately, individuals develop full-blown dementia.

What Is Multiple Sclerosis?

Multiple sclerosis is a disease caused by an autoimmune response through which damage occurs to the covering of the nerves in the central nervous system and also the coverings to the axons of cells in the brain. The damage leads to progressive symptoms involving both physical and cognitive impairments. Early symptoms usually include strange sensations that come and go, such as the feeling of pins and needles or numbness. There can be muscle spasms, problems with balance, and difficulty with coordinating movements. As the disease progresses, there may be difficulty speaking; problems swallowing; and problems with vision, pain, and fatigue.

There are at least four different types of multiple sclerosis. Some individuals with multiple sclerosis experience bouts of typical multiple sclerosis symptoms that persist for a period of time and then disappear with months or years of remission. This type of multiple sclerosis has been called the **relapsing-remitting multiple sclerosis.** Others develop **secondary progressive multiple sclerosis,** which is sometimes referred to as galloping multiple sclerosis. This type begins with a bout of symptoms that appear to go into remission; however, the symptoms then recur with progressively worse symptoms. Future remissions and recurrence involve still progressively worsening symptoms. A third type of multiple sclerosis is **primary progressive multiple sclerosis,** which does not involve periods of remission; rather, the individual experiences progressively debilitating symptoms from the onset. The least common type of multiple sclerosis is progressive relapsing multiple sclerosis; it involves symptoms that steadily worsen, but there are periods in which symptoms flare up and are worse than usual.

In the United States, there are approximately 250,000 to 350,000 people living with multiple sclerosis (NINDS, 2012). Each week, approximately 200 new cases are identified. The age of onset is usually between the ages of 20 and 40. Women are twice as likely to be affected as men. Individuals with European ancestry are twice as likely to be affected as individuals from other ancestries. The prevalence of multiple sclerosis varies geographically; it is five times more common in the northern United States, Canada, and Europe than it is in southern regions of the world.

Problems With Speech and Language

Approximately 23% to 50% of patients with multiple sclerosis experience difficulty producing speech (Beukelman, Kraft, & Freal, 1985; Hartelius & Svensson, 1994). Dysarthria is common because of weakening or paralysis of muscles in the speech tract (Brookshire, 1992; Hartelius & Svensson, 1994). Damage to motor neurons can also lead to problems with articulation, prosody, and other acoustic aspects of speech. Several longitudinal studies of patients with multiple sclerosis provide evidence that there are declines in performance in visuospatial tasks (Canter, 1951), in the ability to comprehend sentences (Grossman et al., 1995), and in word naming (Friend et al., 1999). Other studies have found that individuals with multiple sclerosis experience difficulties understanding ambiguous sentences and figurative language (Lethlean & Murdoch, 1993). They also had trouble repeating sentences. Problems in verbal fluency appear to be reliable indicators of cognitive impairment in multiple sclerosis (Henry & Beatty, 2006).

What Is Huntington's Disease?

Huntington's disease, or Huntington's chorea, is an inherited disease. Individuals with one biological parent with Huntington's disease have a 50% chance of acquiring the disease. The word *chorea* refers to unwanted involuntary movements. Huntington's disease is characterized by a progressive loss of one's ability to control movements as well as cognitive impairments and psychiatric symptoms, such as depression, anxiety, compulsivity, and aggression. Symptoms typically emerge during midlife between the ages of 30 and 50; however, they can develop at any age, including during childhood. Children who develop symptoms of Huntington's disease are unlikely to survive into adulthood. Adults typically live 10 to 20 years following the onset of the disease. The disease progresses more rapidly when symptoms emerge earlier in life. Huntington's disease is more often observed in individuals with European ancestry than in those with African or Asian roots.

In the United States, approximately 30,000 people live with Huntington's disease. Another 200,000 people are living with the possibility of developing the disease, because they have at least one biological parent with the genetic marker associated with the disease. It should be noted that approximately 20% of cases do not involve family inheritance. There is a genetic test that can determine whether an individual will develop the disease. However, because there is currently no cure for Huntington's disease, many individuals who are at risk for Huntington's disease elect not to take the test. Taking the test and receiving a positive result also places the individual at risk of losing health insurance, jobs, and friends. Families affected by Huntington's disease often experience prejudice and discrimination. One of the most famous cases of Huntington's disease was that of the American folk singer Woody Guthrie, who passed away in 1967. Text Box 12.3 provides more information about this extraordinary individual.

Text Box 12.3 Extraordinary Individuals: Woody Guthrie (1912–1967)

Woody Guthrie was an American folk singer and songwriter who became popular for his song "This Land Is Your Land." He inspired many other artists, such as Bob Dylan, Bruce Springsteen, Joe Strummer, and Billy Bragg. He was born in Okemah, Oklahoma, but lived in Texas and later in California, which became the destination of many Oklahomans during the Great Depression (Cray, 2004). His early songs reflected themes familiar to those who had been displaced by the Dust Bowl. Guthrie began to show symptoms of Huntington's disease in the late 1940s. Many assumed that his frequent lack of physical coordination during walking and slurred speech was due to excessive alcohol consumption. His symptoms were also mistaken for schizophrenia. It was not until 1952 that an accurate diagnosis was received. He had inherited the disease from his mother, Nora, who died in 1930 after being committed to a hospital for the insane in Oklahoma. His condition deteriorated so much by 1956 that he required round-the-clock hospitalization. At the height of his illness, there was widespread ignorance about the disease. He and his family helped to bring attention to it. His second wife, Marjorie, helped to establish the Committee to Combat Huntington's Disease. Later, the group was renamed to the Huntington's Disease Society of America. Of Guthrie's eight children, two went on to develop the disease. His son Arlo Guthrie, who did not develop the disease, became well known for his hit song "Alice's Restaurant."

Problems With Cognition and Language

Problems with cognition and language are common with Huntington's disease. During the early stages of the disease, forgetfulness is common as are problems sustaining focused attention. Later, one may experience greater difficulty involving memory, decision making, and problem solving. Individuals may have trouble producing single words as well as sentences (Ho et al., 2002). Recent studies indicate that language deficits specifically involve morphology and syntax because of the gradual destruction of a part of the brain known as the striatum (Teichmann et al., 2005; Teichmann, Dupoux, Cesaro, & Bachoud-Levi, 2008).

Negative Attitudes About Aging

What Is Ageism?

Despite the fact that the average life span has increased and the number of older adults living healthy, productive lives is the highest in human history, our attitudes toward older adults and toward aging remain quite negative. Ageism refers to negative attitudes toward older adults (Butler, 1969). Some of the most common negative stereotypes of older adults is that they are unhealthy, unable to think clearly, and disagreeable. Table 12.4 displays a commonly used quiz to assess ageism. Those who have used this quiz have found that many people hold misconceptions about aging (Nelson, 2004). Interestingly, ageism is the only form of prejudice that we hold toward a group of others that, if all goes well, we will one day join. Many of the beliefs and fears that we have about aging are not accurate. Harris and Clancy Dollinger (2001) investigated the relationship between ageism in young adults and their anxiety about aging; they found a strong, positive relationship. Those with higher levels of anxiety about aging held more negative attitudes toward aging and also rated themselves at the age of 70 more negatively. In a study with adolescents, Haught, Walls, Laney, Leavell, and Stuzen (1999) found that boys held more negative attitudes about aging than girls. In a study with first-year graduate students, Gellis, Sherman, and Lawrence (2003) found that men held more negative attitudes about aging than women. However, girls and women may experience more anxiety about getting older (Cummings, Kropf, & DeWeaver, 2000).

What Is Elderspeak?

One form of subtle ageism can occur when a young adult speaks to an older adult. Sometimes, when young adults and teenagers speak to older adults, they tend to use the same intonation and phrasing as one uses with small children (Caporael, 1981; Kemper, Vendeputte, Rice, Cheung, & Gubarchuk, 1995) or pets. In Chapter 3, you learned that adults' special way of talking to young children is called motherese, or child-directed speech. When this type of speech is used with older adults, it is called **elderspeak,** or infantilizing talk. As with motherese, elderspeak involves a singsong tone of voice that is more high-pitched than one uses when speaking with same-age peers. One might address the older person as *honey, dear,* or *sweetie.* Those who use this manner of speaking may aim for their interaction with the older adult to be caring, compassionate, and effective; however, their actions inadvertently harm rather than help. Elderspeak sends the message that the speaker views the older adult as childlike and incompetent.

Table 12.4 Palmore's Facts of Aging Quiz

1. The majority of old people—age 65-plus—are senile.
2. The five senses (sight, hearing, taste, touch, smell) all tend to weaken in old age.
3. The majority of old people have no interest in, nor capacity for, sexual relations.
4. Lung vital capacity tends to decline with old age.
5. The majority of old people feel miserable most of the time.
6. Physical strength tends to decline with age.
7. At least one-tenth of the aged are living in long-stay institutions such as nursing homes, mental hospitals and homes for the aged.
8. Aged drivers have fewer accidents per driver than those under age 65.
9. Older workers usually cannot work as effectively as younger workers.
10. More than three-fourths of the aged are healthy enough to do their normal activities without help.
11. The majority of old people are unable to adapt to change.
12. Older people usually take longer to learn something new.
13. Depression is more frequent among the elderly than among younger people.
14. Older people tend to react slower than younger people.
15. In general, old people tend to be pretty much alike.
16. The majority of old people say they are seldom bored.
17. The majority of older people are socially isolated.
18. Older workers have fewer accidents than younger workers.
19. More than 20% of the population is now 65 and older.
20. The majority of medical practitioners tend to give low priority to the aged.
21. The majority of old people have incomes below the poverty line, as defined by the U.S. federal government.
22. The majority of old people are working or would like to have some kind of work to do, including housework and volunteer work.
23. Old people tend to become more religious as they age.
24. The majority of old people say they are seldom irritated or angry.
25. The health and economic status of old people will be about the same or worse in the year 2010, compared with younger people.

Answers: All odd-numbered statements are *False*. All even-numbered statements are *True*.

Source: Palmore (2004).

Research by Kemper and colleagues (Kemper et al., 1995; Kemper, Othick, Warren, Gubarchuk, & Gehring, 1996) has shown that younger adults will use elderspeak spontaneously when interacting with older adults. A commonly used task in these laboratory studies has two participants in rooms where they are able to hear each other; however, they are not able to see each other. Both participants are given maps for a town. Each person takes turns serving as the speaker and the listener. The speaker is instructed to describe the map so that the listener can draw it. Young adults in these studies also used more utterances, shorter sentences, less complex sentences, and slower speaking rates when talking to older adults than when they talked to other young adults. One possibility is that young adults use elderspeak because there is some aspect of the interaction with the older adult that the person is having difficulty understanding. Kemper and colleagues (1995) provided evidence against this view, showing that elderspeak occurs in laboratory studies both when older adults are able to interrupt the speaker, ask questions, and make comments and when they are instructed not to interrupt, ask questions, or make comments (Kemper et al., 1996).

In 1981, Caporeal shined a light on the issue of elderspeak, showing that in long-term care settings, health care providers and other staff members frequently use elderspeak with older patients. In such settings, elderspeak may be more accepted by older adults. Nevertheless, research suggests that older adults who experience elderspeak may become frustrated and, over time, begin to experience low self-esteem (Caporael, Lukaszewski, Culbertson, 1983; Coupland, Coupland, Giles, & Henwood, 1998; Ryan, Meredith, MacLean, & Orange, 1995). On the other hand, O'Connor and Rigby (1996) found that there are some older adults who indicate that they like being addressed in elderspeak and like being treated in ways that others describe as infantilizing. Self-esteem was found to be lowest for older adults who did not like elderspeak but who experienced it frequently and highest for older adults who liked elderspeak and who experienced it frequently.

Williams and colleagues (Williams, 2006; Williams, Kemper, & Hummert, 2003) have shown that short-term interventions with professional caregivers are effective in reducing how much elderspeak is used in health care settings. They recorded the speech of caregivers before and after the intervention and coded each example of elderspeak. The intervention educated caregivers about elderspeak and how it can negatively affect older adults' self-esteem. The intervention also provided caregivers with examples of elderspeak and with alternatives that did not contain elderspeak. For example, caregivers were encouraged to avoid the terms *honey, sweetie,* and *darling* and instead use the person's name.

Wingfield and colleagues (Stine, Wingfield, & Myers, 1990; Wingfield, Wayland, & Stine, 1992) have shown that older adults' comprehension suffers when speakers speak quickly and when they use unusual intonational patterns. While older adults prefer others to speak slowly versus quickly, studies have shown that comprehension can be impaired by speech that is too slow as well as by speech that is too fast (Schmitt & Moore, 1989). Wingfield (1998) found that the comprehension of older adults could be improved by a procedure that slowed a sample of recorded speech by splicing in pauses at the ends of clauses and sentences. Other methods of slowing speech, such as splicing in pauses at random points in the speech sample or slowing the speech rate overall, were found to impair comprehension.

What Is Elder Abuse?

The darker side of ageism includes the victimization of older adults. Studies have shown that many older adults suffer from abuse at the hands of others. Elder abuse includes a wide range of circumstances, including physical abuse, emotional abuse, and financial abuse. Table 12.5 displays the different types of abuse that older adults might experience. Older adults are particularly vulnerable financially. They are frequently victims of financial scams, occurring either in person or via telephone. A study by the Association for the Advancement of Retired Persons (AARP) (2003) suggests that between 56% and 80% of telemarketing calls are directed to older adults.

Some surveys have found that 23% of nonspousal caregivers were perpetrators of elder abuse (NCEA, 2012). Another survey of individuals who called a caregiver help line found that 12% of respondents indicated that they engaged in some form of physical abuse of the person that they were caring for at least one time (Coyne, Reichman, & Berbig, 1993). Pillemer and Suitor (1992) found that 6% of caregivers revealed that they had abused the person in their care (Pillemer & Suitor, 1992). Paveza, Cohen, Eisdorfer, and Freels (1992) found that 5% of caregivers abused someone in their care. Compton, Flanagan, and Gregg (1997) found that 10.5% of caregivers indicated that they were abusers of an older adult. Research suggests that elder abuse is more common when one is being cared for by a family member (e.g., spouse, child) than when one is cared for by a nonrelative (Paveza et al., 1992; Pillemer & Suitor, 1992). Among caregivers who are relatives, elder abuse may be more common when the past relationship between the caregiver and care receiver was predictive of elder abuse. Elder abuse was more likely if the care receiver had been aggressive or violent with the caregiver in the past (Coyne et al., 1993; Hamel et al., 1990).

The tremendous stress that some caregivers experience may make elder abuse more likely. Caregiver stress can lead to serious depression (Coyne et al., 1993; Paveza et al., 1992). Researchers do not yet know exactly how stress, depression, and the violence involved in elder abuse are linked. Bendik (1992) has argued that caregiver stress does not

Table 12.5 Examples of Elder Abuse

Physical Abuse	Hitting, kicking, misuse of medications, restraints
Sexual Abuse	Rape, sexual assault, sex without consent
Verbal Abuse	Profanity, name-calling, threats
Emotional Abuse	Manipulation, coercion, isolation
Financial Exploitation	Theft, fraud, misappropriation of funds
Neglect	Physical, nutritional, and medical
Harassment	Racial, sexual, sexual orientation, national origin

Source: National Center for Elder Abuse (NCEA) (2012).

always lead to elder abuse; rather, caregiver stress may cause mood disturbances, which make abuse more likely. Garcia and Kosberg (1992) suggested that anger may be the critical link; mood disturbances that include anger may be particularly prone to leading to elder abuse. Research has shown that elder abuse is more likely when caregivers provide larger amounts of care (in terms of number of years and in terms of hours per day) than when they provide less care (Coyne et al., 1993).

Some daily activities of older adults and caregivers appear to be more stressful and, thus, more likely to lead to an abuse situation than others (Anetzberger, 1987; Compton et al., 1997; Pillemer & Suitor, 1992). For example, mealtimes and the routines centering on the taking of medications are particularly stressful. The personal habits and the privacy concerns for both the caregiver and care receiver can lead to stressful interactions. Public interactions that do not go as planned or that are embarrassing can also be triggers for stress and violence. Older adults who suffer from health problems and/or cognitive impairments may be more likely to be victims of elder abuse.

How Can Ageism Affect Adolescents and Young Adults?

Research suggests that negative attitudes about aging can influence the choices that we make. Kennison and colleagues (Kennison & Ponce-Garcia, 2012; Popham, Kennison, & Bradley, 2011a, 2011b) found that young adults who hold more ageist attitudes and who report more ageist behaviors take more risks in daily life than others. The risk-taking included using alcohol, tobacco, and illegal drugs and also engaging in risky sexual behavior. Prior research has shown that risk-taking is strongly related to an individual difference variable called novelty-seeking or sensation-seeking (Zuckerman, Eysenck, & Eysenck, 1978). Popham, Kennison, and Bradley (2011b) showed that even after taking into account sensation-seeking, participants' ageism explained a significant portion of the variance in risk-taking. Kennison and Ponce-Garcia (2012) found that young adults who had close, positive relationships with an older adult, such as a grandparent, during childhood also reported lower levels of ageism and lower levels of risk-taking in their current daily life. Kennison and colleagues (Kennison & Ponce-Garcia, 2012; Popham, Kennison, & Bradley, 2011a, 2011b) suggested that the link between ageism and adult risk-taking involves death anxiety in a way that is consistent with terror management theory (TMT) (Greenberg, Pyszczynski, &

▲ Photo 12.5 A man climbs a mountain without ropes or other gear. Do you have any hobbies that others would consider dangerous?

Solomon, 1986; Solomon, Greenberg, & Pyszczynski, 1991). TMT claims that people normally have to deal with fear of death and may use multiple strategies to buffer their fears. The theory evolved out of the views of the cultural anthropologist Ernest Becker (1973). Kennison and colleagues proposed that young adults may attempt to reduce their death anxiety by engaging in activities that are invigorating and make them feel fearless, invulnerable, and physically strong.

Medical research has shown that holding negative attitudes about aging may be a risk to one's own health. Levy and colleagues (Levy, Kasl, & Gill, 2004; Levy, Slade, Kunkel, & Kasl, 2002; Levy, Zonderman, Slade, & Ferruci, 2009) found that those who hold ageist views are more likely to suffer from negative health conditions linked to lifestyle (e.g., diet, use of alcohol, use of tobacco, lack of daily exercise), such as cardiovascular disease (Levy et al., 2009). Levy and colleagues (2002) carried out a study of 660 people living in a small town and found that those who had positive attitudes about getting older lived on average 7.5 years longer than others in the sample who did not have positive attitudes about getting older.

Policymakers around the world are likely to be particularly interested in how the misconceptions that people have about aging and the anxiety that they have about aging may lead individuals to make poor lifestyle choices. Risk-taking in adolescents and young adults is a major public health issue not only because of the link between risk-taking and mortality (Eaton et al., 2010) but also because of the cost to society (DiClemente, Santelli, & Crosby, 2009). There is also a cost to society when adults of any age take risks with their health through smoking, drinking, and eating. Large-scale efforts to decrease anxiety about aging among adults of all ages promise to have far-reaching benefits for society.

Summary and Theoretical Implications

Older adults experience physical changes, such as declines in hearing, vision, working memory, and reaction time. However, they also experience increases in crystallized intelligence. The diseases that are most closely associated with aging are generally experienced in a relatively small percentage of older adults. These illnesses include dementia, Alzheimer's disease, Parkinson's disease, multiple sclerosis, and Huntington's disease. In each of these illnesses, language and cognition are adversely affected. Despite the fact that average life spans are increasing and the number of people over 65 is increasing worldwide, many people fear getting older and have negative attitudes about those who are older and about aging in general. Negative attitudes toward older adults can lead to prejudice, discrimination, and also abuse. The fear of aging may also be related to how likely young adults are to engage in risky activities.

In terms of the four theoretical approaches to language development, the research that has been conducted exploring language in older adults does not help us in determining which theory is most accurate, because none of the theories has directly addressed this topic. As the average age of the population in the United States as well as around the world continues to increase, the topic of aging will likely attract more research. As more research is conducted, it is possible that advocates of the different theoretical perspectives will be able to extend their theories to make specific predictions about language in older adults.

KEY TERMS

ageism
Alzheimer's disease
anecdotal evidence
attrition
centenarians
circumlocution
cross-sectional design
crystallized intelligence
delirium
dementia
digit span task
elderspeak
fluid intelligence
fraternal twins
gerontologist
gerontology
Huntington's disease
hyperpriming
identical twins
longitudinal design
multiple sclerosis
nonagenarians
octogenarians
osteoporosis
Parkinson's disease
perceptual normalization
primary progressive multiple sclerosis
quasi-experiment
reading span test
relapsing-remitting multiple sclerosis
secondary progressive multiple sclerosis
senility
septuagenarians
sexagenarians
subject variable
testing confound
true experiment

REVIEW QUESTIONS

1. How have average life spans changed over the past 100 years? What changes are expected in the coming decades regarding the percentage of the population made up of older adults?
2. What is gerontology? How long has it existed as a field of study?
3. What research designs are used to study aging? What are some disadvantages of these research designs?
4. What are some of the physical changes that occur during normal aging?
5. How do fluid intelligence, crystallized intelligence, and memory processing change across the life span?
6. What has research shown about how comprehension processes change as people age?
7. What has research shown about how speech production processes change as people age?
8. What is ageism? What has research shown about the relationship between ageism and health as we age?

9. What is the relationship between ageism and risk-taking in young adults?
10. What factors are predictive of elder abuse among those who care for older adults?
11. What is elderspeak? Why do people use this form of language?
12. What are the most common causes of mortality in adults 65 years of age or older?
13. What is dementia? How is it different from Alzheimer's disease?
14. How are one's language abilities affected by dementia?
15. What are the symptoms of Alzheimer's? How is it diagnosed? What are the current treatments?
16. How are one's language abilities affected by Alzheimer's disease?
17. How are one's language abilities affected by Parkinson's disease?
18. How are one's language abilities affected by multiple sclerosis?
19. How are one's language abilities affected by Huntington's disease?
20. Why are those at risk for developing Huntington's disease often uninterested in obtaining the genetic test for the disease?

RECOMMENDED READING

Buettner, D. (2008). *The blue zones: Lessons for living longer from the people who've lived the longest*. New York: National Geographic.

Cozolino, L. (2008). *The healthy aging brain: Sustaining attachment, attaining wisdom*. New York: W. W. Norton & Company.

Friedman, H. S., & Friedman, L. R. (2012). *The longevity project: Surprising discoveries for health and long life from the landmark eight-decade study*. New York: Plume.

Hooyman, N. R., & Kiyak, H. A. (2011). *Social gerontology: A multidisciplinary perspective* (9th ed.). Boston: Allyn & Bacon.

Snowdon, D. (2002). *Aging with grace: What the Nun Study teaches us about leading longer, healthier, and more meaningful lives*. New York: Bantam.

RECOMMENDED FILMS

Arledge, E. (Director). (2004). *Forgetting: A portrait* [DVD]. United States: Public Broadcasting Service.

Iverson, D. (Director). (2009). *My father my brother and me* [DVD]. United States: Public Broadcasting Service.

Polley, S. (Director). (2006). *Away from her* [Motion picture]. Canada: Foundry Films.

SUGGESTIONS FOR CLASS PROJECTS

1. Administer Palmore's (2004) Facts on Aging Quiz to peers in order to investigate how well-informed people are about growing older. Compare quiz results of different groups of people: men versus women, young adults versus older adults, American students versus international students. A minimum of 10 to 12 participants per type of group is recommended. Report your results in a written report and/or a class presentation.

2. Poll friends and family about the top 10 leading causes of death in older adults. Use multiple formats for the poll. One format could be having participants list the 10 causes of death and then report the results in a ranked order, starting with the most commonly listed disease. A second format could be a checklist in which students create a list of 50 common diseases, which include the 10 leading causes of death among older adults. Participants can be asked to check their best guesses for the leading causes of death for those over 65. The results can be summarized in terms of correct responses and errors.

3. Investigate public awareness about Alzheimer's disease. Do people know what it is? Do people know anyone personally who has been diagnosed with Alzheimer's disease? Do people believe that all or most older adults will go on to develop Alzheimer's disease? Do people think that senility, dementia, and Alzheimer's disease all refer to the same thing? Do they fear having Alzheimer's disease one day? Work alone or in groups to develop a brief questionnaire, and present your results in either a class presentation or a written report.

4. Investigate public awareness about Huntington's disease. Do people know what it is? Investigate views about genetic testing for Huntington's disease. The genetic test for Huntington's disease is one of the few genetic tests that can tell a person that they will definitely develop or definitely not develop a disease. Ask individuals the following question: If one of your parents or grandparents suffered from Huntington's disease, would you want to take the genetic test for the disease? The genetic test would be able to reveal whether you definitely will or will not develop the disease. Ask individuals who respond with no whether they would change their mind about taking the test if they were planning to have a child. A child born to a parent who has the genetic marker for Huntington's disease has a 50% chance of having the genetic marker that will cause the disease.

For additional ancillary resources, please visit the companion website at www.sagepub.com/kennison.

- Video Links
- Audio Links
- Web Resources
- Internet Activities
- Flashcards
- Web Quizzes

Glossary

Agraphia is a disorder in which one has an inability to write.

Accommodation refers to a type of speech error in which a phoneme appears to be produced in a different location within the utterance than intended; the phoneme takes on the phonetic characteristics of the location to which it moved.

Acquired dyslexia refers to dyslexia that arises following a brain injury.

Active sentence is a sentence in which the subject of the sentence serves the semantic role of agent.

Additions are a type of speech error in which an unintended word or morpheme occurs in an utterance.

African American Vernacular English (AAVE) is a dialect of American English that is spoken by African American communities throughout the United States.

Age of acquisition refers to the age at which one learns a language or the age at which one learns a particular word in a language.

Age of acquisition effect is the fact that words learned earlier in life are responded to faster and remembered better than words that are learned later in life.

Ageism refers to having negative attitudes or prejudice against older adults.

Agenesis of the corpus callosum (ACC) is a congenital disorder characterized by the absence of the corpus callosum in the brain. The condition has also been called the *natural split brain*.

Agent is the semantic role of a noun that serves as the performer or doer of the action denoted by a verb.

Agrammatism can occur following brain damage to Broca's area and is characterized by one producing speech that lacks syntactic function words and morphological word endings.

Alexia is a language disorder in which one has an inability to read.

Alphabetic writing systems are ones in which symbols represent individual phonemes.

Alzheimer's disease is a type of dementia in which individuals have trouble with memory, cognition, and language; it is most often observed in older adults but can also affect younger adults.

American Sign Language (ASL) is a language used by deaf individuals primarily in the United States, involving the use of gestures and facial expressions.

Anaphoric pronoun is a pronoun that refers back in a sentence or discourse to a preceding noun or pronoun.

Anecdotal evidence is the type of evidence contained in personal stories, experiences, and observations; it is not considered a strong form of evidence.

Anomic aphasia is a type of expressive aphasia in which one experiences word-finding difficulty when speaking but has relatively good comprehension.

Antecedents are nouns or proper names that precede pronouns or other referential expressions; antecedents and referents refer to the same discourse entity.

Anticipations are a type of speech error in which a word or morpheme is produced earlier in an utterance than originally intended.

Aphasia is a language deficit caused by an injury to the brain, which can include brain diseases, stroke, or trauma-based injury.

Apraxia is a speech disorder involving the inability to plan speech; individuals with apraxia have difficulty producing speech sounds, syllables, and words.

Arcuate fasciculus is an area of the brain that connects Wernicke's area to the premotor area and is involved in controlling movements.

Artificial intelligence is a subfield within cognitive science that is devoted to implementing human-like intelligence in machines.

Asperger's syndrome is a condition classified as on the autism spectrum in which one is relatively high functioning but has difficulty with social relationships and understanding and conforming to social norms.

Attention deficit disorder (ADD) is a disorder associated with inattention.

Attention deficit/hyperactivity disorder (ADHD) is a disorder associated with both inattention and overactivity, such as fidgeting, talking a great deal, and having difficulty sitting still for long periods.

Attrition is a methodological issue that is common in longitudinal studies. Participants drop out of the study over time. The longer the study, the greater the chances of attrition are.

Auditory brainstem response (ABR) is one of the techniques used to test infants' hearing before they leave the hospital after birth. Small electrodes are placed on the infant's head, and the electrical activity is recorded as the infant listens to sounds.

Autism is a developmental disorder in which one displays social and language deficits.

Babbling is a form of language production produced by infants starting at around 4 months. It is characterized by the random production of speech sounds.

Babbling drift refers to the fact that infants' babbling is composed increasingly of the sounds from the language(s) of the home.

A **baby biography** is an early methodology for studying child development; someone, usually a parent, observes the child's behavior and records important milestones in a diary.

Balanced bilinguals are bilinguals who are equally skilled and fluent in both of the languages that they know.

Basic level category is the intermediate level of categorization for concepts that have three levels (e.g., animal—bird—robin; plant—fruit—apple).

Behavioral modification refers to the use of operant conditioning principles (e.g., reinforcement and punishment) to shape behavior; speech therapy techniques may involve reinforcing target speech sounds.

Bilingual education refers to education that provides, at least, a portion of the education in a bilingual's first language in order to facilitate learning.

Bilinguals are those who are fluent in more than one language.

Bimodal bilinguals are those who know two languages: one that is spoken and one that is signed.

Blends are a type of speech error in which words are combined, often resulting in a nonword.

Bottom-up processing is processing that only uses information from the stimulus itself to interpret the stimulus.

Bound morphemes are morphemes that cannot appear as a word on their own; in order to be used, they must be added to another morpheme. Examples of bound morphemes include prefixes and suffixes.

Broca's aphasia is a type of aphasia characterized by difficulty producing speech and also reduction in the use of syntactic words (e.g., helping verbs, prepositions, and pronouns).

Canonical babbling is the first stage of babbling, typically occurring between 4 and 6 months and involving the repetition of a single syllable at a time (e.g., *babababa* or *gagagaga*).

Cardinality is one of the five principles of counting; the number of the last object in a series represents the total number of items.

Cataphoric pronoun is a pronoun that refers forward in a sentence or discourse to a following noun or pronoun.

Categorical perception refers to the ability to distinguish different speech sounds.

Caudate nucleus is part of the basal ganglia, which has been involved in motor disorders, such as Parkinson's disease and Huntington's disease.

Centenarians are individuals who are 100 years of age or older.

Chatterbox syndrome is a rare condition in which one has severe cognitive deficits as well as near normal language ability.

Childhood amnesia refers to the fact that individuals typically have few memories of events that occurred before the age of 5 years.

Circumlocution is a manner of speaking in which one uses roundabout ways of explaining meaning (i.e., using more words and phrases than are needed).

Cleft palate occurs when one is born with a hole in the roof of the mouth, which is also the floor of the nose; the hole involves missing or misplaced bone and muscle.

Closed class words are those for which new words cannot be created, such as prepositions, pronouns, determiners (e.g., *the, an, a, those, these*), and bound morphemes.

Cluttering is a rare speech problem in which speakers are difficult to understand because they speak quickly and may use erratic speech rhythms; they may also produce utterances that are ungrammatical and/or contain words that are nonsensical.

Code-switching occurs when bilinguals produce utterances that contain words from both of the languages that they know; the listeners are typically also bilingual and know the same two languages.

Cognates refer to words in one of a bilingual's two languages that sound and look similar to a word in the bilingual's other language; the two words have similar meanings.

Cohort model is a model of word recognition developed in the early 1980s; the model emphasized the fact that words sharing phonemes are activated in parallel during word recognition and that as the input is processed, the number of words activated decreases until the target word is identified.

Collective monologues occur when children communicate with one another but do so without talking about the same topic; toddlers will often talk to each other with each speaking about her own topic yet taking turns as in a conversation.

Color term is a word that refers to a color (e.g., *blue, red,* and *green*).

Communicative competence refers to achieving the ability to use a language appropriately in a variety of social settings.

Comorbidity refers to the condition of having more than one disease or disorder.

Complement clause is a type of subordinate clause, as those introduced by the word *that* (e.g., *Barbara exclaimed that the soup was cold.*).

Conduction aphasia is a type of aphasia characterized by poor word repetition; good comprehension; and fluent, nonsensical speech. It occurs following damage to the arcuate fasciculus.

Consonants are one of two types of phonemes. Consonants are produced with an interruption of airflow, such as a closure of the lips.

Continuity hypothesis is the view that the mechanism that is used for syntactic parsing is the same for children and for adults.

Conversation is an exchange of utterances between two or more people.

Cooing is pleasant-sounding vocalizations, usually involving one elongated vowel (e.g., *oooooo* or *aaahhh*).

Coordinated clauses are clauses that are connected using the conjunctions *and, but,* and *or* (e.g., *Darla ordered the burger; Stan ordered the salad.*).

Copula is a word that links a subject and a predicate, such as *is* and *was.*

Corpus callosum is the bundle of fibers that connect the left and right hemispheres of the brain.

Counterfactual is a statement expressing a state that is counter-to-fact or hypothetical (e.g., *If I were the president of the United States, I would create more national holidays.*).

Critical period hypothesis is the view that there is a period in development in which language is learned, and after that window of development passes, language learning is more difficult.

Cross modal task is a methodology developed to investigate language comprehension; participants process spoken language while also having to respond to visually presented stimuli.

Crossed aphasia refers to cases in which aphasia results following damage to the right hemisphere.

Cross-sectional design is a design in which multiple groups of individuals representing different ages are compared during the same time period (e.g., a study in which first, second, and third graders are compared on measures of mathematical ability).

Crystallized intelligence refers to the type of intelligence that tends to increase with age, specifically stored knowledge of word meanings and facts.

Curse of knowledge is the tendency to attribute one's own state of understanding to another's state of mind or even one's own previous state of mind.

Deep dyslexia is a type of developmental dyslexia that is characterized by semantic errors involving individual words in reading, such as seeing the word *ape* but experiencing the word as *monkey.*

Deletions refer to a type of speech error in which a phoneme or word is missing from an utterance.

Delirium is used to refer to such cognitive problems that manifest and resolve themselves over a shorter period.

Dementia refers to the loss of general cognitive ability that can arise from a variety of circumstances.

Derivational morphemes are bound morphemes (e.g., suffixes or prefixes) that change the part of speech of the word to which they are added.

Developmental dyslexia is a type of dyslexia that appears when children begin learning to read.

Dialect is a variant of a language, spoken by individuals in the same geographic region and/or social group.

Diary method is a technique for studying speech errors; one observes speech errors in everyday life and records instances in a diary.

Digit span task is a method for measuring working memory; one listens to a series of numbers (or digits) and is asked to recall the numbers.

Direct method is a method for teaching second languages that involves students learning the second language without the use of the first language. Grammar of L2 is learned inductively through exposure to L2 sentences.

Dominant language is the language that is most often used by a bilingual or trilingual and also the language used with the greatest proficiency.

Double dissociation is observed when one is studying two processes, A and B, and one can find a factor that influences A but not B, and a second factor that influences B but not A.

Down syndrome is a developmental disorder caused when an extra copy of chromosome 21 becomes part of the embryo. The disorder results in distinctive physical features, organ defects, severe intellectual impairment, and delays in language development.

Dual-route model is Coltheart's model of word recognition that describes two paths to word identification; one path involves whole word recognition, and the other path involves using spelling-to-sound rules.

Dysarthria is difficulty producing speech because of weakened or paralyzed muscles in the speech tract.

Dyscalculia is a type of disability in which one specifically has trouble with mathematical processes; the disability may occur developmentally or be the result of a brain injury.

Dyseidetic dyslexia is a type of developmental dyslexia in which one cannot identify words as wholes but can recognize words using the spelling-to-sound rules.

Dysgraphia refers to a condition in which one has difficulty with writing.

Dyslexia is a disorder in which reading is impaired. Dyslexia may occur developmentally or arise following brain injury.

Dysphonetic dyslexia is a type of developmental dyslexia in which one cannot identify words using the spelling-to-sound rules but can recognize words as wholes.

Early left anterior negativity (ELAN) is a type of waveform pattern observed during event-related brain potentials (ERP) studies on the left hemisphere anterior area of the scalp occurring relatively early in processing.

Elderspeak is a form of speech used with older adults; though it is often perceived as demeaning and disrespectful. The characteristics of elderspeak are similar to infant-directed speech.

Electrical brain stimulation (EBS) is a methodology developed to study brain processing; electrical stimulation is applied to brain tissue in order to observe the effect.

End-of-clause wrap-up effect refers to the fact that readers tend to spend more time processing words occurring at the ends of clauses and sentences as compared with the same words that are preceded by the same information but do not occur at the end of the clause or sentence.

Exchanges are a type of speech error in which two phonemes or words appear in each other's location in an utterance.

Expressive aphasia is a broad class of aphasias in which people have trouble producing speech.

Expressive language refers to speech or other forms of language production.

False belief test is a test developed to investigate children's theory of mind.

False cognates are words in one of a bilingual's two languages that sound and look similar to words in the bilingual's other language, but the words have different meanings.

Fast mapping refers to children's ability to learn new words after one exposure.

Feral children are children who spend a significant portion of their early lives without nurturing and/or physical care by adults.

First language (L1) refers to one's mother tongue, which is the language of one's childhood home and the first language acquired in life.

Fis phenomenon refers to children's common mispronunciation of words, such as when a child says *fis* when trying to say *fish*.

Fluid intelligence is a type of intelligence that tends to decline as one is aging; fluid intelligence involves the ability to think logically, reason, and solve problems.

Focal color is the shade of a color that represents the prototypical shade.

FOXP2 is a protein regulated by the FOXP2 gene located on chromosome 7 that has been found to be involved in human language and also in animal communication.

Fragile X is a genetic condition that results in congenital intellectual impairment. It is the leading cause of intellectual impairment in children.

Fraternal twins are twins who result from two eggs fertilized by two different sperm. Although the twins develop simultaneously in the same womb, they are not more genetically similar than regular siblings.

Free morphemes are morphemes that can stand alone as words.

Free morpheme constraint refers to the fact that bilingual speakers do not produce multimorphemic words in which some morphemes are in one language and the other morphemes are in another language.

Frenulum is connective tissue that connects the tongue to the mouth.

Freudian slip is a speech error that is perceived to reveal some unspoken thought or intention of the speaker.

Fricatives are consonants that are produced when airflow is partially stopped between any two places of articulation (e.g., lips, teeth, or palate).

Frontier words are words whose meanings are only partially known.

Function words are words that serve a syntactic role in sentences, such as helping verbs (e.g., *was, had, could,* and *should*), prepositions (e.g., *in, on,* and *under*), and pronouns (e.g., *he* and *she*).

Fusion hypothesis is the view that young bilingual children do not initially differentiate their two languages but do so at some point later in childhood.

Genderlect refers to the different dialects used by men and women, involving differences in word choice, amount of polite speech, and sentence structures.

Gerontologist is one who specializes in the field of gerontology.

Gerontology is the specialty dedicated to the study of older adults and the conditions affecting older adults.

Global aphasia occurs following extensive damage to the perisylvian area, which includes Broca's and Wernicke's areas and the area in between; individuals lose their ability to speak and comprehend language.

Grammar refers to the rules of a language that must be learned in order for a speaker to produce and to comprehend the acceptable sentences of the language. Grammar includes phonological, morphological, semantic, and syntactic rules.

Grammar-translation method is a method for teaching a second language; it utilizes the first language to explicitly teach students the grammatical rules of L2.

Grapheme refers to the smallest unit of writing in a language. In English, a grapheme corresponds to a letter of the alphabet.

Grapheme-to-phoneme correspondences refer to letter-to-sound regularities in languages written using an alphabetic writing system.

Gricean maxims describe norms of conversation, which may be universal.

Habituation paradigm is a technique that was developed to investigate infant cognition. Infants are presented with a stimulus repeatedly and also novel stimuli. Infants tend to respond less and less to stimuli as they are processed repeatedly. Differences in responding to stimuli are interpreted as reflecting familiarity or memory for the stimuli.

Head-turn technique is a methodology that was developed to investigate infant cognition; infants are conditioned to turn their heads when they discriminate one phoneme from another.

Heteronym words share the same root but have different pronunciations for the different meanings.

Holophrase is a word that is used to express more meaning than a single word, such as saying *banana* to express *please, give me the banana.*

Homograph refers to a word that is spelled the same but has a different meaning.

Homophones are words that are spelled differently but pronounced the same.

Huntington's disease is a rare, genetically based terminal disorder that involves the gradual loss of the ability to control movements as well as cognitive decline.

Hyperpriming is larger than typical priming effects, which are observed in older adults.

Identical twins are twins who share the same DNA because they are formed by the same egg and sperm.

Idiomorph is a word that children use when they are first learning to speak. The word is made up by the child but is used consistently to refer to the same object or event.

Immersion programs are language learning environments in which a second language is taught without reliance on a shared first language. Students are forced to use only the second language in the language learning environment.

Index of Productive Syntax (IPSyn) is a method for measuring the syntactic complexity of children's spontaneous utterances.

Indirect meaning refers to meaning that is not directly reflected in the grammatical structure of the statement.

Indo-European language is a language that belongs to the Indo-European language family.

Infant-directed speech (or motherese) is the way of speaking that adults use with infants. It is characterized by variable pitch, elongation of some sounds or words, and repetition of words.

Inflectional morphemes are bound morphemes (e.g., suffix or prefix) that do not change the part of speech of the word to which they are added.

Instrument refers to a phrase that specifies an object that is used to carry out an action.

Intellectual disability is the contemporary term to refer to intellectual impairment (low IQ).

Intransitive verb is a verb that cannot be used with a direct object (e.g., *Sue hoped.*).

Irregular word refers to a word that does not follow the morphological rules in a language, such as English plural nouns that are not formed with the addition of the suffix *-s* and English past tense verbs that are not formed with the addition of the suffix *-ed*.

Irreversible passives are passive sentences in which the words that are in the position of subject and object cannot be exchanged and still result in an interpretable sentence.

Isolating languages are languages in which most words are composed of a single morpheme. Such languages typically have relatively fixed word order.

Jargon is a type of slang that develops within professional groups (e.g., medical jargon, legalese, military speak).

Joint attention occurs when one interacts with another person, and together, they focus on some object or other person.

Labial consonants are consonants that are produced when the airflow is interrupted at the lips.

Language acquisition device (LAD) refers to Noam Chomsky's concept of the organ in the brain that is responsible for the rapid acquisition of language by children.

Language competence is a term created by Noam Chomsky to refer to the knowledge of language in its abstract form, separate from the psychological mechanisms that are involved in language production and comprehension (i.e., language performance).

Language family is a group of languages that share a common ancestor language; such languages share some characteristics and root words.

Language isolates are languages that are highly dissimilar from other languages of the world and do not appear to be descendants of any known languages.

Language performance is a term created by Noam Chomsky to refer to the use of language, which involves psychological processes as well as knowledge of language (i.e., language competence).

Language transfer refers to the situation in which one's use of a second language is influenced by some aspect of one's first language.

Late closure is a syntactic parsing strategy that is applied in cases in which a phrase has two possible attachments that are comparable in complexity; late closure predicts that the phrase will be attached to the most recently processed part of the sentence.

Late talker refers to a child who produces first words and other language milestones later than the typical child.

Lexeme is the phonological level of representation of a word, which is separate from the meaning representation of a word.

Lexical ambiguity refers to a word that has more than one meaning (e.g., *bank* as in *riverbank* or *money bank*).

Lexical bias effect refers to the fact that speech errors result in real words more often than they result in a nonword.

Lexical decision task is a task used in language research in which participants view a series of letter strings and are asked to judge whether each string is or is not a real word.

Lexical differentiation refers to the extent to which a language has words to specify a particular meaning. A language with a great number of words for a concept would be described as having greater lexical differentiation than a language with fewer words for a concept.

Lexicon is the term that is used to refer to one's knowledge of words and word meaning.

Linguistic determinism refers to the strong version of the Sapir–Whorf hypothesis, which claims that the language one speaks determines all aspects of thought, including perception.

Linguistic relativity refers to the weak version of the Sapir–Whorf hypothesis, which claims that the language that one speaks can influence one's cognitive processing (e.g., memory and/or decision-making processes).

Linguistics is the scientific study of language and languages.

Lisp is a speech problem in which one pronounces /s/ as the *th* in *thin* and pronounces /z/ as the *th* in *that*.

Localization hypothesis is the view that specific parts of the brain can be associated with specific functions.

Logogen model is one of the earliest models of word recognition.

Logographic writing systems are writing systems in which symbols represent entire words.

Longitudinal design is a type of research design in which one group of participants are tested multiple times—sometimes over a period of years.

Malapropism is an error in speaking that is characterized by using a similar-sounding word instead of the word that is intended (e.g., saying *prostrate* instead of *prostate*).

Manner of articulation refers to the extent to which the airflow is interrupted during the production of a consonant.

Math disability is a type of learning disability that specifically affects mathematical processing.

Matthew effect refers to the notion that children who read poorly will make less progress in reading than children who start out reading well (as in the poor get poorer and the rich get richer).

McGurk effect demonstrates that auditory perception in humans is influenced by visual processing; one who hears the syllable /ga/ but sees the speaker pronouncing the syllable /ba/ will perceive the syllable /da/.

Mean length utterance (MLU) is a unit of measurement used to study language development. The utterances of a child in a conversation are coded in terms of number of morphemes per utterance. From this, the average number of morphemes per utterance is computed.

Mental number line refers to how individuals imagine the number line with negative numbers, zero, and positive numbers ordered relatively to one another.

Mental retardation (MR) is an old-fashioned term to refer to intellectual impairment (low IQ).

Metalinguistic awareness is awareness of one's own language use, such as the extent to which one's intentions have been understood by a listener.

Metaphors are a type of figurative language use in which two dissimilar concepts are compared (e.g., *The exam was a massacre.*).

Mind blindness is Baron-Cohen's term to refer to autism; he claims that individuals with autism have profound deficits in understanding the perspectives or minds of others.

Minimal attachment is a syntactic parsing strategy that is applied in cases in which a phrase has two possible attachments and those attachments differ in complexity; minimal attachment predicts that the syntactically least complex attachment will be favored.

Mirror neurons are specialized cells in the brain that become activated when one performs an action or observes an action performed.

Misinformation effect is the finding that when answering questions about one's memory for events, one generally performs less accurately when the questions contain incorrect details.

Mixed transcortical aphasia is caused by damage to the regions surrounding, but not including, Broca's area, Wernicke's area, and the arcuate fasciculus and characterized by poor language production and comprehension but the ability to repeat words and phrases.

Modularity is the view that the mind is composed of specialized, somewhat independent subsystems or modules.

A **monolingual** is a person who knows only one language.

A **morpheme** is the smallest unit of meaning in language; morphemes may be whole words, prefixes, or suffixes.

Morphological complexity effect refers to the fact that the time taken to recognize a word increases as the number of morphemes in the word increases.

Morphological rules are the rules in a language that govern the formation of new words through the combination of morphemes.

Motherese (also called baby talk or infant-directed speech) is the way of speaking that adults use with infants. It is characterized by variable pitch, elongation of some sounds or words, and repetition of words.

Multiple sclerosis is an autoimmune disease that causes damage to the covering of the nerves in the central nervous system and also the coverings to the axons of cells in the brain. Symptoms include both physical and cognitive impairments.

Mutual exclusivity principle refers to a bias in word learning that children have; if an object has already been associated with one label (i.e., a word), the child will not associate it with a second lable (i.e., if a toy is called a *blick,* it cannot also be called something else).

Nativism is the view that knowledge of language is innate, accounting for children's rapid acquisition of language.

Naturalistic observation refers to a research methodology in which a researcher observes a naturally occurring behavior in the setting in which it typically occurs.

Negative language transfer is a type of language transfer in which one makes errors in a second language because of rules from one's first language.

Negative sentences are sentences that contain the word *not* or other negative words, such as *never* (e.g., *Sue did not go to the party.*).

Neologism is an utterance that represents a made-up word or nonexistent word.

Nonagenarians are individuals who are between 90 and 99 years of age.

Nonfocal colors are examples of colors but not the shade of the colors that are thought of as the prototypical shade.

Occipital lobe is the area of the brain that is responsible for visual processing; it is located at the back of the head.

Octogenarians are individuals who are between 80 and 89 years of age.

Omissions are a type of speech error in which a word or morpheme is missing from the utterance.

Onomatopoeia occurs when a word's pronunciation is similar to the meaning of the word, such as *zip, buzz,* or *drip.*

Onset is the part of the syllable that contains the initial consonant or consonant cluster.

Open class words are those for which new words can be created, such as nouns, verbs, and adjectives.

Operant conditioning refers to a form of learning popularized by B. F. Skinner in which the frequency of a naturally occurring behavior (or operant) can be increased or decreased through reinforcement or punishment, respectively.

Original word game is the term used to describe the interactions between children and caregivers in which the child points to an object and asks what it is called.

Orthographic neighborhood effect refers to the fact that the time taken to process a written word is influenced by the number of similarly spelled words there are in the language. Words with similar spellings are called *neighbors*.

Osteoporosis is a condition that affects older adults, mostly women; bones are weakened due to calcium loss that occurs because of the aging process or dietary issues.

Otitis media is a condition involving the buildup of fluid in the ear, sometimes caused by repeated ear infections and associated with language problems in children.

Otoacoustic emissions test (OAE) is a common hearing test administered to infants before they leave the hospital after birth. A small earphone and microphone are placed in the infant's ear canal. Sounds are played. When an infant hears normally, an echo is produced. If an infant is hearing impaired, then no echo will be observed.

Overextension is a type of error produced by children learning words; a word is used more broadly than adults would use the word (e.g., cow is used to refer to all four-legged animals).

Overregularization errors are errors that most children typically make during the period of time in which they are learning new language rules; they apply the rule to words that are irregular, which results in an ungrammaticality.

Parkinson's disease is a disorder of the central nervous system caused by the degeneration of cells in the part of the brain called the substantia nigra. The disease causes tremors, involuntary shaking, difficulty controlling movements, and cognitive problems.

Pars opercularis is a region within Broca's area associated with phonological processing and syntactic processing.

Pars orbitalis is a region within Broca's area associated with semantic processing.

Pars triangularis is a region within Broca's area associated with semantic processing and syntactic processing.

Passive sentences are sentences in which the subject serves in the semantic role of patient (e.g., *The cookie was eaten.*).

Patient is a type of thematic role, which specifies which noun in the sentence is affected by the action.

Perceptual normalization is the process of adjusting to others' speech patterns.

Perseverations are a type of speech error in which a phoneme that occurs in an utterance appears again later in the utterance in an incorrect location.

Phoneme is the smallest unit of sound in a language.

Phoneme-monitoring task is a task that was developed to investigate language processing; a participant listens to a sentence while attempting to identify a target phoneme specified by the experimenter.

Phonics method is a method for teaching reading that emphasizes the learning of spelling-to-sound rules and other aspects of the phonological aspects of words.

Phonological awareness is the understanding that children acquire regarding the sounds and the sound structure of their language.

Phonological bias technique is a methodology developed to study speech errors in the laboratory. Participants are presented with word pairs having similar phonological composition. The last word pair of the series is varied in an attempt to induce a speech error.

Phonological bootstrapping hypothesis refers to the view that children's language learning is aided by their analysis of the characteristics of the speech that they hear.

Phonological dyslexia is a type of acquired dyslexia characterized by difficulty in applying the spelling-to-graphemes rules to identify words.

Phonological rules are the rules of language that govern the basic sounds of words and sentences.

Phrase marker is a tree diagram that is commonly used to represent the syntactic structure of a sentence.

Phrase structure rules describe the word order norms in a language in terms of what types of phrases are contained within sentences and what types of words are contained in phrases.

Picture-naming task is a task that is used in language research in which participants are shown a series of pictures and asked to say the word that describes the picture. Response times are often recorded.

Pig Latin is a word game in English commonly used by children.

Pinyin is a phonologically transparent writing system that is used with children learning to read Chinese; those learning Chinese can use pinyin to read and write words based on their sound before they go on to learn the complex, idiographic Chinese symbols.

Place of articulation is a feature of consonants, describing where in the vocal tract the airflow is impeded or completely disrupted. For example, the place of articulation for /m/ is the lips.

Planum temporale is a triangular region within Wernicke's area that likely plays some role in language learning.

Politeness refers to a form of manner and accompanying language use that displays respect for the listener as well as a desire not to offend.

A **polyglot** is an individual who knows many languages.

Positive language transfer is a type of language transfer in which one's second language usage is facilitated because the rules of the first and second languages are similar.

Pragmatic rules are those rules in a language that govern the social conventions of language use.

Preferential looking paradigm is a technique that is developed to investigate infant cognition. Infants are presented with speech in the context of two visual displays (i.e., TV screens). Differences in looking time at the two visual displays are analyzed.

Primary progressive aphasia is a form of dementia caused by degenerative disease; patients have increasing difficulty finding words, reading, writing, and understanding speech.

Primary progressive multiple sclerosis is a type of multiple sclerosis that does not involve periods of remission; rather, the individual experiences progressively more debilitating symptoms from the onset.

Priming refers to the facilitation of processing that may occur for a stimulus when it is preceded by a related stimulus.

Principle of arbitrariness refers to the property of human language related to the fact that the spoken and written forms of most words are not related to the meaning of the words.

Principle of displacement refers to the property of human language related to the fact that the speaker can refer to events in the past or future and not exclusively events from the present.

Principle of productivity refers to the fact that in human languages, speakers can create new forms (e.g., words and sentences) that have never been produced before.

Private speech refers to speech directed toward oneself.

Productive language ability is one's ability to produce speech. In the case of one who knows signed language, productive language ability may also refer to one's signing ability.

Pronouns are referential words (e.g., *he, she, it, they*) that refer to the same discourse entity that has been referred to by the use of a noun or proper name.

Prosodic bootstrapping hypothesis is the view that infants use the prosody or melody of the speech stream to extract information about language that aids them in learning language.

Protowords are among the first words that children use; however, they are words that only the child uses. To be considered a protoword, it must be one that the child uses consistently to refer to objects or actions and it must not be one that adults in the environment introduced to the child.

Psycholinguistics is the study of language use, including language processing and language acquisition.

Punishments is a term from operant conditioning that refers to any acts that are designed to decrease the frequency of a particular behavior.

Pure aphasia is a type of aphasia where an individual experiences a deficit in only one area of language use, such as reading, writing, or speaking.

Pure word deafness is characterized by the inability to comprehend spoken language despite adequate hearing and also retaining the ability to speak, read, and write.

Quasi-experiment is a design in which different groups of participants are compared, but random assignment is not used to place participants in groups.

Reading span test is a test to assess individual differences in working memory capacity.

Receptive aphasia is a type of aphasia that is characterized by difficulties understanding language (i.e., listening and reading), such as Wernicke's aphasia.

Receptive language ability refers to comprehension ability.

Register is a form of language that is used in a particular social setting.

Regular word refers to a word that follows the morphological rules in a language, such as English plural nouns that are formed with the adding of the suffix *-s* and English past tense verbs with the adding of the suffix *-ed*.

Reinforcements are any actions that aim to increase the frequency of a behavior, such as the application of a reward.

Relapsing-remitting multiple sclerosis is a type of multiple sclerosis that begins with symptoms that persist for a period of time and then disappear within months or years of remission.

Repair occurs during speech when a speech error is made and the speaker corrects the error in the following utterance.

Request is a type of utterance in which one asks another for something.

Reversible passives are passive sentences in which the words that are in the position of subject and object could be exchanged and result in an interpretable sentence.

Rhotacism is a type of speech problem in which the speaker consistently replaces the /r/ sound with /w/ sounds.

Right ear advantage (REA) for language refers to the fact that language stimuli are perceived better when listened to by the right ear than by the left ear.

Rime is the part of the syllable that does not contain the initial consonant or consonant cluster.

Romance languages are those languages that descended from Latin (e.g., Spanish, French, Italian, Portuguese, and Romanian).

Sapir–Whorf hypothesis is the view proposed by Edward Sapir and Benjamin Whorf that the language one speaks influences what one can think about and how one thinks.

Savantism occurs when an individual with one or more developmental disorders also has an exceptionally developed skill or ability, such as amazing memory, musical ability, or artistic talent.

School psychologists are psychologists who specialize in the diagnosis and treatment of skills that are used in academic settings.

Second language (L2) is a language that is acquired in addition to one's first language (L1).

Secondary progressive multiple sclerosis is a type of multiple sclerosis that begins with a bout of symptoms that appear to go into remission; however, the symptoms then recur with progressively worse symptoms. Future remissions and recurrence involve still progressively worsening symptoms.

Semantic associate is a word that is related in meaning to another word.

Semantic bootstrapping refers to the view that children learn syntactic rules of language through the acquisition of word meanings.

Semantic complexity refers to the complexity of a word or phrase in terms of meaning (i.e., how many meanings or changes in meaning are involved).

Semantic memory is the type of memory used to store the meanings of words and world knowledge, such as facts learned in school.

Semantic network is a metaphor used to describe how concepts are stored in memory—with related concepts located closer together than unrelated concepts. When one concept is activated during processing, the activation can spread to related concepts through pathways between concepts in the network.

Semantic priming effect refers to the fact that processing is facilitated on a stimulus if the preceding stimulus is related to the second in terms of meaning (e.g., *shoe* is processed faster following *sock* than following *rock*).

Semantic priming paradigm is a technique that was developed to investigate semantic processing; processing is measured for words preceded by semantically related or unrelated words (e.g., *shoe–sock* versus *bird–sock*).

Semantic rules are the rules of a language that govern how meaning is derived from combinations of words within sentences and the interpretation of sentences within longer discourses.

Semantic verification task is used to investigate the organization of semantic memory; participants are presented with statements and asked to press a key on a keyboard as soon as they can determine whether the statement is true or false.

Senility is an old-fashioned term used to refer to dementia.

Septuagenarians are individuals between the ages of 70 and 79 years.

Sequential bilingualism occurs when one learns a second language later in life than one learns a first language.

Sexagenarians are individuals between the ages of 60 and 69 years.

Shifts are speech errors in which a phoneme or morpheme shows up in the wrong location in an utterance, such as later in the utterance than intended.

Similes are a form of figurative language in which two dissimilar things are compared using the word *like* (e.g., the sunset was like a symphony).

Simultaneous bilingualism occurs when one learns a second language at the same time one is learning a first language.

Single dissociation is observed when one is studying two processes and a factor influences one process but not the other.

Slang refers to informal language use within a culture, most typically involving newly innovated words and expressions.

Slip of the ear refers to when one mishears spoken language.

Slip of the tongue refers to a speech error.

Slips of the hand are signing errors that are made by a user of a signed language, such as American Sign Language (ASL).

Social smile is a communicative smile displayed by infants around the 6th week to caregivers.

Southern American English (SAE) is a dialect of American English spoken in the southern regions of the country. It differs in both phonological and morphological features from standard American English.

Specific language impairment (SLI) is a genetically based disorder affecting language; affected individuals have IQ in the normal range but have persistent difficulties with grammatical aspects of language.

Speech acts fulfill a specific intention of a speaker, such as issuing a greeting or making a request.

Spoonerism is a speech error. The term gets its name from the Reverend Spooner, who frequently made speech errors when he spoke.

Spreading activation is the metaphor that is used to describe why a concept in a semantic network is easier to process when a related concept has been processed recently.

Standard dialect is the dialect that is considered to be the ideal. It is the dialect used for most official written documents and for spoken programs, such as national news broadcasts.

Stop consonants are types of consonants that involve the complete interruption of airflow through the vocal tract, such as at the lips, in the throat, or by the placement of the tongue on the roof of the mouth. The following phonemes in English are stop consonants: /b/, /p/, /d/, /t/, /g/, and /k/.

Strong version of the Sapir–Whorf hypothesis is the same as linguistic determinism, which claims that the language one speaks determines all aspects of thought, including perception.

Stuttering is a rare speech problem in which individuals have difficulty producing speech in a consistently fluent manner.

Subject variable refers to a variable in a study that is associated with individual participants and cannot be changed or randomly assigned, such as participants' age, gender, smoking status, etc.

Subordinate categories are the lowest level of categorization for concepts that have three levels (e.g., animal–bird–robin and plant–fruit–apple).

Substitutions are a type of speech error in which a word or morpheme is produced in an utterance in place of the intended word or morpheme (e.g., *The coffee is too cold, I mean, too hot.*).

Subvocalization refers to silent speech that is sometimes produced when we hear speech in our minds, such as when we read silently.

Superordinate categories are the highest level of categorization for concepts that have three levels (e.g., animal–bird–robin and plant–fruit–apple).

Surface dyslexia is a type of acquired dyslexia characterized by difficulty in recognizing words as wholes; consequently, individuals have trouble recognizing irregularly spelled words.

Syllabic writing systems are those in which symbols represent individual syllables.

Syntactic ambiguity refers to a word or phrase that can serve more than one syntactic role in a sentence.

Syntactic bootstrapping refers to the view that children's knowledge of grammar can help them learn new words.

Syntactic category ambiguity refers to words that can function as more than one part of speech (i.e., noun, verb, adjective, preposition, determiner), such as *kick*, *dance*, and *burn*; each of these words can be used as either a noun or a verb.

Syntactic complexity refers to the complexity of a word or phrase in terms of syntactic structure (i.e., how many syntactic constituents are involved, how many nodes or branches must be constructed for the structure).

Syntactic parsing refers to the processing of words in a sentence in terms of their syntactic structure.

Syntactic reanalysis occurs when comprehenders process sentences containing a temporary syntactic ambiguity and the first analysis of the ambiguity turns out to be incorrect; the following sentence context disambiguates the ambiguity as being different from the comprehender's initial interpretation.

Syntactic rules are the rules of language that govern basic word order, such as the positioning of the subject, verb, and object in a sentence.

Synthetic languages tend to have more than one morpheme per word on average. Some synthetic languages have a large number of morphemes per word, and word order is much more flexible than in English.

Tachylalia refers to extremely rapid speech.

Taxonomy bias is displayed by children learning words; they tend to assume that a word refers to a type of an object or whole category rather than a specific example of the category.

Testing confound is a methodological problem that can occur in studies in which participants are tested multiple times, such as longitudinal studies. The participants' performance during subsequent testing sessions may be influenced by earlier testing sessions.

Thematic roles are the semantic functions that phrases in a sentence satisfy, such as agent, patient, and instrument.

Theory of mind refers to the understanding that develops in children typically between 3 and 4 years resulting in their having an understanding that others may have different thoughts and perspectives than their own.

Tip-of-the-tongue state (TOT state) occurs when one has difficulty producing a word during speaking despite the fact that the person knows the word and, sometimes, can identify the first sound of the word, number of syllables, and stress pattern.

Tongue twister effect refers to the fact that readers take longer to read silently sentences that contain words sharing the same phoneme.

Tongue-tied is an old-fashioned way of describing individuals with lisps or other speech impediments; however, the expression stems from a particular condition in which the tongue is abnormally connected to the floor of the mouth.

Top-down processing is processing that uses information from memory in addition to information from a stimulus itself to interpret the stimulus.

Transcortical motor aphasia occurs following brain damage outside of Broca's area, specifically regions connected to Wernicke's area, and is characterized by impaired speech but good comprehension.

Transcortical sensory aphasia occurs in patients with damage posterior to Wernicke's area; it is characterized by poor comprehension but fluent, grammatical speech and good word repetition.

Transitive verb is a verb that occurs with a direct object (e.g., *John broke the glass.*).

Translation equivalents refer to the words from different languages that share the same meaning (e.g., *apple* and *manzana*).

Trilingual is one who is fluent in three languages.

Trisomy 21 is the term used to refer to Down syndrome, as the disorder involves one having three copies of chromosome 21.

True experiment is an experiment in which at least one variable is manipulated, and members of the sample are randomly assigned to conditions.

Truncated passives are passive sentences that do not specify the agent of the action in a by-phrase (e.g., *The cookies were eaten.*).

Turn-taking is a characteristic of conversations; one person speaks, then the other person speaks, and so on.

Typicality effect refers to the fact that one can confirm typical members of a category faster than atypical members (e.g., *A robin is a bird.* vs. *A penguin is a bird.*).

Underextension is a common error that children make when learning new words. They use the word on a more limited basis than adults would, specifically using the word to refer to fewer members of a category than adults would.

Universal grammar (UG) refers to Chomsky's concept of the knowledge that is innate and allows children to learn any human language to which they receive adequate exposure; the knowledge must contain information about what makes a possible human language.

Variegated babbling is the second stage of babbling occurring around 11 months of age; infants produce sequences of speech sounds involving different consonants and vowels, such as *bagadabaga* or *mabadagama*.

Vegetative sounds are nonlanguage sounds made by infants early in life, such as sounds related to sucking, sneezing, and breathing.

Visual world paradigm is used to investigate language comprehension; participants' eye movements are recorded as they interact with objects as they listen to spoken instructions over headphones.

Voice-onset time refers to the duration between the start of the production of a phoneme and the start of voicing.

Voicing occurs during speaking when the vocal cords are vibrating.

Vowel refers to one of two types of phonemes. Vowels are produced without any interruption of airflow along the vocal tract.

Wada testing involves using an anesthetic to paralyze temporarily one hemisphere of the brain at a time in order to investigate the processing carried out by the hemisphere that is not paralyzed.

Weak version of the Sapir–Whorf hypothesis states that the language one speaks can influence one's cognitive processing (e.g., memory and/or decision-making processes).

Wernicke's aphasia is a language deficit resulting from damage to the area of the brain known as Wernicke's area; one can produce speech relatively fluently, but the speech lacks content words and may contain nonsense words that result with morphemes that are combined incorrectly.

Whole language method is a method for teaching reading that emphasizes the whole reading experience and the use of reading strategies to extract meaning. Instruction involves students reading in groups, reading aloud, and encouraging students to develop a love of reading.

Whole object bias refers to the fact that when one hears a new noun in the context of the object to which it refers, one typically infers that the noun refers to the entire object rather than some part or aspect of the object.

Whorfian hypothesis is the view proposed by Edward Sapir and Benjamin Whorf that the language that one speaks influences what one can think about and how one thinks.

Wh-word refers to words that are used in the formation of questions in English, such as *what, which, where, when*, and *how*.

Williams syndrome is a rare, genetically based disorder characterized by elfin facial features and low IQ but highly developed (or near normal) language ability.

Word frequency effect refers to the fact that words that are used more frequently can be processed faster and remembered more easily than words that occur less frequently.

Word length effect refers to the fact that the time taken to read a word is generally longer for words with a greater number of letters.

Word-naming task is a task in which a research participant is shown a list of words, one at a time, and asked to pronounce the word as quickly as possible. Response time is typically recorded.

Word spurt is a period during a child's life (typically between 18 and 24 months) when vocabulary expands at a rapid rate.

Word superiority effect refers to the fact that participants can recognize a whole word faster than they can recognize any single letter within the word.

Writing disability is a type of learning disability in which one has problems specifically with writing—either the physical movements involved in writing or the cognitive processes involved in writing, or both.

References

Aarnoutse, C., van Leeuwe, J., Voeten, M., & Oud, H. (2001). Development of decoding, reading comprehension, vocabulary and spelling during the elementary school years. *Reading and Writing: An Interdisciplinary Journal, 14*(1–2), 61–89.

Abrahams, B. S., & Geschwind, D. H. (2008). Advances in autism genetics: On the threshold of a new neurobiology. *Nature Reviews Genetics, 9,* 341–355.

Abrams, L., White, K. K., & Eitel, S. L. (2003). Isolating phonological components that increase tip-of-the-tongue resolution. *Memory and Cognition, 31,* 1153–1162.

Abutalebi, J., Cappa, S. F., & Perani, D. (2005). Functional neuroimaging of the bilingual brain. In J. F. K. Kroll & A. M. de Groot (Eds.), *Handbook of bilingualism: Psycholinguistic approaches.* Oxford, UK: Oxford University Press.

Ackerman, B. P. (1981). Encoding specificity in the recall of pictures and words in children and adults. *Journal of Experimental Child Psychology, 31,* 193–211.

Acosta, M. T., Arcos-Burgos, M., & Muenke, M. (2004). Attention deficit/hyperactivity disorder (ADHD): Complex phenotype, simple genotype? *Genetic Medicine, 6*(1), 1–15.

Acredolo, L. P., Goodwyn, S. W., Horobin, K., & Emmons, Y. (1999). The signs & sounds of early language development. In L. Balter & C. Tamis-LeMonda (Eds.), *Child psychology* (pp. 116–139). New York: Psychology Press.

Adams, C. (2002). Practitioner review: The assessment of language pragmatics. *Journal of Child Psychology and Psychiatry, 43,* 973–987.

Adelman, J. S., & Brown, G. D. A. (2007). Phonographic neighbors, not orthographic neighbors, determine word naming latencies. *Psychonomic Bulletin & Review, 14,* 455–459.

Aguirre, A., Jr. (1980). In search of a paradigm for bilingual education. In R. Padilla (Ed.), *Theory in bilingual education* (pp. 3–13). Ypsilanti: Department of Foreign Languages and Bilingual Studies, Eastern Michigan University.

Aitchison, J. (1998). On discontinuing the continuity-discontinuity debate. In J. R. Hurford, M. Studdert-Kennedy, & C. Knight (Eds.), *Approaches to the evolution of language* (pp. 17–29). Cambridge, UK: Cambridge University Press.

Aitchison, J. (2003). *Words in the mind: An introduction to the mental lexicon* (3rd ed.). Oxford and New York: Basil Blackwell.

Alario, F. X., & Caramazza, A. (2002). The production of determiners: Evidence from French. *Cognition, 82,* 179–223.

Albert, M. S., Moss, M. B., Tanzi, R., & Jones, K. (2001). Preclinical prediction of AD using neuropsychological tests. *Journal of the International Neuropsychological Society, 7*(5), 631–639.

Alegre, M. A., & Gordon, P. (1996). Red rats eater exposes recursion in children's word formation. *Cognition, 60,* 65–82.

Alex the African grey: Science's best known parrot died on September 6th, aged 31. (2007). *The Economist.* Retrieved May 6, 2012, from www.economist.com/node/9828615

Alexander, D., Wetherby, A., & Prizant, B. (1997). The emergence of repair strategies in infants and toddlers. *Seminars in Speech and Language, 18,* 197–212.

Alexander, M. P., Fischette, M. R., & Fischer, R. S. (1989). Crosse aphasias can be mirror image or anomalous. *Brain, 112*, 953–973.

Alexander-Passe, N. (2006). How dyslexic teenagers cope: An investigation of self-esteem, coping and depression. *Dyslexia, 12*(4), 256–275.

Alibali, M. W., Heath, D. C., & Myers, H. J. (2001). Effects of visibility between speaker and listener on gesture production: Some gestures are meant to be seen. *Journal of Memory and Language, 44*, 169–188.

Allemand, M., Zimprich, D., & Hendriks, A. A. J. (2008). Age differences in five personality domains across the life span. *Developmental Psychology, 44*, 758–770.

Allen, S., & Crago, M. (1996). Early passive acquisition in Inuktitut. *Journal of Child Language, 23*, 129–155.

Allen, W. H. (1977). *Spooner: A biography*. London: Virgin Books.

Allum, P. H., & Wheeldon, L. R. (2007). Planning scope in spoken sentence production: The role of grammatical units. *Journal of Experimental Psychology: Learning, Memory, and Cognition, 33*, 791–810.

American Psychiatric Association. (2000). *Diagnostic and statistical manual of mental disorders* (Rev. ed.). Washington, DC: Author.

American Psychiatric Association. (2013). *Diagnostic and statistical manual of mental disorders* (5th ed.). Washington, DC: Author.

Ames, L. B. (1946). The development of the sense of time in the young child. *Journal of Genetic Psychology, 68*, 97–126.

Amunts, K., Schleicher, A., & Zilles, K. (2004). Outstanding language competence and cytoarchitecture in Broca's speech region. *Brain and Language, 89*, 346–353.

Andersen, E. S. (1996). A cross-cultural study of children's register knowledge. In D. I. Slobin, J. Gerhardt, A. Kyratzis, & J. Guo (Eds.), *Social interaction, social context, and language: Essays in honor of Susan Ervin-Tripp* (pp. 125–142). Mahwah, NJ: Erlbaum.

Anderson, L. S., Shin, C., Fullilove, M. T., Scrimshaw, S. C., Fielding, J. E., Normand, J., et al. (2003). The effectiveness of early childhood development programs: A systematic review. *American Journal of Preventive Medicine, 24*(Suppl. 3), 32–46.

Anderson, M. C., Bjork, R. A., & Bjork, E. L. (1994). Remembering can cause forgetting: Retrieval dynamics in long-term memory. *Journal of Experimental Psychology: Learning, Memory, and Cognition, 20*, 1063–1087.

Anderson, R. N., & Smith, B. L. (2005). *Deaths: Leading causes for 2002. National vital statistics reports* (Vol. 53, No. 17). Hyattsville, MD: National Center for Health Statistics.

Andrade, C., Kretschmer, R. R., & Kretschmer, L. W. (1989). Two languages for all children: Expanding to low achievers and the handicapped. In K. E. Muller (Ed.), Languages in the elementary schools (pp. 177–203). New York: The American Forum.

Andrews, S. (1989). Frequency and neighborhood effects on lexical access: Activation or search? *Journal of Experimental Psychology: Learning, Memory and Cognition, 15*, 802–814.

Anetzberger, G. (1987). *The etiology of elder abuse by adult offspring*. Springfield, IL: Charles C Thomas.

Anglin, J. M. (1986). Semantic and conceptual knowledge underlying the child's words. In S. A. Kuczaj & M. D. Barrett (Eds.), *The development of word meaning* (pp. 83–97). New York: Springer-Verlag.

Anthoni, H., Zucchelli, M., Matsson, H., Muller-Myhsok, B., Fransson, I., Schumacher, J., et al. (2007). A locus on 2p12 containing the co-regulated MRPL19 and C2ORF3 genes is associated to dyslexia. *Human Molecular Genetics, 16*, 667–677.

Antinucci, F., & Miller, R. (1976). How children talk about what happened. *Journal of Child Language, 3*, 167–189.

Appell, J., Kertesz, A., & Fisman, M. (1982). A study of language functioning in Alzheimer patients. *Brain and Language, 17,* 73–91.

Applebee, A. N. (1978). *The child's concept of story: Ages two to seventeen.* Chicago: University of Chicago Press.

Ardila, A., & Rosselli, M. (2002). Acalculia and dyscalculia. *Neuropsychology Review, 12,* 179–232.

Ariel, M. (1990). *Accessing noun phrase antecedents.* London: Routledge.

Aristotle, & Barnes, J. (1971). *The complete works of Aristotle.* Princeton, NJ: Princeton University Press.

Arlitt, A. H. (1946). *Psychology of infancy and early childhood* (3rd ed.). McGraw-Hill home economics series (pp. 1–15). New York: McGraw-Hill.

Armstrong, P. W., & Rogers, J. D. (1997, Spring). Basic skills revisited: The effects of foreign language instruction on reading, math and language arts. *Learning Languages,* 20–31.

Arndt, T. L., Stodgell, C. J., & Rodier, P. M. (2005). The teratology of autism. *International Journal of Developmental Neuroscience, 23*(2–3), 189–199.

Arnott, D. W. (1970). *The nominal and verbal systems of Fula.* Oxford, UK: Clarendon Press.

Ashby, J., & Rayner, K. (2007). Literacy development: Insights from research on skilled reading. In D. Dickinson & S. B. Neuman (Eds.), *Handbook of early literacy research* (Vol. 2, pp. 52–63). New York: Guilford Press.

Ashcraft, M. H., & Radvansky, G. A. (2009). *Cognition* (5th ed.). New York: Pearson.

Asperger, H. (1944). Die "Autistischen Psychopathen" im Kindesalter [Autistic psychopathy in childhood] (in German). *Archiv für psychiatrie und nervenkrankheiten, 117,* 76–136.

Aspies for Freedom. (n.d.). Homepage. Retrieved February 7, 2013, from www.aspiesforfreedom.com/

Association for the Advancement of Retired Persons. (2003). National fraud victim study. Retrieved February 7, 2013, from http://assets.aarp.org/rgcenter/econ/fraud-victims-11.pdf

Au, T. K. (1983). Chinese and English counterfactuals: The Sapir-Whorf hypothesis revisited. *Cognition, 15,* 155–187.

Au, T. K. (1984). Counterfactuals: In reply to Alfred Bloom. *Cognition, 17,* 289–302.

Au, T. K., & Markman, E. (1987). Acquiring word meaning via linguistic contrast. *Cognitive Development, 3,* 217–236.

August, D., & Shanahan, T. (2006). Executive summary of developing literacy in second-language learners: Report of the National Literacy Panel on Language-Minority Children and Youth. Mahwah, NJ: Erlbaum.

Austin, J. L. (1962). *How to do things with words.* Cambridge, MA Harvard University Press.

Austin, P., & Sallaback, J. (2011). *Cambridge encylopedia of endangered languages.* Cambridge, UK: Cambridge University Press.

Axia, G., & Baroni, M. R. (1985). Linguistic politeness at different age levels. *Child Development, 56*(4), 918–927.

Azevedo, F. A. C., Carvalho, L. R. B., Grinberg, L. T., Farfel, J. M., Ferretti, R. E. L., Leite, R. E. P., et al. (2009). Equal numbers of neuronal and nonneuronal cells make the human brain an isometrically scaled-up primate brain. *Journal of Comparative Neurology, 513,* 532–541.

Baars, B. J. (1980). On eliciting predictable speech errors in the laboratory. In V. A. Fromkin (Ed.), *Errors in linguistic performance* (pp. 307–318). New York: Academic.

Baars, B. J., Motley, M. T., & MacKay, D. (1975). Output editing for lexical status in artificially elicited slips of the tongue. *Journal of Verbal Learning & Verbal Behavior, 14,* 382–391.

Bächtold, D., Baumüller, M., & Brugger, P. (1998). Stimulus-response compatibility in representational space. *Neuropsychologia, 36,* 731–735.

Bäckman, I., Small, B., & Fratiglioni, L. (2001). Stability of the preclinical episodic memory deficit in Alzheimer's disease. *Brain, 124,* 96–102.

Baddeley, A. D. (1990). *Human memory: Theory and practice.* London: Erlbaum.

Baddeley, A. D. (2003). Working memory: Looking back and looking forward. *Nature Reviews: Neuroscience, 4,* 829–839.

Baddeley, A. D., & Hitch, G. (1974). Working memory. In G. H. Bower (Ed.), *The psychology of learning and motivation: Advances in research and theory* (Vol. 8, pp. 47–89). New York: Academic Press.

Baddeley, A. D., Logie, R., Nimmo-Smith, I., & Brereton, N. (1985). Components of fluent reading. *Journal of Memory & Language, 24,* 119–131.

Bahrick, H. P., Hall, L. K., Goggin, J. P., Bahrick, L. E., & Berger, S. A. (1994). Fifty years of language maintenance and language dominance in bilingual Hispanic immigrants. *Journal of Experimental Psychology: General, 123*(3), 264–283.

Baker, C. (1993). *Foundations of bilingual education and bilingualism*. Clevedon, UK: Multilingual Matters.

Bakker, P., & Mous, M. (1994). Mixed languages: 15 case studies in language intertwining. *Studies in Language and Language Use 13*. Amsterdam: Institute for Functional Research Into Language and Language Use.

Baldwin, D. A., Markman, E. M., Bill, B., Desjardins, N., Irwin, J. M., & Tidball, G. (1996). Infants' reliance on a social criterion for establishing word–object relations. *Child Development, 67,* 3135–3153.

Balota, D. A., & Chumbley, J. I. (1984). Are lexical decisions a good measure of lexical access? The role of word frequency in the neglected decision stage. *Journal of Experimental Psychology: Human Perception and Performance, 10,* 340–357.

Balota, D. A., & Chumbley, J. I. (1985). The locus of word-frequency effects in the pronunciation task: Access and/or production? *Journal of Memory and Language, 24,* 89–106.

Banaji, M. R., & Hardin, C. D. (1996). Automatic stereotyping. *Psychological Science, 7,* 136–141.

Barkley, R. A. (2006). *Attention deficit hyperactivity disorder: A handbook for diagnosis and treatment* (3rd ed.). New York: Guilford Press.

Barnhart, R. K. (1988). *The Barnhart dictionary of etymology*. Bronx, NY: H. W. Wilson Co.

Baron, J., & Strawson, C. (1976). Use of orthographic and word-specific knowledge in reading words aloud. *Journal of Experimental Psychology: Human Perception and Performance, 2,* 386–393.

Baron-Cohen, S. (1990). Autism: A specific cognitive disorder of mind-blindness. *International Review of Psychiatry, 2,* 81–90.

Baron-Cohen, S. (2009). Autism: The emphathizing-systemizing (E-S) theory. *Annals of New York Academy of Science, 1156,* 68–80.

Baron-Cohen, S., Leslie, A. M., & Frith, U. (1985). Does the autistic child have a "theory of mind"? *Cognition, 21,* 37–46.

Barr, L., & Couto, M. (2007). Molecular genetics of reading. In E. Grigorenko & A. J. Naples (Eds.), *Single word reading: Behavioural and biological perspectives* (pp. 255–281). New York: Psychology Press.

Barth, A. E., Denton, C. A., Stuebing, K. K., Fletcher, J. M. Cirino, P. T., Francis, D. J., et al. (2010). A test of the cerebellar hypothesis of dyslexia in adequate and inadequate responders to reading intervention. *Journal of the International Neuropsychological Society, 16,* 526–536.

Barth, H., Starr, A., & Sullivan, J. (2009). Children's mappings of large number words to numerosities. *Cognitive Development, 24,* 248–264.

Bartlett, J. C. (1920, March 30). Some experiments on the reproduction of folk-stories. *Folklore, 31*(1), 30–47.

Bast, J., & Reitsma, P. (1998). Analyzing the development of individual differences in terms of Matthew effects in reading: Results from a Dutch longitudinal study. *Developmental Psychology, 34*(6), 1373–1399.

Bates, E. (1976). Pragmatics and sociolinguistics in child language. In D. Morehead & A. Morehead (Eds.), *Normal and deficient child language* (pp. 247–307). Baltimore: University Park Press.

Bates, E. (1994). Modularity, domain specificity and the development of language. In D. C. Gajdusek, G. M. McKhann, & C. L. Bolis (Eds.), Evolution and neurology of language. *Discussions in Neuroscience, 10*(1–2), 136–149.

Bates, E., Marchman, V., Thal, D., Fenson, L., Dale, P., Reznick, S., Reilly, J., & Hartung, J. (1994). Developmental and stylistic variation in the composition of early vocabulary. *Journal of Child Language, 21*(1), 85–124.

Bates, E., & Silvern, L. (1977). Social adjustment and politeness in preschoolers. *Journal of Communication, 27*(2), 104–111.

Battro, A. M. (2001). *Half a brain is enough: The story of Nico*. Cambridge, England: Cambridge University Press.

Baugh, A. C., & Cable, T. (2002). *A history of the English language*. New York: Psychology Press.

Baugh, J. (2000). *Beyond Ebonics: Linguistic pride and racial prejudice*. New York: Oxford University Press.

Baum, W. M. (2005). *Understanding behaviorism: Behavior, culture & evolution*. London: Blackwell.

Baydar, N., Brooks-Gunn, J., & Furstenberg, F. F., Jr. (1993). Early warning signs of functional illiteracy: Predictors in childhood and adolescence. *Child Development. 64,* 815–829.

Bayles, K. A. (1979). *Communication profiles in a geriatric population*. Unpublished doctoral dissertation, University of Arizona.

Beattie, G., & Butterworth, B. (1979). Contextual probability and word-frequency as determinants of pauses in spontaneous speech. *Language & Speech, 22,* 201–221.

Beaumont, J. G. (1983). Methods for studying cerebral hemispheric function. In A. W. Young (Ed.), *Functions of the right cerebral hemisphere* (pp. 114–146). London: Academic Press.

Beauvillain, C., & Grainger, J. (1987). Accessing interlexical homographs: Some limitations of a language-selective access. *Journal of Memory and Language, 26,* 658–672.

Becker, E. (1973). *The denial of death*. New York: The Free Press.

Becker, J. A. (1986). Bossy and nice request: Children's production and interpretation. *Merrill-Palmer Quarterly, 32,* 393–413.

Behar, D. M., Harmant, C., Manry, J., van Oven, M., Haak, W., Martinez-Cruz, B., et al. (2012). The basque paradigm: Genetic evidence of a maternal continuity in the Franco-Cantabrian region since pre-Neolithic times. *The American Journal of Human Genetics, 90,* 1–8.

Behrend, D., Rosengren, K. S., & Perlmutter, M. A. (1992). Parental scaffolding and children's private speech: Differing sources of cognitive regulation. In R. M. Diaz & L. E. Berk (Eds.), *Private speech: From social interaction to self-regulation*. Hillsdale, NJ: Erlbaum.

Bellugi, U., Lichtenberger, L., Mills, D., Galaburda, A., & Korenberg, J. R. (1999). Bridging cognition, the brain and molecular genetics: Evidence from Williams syndrome. *Trends in Neuroscience, 22,* 197–207.

Bellugi, U., Marks, S., Bihrle, A., & Sabo, H. (1988). Dissociation between language and cognitive functions in Williams syndrome. In D. Bishop & K. Mogford (Eds.), *Language development in exceptional circumstances* (pp. 177–189). London: Churchill Livingstone.

Bendik, M. F. (1992). Reaching the breaking point: Dangers of mistreatment in elder caregiving situations. *Journal of Elder Abuse & Neglect, 4*(3), 39–60.

Benjamin, L. T., Jr. (2007). *A brief history of modern psychology*. Malden, MA: Blackwell Publishing.

Benton, A. L., & Joynt, R. J. (1960). Early descriptions of aphasia. *Archives of Neurology, 3*(2), 205–222.

Ben-Zeev, S. (1977). Mechanisms by which childhood bilingualism affects understanding of language and cognitive structures. In P. A. Hornby (Ed.), *Bilingualism* (pp. 29–55). New York: Academic Press.

Berch, D. B., Foley, E. J., Hill, R. J., & Ryan, P. M. (1999). Extracting parity and magnitude from Arabic numerals: Developmental changes in number processing and mental representation. *Journal of Experimental Child Psychology, 74,* 286–308.

Berk, L. (1986). Relationship of elementary school children's private speech to behavioral accompaniment to task, attention, and task performance. *Developmental Psychology, 22,* 671–680.

Berk, L. (1992). Children's private speech: An overview of theory and the status of research. In R. M. Diaz & L. E. Berk (Eds.), *Private speech: From social interaction to self regulation* (pp. 17–53). Hillsdale, NJ: Erlbaum.

Berk, L., & Spuhl, S. (1995). Maternal interaction, private speech, and task performance in preschool children. *Early Childhood Research Quarterly, 10,* 145–169.

Berko, J. (1958). The child's learning of English morphology. *Word, 14,* 150–177.

Berko, J., & Brown, R. (1960). Psycholinguistic research methods. In P. Mussen (Ed.), *Handbook of research methods in child development* (pp. 517–557). New York: John Wiley.

Berkout, O. V., Young, J. N., & Gross, A. M. (2011). Mean girls and bad boys: Recent research on gender differences in conduct disorder. *Aggression and Violent Behavior, 16*(6), 503–511.

Berlin, B., & Kay, P. (1969). *Basic color terms: Their universality and evolution.* Berkeley: University of California Press.

Berman, R. (1985). The acquisition of Hebrew. In D. Slobin (Ed.), *The crosslinguistic study of language acquisition 1* (pp. 255–371). Hillsdale, NJ: Erlbaum.

Bermejo, V. (1996). Cardinality development and counting. *Developmental Psychology, 32*(2), 263–268.

Bernardo, A. B. I. (1998). Language format and analogical transfer among bilingual problem solvers in the Philippines. *International Journal of Psychology, 33*(1), 33–44.

Berninger, V., & Wolf, B. (2009). *Teaching students with dyslexia and dysgraphia: Lessons from teaching and science.* Baltimore: Paul H. Brookes.

Berrington de González, A., Mahesh, M., Kim, K. P., Bhargavan, M., Lewis, R., Mettler, F., et al. (2009). Projected cancer risks from computed tomographic scans performed in the United States in 2007. *Archives of Internal Medicine, 169,* 2071–2077.

Beukelman, D., Kraft, G., & Freal, J. (1985). Expressive communication disorders in persons with multiple sclerosis: A survey. *Archives of Physical Medicine and Rehabilitation, 66,* 675–677.

Bever, T. G. (1968). Associations to stimulus-response theories of language. *Verbal Behavior and General Behavior Theory,* 418–494.

Bever, T. G. (1970). The cognitive basis for linguistic structures. In J. R. Hayes (Ed.), *Cognition and the development of language* (pp. 279–362). New York: Wiley.

Bialystok, E. (1986). Children's concept of word. *Journal of Psycholinguistic Research, 15*(1), 13–32.

Bialystok, E. (1988). Levels of bilingualism and levels of linguistic awareness. *Developmental Psychology, 24*(4), 560–567.

Bialystok, E. (1999). Cognitive complexity and attentional control in the bilingual mind. *Child Development, 70,* 636–644.

Bialystok, E. (2005). Consequences of bilingualism for cognitive development. In J. F. Kroll & A. M. B. de Groot (Eds.), *Handbook of bilingualism: Psycholinguistic approaches* (pp. 417–432). New York: Oxford University Press.

Bialystok, E., Barac, R., Blaye, A., & Poulin-Dubois, D. (2010). Word mapping and executive functioning in young monolingual and bilingual children. *Journal of Cognition and Development, 11,* 485–508.

Bialystok, E., Craik, F., & Freedman, M. (2007). Bilingualism as a protection against the onset of symptoms of dementia. *Neuropsychologia, 45,* 459–464.

Bialystok, E., Craik, F., Klein, R., & Viswanathan, M. (2004). Bilingualism, aging, and cognitive control: Evidence from the Simon task. *Psychology and Aging, 19*(2), 290–303.

Bialystok, E., Craik, F., & Luk, G. (2008). Cognitive control & lexical access in younger & older bilinguals. *Journal of Experimental Psychology: Learning, Memory, & Cognition, 34,* 859–873.

Bialystok, E., Luk, G., Peets, K. F., & Yang, S. (2010). Receptive vocabulary differences in monolingual and bilingual children. *Bilingualism: Language and Cognition, 13,* 525–531.

Bialystok, E., & Majumder, S. (1998). The relationship between bilingualism and the development of cognitive processes in problem solving. *Applied Psycholinguistics, 19,* 69–85.

Bialystok, E., & Martin, M. M. (2004). Attention and inhibition in bilingual children: Evidence from the dimensional change card sort task. *Developmental Science, 7,* 325–339.

Bickerton, D. (1981). *Roots of language*. Ann Arbor, MI: Karoma.

Bickerton, D. (1983). Creole languages. *Scientific American, 249*(8), 116–122.

Bickerton, D. (1984). The language bioprogram hypothesis. *Behavioral Brain Sciences, 7,* 173–221.

Bickerton, D. (1991). Creole languages. In W. S. Y. Wang (Ed.), *The emergence of language: Development & evolution; Readings from Scientific American Magazine* (pp. 59–69). New York: W. H. Freeman.

Binder, J. R., Westbury, C. F., McKiernan, K. A., Possing, E. T., & Medler, D. A. (2005). Distinct brain systems for processing concrete and abstract concepts. *Journal of Cognitive Neuroscience, 17*(6), 905–917.

Birch, S. A. J., & Bloom, P. (2003). Children are cursed: An asymmetric bias in mental state attribution. *Psychological Science, 14,* 283–286.

Birdsong, D. (2005). Interpreting age effects in second language acquisition. In J. Kroll & A. de Groot (Eds.), *Handbook of bilingualism: Psycholinguistic perspectives* (pp. 109–127). Oxford, UK: Oxford University Press.

Bishop, D. V. M. (1998). Development of the children's communication checklist (CCC): A method for assessing qualitative aspects of communicative impairment in children. *Journal of Child Psychology and Psychiatry, 39,* 879–892.

Bishop, D. V. M. (2002). Cerebellar abnormalities in developmental dyslexia: Cause, correlate or consequence? *Cortex, 38,* 491–498.

Bishop, D. V. M. (2004). Specific language impairment: Diagnostic dilemmas. In L. Verhoeven & H. Van Balkom (Eds.), *Classification of developmental language disorders* (pp. 309–326). Mahwah, NJ: Erlbaum.

Bishop, D. V. M. (2007). Curing dyslexia and attention-deficit hyperactivity disorder by training motor co-ordination: Miracle or myth? *Journal of Paediatrics and Child Health, 43,* 653–655.

Bishop, D. V. M., & Adams, C. (1989). Conversational characteristics of children with semantic-pragmatic disorder II: What features lead to a judgement of inappropriacy? *British Journal of Disorders of Communication, 24,* 241–263.

Bishop, D. V. M., & Edmundson, A. (1987). Specific language impairment as a maturational lag: Evidence from longitudinal data on language and motor development. *Developmental Medicine and Child Neurology, 29,* 442–459.

Bishop, D. V. M., North, T., & Donlan, C. (1995). Genetic basis of specific language impairment: Evidence from a twin study. *Developmental Medicine & Child Neurology, 37*(1), 56–71.

Bivens, J. A., & Berk, L. E. (1990). A longitudinal study of the development of elementary school children's private speech. *Merrill Palmer Quarterly, 36,* 443–463.

Blood, G., Blood, I., Tellis, G., & Gabel, R. (2001). Communication apprehension and self-perceived communication competence in adolescents who stutter. *Journal of Fluency Disorders, 26,* 161–178.

Bloom, A. H. (1981). *The linguistic shaping of thought: A study in the impact of language on thinking in China and the West.* Hillsdale, NJ: Erlbaum.

Bloom, A. H. (1984). Caution—The words you use may affect what you say: A response to Terry Kit-fong Au's "Chinese and English counterfactuals: The Sapir-Whorf hypothesis revisited." *Cognition, 17,* 275–287.

Bloom, L., Lahey, M., Hood, L., Lifter, K., & Fiess, K. (1980). Complex sentences: Acquisition of syntactic connectives and the semantic relations they encode. *Journal of Child Language, 7,* 235–261.

Bloom, P., & German, T. (2000). Two reasons to abandon the false belief task as a test of theory of mind. *Cognition, 77,* B25–B31.

Boas, F. (1911). *Handbook of American Indian languages* (Vol. 1). Bureau of American Ethnology, Bulletin 40. Washington, DC: Government Printing Office (Smithsonian Institution, Bureau of American Ethnology).

Bock, K., & Cutting, J. C. (1992). Regulating mental energy: Performance units in language production. *Journal of Memory and Language, 31,* 99–127.

Bock, K., & Levelt, W. (1994). Language production: Grammatical encoding. In M. A. Gernsbacher (Ed.), *Handbook of psycholinguistics* (pp. 945–984). San Diego: Academic Press.

Bollard, P., Chute P., Popp, A., & Parisier, S. (1999). Specific language growth in young children using the CLARION cochlear implant. *Annals of Otology, Rhinology, & Laryngology, 177,* 119–123.

Boltz, W. G. (1986). Early Chinese writing. *World Archaeology, 17*(3), 420–436.

Bond, M., & Lai, T. (1986). Embarrassment and code-switching into a second language. *The Journal of Social Psychology, 126*(2), 179–186.

Booth, J. R., & Burman, D. D. (2001). Development and disorders of neuro-cognitive systems for oral-language and reading. *Learning Disabilities Quarterly, 24,* 205–215.

Bornstein, M. H. (1985). On the development of color naming in young children: Data and theory. *Brain and Language, 26,* 72–93.

Bornstein, M. H., Kessen, W., & Weiskopf, S. (1976). Color vision and hue categorization in young human infants. *Journal of Experimental Psychology: Human Perception and Performance, 2,* 115–129.

Boroditsky, L. (2001). Does language shape though? English and Mandarin speakers' conceptions of time. *Cognitive psychology, 43*(1), 1–22.

Boroditsky, L. (2008). Do English and Mandarin speakers think differently about time? In B. C. Love, K. McRae, & V. M. Sloutsky (Eds.), *Proceedings of the 30th annual conference of the cognitive science society* (pp. 64–70). Austin, TX: Cognitive Science Society.

Boroditsky, L., Fuhrman, O., & McCormick, K. (2011). Do English and Mandarin speakers think about time differently? *Cognition, 118,* 123–129.

Boroditsky, L., & Gaby, A. (2010). Remembrances of times east: Absolute spatial representations of time in an Australian Aboriginal community. *Psychological Science, 21,* 1635–1639.

Boroditsky, L., Schmidt, L. A., & Phillips, W. (2003). Sex, syntax, and semantics. In D. Gentner & S. Goldin-Meadow (Eds.), *Language in mind: Advances in the study of language and thought* (pp. 61–79). Cambridge, MA: MIT Press.

Bortfeld, H., Leon, S. D., Bloom, J. E., Schober, M. F., & Brennan, S. E. (2001). Disfluency rates in spontaneous speech: Effects of age, relationship, topic, role, and gender. *Language and Speech, 44,* 123–149.

Bosch, L., & Sebastián-Gallés, N. (2003). Simultaneous bilingualism and the perception of a language-specific vowel contrast in the first year of life. *Language and Speech, 46,* 217–243.

Bottéro, J. (1992). *Mesopotamia: Writing, reasoning, and the gods.* Chicago: University of Chicago Press.

Botting, N. (2002). Narrative as a tool for the assessment of linguistic and pragmatic impairments. *Child Language Teaching and Therapy, 18*(1), 1–22.

Bouckaert, R., Lemey, P., Dunn, M., Greenhill, S. J., Alekseyenko, A. V., Drummond, A. J., et al. (2012). Mapping the origins and expansion of the Indo-European language family. *Science, 337,* 957.

Bowers, J. M., & Kennison, S. M. (2011). The role of age of acquisition in bilingual translation. *Journal of Psycholinguistic Research, 40,* 279–289.

Bowers, J. M., Perez-Pouchoulen, M., Edwards, N. S., & McCarthy, M. M. (2013). FOXP2 mediates sex differences in ultrasonic vocalization by rat pups and directs order of maternal retrieval. *The Journal of Neuroscience, 33*(8), 3276–3283.

Bowers, J. S., Davis, C. J., & Hanley, D. A. (2005a). Automatic semantic activation of embedded words: Is there a "hat" in "that"? *Journal of Memory and Language, 52,* 131–143.

Bowers, J. S., Davis, C. J., & Hanley, D. A. (2005b). Interfering neighbours: The impact of novel word learning on the identification of visually similar words. *Cognition, 97,* 45–54.

Bowey, J. A. (1995). Socioeconomic status differences in preschool phonological sensitivity and first-grade reading achievement. *Journal of Educational Psychology, 87*(3), 476–487.

Boysson-Bardies, B. de. (1993). Ontogeny of language-specific syllabic productions. In B. de Boysson-Bardies, S. de Schonen, P. Jusczyk, P. McNeilage, & J. Morton (Eds.), *Developmental neurocognition: Speech and face processing in the first year of life* (pp. 353–363). Dordrecht: Kluwer.

Bradley, J. (1988). Yanyuwa: "Men speak one way, women speak another." *Aboriginal Linguistics, 1,* 126–134.

Bradshaw, J. L., & Nettleton, N. C. (1983). *Human cerebral asymmetry*. Englewood Cliffs, NJ: Prentice Hall.

Brambring, M. (2007). Divergent development of verbal skills in children who are blind or sighted. *Journal of Visual Impairment & Blindness, 101*(12), 749–762.

Bramwell, B. (1899). On "crossed" aphasia and the factors which go to determine whether the "leading" or "driving" speech-centres shall be located in the left or in the right hemisphere of the brain, with notes on a case of "crossed" aphasia (aphasia with right-sided hemiplegia in a left-handed man). *Lancet, 1,* 1473–1479.

Branch, C., Milner, B., & Rasmussen, T. (1964). Intracarotid sodium amytal for the lateralization of cerebral speech dominance. *Journal of Neurosurgery, 21*(5), 399–405.

Branson, R. (2011). *Losing my virginity: How I've survived, had fun, and made a fortune doing business my way.* New York: Crown Business.

Braun, J. M., Kahn, R. S., Froehlich, T., Auinger, P., & Lanphear, B. P. (2006). Exposures to environmental toxicants and attention deficit hyperactivity disorder in U.S. children. *Environmental Health Perspectives, 114*(12), 1904–1909.

Breland, K., & Breland, M. (1961). The misbehavior of organisms. *American Psychologist, 16,* 681–684.

Briganti, A. M., & Cohen, L. B. (2011). Examining the role of social cues in early word learning. *Infant Behavior Development, 34*(1), 211–214.

Brinton, B., Fujiki, M., Loeb, D. F., & Winkler, E. (1986). Development of conversational repair strategies in response to requests for clarification. *Journal of Speech and Hearing Research, 29,* 75–81.

Brkanac, Z., Chapman, N. H., Matsushita, M. M., Chun, L., Nielsen, K., Cochrane, E., et al. (2007). Evaluation of candidate genes for DYX1 and DYX2 in families with dyslexia. *American Journal of Medical Genetics, Part B: Neuropsychiatric Genetics, 144,* 556–560.

Broca, P. (1950). Remarques sur le siège da la faculté du langage articulé, suive d'une observation d'aphémie (J. Kann, Trans.). *Journal of Speech & Hearing Disorders, 15,* 16–20. (Original work published 1861)

Brooks, P., & Tomasello, M. (1999). How children constrain their argument structure constructions. *Language, 75,* 720–738.

Brooks-Gunn, J., Guo, G., & Furstenberg, F. (1993). Who drops out of and who continues beyond high school? *Journal of Research on Adolescence, 3,* 271–294.

Brookshire, R. (1992). *An introduction to neurogenic communication disorders* (4th ed.). St. Louis, MO: Mosby–Year Book.

Brown, G. D. A., & Watson, F. L. (1987). First in, first out: Word learning age and spoken word frequency as predictors of word familiarity and word naming latency. *Memory & Cognition, 15,* 208–216.

Brown, P., & Levinson, S. C. (1987). *Politeness: Some universals in language usage*. Cambridge, UK: Cambridge University Press.

Brown, R. (1958). *Words and things*. New York: Simon & Schuster.

Brown, R. (1970). The first sentences of child and chimpanzee In R. Brown (Ed.), *Psycholinguistics* (pp. 208–231). New York: Free Press.

Brown, R. (1973). *A first language*. Cambridge, MA: Harvard University Press.

Brown, R., & Lenneberg, E. H. (1954). A study in language and cognition. *Journal of Abnormal and Social Psychology, 49,* 454–462.

Brown, R., & McNeill, D. (1966). The "tip of the tongue" phenomenon. *Journal of Verbal Learning and Verbal Behavior, 5,* 325–337.

Brustein, W. I. (1996). *The logic of evil.* New Haven, CT: Yale University Press.

Bryant, G. A., & Barrett, H. C. (2007). Recognizing intentions in infant-directed speech: Evidence for universals. *Psychological Science, 18*(8), 746–751.

Bryant, P. E., Bradley, L., Maclean, M., & Crossland, J. (1989). Nursery rhymes, phonological skills and reading. *Journal of Child Language, 16,* 407–428.

Bryden, M. P. (1988). Does laterality make any difference? Thoughts on the relation between cerebral asymmetry and reading. In D. L. Molfese & S. J. Segalowitz (Eds.), *Brain lateralization in children: Developmental implications* (pp. 509–526). New York: Guilford Press.

Budwig, N. (1990). The linguistic marking of non-prototypical agency: An exploration into children's use of passives. *Linguistics, 28*(6), 1221–1252.

Burke, D. M. (1997). Language, aging, and inhibitory deficits: Evaluation of a theory. *Journal of Gerontology: Psychological Sciences, 52B,* 254–264.

Burke, D. M., Locantore, J. K., Austin, A. A., & Chae, B. (2004). Cherry pit primes Brad Pitt: Homophone priming effects on young and older adults' production of proper names. *Psychological Science, 15,* 164–170.

Burke, D. M., MacKay, D. G., & James, L. E. (2000). Theoretical approaches to language and aging. In T. Perfect & E. Maylor (Eds.), *Models of cognitive aging* (pp. 204–237). Oxford, UK: Oxford University Press.

Burke, D. M., MacKay, D. G., Worthley, J. S., & Wade, E. (1991). On tip-of-the-tongue: What causes word finding failures in young and older adults? *Journal of Memory and Language, 30,* 542–579.

Burke, D. M., White, H., & Diaz, D. L. (1987). Semantic priming in young and old adults: Evidence for age constancy in automatic and attentional processes. *Journal of Experimental Psychology: Human Perception and Performance, 13,* 79–88.

Burns, T. C., Yoshida, K. A., Hill, K., & Werker, J. F. (2007). Bilingual and monolingual infant phonetic development. *Applied Psycholinguistics, 28*(3), 455–474.

Busby, J., & Suddendorf, T. (2005). Recalling yesterday and predicting tomorrow. *Cognitive Development, 20,* 362–372.

Bush, G., Valera, E. M., & Seidman, L. J. (2005). Functional neuroimaging of attention-deficit/hyperactivity disorder: A review and suggested future directions. *Biological Psychiatry, 57,* 1273–1284.

Butler, R. N. (1969). Age-ism: Another form of bigotry. *Gerontologist, 9,* 243–246.

Butterworth, B., Reeve, R., & Reynolds, F. (2011). Using mental representations of space when words are unavailable: Studies of enumeration and arithmetic in Indigenous Australia. *Journal of Cross-Cultural Psychology, 42*(4), 630–638.

Butterworth, B., Reeve, R., Reynolds, F., & Lloyd, D. (2008). Numerical thought with and without words: Evidence from Indigenous Australian children. *PNAS, 105*(35), 13179–13184.

Cafemom.com (n.d.). *20 words that cannot be translated into English (and their close-to meanings).* Retrieved March 14, 2013, from www.cafemom.com/group/33200/forums/read/12655342/20_Words_That_Cannot_Be_Translated_Into_English_and_their_close_to_meanings

Call, J., & Tomasello, M. 1999. A nonverbal theory of mind test: The performance of children and apes. *Child Development, 70,* 381–395.

Camerer, C., Loewenstein, G., & Weber, M. (1989). The curse of knowledge in economic settings: An experimental analysis. *Journal of Political Economy, 97,* 1232–1254.

Canter, A. H. (1951). Direct and indirect measures of psychological deficit in multiple sclerosis. *Journal of General Psychology, 44,* 3–50.

Caporael, L. (1981). The paralanguage of caregiving: Baby talk to the institutionalized aged. *Journal of Personality and Social Psychology, 40,* 876–884.

Caporael, L., Lukaszewski, M. P., & Culbertson, G. H. (1983). Secondary baby talk: Judgments by institutionalized elderly and their caregivers. *Journal of Personality and Social Psychology, 44*(4), 746–754.

Caramazza, A., & Hillis, A. E. (1991). Lexical organization of nouns and verbs in the brain. *Nature, 349*(6312), 788–790.

Caramazza, A., & Miozzo, M. (1997). The relation between syntactic and phonological knowledge in lexical access: Evidence from the "tip-of-the-tongue" phenomenon. *Cognition, 64,* 309–343.

Carey, S. (1978). The child as a word learner. In M. Halle, J. Bresnan, & G. Miller (Eds.), *Linguistic theory and psychological reality* (pp. 264–293). Cambridge, MA: MIT Press.

Carey, S., & Bartlett, E. (1978). Acquiring a single new word. *Proceedings of the Stanford Child Language Conference, 15,* 17–29.

Carli, L. L. (1990). Gender, language, and influence. *Journal of Personality and Social Psychology, 59,* 941–951.

Carlson, S. M., & Meltzoff, A. N. (2008). Bilingual experience and executive functioning in young children. *Developmental Science, 11,* 282–298.

Carreiras, M., Álvarez, C. J., & de Vega, M. (1993). Syllable frequency and visual word recognition in Spanish. *Journal of Memory and Language, 32,* 766–780.

Carreiras, M., Garnham, A., Oakhill, J., & Cain, K. (1996). The use of stereotypical gender information in constructing a mental model: Evidence from English and Spanish. *The Quarterly Journal of Experimental Psychology A: Human Experimental Psychology, 49A*(3), 639–663.

Carroll, J. B. (Ed.). (1956). *Language, thought, and reality: Selected writings of Benjamin Lee Whorf.* Cambridge: Technology Press of Massachusetts Institute of Technology.

Carroll, J. B., & White, M. N. (1973). Word frequency and age of acquisition as determiners of picture-naming latency. *Quarterly Journal of Experimental Psychology, 25,* 85–95.

Casasanto, D., Fotakopoulou, O., & Boroditsky, L. (2010). Space and time in the child's mind: Evidence for a cross-dimensional asymmetry. *Cognitive Science, 34,* 387–405.

Casasola, M., Wilbourn, M. P., & Yang, S. (2006). Can English-learning toddlers acquire and generalize a novel spatial word? *First Language, 26,* 187–205.

Caselli, M. C., Bates E., Casadio, P., Fenson, J., Fenson, L., Sanderl, L., et al. (1995). A cross-linguistic study of early lexical development. *Cognitive Development, 10,* 159–199.

Castles, A., & Coltheart, M. (1993). Varieties of developmental dyslexia. *Cognition, 47,* 149–180.

Cattell, J. M. (1886). The time taken up by cerebral operations. *Mind, 11,* 220–242.

Catts, H. W., Aldolf, S. M., & Weismer, E. S. (2006). Language deficits in poor comprehenders: A case for the simple view of reading. *Journal of Speech, Language, and Hearing Research, 49,* 1278–1293.

Cazden, C. B. (1968). The acquisition of noun and verb inflections. *Child Development, 39,* 433–448.

Centers for Disease Control and Prevention. (n.d.). *Autism spectrum disorders.* Retrieved March 11, 2013, from www.cdc.gov/ncbddd/autism/data.html

Chall, J. S. (1996). *Learning to read: The great debate.* Fort Worth, TX: Harcourt Brace College Publishers.

Chan, T. T., & Bergen, B. (2005). *Writing direction influences spatial cognition.* Proceedings of the twenty-seventh annual conference of the Cognitive Science Society.

Chapey, R. (1994). *Language intervention strategies in adult aphasia* (3rd ed.). Baltimore: Williams and Wilkins.

Chee, M. W., Soon, C. S., Lee, H. L., & Pallier, C. (2004). Left insular activation: A marker for language attainment in bilinguals. *Proceedings of the National Academy of Sciences, USA, 101,* 15265–15270.

Chen, J. Y. (2007). Do Chinese and English speakers think about time differently? Failure of replicating Boroditsky (2001). *Cognition, 104*(2), 427–436.

Cherry-Wilkinson, L. C., & Dollaghan C. (1979). Peer communication in first grade reading groups. *Theory Into Practice, 18*(4), 267–274.

Chesner, C. A., Westgate, J. A., Rose, W. I., Drake, R., & Deino, A. (1991). Eruptive history of Earth's largest quarternary caldera (Toba, Indonesia) clarified. *Geology, 19*, 200–203.

Cheung, H., & Ng, L. (2003). Chinese reading development in some major Chinese societies: An introduction. In C. McBride-Chang & H. C. Chen (Eds.), *Reading development in Chinese children* (pp. 3–17). Westport, CT: Praeger.

Chien, Y.-C., & Wexler, K. (1990). Children's knowledge of locality conditions in binding as evidence for the modularity of syntax and pragmatics. *Language Acquisition: A Journal of Developmental Linguistics, 1*(3), 225–295.

Child, B. (2000). *Boarding school seasons: American Indian families, 1900–1940*. Lincoln: University of Nebraska Press.

Chincotta, D., & Hoosain, R. (1995). Reading rate, articulatory suppression and bilingual digit span. *European Journal of Cognitive Psychology, 7*, 201–211.

Chincotta, D., & Underwood, G. (1996). Mother tongue, language of schooling and bilingual digit span. *British Journal of Psychology, 87*, 193–208.

Choi, S. (2006). Influence of language-specific input on spatial cognition: Categories of containment. *First Language, 26*(2), 207–232.

Choi, S., & Bowerman, M. (1991). Learning to express motion events in English and Korean: The influence of language-specific lexicalization patterns. *Cognition, 41*, 83–121.

Choi, S., & Gopnik, A. (1995). Early acquisition of verbs in Korean. *Journal of Child Language, 22*, 297–531.

Choi, S., McDonough, L., Bowerman, M., & Mandler, J. (1999). Early sensitivity to language-specific spatial categories in English and Korean. *Cognitive Development, 14*, 241–268.

Cholin, J., Dell, G. S., & Levelt, W. J. M. (2011). Planning and articulation in incremental word production: Syllable-frequency effects in English. *Journal of Experimental Psychology: Learning, Memory, and Cognition, 37*, 109–122.

Cholin, J., Levelt, W. J. M., & Schiller, N. O. (2006). Effects of syllable frequency in speech production. *Cognition, 99*, 205–235.

Chomsky, N. (1957). *Syntactic structures*. Cambridge, MA: MIT Press.

Chomsky N. (1959). Review of B. F. Skinner's *Verbal behavior. Language, 35*, 26–58.

Chomsky, N. (1965). *Aspects of the theory of syntax*. Cambridge, MA: MIT Press.

Chomsky, N. (1981). *Lectures on government and binding: The Pisa lectures*. Holland: Foris Publications.

Chomsky, N. (1985). *Knowledge of language: Its nature, origin, and use*. Santa Barbara, CA: Praeger.

Clahsen, H., & Felser, C. (2006). Grammatical processing in language learning. *Applied Psycholinguistics, 27*, 3–42.

Clancy, P. (1985). The acquisition of Japanese. In D. I. Slobin (Ed.), *The crosslinguistic study of language acquisition* (pp. 373–524). Hillsdale, NJ: Erlbaum.

Clark, G. M., Tong, Y. C., & Patrick, J. F. (1990). *Cochlear prostheses*. Melbourne: Churchill Livingstone.

Clark, H. H. (1973). The language-as-fixed-effect fallacy: A critique of language statistics in psychological research. *Journal of Verbal Learning & Verbal Behavior, 12*, 335–359.

Clark, H. H., & Chase, W. G. (1972). On the process of comparing sentences against pictures. *Cognitive Psychology, 3*, 472–517.

Clark, H. H., & Fox Tree, J. E. (2002). Using uh and um in spontaneous speech. *Cognition , 84*, 73–111.

Cleveland Clinic. (n.d.). *Types of dementia*. Retrieved February 7, 2013, from http://my.clevelandclinic .org/disorders/dementia/hic_types_of_dementia.aspx

Clifton, C., Jr., & Duffy, S. (2001). Sentence comprehension: Roles of linguistic structure. *Annual Review of Psychology, 52*, 167–196.

Cohen, A. A., & Harrison, R. P. (1973). Intentionality in the use of hand illustrators in face-to-face communication situations. *Journal of Personality and Social Psychology, 28,* 276–279.

Colcombe, S. J., Erickson, K. I., Raz, N., Webb, A. G., Cohen, N. J., McAuley, E., et al. (2003). Aerobic fitness reduces brain tissue loss in aging humans. *Journal of Gerontology Series A: Biological Sciences and Medical Sciences, 58A,* 176–180.

Colcombe, S. J., Kramer, A. F., Erickson, K. I., Scalf, P., McAuley, E., Cohen, N. J., et al. (2004). Cardiovascular fitness, cortical plasticity, and aging. *Proceedings of the National Academy of Sciences, USA, 101,* 3316–3321.

Collins, A. M., & Loftus, E. F. (1975). A spreading activation theory of semantic processing. *Psychological Review, 82,* 407–428.

Collins, A. M., & Quillian, M. R. (1969). Retrieval time from semantic memory. *Journal of Verbal Learning and Verbal Behavior, 8*(2), 240–247.

Coltheart, M. (1978). Lexical access in simple reading tasks. In G. Underwood (Ed.), *Strategies of information processing.* London: Academic Press.

Coltheart, M., Jonasson, J. T., Davelaar, E., & Besner, D. (1977). Access to the internal lexicon. In S. Dornic (Ed.), *Attention and performance VI.* New York: Academic Press.

Compton, S. A., Flanagan, P., & Gregg, W. (1997). Elder abuse in people with dementia in Northern Ireland: Prevalence and predictors in cases referred to a psychiatry of old age service. *International Journal of Geriatric Psychiatry, 12,* 632–635.

Comrie, B. (1989). *Language universals and linguistic typology* (2nd ed.). Chicago: University of Chicago.

Comrie, B., Matthews, S., & Polinksy, M. (Eds.). (1996). *The atlas of languages: The origin and development of languages throughout the world.* New York: Quarto.

Comrie, B., Matthews, S., & Polinksy, M. (Eds.). (2003). *The atlas of languages: The origin and development of languages throughout the world* (2nd ed.). New York: Quarto.

Conger, D. 2009. Testing, time limits, and English learners: Does age of entry affect how quickly students can learn English? *Social Science Research, 2,* 383–396.

Connine, C. M., & Titone, D. (1996). Phoneme monitoring. *Language and Cognitive Processes, 11*(6), 635–646.

Conti, D. J., & Camras, L. A. (1984). Children's understanding of conversational principles. *Journal of Experimental Child Psychology, 38,* 456–463.

Conti-Ramsden, G., & Botting, N. (1999). Classification of children with specific language impairment: Longitudinal considerations. *Journal of Speech, Language, and Hearing Research, 42*(5), 1195–1204.

Convit, A., de Asis, J., De Leon, M. J., Tarshish, C. Y., De Santi, S., & Rusinek, H. (2000). Atrophy of the medial occipitotemporal, inferior, and middle temporal gyri in non-demented elderly predict decline to Alzheimer's disease. *Neurobiology of Aging, 21,* 19–26.

Conwell, E., & Demuth, K. (2007). Early syntactic productivity: Evidence from dative shift. *Cognition, 103,* 163–179.

Cook, V. J. (1988). *Chomsky's universal grammar: An introduction.* Oxford: Basil Blackwell.

Cook, W. M. (1931). Ability of children in color discrimination. *Child Development, 2,* 303–320.

Coombs, R. H., Chopra, S., Schenk, D., & Yutan, E. (1993). Medical slang and its functions. *Social Science and Medicine: An International Journal, 36*(8), 987–998.

Cooper, R. (1967). The ability of deaf and hearing children to apply morphological rules. *Journal of Speech and Hearing, 10,* 77–86.

Cooper, R. M. (1974). Control of eye fixation by meaning of spoken language: New methodology for real-time investigation of speech perception, memory, and language processing. *Cognitive Psychology, 6,* 84–107.

Cooper, R. P., & Aslin, R. N. (1990). Preference for infant-directed speech in the first month after birth. *Child Development, 61,* 1584–1595.

Cope, N., Harold, D., Hill, G., Moskvina, V., Holmans, P., Owen, M. J., et al. (2005). Strong evidence that KIAA0319 on chromosome 6p is a susceptibility gene for developmental dyslexia. *American Journal of Human Genetics, 76,* 581–591.

Corsaro, W. A. (1979). We're friends, right? Children's use of access rituals in a nursery school. *Language in Society, 8,* 315–336.

Costa, A., Hernández, M., & Sebastián-Gallés, N. (2008). Bilingualism aids conflict resolution: Evidence from the ANT task. *Cognition, 106*(1), 59–86.

Coupland, J., Coupland, N., Giles, H., & Henwood, K. (1988). Accommodating the elderly: Invoking and extending a theory. *Ageing and Society, 11,* 189–208.

Coyne, A., Reichman, W., & Berbig, L. (1993). The relationship between dementia and elder abuse. *American Journal of Psychiatry, 150*(4), 643–646.

Craik, F. I. M. (1968). Two components in free recall. *Journal of Verbal Learning and Verbal Behavior, 7*(6), 996–1004.

Craik, F. I. M., & Byrd, M. (1982). Aging and cognitive deficits: The role of attentional resources. In F. I. M. Craik & S. E. Trehub (Eds.), *Aging and cognitive processes* (pp. 191–211). New York: Plenum.

Cray, E. (2004). *Ramblin man: The life and times of Woody Guthrie.* New York: W. W. Norton & Company.

Crick, F. (2003). Letter to Jacques Monod. Reprinted in *Nature Correspondence, 425,* 15. (Original work published 1961)

Critchley, M. (1975). Specific developmental dyslexia. In E. H. & E. Lenneberg (Eds.), *Foundations of language development* (Vol. 2). New York: Academic Press.

Croft, W. (2002). *Typology & universals* (2nd ed.). Cambridge, UK: Cambridge University Press.

Cromdal, J. (1999). Childhood bilingualism and metalinguistic skills: Analysis and control in young Swedish-English bilinguals. *Applied Psycholinguistics, 20,* 1–20.

Cromer, L. (1902, August 9). Our strange lingo. *Spectator.*

Cromer, R. (1994). A case study of dissociations between language and cognition. In H. Tager-Flusberg (Ed.), *Constraints on language acquisition: Studies of atypical children* (pp. 141–153). Hillsdale, NJ: Erlbaum.

Cross, E. S., & Burke, D. M. (2004). Do alternative names block young and older adults' retrieval of proper names? *Brain and Language, 89,* 174–181.

Crow, J. F. (2002, Winter). Unequal by nature: A geneticist's perspective on human differences. *Dedalus,* 81–88.

Cumming, R. (2007). Language play in the classroom: Encouraging children's intuitive creativity with words through poetry. *Literacy, 41*(2), 93–98.

Cummings, S. M., Kropf, N. P., & DeWeaver, K. L. (2000). Knowledge of and attitudes toward aging among non-elders: Gender and race differences. *Journal of Women & Aging, 12,* 77–91.

Curtain, H., & Dahlberg, C. A. (2004). *Languages and children: Making the match: New languages for young learners, Grades K–8* (3rd ed.). New York: Longman.

Curtiss, S., & de Bode, S. (2003). How normal is grammatical development in the right hemisphere following hemispherectomy? The RI stage and beyond. *Brain and Language, 86,* 193–206.

Curtiss, S., & Schaeffer, J. (2005). Syntactic development in children with hemispherectomy: The I-, D-, and C-systems. *Brain and Language, 94,* 147–166.

Dahlgren, D. J. (1998). Impact of knowledge and age on tip-of-the tongue rates. *Experimental Aging Research, 24,* 139–153.

Damasio, A. R. (1977). Varieties and significance of the alexias. *Archives of Neurology, 34*(6), 325–326.

Damasio, H., Grabowski, T., Frank, R., Galaburda, A. M., & Damasio, A. R. (1994). The return of Phineas Gage: Clues about the brain from the skull of a famous patient. *Science, 264*(5162), 1102–1105.

Damian, M. F., & Dumay, N. (2007). Time pressure and phonological advance planning in spoken production. *Journal of Memory and Language, 57,* 195–209.

Daneman, M., & Carpenter, P. A. (1980). Individual differences in working memory and reading. *Journal of Verbal Learning and Verbal Behavior, 19*(4), 450–466.

Daniels, D. E. (2007). *Recounting the school experiences of adults who stutter: A qualitative analysis.* Unpublished doctoral dissertation, Bowling Green State University, Ohio.

Darwin, C. (1877). Biographical sketch of an infant. *Mind, 2,* 285–294.

Darwin, C. (2007). *The expression of the emotions in man and animals.* New York: Filiquarian. (Original work published 1872)

Davidson, D., Jergovic, D., Imami, Z., & Theodos, V. (1997). Monolingual and bilingual children's use of the mutual exclusivity constraint. *Journal of Child Language, 24,* 3–24.

Dawkins, R. (2004). *The ancestor's tale: A pilgrimage to the dawn of life.* Boston: Houghton Mifflin Company.

Dear John letter. (n.d.). Wiktionary.com. Retrieved March 10, 2013, from http://en.wiktionary.org/wiki/Dear_John_letter

De Beni, R., Borella, E., & Carretti, B. (2007). Reading comprehension in aging: The role of working memory and metacomprehension. *Aging, Neuropsychology, and Cognition, 14,* 189–212.

de Bot, K. (1999). The psycholinguistics of language loss. In G. Extra & L. Verhoeven (Eds.), *Studies on language acquisition* (pp. 345–361). Berlin, Germany: Walter de Gruyter.

DeCasper, A. J., & Fifer, W. P. (1980). Of human bonding: Newborns prefer their mothers' voices. *Science, 208,* 1174–1176.

DeCasper, A. J., & Spence, M. J. (1986). Prenatal maternal speech influences newborns' perception of speech sounds. *Infant Behavior and Development, 9,* 133–150.

Deckers, L., & Kizer, P. (1975). Humor and the incongruity hypothesis. *Journal of Psychology, 90,* 215–218.

DeFrancis, J. (1989). *Visible speech: The diverse oneness of writing systems.* Honolulu: University of Hawaii Press.

DeFries, J. C., Fulker, D. W., & LaBuda, M. C. (1987). Evidence for a genetic etiology in reading disability of twins. *Nature, 329,* 537–539.

de Groot, A. (1992). Determinants of word translation. *Journal of Experimental Psychology: Learning Memory and Cognition, 18,* 1001–1018.

Dehaene, S., Bossini S., & Giraux, P. (1993). The mental representation of parity and numerical magnitude. *Journal of Experimental Psychology: General, 122,* 371–396.

Dehaene, S., Dupoux, E., Mehler, J., Cohen, L., Paulesu, E., Perani, D., et al. (1997). Anatomical variability in the cortical representation of first and second language. *NeuroReport, 8,* 3809–3815.

Dehaene-Lambertz, G., Dehaene, S., & Hertz-Pannier, L. (2002). Functional neuroimaging of speech perception in infants. *Science, 298,* 2013–2015.

Dehaene-Lambertz, G., Hertz-Pannier, L., & Dubois, J. (2006). Nature and nurture in language acquisition: Anatomical and functional brain-imaging studies in infants. *Trends in Neurosciences, 29*(7), 367–373.

Dehaene-Lambertz, G., Montavont, A., Jobert, A., Allirol, L., Dubois, J., Hertz-Pannier, L., et al. (2010). Language or music, mother or Mozart? Structural and environmental influences on infants' language networks. *Brain and Language, 114,* 53–65.

De Houwer, A. (2005). Early bilingual acquisition: Focus on morphosyntax and the separate development hypothesis. In J. Kroll & A. de Groot (Eds.), *The handbook of bilingualism* (pp. 30–48). Oxford, UK: Oxford University Press.

de Kovel, C. G., Franke, B., Hol, F. A., Lebrec, J. J., Maassen, B., Brunner, H., et al. (2008). Confirmation of dyslexia susceptibility loci on chromosomes 1p and 2p, but not 6p in a Dutch sib-pair collection. *American Journal Medical Genetics Part B: Neuropsychiatric Genetics, 147,* 294–300.

Delgado, B., Gómez, J. C., & Sarriá, E. (2009). Private pointing and private speech: Developing parallelisms. In A. Winsler, C. Fernyhough, & I. Montero (Eds.), *Private speech, executive functioning, and the development of verbal self-regulation.* Cambridge, UK Cambridge University Press.

Delgado, P., Guerrero, G., Goggin, J. P., & Ellis, B. B. (1999). Self-assessment of linguistic skills by bilingual Hispanics. *Hispanic Journal of Behavioral Sciences, 21,* 31–46.

Dell, G. S. (1986). A spreading-activation model of retrieval in sentence production. *Psychological Review, 93,* 283–321.

Demuth, K. (1990). Locatives, impersonals and expletives in Sesotho. *The Linguistic Review, 7,* 233–249.

Desalles, J. (2007). *Why we talk.* Oxford, UK: Oxford University Press.

Deutsch, G. K., Dougherty, R. F., Bammer, R., Siok, W. T., Gabrieli, J. D. E., & Wandell, B. (2005). Children's reading performance is correlated with white matter structure measured by diffusion tensor imaging. *Cortex, 41,* 354–363.

Deutscher, G. (2010). *Through the language glass: Why the world looks different in other languages.* New York: Macmillan.

de Villiers, J. G. (1984). *Learning the passive from models: Some contradictory data.* Paper presented at the Ninth Annual BUCLD, Boston.

de Villiers, J. G., & de Villiers, P. A. (1973). A cross sectional study of the acquisition of grammatical morphemes in child speech. *Journal of Psycholinguistic Research, 2,* 267–278.

de Villiers, J. G., & de Villiers, P. A. (1985). The acquisition of English. In D. I. Slobin (Ed.), *The crosslinguistic study of language acquisition.* Hillsdale, NJ: Erlbaum.

Devlin, J. T., Matthews, P. M., & Rushworth, M. F. (2003). Semantic processing in the left inferior prefrontal cortex: A combined functional magnetic resonance imaging and transcranial magnetic stimulation study. *Journal of Cognitive Neurosciences, 15,* 71–84.

DiClemente, R. J., Santelli, J. S., & Crosby, R. A. (Eds.). (2009). *Adolescent health: Understanding and preventing risk behaviors.* San Francisco: Jossey-Bass.

Dijkstra, A., & Van Heuven, W. J. B. (2002). The architecture of the bilingual word recognition system: From identification to decision. *Bilingualism: Language and Cognition, 5,* 175–197.

di Pellegrino, G., Fadiga, L., Fogassi, L., Gallese, V., & Rizzolatti, G. (1992). Understanding motor events: A neurophysiological study. *Experimental Brain Research, 91*(1), 176–180.

Dobel, C., Diesendruck, G., & Bölte, J. (2007). How writing system and age influence spatial representations of actions: A developmental, cross-linguistic study. *Psychological Science, 18*(6), 487–491.

Dore, J. (1974). A pragmatic description of early language development. *Journal of Psycholinguistic Research, 3,* 343–350.

Dore, J. (1975). Holophrases, speech acts, and language universals. *Journal of Child Language, 2,* 21–40.

Dore, W. (2006). *Dyslexia: The miracle cure.* London: John Blake Publishing.

Dowty, D. (1990). Thematic proto-roles and argument selection. *Language, 67*(3), 547–619.

Dromi, E. (1987). *Early lexical development.* London: Cambridge University Press.

Duez, D. (1982). Silent pauses and non-silent pauses in three speech styles. *Language and Speech, 25,* 11–28.

Duffy, S. A., & Keir, J. A. (2004). Violating stereotypes: Eye movement and comprehension processes when text conflicts with world knowledge. *Memory & Cognition, 32*(4), 551–559.

Dumas, L. S. (1999, February). Learning a second language: Exposing your child to a new world of words boosts her brainpower, vocabulary & self-esteem. *Child, 72,* 74, 76–77.

Dunham, F., Dunham, F., & O'Keefe, C. (2000). Two-year-olds' sensitivity to a parent's knowledge state: Mind reading or contextual cues? *British Journal of Developmental Psychology, 18,* 519–532.

Dunn, L. M., & Dunn, D. M. (2007). *Manual: Peabody Picture Vocabulary Test* (4th ed.). Bloomington, MN: Pearson Assessments.

Dupoux, E., & Mehler, J. (1990). Monitoring the lexicon with normal and compressed speech: Frequency effects and the prelexical code. *Journal of Memory and Language, 29*(3), 316–335.

Durgunoğlu, A. Y., & Roediger, H. L., III. (1987). Test differences in accessing bilingual memory. *Journal of Memory & Language, 26,* 377–391.

Durso, F. T., & Shore, W. J. (1991). Partial knowledge of word meanings. *Journal of Experimental Psychology: General, 120*(2), 190–202.

Eaton, D. K., Kann, L., Kinchen, S., Shanklin, S., Ross, J., Hawkins, J., et al. (2010). Youth risk behavior surveillance—United States, 2009. *Morbidity and Mortality Weekly Report Surveillance Summary, 59*(5), 1–142.

Eilers, R. E., Wilson, W. R., & Moore, J. M. (1977). Developmental changes in speech discrimination in infancy. *Journal of Speech and Hearing Research, 20,* 766–780.

Eimas, P. D., & Nygaard, L. C. (1992). Contextual coherence and attention in phoneme monitoring. *Journal of Memory and Language, 31,* 375–395.

Eimas, P. D., Siqueland, E. R., Jusczyk, P. W., & Vigorito, J. (1977). Speech perception in infants. *Science, 171*(968), 303–306.

Ekman, P., & Friesen, W. (1971). Constants across cultures in the face and emotion. *Journal of Personality and Social Psychology, 17*(2), 124–129.

Elfenbein, J. L., Hardin-Jones, M. A., & Davis, J. M. (1994). Oral communication skills of children who are hard of hearing. *Journal of Speech and Hearing Research, 37,* 216–226.

Elgh, E., Domellöf, M., Linder, J., Edström, M., Stenlund, H., & Forsgren, L. (2009). Cognitive function in early Parkinson's disease: A population-based study. *European Journal of Neurology, 16*(12), 1278–1284.

Elia, J., Ambrosini, P., & Rapoport, J. (1999). Treatment of attention-deficit/hyperactivity disorder. *The New England Journal of Medicine, 340,* 780–788.

Ellis, A. W., & Morrison, C. M. (1998). Real age of acquisition effects in lexical retrieval. *Journal of Experimental Psychology: Learning, Memory and Cognition, 24,* 515–523.

Elman, J., Bates, E., Johnson, M., Karmiloff-Smith, A., Parisi, D., & Plunkett, K. (1996). *Rethinking innateness: A connectionist perspective on development.* Cambridge, MA: MIT Press/Bradford Books.

Elowson, A. M., Snowdon, C. T., & Lazaro-Perea, C. (1998). Babbling and social context in infant monkeys: Parallels to human infants. *Trends in Cognitive Science, 2,* 31–37.

Emmorey, K., & McCullough, S. (2009). The bimodal brain: Effects of sign language experience. *Brain and Language, 110*(2), 208–221.

Enard, W., Przeworski, M., Fisher, S. E., Lai, C. S., Wiebe, V., Kitano, T., et al. (2002). Molecular evolution of FOXP2, a gene involved in speech and language. *Nature, 418,* 869–872.

Erickson, K. I., Prakash, R. S., Voss, M. W., Chaddock, L., Hu, L., Morris, K. S., et al. (2009). Aerobic fitness is associated with hippocampal volume in elderly humans. *Hippocampus, 19,* 1030–1039.

Ervin, S. M. (1964). Imitation and structural change in children's language. In E. H. Lenneberg (Ed.), *New directions in the study of language* (pp. 163–189). Cambridge, MA: MIT Press.

Ervin-Tripp, S. M. (1976). Is Sybil there: Some American English directives. *Language in Society, 5,* 25–66.

Ervin-Tripp, S. M. (1977). From conversation to syntax. *Stanford Papers and Reports on Child Language Development, 13,* 1–21.

Ervin-Tripp, S. M., & Gordon, D. P. (1985). The development of requests. In R. L. Schiefelbusch (Ed.), *Communicative competence: Acquisition and intervention.* Beverly Hills, CA: College Hills Press.

Ervin-Tripp, S. M., Guo, J., & Lampert, M. (1990). Politeness and persuasion in children's control acts. *Journal of Pragmatics, 14,* 195–219.

Eskritt, M., Whalen, J., & Lee, K. (2008). Preschoolers can recognize violations of the Gricean Maxims. *British Journal of Developmental Psychology, 26,* 435–443.

Eson, M. E., & Shapiro, A. S. (1982). When "don't" means "do": Pragmatic and cognitive development in understanding an indirect imperative. *First Language, 3,* 83–91.

Espenshade, T. J., & Fu, H. (1997). An analysis of English-language proficiency among U.S. immigrants. *American Sociological Review, 62,* 288–305.

Estes, R. D. (1992). *The behavior guide to African mammals: Including hoofed mammals, carnivores, primates.* Berkeley: University of California Press.

Eurydice. (2005). *How boys and girls in Europe are finding their way with information and communication technology.* Brussels: Author.

Evens, R. G. (1995). Rontgen retrospective: One hundred years of a revolutionary technology. *The Journal of the American Medical Association, 274,* 912–917.

Everett, D. (2005). Cultural constraints on grammar and cognition in Pirahã: Another look at the design features of human language. *Current Anthropology, 46,* 621–646.

Fabbro, F. (1999). *The neurolinguistics of bilingualism.* Hove, UK: Psychology Press.

Fabrigoule, C., Rouch, I., Taberly, A., Letenneur, L., Commenges, D., Mazaux, J. M., et al. (1998). Cognitive process in preclinical phase of dementia. *Brain, 121,* 135–141.

Fagerheim, T., Raeymaekers, P., Tønnessen, F. E., Pedersen, M., Tranebjaerg, L., & Lubs, H. A. (1999). A new gene (DYX3) for dyslexia is located on chromosome 2. *Journal of Medical Genetics, 36,* 664–669.

Farb, P. (1974). *Word play: What happens when people talk.* New York: Knopf.

Faust, M., Kravetz, S., & Babkoff, H. (1993). Hemisphericity and topdown processing of language. *Brain and Language, 53,* 234–259.

Feder, K. P., & Majnemer, A. (2007). Handwriting development, competency, and intervention. *Developmental Medicine & Child Neurology, 49,* 312–317.

Fenson, L., Dale, P. S., Reznick, S., Bates, E., Thal, D., & Pethick, S. (1994). Variability in early communicative development. *Monographs of the Society for Research in Child Development, 59* (Serial No. 242).

Ferguson, C. A., Menn, L., & Stoel-Gammon, C. (Eds.). (1992). *Phonological development: Models, research, implications.* Timonium, MD: York Press.

Fernald, A. (1982). Acoustic determinants of infant preference for "motherese" (Doctoral dissertation, University of Oregon, 1982). *Dissertation Abstracts International, 43,* 545-B.

Fernald, A. (1985). Four-month-old infants prefer to listen to motherese. *Infant Behavior and Development, 8,* 181–195.

Fernald, A. (1989). Intonation and communicative intent in mother's speech to infants: Is the melody the message? *Child Development, 60*(6), 1497–1510.

Fernald, A., Taeschner, T., Dunn, J., Papousek, M., de Boysson-Bardies, B., & Fukui, I. (1989). A cross-language study of prosodic modifications in mothers' and fathers' speech to preverbal infants. *Journal of Child Language, 16,* 477–501.

Fernald, A., Thorpe, K., & Marchman, V. A. (2010). Blue car, red car: Developing efficiency in online interpretation of adjective-noun phrases. *Cognitive Psychology, 60,* 190–217.

Ferreira, F., & Henderson, J. (1990). Use of verb information during syntactic parsing: Evidence from eye tracking and word by word self-paced reading. *Journal of Experimental Psychology: Learning, Memory, & Cognition, 16,* 555–568.

Ferreira, F., & Henderson, J. (1991). Recovery from misanalyses of garden-path sentences. *Journal of Memory & Language, 30,* 725–745.

Ferrer, E., Shaywitz, B. A., Holahan, J. M., Marchione, K., & Shaywitz, S. E. (2010). Uncoupling of reading and IQ over time: Empirical evidence for a definition of dyslexia. *Psychological Science, 21*(1), 93–101.

Ferrier, S., Dunham, P., & Dunham, F. (2000). The confused robot: Two-year-olds' responses to breakdowns in conversation. *Social Development, 9,* 337–347.

Field, L. L., & Kaplan, B. J. (1998). Absence of linkage of phonological coding dyslexia to chromosome 6p23-p21.3 in a large family data set. *American Journal of Human Genetics, 63,* 1448–1456.

Fischer, S. R. (2004). *A history of writing. Globalities.* London: Reaktion Books.

Fisher, C. (1994). Structure and meaning in the verb lexicon: Input for a syntax-aided verb learning procedure. *Language and Cognitive Processes, 9,* 473–517.

Fisher, S. E. (2006). Tangled webs: Tracing the connections between genes and cognition. *Cognition, 101,* 270–297.

Fisher, S. E., & DeFries, J. C. (2002). Developmental dyslexia: Genetic dissection of a complex cognitive trait. *Nature Reviews Neuroscience, 3,* 767–780.

Fisher, S. E., & Scharff, C. (2009). FOXP2 as a molecular window into speech and language. *Trends in Genetics, 25*(4), 166–177.

Fishman, J. (1980). Bilingualism and biculturism as individual and as societal phenomena. *Journal of Multilingual and Multicultural Development, 1*(1), 3–15.

Flege, J., MacKay, I. R. A., & Piske, T. (2002). Assessing bilingual dominance. *Applied Psycholinguistics, 23*(4), 567–598.

Flege, J., Yeni-Komshian, G., & Liu, S. (1999). Age constraints on second language learning. *Journal of Memory and Language, 41,* 78–104.

Fodor, J. A. (1983). *Modularity of mind: An essay on faculty psychology.* Cambridge, MA: MIT Press.

Fogelin, R. (1991). Review of *Studies in the way of words* by H. P. Grice. *Journal of Philosophy, 88,* 213–219.

Forbes, K. E., Shanks, M. F., & Venneri, A. (2004). The evolution of dysgraphia in Alzheimer's disease. *Brain Research Bulletin, 63,* 19–24.

Forster, K. I. (1976). Accessing the mental lexicon. In R. J. Wales & E. Walker (Eds.), *New approaches to language mechanisms*. Amsterdam: North Holland.

Forster, K. I. (1979). Levels of processing and the structure of the language processor. In W. Cooper & E. Walker (Eds.), *Sentence processing psycholinguistics studies presented to Merrill Garrett.* Hillsdale, NJ: Erlbaum.

Foss, D. J. (1969). Decision processes during sentence comprehension: Effects of lexical item difficulty and position upon decision times. *Journal of Verbal Learning and Verbal Behavior, 8,* 457–462.

Foss, D. J., & Blank, M. A. (1980). Identifying the speech codes. *Cognitive Psychology, 12,* 1–31.

Foster, E. M., & Jones, D. E. (2005). The high costs of aggression: Public expenditures resulting from conduct disorder. *American Journal of Public Health, 95,* 1767–1772.

Fouts, R. S., Fouts, D. H., & Van Cantfort, T. E. (1989). The infant Loulis learns signs from cross-fostered chimpanzees. In R. Gardner, B. T. Gardner, & T. Van Cantfort (Eds.), *Teaching sign language to chimpanzees* (pp. 280–292). Albany: SUNY Press.

Fowler, A. (1990). Language abilities in children with Down syndrome: Evidence for a specific syntactic delay. In D. Cicchetti & M. Beeghly (Eds.), *Children with Down syndrome: A developmental perspective*. New York: Cambridge University Press.

Fowler, K. S., Saling, M. M., Conway, E. L., Semple, J. M., & Louis, W. J. (2002). Paired associate performance in the early detection of DAT. *Journal of the International Neuropsychological Society, 8,* 58–71.

Fox, D., & Grodzinsky, Y. (1998). Children's passive: A view from the by-phrase. *Linguistic Inquiry, 29,* 311–332.

Fox, N. C., Warrington, E. K., Seiffer, A. L., Agnew, S. K., & Rossor, M. N. (1998). Presymptomatic cognitive deficits in individuals at risk of familial Alzheimer's disease: A longitudinal prospective study. *Brain, 121,* 1631–1639.

Fox Tree, J. E. (1995). The effects of false starts and repetitions on the processing of subsequent words in spontaneous speech. *Journal of Memory and Language, 34,* 709–738.

Fraisse, P. (1982). The adaptation of the child to time. In W. J. Friedman (Ed.), *The developmental psychology of time* (pp. 113–140). New York: Academic Press.

Francis, W. N., & Kučera, H. (1982). *Frequency analysis of English usage: Lexicon and grammar*. Boston: Houghton Mifflin.

Francis, W. S. (1999). Analogical transfer of problem solutions within and between languages in Spanish-English bilinguals. *Journal of Memory and Language, 40*(3), 301–329.

Frank, M., Everett, D., Fedorenko, E., & Gibson, E. (2008). Number as a cognitive technology: Evidence from Pirahã language and cognition. *Cognition, 108,* 819–824.

Franklin, A., Catherwood, D., Alvarez, J., & Axelsson, E. (2010). Hemispheric asymmetries in categorical perception of orientation in infants and adults. *Neuropsychologia, 48,* 2648–2657.

Franklin, A., Drivonikou, G. V., Clifford, A., Kay, P., Regier, T., & Davies, I. R. L. (2008). Lateralization of categorical perception of color changes with color term acquisition. *Proceedings of the National Academy of Sciences, 105*(47), 18221–18225.

Fraser, B. (1976). *The verb-particle combination in English*. New York: Academic Press.

Frazier, L. (1979). *On comprehending sentences: Syntactic parsing strategies.* Unpublished doctoral dissertation, University of Connecticut, Storrs-Mansfield.

Frazier, L., & Clifton, C., Jr. (1996). *Construal.* Cambridge, MA: MIT Press.

Frazier, L., & Fodor, J. D. (1978). The sausage machine: A new two stage parsing model. *Cognition, 6,* 1–34.

Frazier, L., Pacht, J. M., & Rayner, K. (1999). Taking on semantic commitments II: Collective versus distributive readings. *Cognition, 70,* 87–104.

Frazier, L., & Rayner, K. (1982). Making and correcting errors during sentence comprehension: Eye movements in the analysis of structurally ambiguous sentences. *Cognitive Psychology, 14,* 178–210.

Frazier, L., & Rayner, K. (1987). Resolution of syntactic category ambiguities: Eye movements in parsing lexically ambiguous sentences. *Journal of Memory and Language, 26,* 505–526.

Frazier, L., & Rayner, K. (1990). Taking on semantic commitments: Processing multiple meanings versus multiple senses. *Journal of Memory and Language, 29,* 181–200.

Freud, S. (1971). *The psychopathology of everyday life*. New York: W. W. Norton & Co. (Original work published 1901)

Frick, P. J., Lahey, B. B., Christ, M. A. G., Loeber, R., & Green, S. M. (1991). History of childhood behavior problems in biological relatives of boys with attention-deficit hyperactivity disorder and conduct disorder. *Journal of Clinical Child Psychology, 20,* 445–451.

Fridriksson, J., Kjartansson, O., Morgan, P. S., Hjaltason, H., & Magnusdottir, S. (2010). Impaired speech repetition & left parietal lobe damage. *The Journal of Neuroscience, 30*(33), 11057–11061.

Friedenberg, J. D., & Silverman, G. (2005). *Cognitive science: An introduction to the study of mind.* Thousand Oaks, CA: Sage.

Friedman, W. J. (1991). The development of children's memory for the time of past events. *Child Development, 62,* 139–155.

Friedman, W. J. (2000). The development of children's knowledge of the times of future events. *Child Development, 71,* 913–932.

Friedman, W. J., Gardner, A. G., & Zubin, N. R. E. (1995). Childrens' comparisons of the recency of two events from the past year. *Child Development, 66,* 970–983.

Friel, B. M., & Kennison, S. M. (2001). Identifying German-English cognates, false cognates, and noncognates: Methodological issues and descriptive norms. *Bilingualism: Language and Cognition, 4,* 249–274.

Friend, K. B., Rabin, B. M., Groninger, L., Deluty, R. H., Bever, C., & Grattan, L. (1999). Language functions in patients with multiple sclerosis. *The Clinical Neuropsychologist, 13,* 78–94.

Frings, M., Gaertner, K., Buderath, P., Gerwig, M., Christiansen, H., Schoch, B., et al. (2010). Timing of conditioned eyeblink responses is impaired in children with attention-deficit/hyperactivity disorder. *Experimental Brain Research, 201*(2), 167–176.

Fromkin, V. A. (1971). The non-anomalous nature of anomalous utterances. *Language, 47*(1), 27–52.

Fromkin, V. A. (1988). Grammatical aspects of speech errors. In F. J. Newmeyer (Ed.), *Linguistics: The Cambridge survey: Vol.: Linguistic theory: Extensions and implications* (pp. 117–138). Cambridge, UK: Cambridge University Press.

Fromkin, V. A., Rodman, R., & Hyams, N. (2013). *An introduction to language* (10th ed.). Belmont, CA: Wadsworth.

Frye, D., Braisby, N., Lowe, J., Maroudas, C., & Nicholls, J. (1989). Young children's understanding of counting and cardinality. *Child Development, 60,* 1158–1171.

Fuhrman, O., & Boroditsky, L. (2010). Cross-cultural differences in mental representations of time: Evidence from an implicit non-linguistic task. *Cognitive Science, 34,* 1430–1451.

Fukumine, E., & Kennison, S. M. (2011). *The role of knowledge acquisition in analogical transfer in Spanish-English bilinguals.* Unpublished data.

Galaburda, A. M., Schrott, L. S., Sherman, G. F., Rosen, G. D., & Denenberg, V. H. (1996). Animal models of developmental dyslexia. In C. Chase, G. D. Rosen, & G. F. Sherman (Eds.), *Developmental dyslexia: Neural, cognitive, and genetic mechanisms* (pp. 1–14). Timonium, MD: York Press.

Galaburda, A. M., Sherman, G. F., Rosen, G. D., Aboitiz, F., & Geschwind, N. (1985). Developmental dyslexia: Four consecutive patients with cortical anomalies. *Annals of Neurology, 18*(2), 222–233.

Galambos, S., & Goldin-Meadow, S. (1990). The effects of learning two languages on levels of metalinguistic awareness. *Cognition, 34,* 1–56.

Galambos, S. J., & Hakuta, K. (1988). Subject-specific and task-specific characteristics of metalinguistic awareness in bilingual children. *Applied Psycholinguistics, 9,* 141–162.

Gallagher, R., Jens, K., & O'Donnell, K. (1983). The relationship between physical status and positive affect in handicapped infants. *Infant Behavior and Development, 5,* 73–77.

Gallagher, T. (1977). Revision behaviors in the speech of normal children developing language. *Journal of Speech and Hearing Research, 20,* 303–318.

Gallahorn, G. E. (1971). The use of taboo words by psychiatric ward personnel. *Psychiatry, 34,* 309–321.

Gannon, P. J., Holloway, R. L., Broadfield, D. C., & Braun, A. R. (1998, January 9). Asymmetry of chimpanzee planum temporale: Humanlike pattern of Wernicke's brain language area homolog. *Science, 279,* 220–222.

García, E. (2008). Bilingual education in the United States. In J. Altarriba and R. Heredia (Eds.), *An introduction to bilingualism: Principles and practices* (pp. 321–343). Mahwah, NJ: Erlbaum.

Garcia, J. L., & Kosberg, J. I. (1992). Understanding anger: Implications for formal and informal caregivers. *Journal of Elder Abuse and Neglect, 4*(4), 87–99.

Gardner, B. T., & Gardner, R. A. (1975). Evidence for sentence constituents in the early utterances of child & chimpanzee. *Journal of Experimental Psychology: General, 104,* 244–267.

Garland, A. F., Hough, R. L., McCabe, K. M., Yeh, M., Wood, P. A., & Aarons, G. A. (2001). Prevalence of psychiatric disorders in youths across five sectors of care. *Journal of the American Academy of Child and Adolescent Psychiatry, 40,* 409–418.

Garnsey, S. M., Pearlmutter, N. J., Myers, E., & Lotocky, M. (1997). The contributions of verb bias and plausibility to the comprehension of temporarily ambiguous sentences. *Journal of Memory & Language, 37,* 58–93.

Garrett, M. F. (1975). The analysis of sentence production. In G. H. Bower (Ed.), *The psychology of learning and motivation: Advances in research and theory* (Vol. 9, pp. 133–177). New York: Academic Press.

Garrett, M. F. (1980). The limits of accommodation. In V. Fromkin (Ed.), *Errors in linguistic performance* (pp. 263–271). New York: Academic Press.

Garrity, L. I. (1977). Electromyography: A review of the current status of subvocal speech research. *Memory and Cognition, 5,* 615–622.

Garvey, C. (1975). Requests and responses in children's speech. *Journal of Child Language, 2,* 41–63.

Gazzaniga, M. S. (2005). Forty-five years of split-brain research & still going strong. *Nature Reviews Neuroscience, 6*(8), 653–659.

Geary, D. C. (1993). Mathematical disabilities: Cognitive, neuropsychological, and genetic components. *Psychological Bulletin, 114,* 345–362.

Geary, D. C. (2006). Dyscalculia at an early age: Characteristics and potential influence on socio-emotional development. In R. E. Tremblay & R. D. Peters (Eds.), *Encyclopaedia on early childhood development* [online] (pp. 1–4). Montreal, Quebec: Center of Excellence for Early Childhood Development.

Geary, D. C., Bow-Thomas, C. C., & Yao, Y. (1992). Counting knowledge and skill in cognitive addition: A comparison of normal and mathematically disabled children. *Journal of Experimental Child Psychology, 54,* 372–391.

Geary, D. C., Hamson, C. O., & Hoard, M. K. (2000). Numerical and arithmetical cognition: A longitudinal study of process and concept deficits in children with learning disability. *Journal of Experimental Child Psychology, 77,* 236–263.

Gellis, Z. D., Sherman, S., & Lawrence, F. (2003). First year graduate social work students' knowledge of and attitude toward older adults. *Educational Gerontology, 29,* 1–16.

Gelman, R., & Gallistel, C. R. (1978). *The child's understanding of number*. Cambridge, MA: Harvard University Press.

Gelman, R., & Meck, E. (1983). Preschooler's counting: Principles before skill. *Cognition, 13,* 343–359.

Genesee, F. (1985). Second language learning through immersion: A review of U.S. programs. *Review of Educational Research, 55,* 541–561.

Genesee, F. (1987). *Learning through two languages: Studies of immersion and bilingual education.* Cambridge, MA: Newbury House.

Genesee, F. (1989). Early bilingual development: One language or two? *Journal of Child Language, 16,* 161–179.

Genesee, F. (2003). Rethinking bilingual acquisition. In J. M. Dewaele, A. Housen, & L. Wei (Eds.), *Bilingualism: Beyond basic principles* (pp. 204–228). Clevedon, UK: Multilingual Matters Ltd.

Genesee, F., & Nicoladis, E. (2007). Bilingual first language acquisition. In E. Hoff & M. Shatz (Eds.), *Blackwell handbook of language development* (pp. 324–342). Malden, MA: Blackwell.

Gerber, W., & Routh, D. (1975). Humor response as related to violation of expectancies and to stimulus intensity in a weight-judgment task. *Perceptual and Motor Skills, 41,* 673–674.

Gertner, B. L., Rice, M. L., & Hadley, P. A. (1994). Influence of communicative competence on peer preferences in a preschool classroom. *Journal of Speech and Hearing Research, 37,* 913–923.

Geschwind, N. (1965). Disconnexion syndromes in animals and man. *Brain, 88,* 237–294, 585–644.

Gesell, A. L. (1925). *The mental growth of a preschool child: A psychological outline of normal development from birth to the sixth year, including a system of developmental diagnosis.* New York: Macmillan.

Gesell, A. L. (1940). *The first five years of life: A guide to the study of the preschool child.* New York: Harper & Brothers.

Gesell, A. L. (1948). *Studies in child development.* Westport, CT: Greenwood.

Gibbs, R. (1979). Contextual effects in understanding indirect requests. *Discourse Processes, 2,* 1–10.

Gibson, C. J., & Gruen, J. R. (2008). The human lexinome: Genes of language and reading. *Journal of Communication Disorders, 41*(5), 409–420.

Gibson, E. (1998). Linguistic complexity: Locality of syntactic dependencies. *Cognition, 68,* 1–76.

Gibson, E. (2000). The dependency locality theory: A distance-based theory of linguistic complexity. In Y. Miyashita, A. Marantz, & W. O'Neil (Eds.), *Image, language, brain* (pp. 95–126). Cambridge, MA: MIT Press.

Gick, M. L., & Holyoak, K. J. (1980). Analogical problem solving. *Cognitive Psychology, 12*(3), 306–355.

Gick, M. L., & Holyoak, K. J. (1983). Schema induction and analogical transfer. *Cognitive Psychology, 15*(1), 1–38.

Gilbert, A. L., Regier, T., Kay, P., & Ivry, R. B. (2006). Whorf hypothesis is supported in the right visual field but not the left. *Proceedings of the National Academy of Sciences, 103,* 489–494.

Gilbert, A. L., Regier, T., Kay, P., & Ivry, R. B. (2008). Support for lateralization of the Whorf effect beyond the realm of color discrimination. *Brain and Language, 105,* 91–98.

Gilbert, B., Belleville, S., Bherer, L., & Chouinard, S. (2005). Study of verbal working memory in patients with Parkinson's disease. *Neuropsychology, 19,* 106–114.

Gilhooly, K. J., & Gilhooly, M. L. (1979). Age-of-acquisition effects in lexical and episodic memory tasks. *Memory & Cognition, 7,* 214–223.

Gingras, R. (1974). Problems in the description of Spanish-English intra-sentential code-switching. In G. Bills (Ed.), *Southwest areal linguistics*. San Diego, CA: San Diego State University, Institute for Cultural Pluralism.

Givens, J., Frederick, M., Silverman, L., Anderson, S., Senville, J., Silver, M., et al. (2009). Personality traits of centenarians' offspring. *Journal of the American Geriatric Society, 57,* 683–685.

Gleitman, L., & Gleitman, J. (1992). A picture is worth a thousand words, but that's the problem: The role of syntax in vocabulary acquisition. *Current Directions in Psychological Science, 1,* 1–5.

Glennen, S. L. (2007). The human lexinome: Genes of language & reading. Predicting language outcomes for internationally adopted children. *Journal of Speech, Language, & Hearing Research, 50*(2), 529–548.

Glezerman, T. B. (1983). *Mozgovye disfunktsii u detei* [Minimal brain dysfunctions in children]. Moscow: Nauka.

Glosser, G., Grugan, P. K., & Friedman, R. B. (1999). Comparison of reading and spelling in patients with probable Alzheimer's disease. *Neuropsychology, 13,* 350–358.

Glucksberg, S., & Danks, J. H. (1975). *Experimental psycholinguistics: An introduction*. Hillsdale, NJ: Erlbaum.

Gold, B. T., & Buckner, R. L. (2002). Common prefrontal regions coactivate with dissociable posterior regions during controlled semantic & phonological tasks. *Neuron, 35,* 803–812.

Goldfield, B. A., & Reznick, J. S. (1990). Early lexical acquisition: Rate, content, & the spurt. *Journal of Child Language, 17,* 171–183.

Goldin-Meadow, S., & Butcher, C. (2003). Pointing toward two-word speech in young children. In S. Kita (Ed.), *Pointing: Where language, culture, and cognition meet* (pp. 85–107). Mahwah, NJ: Erlbaum.

Goldin-Meadow, S., Goodrich, W., Sauer, E., & Iverson, J. (2007). Young children use their hands to tell their mothers what to say. *Developmental Science, 10,* 778–785.

Goldin-Meadow, S., & Sandhofer, C. M. (1999). Gesture conveys substantive information about a child's thoughts to ordinary listeners. *Developmental Science, 2,* 67–74.

Goldin-Meadow, S., Seligman, M. E. P., & Gelman, R. (1976). Language in the two-year-old: Receptive and productive stages. *Cognition, 4*(2), 189–202.

Goldman-Eisler, F. (1958a). The predictability of words in context and the length of pauses in speech. *Language and Speech, 1,* 226–231.

Goldman-Eisler, F. (1958b). Speech production and the predictability of words in context. *Quarterly Journal of Experimental Psychology, 10,* 96–106.

Goldman-Eisler, F. (1968). *Psycholinguistics: Experiments in spontaneous speech*. London: Academic Press.

Goldstein, E. B. (2007). *Cognitive psychology* (2nd ed.). Belmont, CA: Wadsworth.

Golinkoff, R. M. (1986). "I beg your pardon?" The preverbal negotiation of failed messages. *Journal of Child Language, 13,* 455–476.

Golinkoff, R. M., Hirsh-Pasek, K., Bailey, L. M., & Wenger, N. R. (1992). Young children & adults use lexical principles to learn new nouns. *Developmental Psychology, 28,* 99–108.

Golinkoff, R. M., Hirsh-Pasek, K., Cauley, K., & Gordon, L. (1987). The eyes have it: Lexical & syntactic comprehension in a new paradigm. *Journal of Child Language, 14,* 23–46.

Gollan, T. H., & Acenas, L. A. (2004). What is ToT? Cognate and translation equivalent effects on tip-of-the-tongue states in Spanish-English and Tagalog-English bilinguals. *Journal of Experimental Psychology: Learning, Memory, and Cognition, 30,* 246–269.

Gollan, T. H., & Silverberg, N. B. (2001). Tip-of-the-tongue states in Hebrew-English bilinguals. *Bilingualism: Language and Cognition, 4,* 63–83.

Goodall, J. (1986). *The chimpanzees of Gombe: Patterns of behavior*. Cambridge, MA: Harvard University Press.

Goodglass, H., & Kaplan, E. (1972). *The assessment of aphasia and related disorders*. Philadelphia: Lea & Febiger.

Goodwin, C. J. (2009). *Research in psychology: Methods and design.* Hoboken, NJ: John Wiley & Sons.

Goodwyn, S., Acredolo, L., & Brown, C. A. (2000). Impact of symbolic gesturing on early language development. *Journal of Nonverbal Behavior, 24,* 81–103.

Gopnik, M. (1990). Feature-blind grammar and dysphasia. *Nature, 344,* 715.

Gopnik, M., & Crago, M. B. (1991). Familial aggregation of a developmental language disorder. *Cognition, 39,* 1–50.

Gordon, P. (2004). Numerical cognition without words: Evidence from Amazonia. *Science, 306,* 496–499.

Gordon-Salant, S., & Fitzgibbons, P. J. (1997). Selected cognitive factors and speech recognition performance among young and elderly listeners. *Journal of Speech Language & Hearing Research, 40,* 423–431.

The Gorilla Foundation. (n.d.). *The gorilla foundation.* Retrieved May 6, 2012, from www.koko.org/

Gottfried, A. W., Gottfried, A. E., Bathurst, K., Guerin, D. W., & Parramore, M. M. (2003). Socioeconomic status in children's development and family environment: Infancy through adolescence. In M. H. Bornstein & R. H. Bradley (Eds.), *Socioeconomic status, parenting and child development* (pp. 189–207). Mahwah, NJ: Erlbaum.

Gough, P. B. (1972). One second of reading. In J. F. Kavanagh & I. G. Mattingly (Eds.), *Language by ear and by eye* (pp. 331–358). Cambridge, MA: MIT Press.

Grainger, J., O'Regan, J. K., Jacobs, A. M., & Segui, J. (1989). On the role of competing word units in visual word recognition: The neighborhood frequency effect. *Perception and Psychophysics, 45,* 189–195.

Grainger, J., & Segui, J. (1990). Neighborhood frequency effects in visual word recognition: A comparison of lexical decision and masked identification latencies. *Perception and Psychophysics, 47,* 191–198.

Grant, C., & Gomez, M. L. (2001). *Campus and classroom: Making schooling multicultural* (2nd ed.). Upper Saddle River, NJ: Merrill Prentice Hall.

Gray, W. S., & Sharp, Z. (1930). *Dick and Jane.* Chicago: Scott Foresman and Company.

Greenberg, J. (1966). *Universals of language*. Cambridge, MA: MIT Press.

Greenberg, J. (1974). *Language typology: A historical & analytic overview.* Amsterdam, The Netherlands: Mouton.

Greenberg, J., Pyszczynski, T., & Solomon, S. (1986). The causes and consequences of a need for self-esteem: A terror management theory. In R. F. Baumeister (Ed.), *Public self and private self* (pp. 189–212). New York: Springer-Verlag.

Greenfield, P. M., & Savage-Rumbaugh, E. S. (1991). Imitation, grammatical development, & the invention of protogrammar by an ape. In N. Krasnegor, D. M. Rumbaugh, M. Studdert-Kennedy, & R. L. Schiefelbusch (Eds.), *Biological & behavioral determinants of language development* (pp. 235–258). Hillsdale, NJ: Erlbaum.

Greenfield, P. M., & Smith, J. H. (1976). *The structure of communication in early language development*. New York: Academic Press.

Gregoire, J., & Van der Linden, M. (1997). Effects of age on forward and backward digit spans. *Aging, Neuropsychology, and Cognition, 4*(2), 140–149.

Grice, E. (2006). Cry of an enfant sauvage. *London Telegraph*. Retrieved May 6, 2012, from www.telegraph.co.uk/culture/tvandradio/3653890/Cry-of-an-enfant-sauvage.html

Grice, H. P. (1957). Meaning. *Philosophical Review, 66*(3), 377–388.

Grice, P. (1975). Logic and conversation. In D. Davidson & G. Harman (Eds.), *The logic of grammar* (pp. 64–75). Encino, CA: Dickenson.

Grice, P. (1989). *Studies in the way of words*. Cambridge, MA: Harvard University Press.

Grieser, D. L., & Kuhl, P. K. (1988). Maternal speech to infants in a tonal language: Support for universal prosodic feature in motherese. *Developmental Psychology, 24,* 14–20.

Griffin, Z. M. (2003). A reversed word length effect in coordinating the preparation and articulation of words in speaking. *Psychonomic Bulletin and Review, 10*(3), 603–609.

Griffin, Z. M., & Bock, K. (2000). What the eyes say about speaking. *Psychological Science, 11,* 274–279.

Grigorenko, E. L. (2001). Developmental dyslexia: An update on genes, brains, and environments. *Journal of Child Psychology and Psychiatry and Allied Disciplines, 42*(1), 91–125.

Grigorenko, E. L. (2005). A conservative meta-analysis of linkage and linkage-association studies of developmental dyslexia. *Scientific Studies in Reading, 9,* 285–316.

Grigorenko, E. L., Naples, A., Chang, J., Romano, C., Ngorosho, D., Kungulilo, S., et al. (2007). Back to Africa: Tracing dyslexia genes in East Africa. *Reading and Writing, 20,* 27–49.

Grigorenko, E. L., Ngorosho, D., Jukes, M., & Bundy, D. (2006). Reading in able and disabled readers from around the word: Same or different? An illustration from a study of reading-related processes in a Swahili sample of siblings. *Journal of Research in Reading, 29,* 104–123.

Grimshaw, J., & Rosen, S. T. (1990). Obeying the binding theory. In L. Frazier & J. de Villiers (Eds.), *Language processing and acquisition*. Boston: Kluwer Academic.

Grober, E., & Kawas, C. (1997). Learning and retention in preclinical and early Alzheimer's disease. *Psychology of Aging, 12,* 183–188.

Grossman, M., Kalmanson, J., Bernhardt, N., Stern, M. B., & Hurtig, H. I. (2000). Cognitive resource limitations during sentence processing in Parkinson's disease. *Brain and Language, 73,* 1–16.

Grossman, M., Robinson, K. M., Onishi, K., Thompson, H., Cohen, J., & D'Esposito, M. (1995). Sentence comprehension in multiple sclerosis. *Acta Neurolgica Scandinavica, 92*(4), 324–331.

Gross-Tsur, V., Manor, O., & Shalev, R. S. (1996). Developmental dyscalculia: Prevalence and demographic features. *Developmental Medicine and Child Neurology, 38,* 25–33.

Grotjahn, M. (1957). *Beyond laughter: Humor and the subconscious*. New York: McGraw-Hill.

Groves-Wright, K., Neils-Strunjas, J., Burnett, R., & O'Neill, M. J. (2004). A comparison of verbal and written language in Alzheimer's disease. *Journal of Communication Disorders, 37,* 109–130.

Gu, D., Dupre, M. E., Sautter, J., Zhu, H., Liu, Y., & Yi, Z. (2009). Frailty and mortality among Chinese at advanced ages. *The Journals of Gerontology Series B: Psychological Sciences and Social Sciences, 64*(2), 279.

Gumperz, J. J. (1971). *Language in social groups*. Stanford, CA: Stanford University Press.

Gumperz, J. (1976). *The sociolinguistic significance of conversational code-switching*. Working papers of the Language Behavior Research Laboratory, No. 46. Berkeley: University of Califomia.

Hagerman, R. J., & Silverman, A. C. (1996). *Fragile X syndrome: Diagnosis, treatment, & research* (2nd ed.). Baltimore: Johns Hopkins University Press.

Hagoort, P. (2005). Broca's complex as the unification space for language. In A. Cutler (Ed.), *Twenty-first century psycholinguistics: Four cornerstones* (pp. 157–173). Mahwah, NJ: Erlbaum.

Hahne, A., Eckstein, K., & Friederici, A. D. (2004). Brain signatures of syntactic and semantic processes during children's language development. *Journal of Cognitive Neuroscience, 16*(7), 1302–1318.

Hahne, A., & Friederici, A. D. (2001). Processing a second language: Late learners' comprehension mechanisms as revealed by event-related brain potentials. *Bilingualism: Language and Cognition, 4,* 123–141.

Hakuta, K., Bialystok, E., & Wiley, E. (2003). Critical evidence: A test of the critical period hypothesis for second language acquisition. *Psychological Science, 14,* 31–38.

Hakuta, K., Butler, Y. G., & Witt, D. (2000). *How long does it take English learners to attain proficiency?* University of California Linguistic Minority Research Institute Policy Report 2000-1.

Hakuta, K., & Diaz, R. (1985). The relationship between degree of bilingualism and cognitive ability: A critical discussion and some new longitudinal data. In K. E. Nelson (Ed.), *Children's language* (Vol. 5, pp. 319–344). Hillsdale, NJ: Erlbaum.

Halberda, J. (2003). The development of a word-learning strategy. *Cognition, 87,* B23–B34.

Hamel, M., Gold, D., Andres, D., Reis, M., Dastoor, D., Grauer, H., et al. (1990). Predictors and consequences of aggressive behavior by community-based dementia patients. *The Gerontologist, 30*(2), 206–211.

Hammes, D. M., Novak, M. A., Rotz, L. A., Willis, M., Edmondson, D. M., & Thomas, J. (2002). Early identification in cochlear implantation: Critical factors for spoken language development. *Annals of Otology, Rhinology and Laryngology Supplement, 189,* 74–78.

Hamstra-Bletz, L., & Blote, A. W. (1993). A longitudinal study on dysgraphic handwriting in primary school. *Journal of Learning Disabilities, 26*(10), 689–699.

Han, J. J., Leichtman, M. D., & Wang, Q. (1998). Autobiographical memory in Korean, Chinese and American children. *Developmental Psychology, 34*(4), 701–713.

Hanna, P. R., Hodges, R. E., & Hanna, J. S. (1971). *Spelling: Structure and strategies.* Boston: Houghton Mifflin.

Hannula-Jouppi, K., Kaminen-Ahola, N., Taipale, M., Eklund, R., Nopola-Hemmi, J., Kaariainen, H., et al. (2005). The axon guidance receptor gene ROBO1 is a candidate gene for developmental dyslexia. *PLoS Genetics, 1*(4), e50.

Harner, L. (1975). Yesterday and tomorrow: Development of early understanding of the terms. *Developmental Psychology, 11,* 864–865.

Harner, L. (1980). Comprehension of past and future reference revisited. *Journal of Experimental Child Psychology, 29,* 170–182.

Harris, M., Barrett, M., Jones, D., & Brookes, S. (1988). Linguistic input and early word meaning. *Journal of Child Language, 15,* 77–94.

Harris, L. A., & Clancy Dollinger, S. M. (2001). Participation in a course on aging: Knowledge, attitudes and anxiety about aging in oneself and others. *Educational Gerontology, 27,* 657–667.

Hart, B., & Risley, T. (1995). *Meaningful differences in the everyday experience of young American children.* Baltimore: Paul H. Brookes.

Hartelius, L., & Svensson, P. (1994). Speech and swallowing symptoms associated with Parkinson's disease and multiple sclerosis: A survey. *Folia Phoniatrica et Logopaedica, 46*(1), 9–17.

Hartman, M. A. (1976). Descriptive study of the language of men and women born in Maine around 1900 as it reflects the Lakoff hypotheses in "Language and Woman's Place." In B. L. Dubois & I. Crouch (Eds.), *The sociology of the languages of American women.* San Antonio, TX: Trinity University Press.

Hasher, L., & Zacks, R. T. (1988). Working memory, comprehension, and aging: A review and a new view. In G. H. Bower (Ed.), *The psychology of learning and motivation* (Vol. 22, pp. 193–225). New York: Academic Press.

Hasselhorn, M., & Schuchardt, K. (2006). Learning disorders: A critical sketch of the epidemiology. *Childhood and Development, 15,* 208–215.

Haught, P. A., Walls, R. T., Laney, J. D., Leavell, A., & Stuzen, S. (1999). Child and adolescent knowledge and attitudes about older adults across time and states. *Educational Gerontology, 25,* 501–517.

Havens, L. L., & Foote, W. E. (1963). The effect of competition on visual duration thresholds and its independence of stimulus frequency. *Journal of Experimental Psychology, 65,* 6–11.

Hayes, B. (2009). *Introducing phonology*. New York: Wiley-Blackwell.

Hayter, W. (1977). *Spooner: A biography.* London: W. H. Allen.

Heaven, P. C. L. (2001). *The social psychology of adolescence.* New York: Palgrave Macmillan.

Hecht, S. A., Burgess, S. R., Torgesen, J. K., Wagner, R. K., & Rashotte, C. A. (2000). Explaining social class differences in growth of reading skills from beginning kindergarten through fourth grade: The role of phonological awareness, rate of access, and print knowledge. *Reading and Writing, 12*(1–2), 99–127.

Hedberg, N. L., & Stoel-Gammon, C. (1986). Narrative analysis: Clinical procedures. *Topics in Language Disorders, 7,* 58–69.

Hedberg, N. L., & Westby, C. E. (1993). *Analyzing storytelling skills: Theory to practice.* Tucson, AZ: Communication Skill Builders.

Heibeck, T. H., & Markman, E. M. (1987). Word learning in children: An examination of fast mapping. *Child Development, 58,* 1021–1034.

Heider, E. (1972). Universals in color naming and memory. *Journal of Experimental Psychology, 93,* 10–20.

Heim, S., Tschierse, J., Amunts, K., Wilms, M., Vossel, S., Willmes, K., et al. (2008). Cognitive subtypes of dyslexia. *Acta Neurobiologiae Experimentalis, 68*(1), 73–82

Heine, M. K., Ober, B. A., & Shenaut, G. K. (1999). Naturally occurring and experimentally induced tip-of-the-tongue experiences in three adult age groups. *Psychology & Aging, 14,* 445–457.

Helmers, H. (1965). *Sprache und Humor des Kindes*. Stuttgart: Klett.

Hennessee, J. A., & Nicholson, J. (1972, May 28). NOW says: Commercials insult women. *New York Times Magazine,* p. 12.

Henry, J. D., & Beatty, W. W. (2006). Verbal fluency deficits in multiple sclerosis. *Neuropsychologia, 44*(7), 1166–1174.

Heredia, R. R., & Altarriba, J. (2001). Bilingual language mixing: Why do bilinguals code-switch? *Current Directions in Psychological Science, 10,* 164–168.

Herman, G. T. (2009). *Fundamentals of computerized tomography: Image reconstruction from projection* (2nd ed.). New York: Springer.

Herman, L. M. (1980). Cognitive characteristics of dolphins. In L. M. Herman (Ed.), *Cetacean behavior: Mechanisms & functions* (pp. 363–429). New York: Wiley Interscience.

Herman, L. M., Richards, D. G., & Wolz, J. P. (1984). Comprehension of sentences by bottlenosed dolphins. *Cognition, 16,* 129–219.

Hernandez, A. E., Martinez, A., & Kohnert, K. (2000). In search of the language switch: An fMRI study of picture naming in Spanish–English bilinguals. *Brain and Language, 73,* 421–431.

Hespos, S., & Spelke, E. (2004). Conceptual precursors to language. *Nature, 430,* 453–456.

Hick, R., Joseph, K., Conti-Ramsden, G., Serratrice, L., & Faragher, B. (2002). Vocabulary profiles of children with specific language impairment. *Child Language Teaching and Therapy, 18*(2), 165–180.

Hier, D. B., Hagenlocker, K., & Shindler, A. G. (1985). Language disintegration in dementia: Effects of etiology and severity. *Brain and Language, 25,* 117–133.

Hiramatsu, K. (2003). Children's judgments of negative questions. *Language Acquisition, 11*(2), 99–126.

Hirsh-Pasek, K., & Golinkoff, R. (1991). Language comprehension: A new look at some old themes. In N. Krasnegor, D. Rumbaugh, M. Studdert-Kennedy, & R. Schiefelbusch (Eds.), *Biological and behavioral aspects of language acquisition*. Hillsdale, NJ: Erlbaum.

Hirsh-Pasek, K., & Golinkoff, R. (1996). *The origins of grammar.* Cambridge, MA: MIT Press.

Hjelmborg, J. V., Iachine, I., Skytthe, A., Vaupel, J. W., McGue, M., Koskenvuo, M., et al. (2006). Genetic influence on human lifespan and longevity. *Human Genetics,119*(3), 312–321.

Ho, A. K., Iansek, R., Marigliani, C., Bradshaw, J. L., & Gates, S. (1998). Speech impairment in a large sample of patients with Parkinson's disease. *Behavioural Neurology, 11,* 131–137.

Ho, A. K., Sahakian, B. J., Robbins, T. W., Barker, R. A., Rosser, A. E., & Hodges, J. R. (2002). Verbal fluency in Huntington's disease: A longitudinal analysis of phonemic and semantic clustering and switching. *Neuropsychologia, 40,* 1277–1284.

Hockett, C. F. (1966). The problem of universals in language. In J. H. Greenberg (Ed.), *Universals of language* (pp. 1–22). Cambridge, MA: MIT Press.

Hoff, E. (2003). The specificity of environmental influence: Socioeconomic status affects early vocabulary development via maternal speech. *Child Development, 74,* 1368–1378.

Holler, J., & Stevens, R. (2007). An experimental investigation into the effect of common ground on how speakers use gesture and speech to represent size information in referential communication. *Journal of Language and Social Psychology, 26,* 4–27.

Holm, J. (1988). *Pidgins and creoles: Volume 1.* Cambridge, UK: Cambridge University Press.

Holm, J. (1989). *Pidgins and creoles: Volume 2.* Cambridge, UK: Cambridge University Press.

Holmes, A., Franklin, A., Clifford, A., & Davies, I. R. L. (2009). Neurophysiological evidence for categorical perception of color. *Brain and Cognition, 69,* 426–434.

Holt, R. F., & Svirsky, M. A. (2008). An exploratory look at pediatric cochlear implantation: Is earliest always best? *Ear and Hearing, 29*(4), 492–511.

Holtgraves, T., McNamara, P., Durso, R., & Cappaert, K. (2010). Linguistic correlates of asymmetric motor symptom severity in Parkinson's disease. *Brain and Cognition, 72*(2), 189–196.

Holyoak, K. J., & Koh, K. (1987). Surface and structural similarity in analogical transfer. *Memory and Cognition, 15*(4), 332–340.

Holyoak, K. J., & Thagard, P. (1989). Analogical mapping by constraint satisfaction. *Cognitive Science, 13*(3), 295–355.

Hook, E. B. (1982). Epidemiology of Down syndrome. In S. M. Pueschel & J. E. Rynders (Eds.), *Down syndrome: Advances in biomedicine and the behavioral sciences.* Cambridge, MA: Ware Press.

Hoosain, R. (1982). Correlation between pronunciation speed and digit span size. *Perceptual and Motor Skills, 55,* 1128.

Hooyman, N. R., & Kiyak, H. A. (2011). *Social gerontology* (9th ed.). Boston: Allyn & Bacon.

Horgan, D. (1978). The development of the full passive. *Journal of Child Language, 5*(1), 65–80.

Hostetter, A. B. (2011). When do gestures communicate? A meta-analysis. *Psychological Bulletin, 137*(2), 297–315.

Hostetter, A. B., & Alibali, M. W. (2008). Visible embodiment: Gestures as simulated action. *Psychonomic Bulletin & Review, 15,* 495–514.

Houston-Price, C., Caloghiris, Z., & Raviglione, E. (2010). Language experience shapes the development of the mutual exclusivity bias. *Infancy, 15*(2), 125–150.

Hubel, D. H., & Wiesel, T. N. (1963). Receptive fields of cells in striate cortex of very young, visually inexperienced kittens. *Journal of Neurophysiology, 26,* 994–1002.

Huey, E. (1968). *The psychology and pedagogy of reading.* Cambridge, MA: MIT Press. (Original work published 1908)

Hugh-Jones, S., & Smith, P. K. (1999). Self-reports of short and long term effects of bullying on children who stammer. *British Journal of Educational Psychology, 69,* 141–158.

Human Genome Project. (n.d.). *Human Genome Project information.* Retrieved February 2, 2013, from www.ornl.gov/sci/techresources/Human_Genome/home.shtml

Humes, L. E. (1996). Speech understanding in the elderly. *Journal of the American Academy of Audiology, 7,* 161–167.

Hunt, E., & Agnoli, F. (1991). The Worfian hypothesis: A cognitive psychology perspective. *Psychological Review, 98*(3), 377–389.

Hutchison, K. A. (2003). Is semantic priming due to association strength or featural overlap? A micro-analytic review. *Psychonomic Bulletin & Review, 10,* 785–813.

Huttenlocher, J., Vasilyeva, M., Cymerman, E., & Levine, S. C. (2002). Language input at home and at school: Relation to syntax. *Cognitive Psychology, 45*(3), 337–374.

Hyltenstam, K., & Abrahamsson, N. (2003). Maturational constraints in SLA. In C. J. Doughty & M. H. Long (Eds.), *The handbook of second language acquisition* (pp. 539–588). Oxford: Blackwell.

Hymes, D. H. (1966). Two types of linguistic relativity. In W. Bright (Ed.), *Sociolinguistics* (pp. 114–158). The Hague: Mouton.

Hyönä, J., & Pollatsek, A. (1998). Reading Finnish compound words: Eye fixations are affected by component morphemes. *Journal of Experimental Psychology. Human Perception & Performance, 24,* 1612–1627.

Iaccino, J. F. (1993). *Left brain–right brain differences.* Hillsdale, NJ Erlbaum.

International Phonetic Association. (1999). *Handbook of the International Phonetic Association: A guide to the use of the International Phonetic Alphabet.* Cambridge, UK: Cambridge University Press.

Irigaray, L. (1973). *Le langage des dements.* The Hague: Mouton.

Irwin, M. L., Tworoger, S. S., Yasui, Y., Rajan, B., McVarish, L., LaCroix, K., et al. (2004). Influence of demographic, physiologic, and psychosocial variables on adherence to a yearlong moderate-intensity exercise trial in postmenopausal women. *Preventive Medicine, 39,* 1080–1086.

Isurin, L. (2000). Deserted island or a child's first language forgetting. *Bilingualism: Language and Cognition, 3*(2), 151–166.

Ivanova, I., & Costa, A. (2008). Does bilingualism hamper lexical access in speech production? *Acta Psychologica, 127,* 277–288.

Iverson, J. M., Capirci, O., Volterra, V., & Goldin-Meadow, S. (2008). Learning to talk in a gesture-rich world: Early communication in Italian vs. American children. *First Language, 28,* 164–181.

Iverson, J. M., & Goldin-Meadow, S. (2005). Gesture paves the way for language development. *Psychological Science, 16,* 367–371.

Ives, K. (1979). *Written dialects & spelling reform.* Chicago: Progressive Publishers.

Izard, C. E., King, K. A., Trentacosta, C. J., Laurenceau, J. P., Morgan, J. K., Krauthamer-Ewing, E. S., et al. (2008). Accelerating the development of emotion competence in Head Start children. *Development & Psychopathology, 20,* 369–397.

Jacobs, N., & Garnham, A. (2007). The role of conversational hand gestures in a narrative task. *Journal of Memory and Language, 56*(2), 291–303.

Jaeger, J. J. (2005). *Kids' slips: What young children's slips of the tongue reveal about language development.* Mahwah, NJ: Erlbaum.

James, L. E. (2004). Meeting Mr. Farmer versus meeting a farmer: Specific effects of aging on learning proper names. *Psychology and Aging, 19,* 515–522.

James, L. E., & Burke, D. M. (2000). Phonological priming effects on word retrieval and tip-of-the-tongue experiences in young and older adults. *Journal of Experimental Psychology, 26,* 1378–1391.

January, D., & Kako, E. (2007). Re-evaluating evidence for linguistic relativity: Reply to Boroditsky (2001). *Cognition, 104,* 417–426.

Jarvella, R. J. (1971). Syntactic processing of connected speech. *Journal of Verbal Learning & Verbal Behavior, 10,* 409–416.

Jasper, H., & Penfield, W. (1954). *Epilepsy & the functional anatomy of the human brain* (2nd ed.). New York: Little, Brown.

Javier, R. (1996). In search of repressed memories in bilingual individuals. In R. Foster, M. Moskowitz, & R. Javier (Eds.), *Reaching across boundaries of culture and class: Widening the scope of psychotherapy* (pp. 225–241). Northvale, NJ: Jason Aronson, Inc.

Jerger, J., Jerger, S., Oliver, T., & Pirozzolo, F. (1989). Speech understanding in the elderly. *Ear and Hearing, 10,* 79–89.

Johnson, J. S., & Newport, E. L. (1989). Critical period effects in second language learning: The influence of maturational state on the acquisition of English as a second language. *Cognitive Psychology, 21,* 60–99.

Johnston, J., Durieux-Smith, A., & Bloom, K. (2005). Teaching gestural signs to infants to advance child development. *First Language, 25,* 235–251.

Jones, G. V. (1989). Back to Woodworth: Role of interlopers in the tip-of-the-tongue phenomenon. *Memory & Cognition, 17,* 69–76.

Jones, M. C. (1926). The development of early behavior patterns in young children. *Pedagogical Seminary, 33,* 537–585.

Jordan, G. E., Snow, C. E., & Porsche, M. V. (2000). Project EASE: The effect of a family literacy project on kindergarten students' early literacy skills. *Reading Research Quarterly, 35,* 524–546.

Joseph, H. S. S. L., Liversedge, S. P., Blythe, H. I., White, S. J., Gathercole, S. E., & Rayner, K. E. (2008). Children's and adults' processing of anomaly and implausibility during reading: Evidence from eye movements. *Quarterly Journal of Experimental Psychology, 61*(5), 708–723.

Juel, C. (1988). Learning to read and write: A longitudinal study of 54 children from first through fourth grades. *Journal of Educational Psychology, 80*(4), 437–447.

Juhasz, B. J., & Rayner, K. (2003). Investigating the effects of a set of intercorrelated variables on eye fixation durations in reading. *Journal of Experimental Psychology: Learning, Memory, and Cognition, 29,* 1312–1318.

Junker, D. A., & Stockman, I. J. (2002). Expressive vocabulary of German-English bilingual toddlers. *American Journal of Speech-Language Pathology, 11*(4), 381–395.

Jusczyk, P. W. (1997). *The discovery of spoken language.* Cambridge, MA: MIT Press.

Jusczyk, P. W. (1999). How infants begin to extract words from speech. *Trends in Cognitive Sciences, 3*(9), 323–328.

Jusczyk, P. W. (2000). *The discovery of language.* Cambridge, MA: The MIT Press.

Jusczyk, P. W., & Aslin, R. N. (1995). Infants' detection of the sound patterns of words in fluent speech. *Cognitive Psychology, 28,* 1–23.

Just, M. A., & Carpenter, P. A. (1980). A theory of reading: From eye fixations to comprehension. *Psychological Review, 87,* 329–354.

Just, M. A., & Carpenter, P. A. (2002). A capacity theory of comprehension: Individual differences in working memory. In T. A. Polk & C. M. Seifert (Eds.), *Cognitive modeling* (pp. 131–178). Cambridge, MA: MIT Press.

Just, M. A., Carpenter, P. A., & Masson, M. E. (1982). What eye fixations tell us about speed reading and skimming. Unpublished technical report, Department of Psychology, Carnegie Mellon University, Pittsburgh, PA.

Katz, A. N. (1998). *Figurative language and thought.* Oxford, UK: Oxford University Press.

Kauschke, C., & Hofmeister, C. (2002). Early lexical development in German: A study on vocabulary growth and vocabulary composition during the second and third year of life. *Journal of Child Language, 29,* 735–757.

Kay, D. A., & Anglin, J. M. (1982). Overextension and underextension in the child's expressive and receptive speech. *Journal of Child Language, 9,* 83–98.

Kay, P., & Kempton, W. (1984). What is the Sapir–Whorf hypothesis? *American Anthropologist, 86,* 65–79.

Kegl, J., Senghas, A., & Coppola, M. (1999). Creation through contact: Sign language emergence & sign language change in Nicaragua. In M. DeGraff (Ed.), *Comparative grammatical change: The intersection of language acquisistion, creole genesis, & diachronic syntax* (pp. 179–237). Cambridge, MA: MIT Press.

Keller, H., Shattuck, R., & Herrmann, D. (2004). *The story of my life*. New York: W. W. Norton.

Kellogg, W. N., & Kellogg, L. A. (1933). *The ape & the child: A comparative study of the environmental influence upon early behavior.* New York: Hafner Publishing Co.

Kelly, D. J. (1998). A clinical synthesis of the "late talker" literature: Implications for service delivery. *Language, Speech, and Hearing Services in Schools, 29,* 76–84.

Kemmerer, D. (1999). "Near" and "far" in language and perception. *Cognition, 73,* 35–63.

Kempe, C., Eriksson-Gustavsson, A., & Samuelsson, S. (2011). Are there any Matthew effects in literacy and cognitive development? *Scandinavian Journal of Educational Research, 55*(2), 181–196.

Kemper, A. R., & Downs, S. M. (2000). A cost-effectiveness analysis of newborn hearing screening strategies. *Archives of Pediatrics and Adolescent Medicine, 154*(5), 484–488.

Kemper, S. (1992). Language and aging. In F. I. M. Craik & T. A. Salthouse (Eds.), *Handbook of aging and cognition* (pp. 213–270). Hillsdale, NJ: Erlbaum.

Kemper, S., & Edwards, L. L. (1986). Children's expression of causality and their construction of narratives. *Topics in Language Disorders, 7*(1), 11–20.

Kemper, S., Greiner, L., Marquis, J., Prenovost, K., & Mitzner, T. (2001). Language decline across the life span: Findings from the Nun Study. *Psychology and Aging, 16,* 227–239.

Kemper, S., Othick, M., Warren, J., Gubarchuk, J., & Gerhing, H. (1996). Facilitating older adults' performance on a referential communication task through speech accommodations. *Aging, Neuropsychology, and Cognition, 3,* 37–55.

Kemper, S., Vandeputte, D., Rice, K., Cheung, H., & Gubarchuk, J. (1995). Speech adjustments to aging during a referential communication task. *Journal of Language and Social Psychology, 14,* 40–59.

Kemper, T. (1984). Neuroanatomical and neuropathological changes in normal aging and dementia. In M. L. Albert (Ed.), *Clinical neurology of aging* (pp. 9–52). New York: Oxford University Press.

Kempler, D., & Zelinski, E. M. (1994). Language function in dementia and normal aging. In F. A. Huppert, C. Brayne, & D. O'Connor (Eds.), *Dementia and normal aging* (pp. 331–365). Cambridge, UK: Cambridge University Press.

Kennedy, A., & Murray, W. S. (1984). Inspection times for words in syntactically ambiguous sentences under three presentation conditions. *Journal of Experimental Psychology: Human Perception & Performance, 10,* 833–847.

Kennison, S. M. (2001). Limitations on the use of verb information in sentence comprehension. *Psychonomic Bulletin & Review, 8,* 132–138.

Kennison, S. M. (2004). The effect of phonemic repetition on syntactic ambiguity resolution: Implications for models of working memory. *Journal of Psycholinguistic Research, 33,* 493–516.

Kennison, S. M. (2005). Different time courses of integrative semantic processing for plural and singular nouns: Implications for theories of sentence processing. *Cognition, 97,* 269–294.

Kennison, S. M. (2007). Why children in the U.S. read less well than children in other industrialized nations: Recommendations for future research. *Public Policy Forum, 3,* 24–37.

Kennison, S. M. (2009). The use of verb information in parsing: Different statistical analyses lead to contradictory conclusions. *Journal of Psycholinguistic Research, 38,* 363–378.

Kennison, S. M., & Bowers, J. M. (2011). Illustrating brain lateralization in a naturalistic observation of cell phone use. *Psychology Learning & Teaching, 10,* 46–51.

Kennison, S. M., & Clifton, C. (1995). Determinants of parafoveal preview benefit in high and low working memory capacity readers: Implications for eye movement control. *Journal of Experimental Psychology: Learning, Memory, and Cognition, 21,* 68–81.

Kennison, S. M., Fernandez, E. C., & Bowers, J. M. (2009). Processing differences for anaphoric and cataphoric pronouns: Implications for theories of referential processing. *Discourse Processes, 46,* 25–35.

Kennison, S. M., Fernandez, E. C., & Bowers, J. M. (in press). The role of semantic and phonological information in word production: Evidence from Spanish-English bilinguals. *Journal of Psycholinguistic Research.*

Kennison, S. M., & Ponce-Garcia, E. (2012). The role of childhood relationships with older adults in reducing risk-taking by young adults. *Journal of Intergenerational Relationships, 10,* 22–33.

Kennison, S. M., Sieck, J. P., & Briesch, K. A. (2003). Evidence for a late occurring effect of phoneme repetition in silent reading. *Journal of Psycholinguistic Research, 32,* 297–312.

Kennison, S. M., & Trofe, J. L. (2003). A role for gender stereotype information in language comprehension. *Journal of Psycholinguistic Research, 32,* 355–378.

Kidd, E., & Bavin, E. L. (2005). Lexical and referential cues to interpretation: An investigation of children's interpretations of ambiguous sentences. *Journal of Child Language, 32,* 855–876.

Kidd, E., Stewart, A., & Serratrice, L. (2011). Children do not overcome lexical biases where adults do: The role of the referential scene in garden path recovery. *Journal of Child Language, 38,* 222–234.

Kim, H. S., Relkin, N. R., Lee, K. M., & Hirsch, J. (1997). Distinct cortical areas associated with native and second languages. *Nature, 388,* 171–174.

Kimball, J. (1973). Seven principles of surface structure parsing in natural language. *Cognition, 2,* 15–47.

Kim-Renaud, Y. (1997). *The Korean alphabet: Its history and structure.* Honolulu, HI: University of Hawaii Press.

Kimura, D. (1961). Cerebral dominance and the perception of verbal stimuli. *Canadian Journal of Psychology, 15,* 166–171.

Kircher, T. T., Brammer, M. J., Levelt, W., Bartels, M., & McGuire, P. K. (2004). Pausing for thought: Engagement of left temporal cortex during pauses in speech. *NeuroImage, 21,* 84–90.

Kirsch, I. S., Jungeblut, A., Jenkins, L., & Kolstad, A. (1993). *Adult literacy in America: A first look at the results of the National Adult Literacy Survey.* Washington, DC: Office of Educational Research and Improvement.

Kirton, J. F. (1988). Men's and women's dialect. *Aboriginal Linguistics, 1,* 111–125.

Kisilevsky, B. S., Hains, S. M., Lee, K., Xie, X., Huang, H., Ye, H. H., et al. (2003). Effects of experience on fetal voice recognition. *Psychological Science, 14*(3), 220–224.

Kisilevsky, B. S., Muir, D. W., & Low, J. A. (1992). Maturation of human fetal responses to vibroacoustic stimulation. *Child Development, 63,* 1497–1508.

Klein, D., Milner, B., Zatorre, R. J., Zhao, V., & Nikelski, J. (1999). Cerebral organization in bilinguals: A PET study of Chinese–English verb generation. *NeuroReport, 10,* 2841–2845.

Klein, R. G., & Edgar, B. (2002). *The dawn of human culture.* New York: John Wiley.

Klima, E. S., & Bellugi, U. (1966). Syntactic regularities in the speech of children. In J. Lyons & R. J. Wales (Eds.), *Psycholinquistic papers: The proceedings of the 1966 Edinburgh Conference* (pp. 183–219). Edinburgh: Edinburgh University Press.

Klima, E. S., & Bellugi, U. (1979). *The signs of language.* Boston: Harvard University Press.

Koerner, E. F. K., & Koerner, K. (1985). *Edward Sapir: Appraisals of his life and work.* Amsterdam: John Benjamins.

Kolb, B., & Whishaw, I. Q. (2003). *Fundamentals of human neuropsychology.* New York: Worth.

Konner, M. (2002). *The tangled wing: Biological constraints on the human spirit.* New York: Times Books.

Köppe, R. (1996). Language differentiation in bilingual children: The development of grammatical and pragmatic competence. *Linguistics, 34,* 927–954.

Kornhaber-Le Chanu, M. (1995). *La communication du jeune enfant avec son père et sa mère: Adaptation à l'interlocuteur.* Thèse de Doctorat, Université Paris V.

Kot, A., & Law, J. (1995). Intervention with preschool children with specific language impairments: A comparison of two different approaches to treatment. *Child Language Teaching and Therapy, 11*(2), 144–162.

Kovács, A. M., & Mehler, J. (2009). Cognitive gains in 7-month-old bilingual infants. *Proceedings of the National Academy of Sciences of the United States of America, 106*, 6556–6560.

Kovelman, I., Baker, S. A., & Petitto, L. A. (2008). Age of first bilingual language exposure as a new window into bilingual reading development. *Bilingualism: Language and Cognition, 11*(2), 203–223.

Kowalski, K., & Zimiles, H. (2006). The relation between children's conceptual functioning with color and color term acquisition. *Journal of Experimental Child Psychology, 94,* 301–321.

Kramer, A. F., Larish, J. F., & Strayer, D. L. (1999). Training for executive control: Task coordination strategies and aging. In D. Gopher & A. Koriat (Eds.), *Attention and performance XVII*. Cambridge, MA: MIT Press.

Krause, C. A. (1916). *The direct method in modern languages.* New York: Scribner's Sons.

Krause, J., Lalueza-Fox, C., Orlando, L., Enard, W., Green, R. E., Burbano, H. A., et al. (2007). The derived FOXP2 variant of modern humans was shared with Neandertals. *Current Biology, 17*, 1–5.

Krauss, R. (1998). Why do we gesture when we speak? *Current Directions in Psychological Science, 7*(2), 54–60.

Krauss, R. M., Chen, Y., & Chawla, P. (1996). Nonverbal behavior and nonverbal communication: What do conversational hand gestures tell us? In M. Zanna (Ed.), *Advances in experimental social psychology* (pp. 389–450). San Diego, CA: Academic Press.

Krauss, R. M., Chen, Y., & Gottesman, R. F. (2000). Lexical gestures and lexical access: A process model. In D. McNeill (Ed.), *Language and gesture* (pp. 261–283). New York: Cambridge University Press.

Kretschmer, R., & Kretschmer, L. (1989). *Second language teaching and the handicapped child.* In D. Nielsen (Ed.), *Two languages for all children*. New York: Addison.

Kroll, J. F., & Stewart, E. (1994). Category interference in translation and picture naming: Evidence for asymmetric connection between bilingual memory representations. *Journal of Memory & Language, 33,* 149–174.

Kroll, J. F., Van Hell, J. G., Tokowicz, N., & Green, D. W. (2010). The revised hierarchical model: A critical review and assessment. *Bilingualism: Language and Cognition, 13,* 373–381.

Kuhl, P. K. (1987). The special-mechanisms debate in speech perception: Nonhuman species & nonspeech signals. In S. Harnad (Ed.), *Categorical perception: The groundwork of cognition* (pp. 355–386). New York: Cambridge University Press.

Kuhl, P. K. (1999). Speech, language, and the brain: Innate preparation for learning. In M. D. Hauser & M. Konishi (Eds.), *The design of animal communication* (pp. 419–450). Cambridge, MA: MIT Press.

Kuhl, P. K. (2007). Is speech learning "gated" by the social brain? *Developmental Science, 10,* 110–120.

Kuhl, P. K., Conboy, B. T., Padden, D., Nelson, T., & Pruitt, J. (2005). Early speech perception & later language development: Implications for the "critical period." *Language Learning & Development, 1,* 237–264.

Kuhl, P. K., & Meltzoff, A. N. (1982). The bimodal perception of speech in infancy. *Science, 218,* 1138–1141.

Kuhl, P. K., Stevens, E., Hayashi, A., Deguchi, T., Kiritani, S., & Iverson, P. (2006). Infants show a facilitation effect for native language perception between 6 and 12 months. *Developmental Science, 9,* F1–F9.

Kuhl, P. K., Tsao, F.-M., & Liu, H.-M. (2003). Foreign-language experience in infancy: Effects of short-term exposure & social interaction on phonetic learning. *Proceedings of the National Academy of Sciences, USA, 100,* 9096–9101.

Kurinski, E., & Sera, M. D. (2011). Does learning Spanish grammatical gender change English adults' concepts of inanimate objects? *Bilingualism: Language and Cognition, 14*(2), 203–220.

Labov, W. (1969). Contraction, deletion, and inherent variability of the English copula. *Language, 45,* 715–762.

Ladd, G. W., Kochenderfer, B. J., & Coleman, C. C. (1996). Friendship quality as a predictor of young children's early school adjustment. *Child Development, 67,* 1103–1118.

Ladefoged, P. (2001). Vowels and consonants: An introduction to the sounds of languages. Oxford, UK: Blackwell.

Lakoff, R. (1973). Language and woman's place. *Language in Society, 2,* 45–79.

Lakoff, R. (1975). *Language and woman's place*. New York: Harper & Row.

Lam, K., & Dijkstra, T. (2010). Modeling code-interactions in bilingual word recognition: Recent empirical studies and simulations with BIA+. *International Journal of Bilingual Education and Bilingualism, 13,* 487–503.

Lane, H. (1976). *The wild boy of Aveyron*. Cambridge, MA: Harvard University Press.

LaPointe, L. B., & Engle, R. W. (1990). Simple and complex word spans as measures of working memory capacity. *Journal of Experimental Psychology: Learning, Memory, and Cognition, 16,* 1118–1133.

Lass, K. (1997). *Historical linguistics & language change.* Cambridge, UK: Cambridge University Press.

Lass, N. J. (1988). *Handbook of speech-language pathology & audiology*. St. Louis, MO: Mosby–Year Book.

Lass, R., & Hogg, R. M. (2000). *The Cambridge history of the English language*. Cambridge, UK: Cambridge University Press.

Laver, G. D., & Burke, D. M. (1993). Why do semantic priming effects increase in old age? A meta-analysis. *Psychology of Aging, 8*(1), 34–43.

Laws, G., & Bishop, D. V. M. (2003). A comparison of language abilities in adolescents with Down syndrome and children with specific language impairment. *Journal of Speech, Language and Hearing Research, 46,* 1324–1339.

Le Corre, M., & Carey, S. (2007). One, two, three, four, nothing more: An investigation of the conceptual sources of the verbal counting principles. *Cognition, 105,* 395–438.

Lederberg, A. R., & Morales, C. (1985). Code switching by bilinguals: Evidence against a third grammar. *Journal of Psycholinguistic Research, 14*(2), 113–136.

Lee, C. D., & Smagorinsky, P. (Eds.). (2000). *Vygotskian perspectives on literacy research: Constructing meaning through collaborative inquiry.* New York: Cambridge University Press.

Lee, K. (2011). Impacts of the duration of Head Start enrollment on children's academic outcomes: Moderation effects of family risk factors and earlier outcomes. *Journal of Community Psychology, 39,* 698–716.

Le Maire, L. (Composer). (1835). *The A.B.C., a German air with variations for the flute with an easy accompaniment for the piano forte* [Sheet music]. Boston: Charles Bradlee.

Lenneberg, E. H. (1964). The capacity for language acquisition. In J. A. Fodor & J. J. Katz (Eds.), *The structure of language: Readings in the philosophy of language* (pp. 579–603). Englewood Cliffs, NJ: Prentice Hall.

Lenneberg, E. H. (1967). *Biological foundations of language*. New York: Wiley.

Leonard, L. B. (1998). *Children with specific language impairment*. Cambridge, MA: MIT Press.

Lesch, M. F., & Pollatsek, A. (1998). Evidence for the use of assembled phonology in accessing the meaning of printed words. *Journal of Experimental Psychology: Learning, Memory, and Cognition, 24,* 573–592.

Lethlean, J. B., & Murdoch, B. E. (1993). Language problems in multiple sclerosis. *Journal of Medical Speech-Language Pathology, 1*(1), 47–59.

Leung, E. H. L., & Rheingold, H. L. (1981). Development of pointing as a social gesture. *Developmental Psychology, 17,* 215–220.

Levelt, W. J. M. (1983). Monitoring and self-repair in speech. *Cognition, 14,* 41–104.

Levelt, W. J. M. (1989). *Speaking: From intention to articulation.* Cambridge, MA: MIT Press.

Levelt, W. J. M., Roelofs, A., & Meyer, A. S. (1999). A theory of lexical access in speech production. *Behavioral and Brain Sciences, 22,* 1–38.

Levelt, W. J. M., & Wheeldon, L. R. (1994). Do speakers have access to a mental syllabary? *Cognition, 50,* 239–269.

Levin, I., & Wilkening, F. (1989). Measuring time via counting: The development of children' conceptions of time as a quantifiable dimension. In I. Levin & D. Zakay (Eds.), *Time and human cognition: A life-span perspective* (pp. 119–143). Amsterdam: Elsevier.

Levine, L. E., & Munsch, J. (2011). *Child development: An active learning approach.* Thousand Oaks, CA: Sage.

Levinson, B. (Director). (1988). *Rain main.* United States: United Artists.

Levinson, S. (1996). Language and speech. *Annual Review of Anthropolgy, 25,* 353–382.

Levitt, S., & Dubner, S. J. (2005). *Freakonomics: A rogue economist explores the hidden side of everything.* New York: William Morrow/HarperCollins.

Levy, B. J., McVeigh, N. D., Marful, A., & Anderson, M. C. (2007). Inhibiting your native language: The role of retrieval-induced forgetting during second language acquisition. *Psychological Science, 18,* 29–34.

Levy, B. R., Kasl, S., & Gill, T. (2004). Image of aging scale. *Journal of Perceptual and Motor Skills, 99,* 208–210.

Levy, B. R., Slade, M., Kunkel, S., & Kasl, S. (2002). Longitudinal benefit of positive self-perceptions of aging on functioning health. *Journal of Personality and Social Psychology, 83,* 261–270.

Levy, B. R., Zonderman, A. B., Slade, M. D., & Ferruci, L. (2009). Age stereotypes held earlier in life predict cardiovascular events in later life. *Psychological Science, 20,* 296–298.

Levy, L. M., Reis, I. L., & Grafman, J. (1999). Metabolic abnormalities detected by 1H-MRS in dyscalculia and dysgraphia. *Neurology, 53*(3), 639–641.

Lewis, M. P. (Ed.). (2009). *Ethnologue: Languages of the world* (16th ed.). Dallas, TX: SIL International.

Li, P. W., & Gleitman, L. R. (2002). Turning the tables: Language and spatial reasoning. *Cognition, 83*(3), 265–294.

Li, X., Harbottle, G., Zhang, J., & Changsui, W. (2003). The earliest writing? Sign use in the seventh millennium BC at Jiahu, Henan Province, China. *Antiquity, 77*(295), 31–45.

Liaw, F., & Brooks-Gunn, J. (1994). Cumulative familial risks and low birth weight children's cognitive and behavioral development. *Journal of Clinical Child Psychology, 23,* 360–372.

Liberman, I. Y., Shankweiler, D., Fischer, F. W., & Carter, B. (1974). Explicit syllable and phoneme segmentation in young children. *Journal of Experimental Child Psychology, 18,* 201–212.

Lieberman, D. E., McCarthy, R. C., Hiiemae, K. M., & Palmer, J. B. (2001). Ontogeny of larynx and hyoid descent in humans: Implications for deglutition and vocalization. *Archives of Oral Biology, 46,* 117–128.

Lieberman, P. (2007). The evolution of human speech: Its anatomical and neural bases. *Current Anthropology, 48*(1), 39–66.

Lieberman, P., Kako, E. T., Friedman, J., Tajchman, G., Feldman, L. S., & Jiminez, E. B. (1992). Speech production, syntax comprehension, and cognitive deficits in Parkinson's disease. *Brain and Language, 43,* 169–189.

Light, L. L. (1992). The organization of memory in old age. In F. I. M. Craik & T. A. Salthouse (Eds.), *The handbook of aging and cognition* (pp. 111–165). Hillsdale, NJ: Erlbaum.

Lima, S. D. (1987). Morphological analysis during sentence reading. *Journal of Memory and Language, 26,* 84–99.

Lin, D., McBride-Chang, C., Shu, H., Zhang, Y. P., Li, H., Zhang, J., et al. (2010). Small wins big: Analytic Pinyin skills promote Chinese word reading. *Psychological Science, 21,* 1117–1122.

Lin, F. R., Thorpe, R., Gordon-Salant, S., & Ferrucci, L. (2011). Hearing loss prevalence and risk factors among older adults in the United States. *Journal of Gerontology Series A: Biological Sciences and Medical Sciences, 66*(5), 582–590.

Linacre, J. M. (1996). The prison literacy problem. *Rasch Measurement Transactions, 10*(1), 473–474.

Linck, J., Kroll, J. F., & Sunderman, G. (2009). Losing access to the native language while immersed in a second language: Evidence for the role of inhibition in second language learning. *Psychological Science, 20,* 1507–1515.

Lindholm, K. J., & Padilla, A. M. (1978). Language mixing in bilingual children. *Journal of Child Language, 5,* 327–335.

Lindwall, M., Rennemark, M., & Berggren, T. (2008). Movement in mind: The relationship of exercise with cognitive status for older adults in the Swedish National Study on Aging and Care (SNAC). *Aging and Mental Health, 12,* 212–220.

Lipton, J. S., & Spelke, E. S. (2006). Preschool children master the logic of number word meanings. *Cognition, 98,* B57–B66.

Liu, L. G. (1985). Reasoning counterfactually in Chinese: Are there any obstacles? *Cognition, 21,* 239–270.

Loftus, E. F. (1973). Category dominance, instance dominance and categorization time. *Journal of Experimental Psychology, 97,* 70–74.

Loftus, E. F. (1996). *Eye witness testimony.* Cambridge, MA: Harvard University Press.

Loftus, E. F. (2005). Planting misinformation in the human mind: A 30-year investigation of the malleability of memory. *Learning and Memory, 12,* 361–366.

Logemann, J. A., Fisher, H. B., Boshes, B., & Blonsky, E. R. (1978). Frequency and co-occurrence of vocal tract dysfunctions in a large sample of Parkinson's patients. *Journal of Speech and Hearing Disorders, 43,* 47–57.

Longoria, A. D., Page, M., Hubbs-Tait, L., & Kennison, S. M. (2009). The relationship between social competence and verbal skills in young children. *Early Childhood Development and Care, 179,* 919–929.

Longstaff, M. G., & Heath, R. A. (1997). Space–time invariance in adult handwriting. *Acta Psychologica, 97,* 201–214.

López, A., Atran, S., Coley, J., Medin, D., & Smith, E. (1997). The tree of life: Universals of folkbiological taxonomies and inductions. *Cognitive Psychology, 32,* 251–295.

López, A., Gelman, S. A., Gutheil, G., & Smith, E. E. (1992). The development of category-based induction. *Child Development, 63,* 1070–1090.

Lotto, L., & de Groot, A. M. B. (1998). Effects of learning method and word type on acquiring vocabulary in an unfamiliar language. *Language Learning, 48,* 31–69.

Lucas, M. (2000). Semantic priming effects without association: A meta-analytical review. *Psychonomic Bulletin and Review, 7,* 618–630.

Lucy, J. A., & Shweder, R. A. (1979). Whorf and his critics: Linguistic and nonlinguistic influences on color memory. *American Anthropologist, 81,* 581–615.

Luria, A. R. (1968). *The mind of a mnemonist: A little book about a vast memory.* London: Jonathan Cape.

Maas, A., & Russo, A. (2003). Directional bias in the mental representation of spatial events: Nature or culture? *Psychological Science, 14*(4), 296.

MacDonald, M. C., Just, M. A., & Carpenter, P. A. (1992). Working memory constraints on the processing of syntactic ambiguity. *Cognition, 24,* 56–98.

MacDonald, M., Pearlmutter, N. J., & Seidenberg, M. S. (1994). The lexical nature of syntactic ambiguity resolution. *Psychological Review, 101,* 676–703.

MacGuinness, D. (2004). Early reading instruction: What science really tells us about how to teach reading. Cambridge, MA: MIT Press.

MacKay, D. G., Abrams, L., & Pedroza, M. J. (1999). Aging on the input versus output side: Theoretical implications of age-linked asymmetries between detecting versus retrieving orthographic information. *Psychology and Aging, 14,* 3–17.

MacKay, D. G., & James, L. E. (2004). Sequencing, speech production, and selective effects of aging on phonological and morphological speech errors. *Psychology and Aging, 19,* 93–107.

Maclay, H., & Osgood, C. E. (1959). Hesitation phenomena in spontaneous English speech. *Word, 15,* 19–44.

Maclean, M., Bryant, P., & Bradley, L. (1987). Rhymes, nursery rhymes, and reading in early childhood. *Merrill-Palmer Quarterly, 33,* 25–282.

MacWhinney, B. (2000). *The CHILDES project: Tools for analyzing talk* (3rd ed.). Mahwah, NJ: Erlbaum.

Madden, D. J. (1988). Adult age differences in the effects of sentence context and stimulus degradation during visual word recognition. *Psychology and Aging, 3,* 167–172.

Mahon, M., & Crutchley, A. (2006). Performance of typically-developing school-age children with English as an additional language on the British Picture Vocabulary Scales II. *Child Language Teaching and Therapy, 22*(3), 333–351.

Mallinson, G., & Blake, B. (1981). *Language typology: Cross-linguistic studies in syntax.* Amsterdam: North Holland.

Mandel, D. R., Jusczyk, P. W., & Pisoni, D. B. (1995). Infants' recognition of the sound patterns of their own names. *Psychological Science, 6,* 315–318.

Mandler, G. (2007). A history of modern experimental psychology: From James & Wundt to cognitive science. Cambridge, MA: MIT Press.

Mansson, H. (2000). Childhood stuttering: Incidence and development. *Journal of Fluency Disorders, 25*(1), 47–57.

Maratsos, M. P. (1974). Children who get worse at understanding the passive: A replication of Bever. *Journal of Psycholinguistic Research, 3*(1), 65–74.

Marian, V., Blumenfeld, H. K., & Kaushanskaya, M. (2007). The Language Experience and Proficiency Questionnaire (LEAP-Q): Assessing language profiles in bilinguals and multilinguals. *Journal of Speech, Language, and Hearing Research, 50,* 940–967.

Marian, V., Spivey, M., & Hirsch, J. (2003). Shared and separate systems in bilingual language processing: Converging evidence from eyetracking and brain imaging. *Brain and Language, 86,* 70–82.

Marino, C., Giorda, R., Lorusso, M. L., Vanzin, L., Salandi, N., Nobile, M., et al. (2005). A family-based association study of the DYX1C1 gene on 15q21.1 in developmental dyslexia. *European Journal of Human Genetics, 13,* 491–499.

Markman, E. M. (1990). Constraints children place on word meanings. *Cognitive Science, 14,* 154–173.

Markman, E. M., & Hutchinson, J. E. (1984). Children's sensitivity to constraints on word meaning: Taxonomic vs. thematic relations. *Cognitive Psychology, 16,* 1–27.

Markman, E. M., & Wachtel, G. F. (1988). Children's use of mutual exclusivity to constrain the meaning of words. *Cognitive Psychology, 20,* 121–157.

Marslen-Wilson, W. (1987). Functional parallelism in spoken word recognition. *Cognition, 25,* 71–102.

Marslen-Wilson, W. (1990). Activation, competition, and frequency in lexical access. In G. T. M. Altmann (Ed.), *Cognitive models of speech processing* (pp. 148–172). Cambridge, MA: MIT Press.

Marslen-Wilson, W., & Tyler, L. K. (1980). The temporal structure of spoken language understanding. *Cognition, 8,* 1–71.

Marslen-Wilson, W., Tyler, L. K., Waksler, R., & Older, L. (1994). Morphology and meaning in the English mental lexicon. *Psychological Review, 101,* 3–33.

Martensen, H., Dijkstra, A., & Maris, E. (2005). A werd is not quite a word: On the role of sublexical phonological information in visual lexical decision. *Language and Cognitive Processes, 20,* 513–552.

Martin, L. (1986). Eskimo words for snow: A case study in the genesis and decay of an anthropological example. *American Anthropologist, 88*(2), 418.

Martin, R. C., Crowther, J. E., Knight, M., Tamborello, F. P., II, & Yang, C. L. (2010). Planning in sentence production: Evidence for the phrase as a default planning scope. *Cognition, 116,* 177–192.

Martin, R. C., & Freedman, M. (2001). Short-term retention of lexical-semantic representations: Implications for speech production. *Memory, 9,* 261–280.

Martin, R. C., Miller, M. D., & Vu, H. (2004). Lexical-semantic retention and speech production: Further evidence from normal and brain-damaged participants for a phrasal scope of planning. *Cognitive Neuropsychology, 21,* 625–644.

Martinez, I., & Shatz, M. (1996). Linguistic influences on categorization in preschool children: A cross-linguistic study. *Journal of Child Language, 23,* 529–545.

Masataka, N. (1992). Pitch characteristics of Japanese maternal speech to infants. *Journal of Child Language, 19,* 213–223.

Masciantonio, R. (1977). Tangible benefits of the study of Latin: A review of research. *Foreign Language Annals, 10,* 375–382.

Mason, T. (2010). *The complete idiot's guide to ventriloquism*. New York: Alpha.

Massaro, D. W., & Cohen, M. M. (2000). Fuzzy logical model bimodal emotion perception: Comment on The perception of emotions by ear and by eye by de Gelder and Vroomen. *Cognition and Emotion, 14*(3), 313–320.

Masson, M. E. J., & Miller, J. (1983). Working memory and individual differences in comprehension and memory of text. *Journal of Educational Psychology, 75,* 314–318.

Matsumoto, Y. (1997). The rise and fall of Japanese nonsubject honorifics: The case of "o-verb-suru." *Journal of Pragmatics, 28*(6), 719–740.

Matzat, W. (2000). Emil Krebs (1867–1930), das Sprachwunder, Dolmetscher in Peking und Tsingtau. Eine Lebensskizze. *Bulletin of the German China Association, 1,* 31–47.

Mayer, E., Martory, M., Pegna, A. J., Landis, T., Delavelle, J., & Annoni, J. (1999). A pure case of Gerstmann syndrome with a subangular lesion. *Brain, 122,* 1107–1120.

Maylor, E. A. (1990). Age and prospective memory. *The Quarterly Journal of Experimental Psychology, 42A,* 471–493.

McCardle, P., & Chhabra, V. (Eds.). (2004). *The voice of evidence in reading research*. Baltimore: Paul H. Brookes.

McCarthy, D. (1930). *The language development of the preschool child* (No. 4). Minneapolis: Institute of Child Welfare, Monograph Series.

McClelland, J. L., & Rumelhart, D. E. (1981). An interactive activation model of context effects in letter perception: I. An account of basic findings. *Psychological Review, 88,* 375–407.

McCutchen, D., & Perfetti, C. A. (1982). The visual tongue-twister effect: Phonological activation in silent reading. *Journal of Verbal Learning and Verbal Behavior, 21,* 672–687.

McDaniel, D., & Cairns, H. S. (1996). Eliciting judgments of grammaticality and reference. In D. McDaniel, C. McKee, & H. S. Cairns (Eds.), *Methods for assessing children's syntax*. Cambridge, MA: MIT Press.

McDonough, L., Choi, S., & Mandler, J. M. (2003). Understanding spatial relations: Flexible infants, lexical adults. *Cognitive Psychology, 46*(3), 229–259.

McGhee, P. E. (1979). *Humor—its origin and development.* San Francisco: W. H. Freeman & Co.

McGhee, P. E. (1989). *Humor and children's development: A guide to practical applications*. Binghamton, NY: Haworth Press.

McGuigan, F. J. (1971). Covert linguistic behavior in deaf subjects during thinking. *Journal of Comparative and Physiological Psychology, 75,* 417–420.

McGurk, H., & MacDonald, J. (1976). Hearing lips and seeing voices. *Nature, 264*(5588), 746–748.

McHenry, H. M. (2009). Human evolution. In M. Ruse & J. Travis (Eds.), *Evolution: The first four billion years*. Cambridge, MA: The Belknap Press.

McMillan, M. (2002). *An odd kind of fame: Stories of Phineas Gage*. Cambridge, MA: MIT Press.

McNeill, D. (1985). So you think gestures are non-verbal? *Psychological Review, 92,* 350–371.

McNeill, D. (2003). Pointing and morality in Chicago. In S. Kita (Ed.), *Pointing: Where language, culture and cognition meet*. Mahwah, NJ: Erlbaum.

McNeill, D. (2005). *Gesture and thought*. Chicago: University of Chicago Press.

McTear, M. F., & Conti-Ramsden, G. (1992). *Pragmatic disability in children*. London: Whurr Publishers.

Mechelli, A., Crinion, J. T., Noppeney, U., O'Doherty, J., Ashburner, J., Frackowiak, R. S., et al. (2004). Structural plasticity in the bilingual brain. *Nature, 431,* 757.

Medler, D. A., & Binder, J. R. (2005). *MCWord: An on-line orthographic database of the English language.* Available from www.neuro.mcw.edu/mcword/

Mehl, M. R., & Pennebaker, J. W. (2003). The sounds of social life: A psychometric analysis of students' daily social environments and natural conversations. *Journal of Personality and Social Psychology, 84*(4), 857–870.

Mehler, J., Jusczyk, P. W., Lambertz, G., Halsted, N., Bertoncini, J., & Amiel-Tison, C. (1988). A precursor of language acquisition in young infants. *Cognition, 29,* 143–178.

Meier, R. P., & Newport, E. L. (1990). Out of the hands of babes: On a possible sign advantage in language acquisition. *Language, 66,* 1–23.

Meisel, J. (1989). Early differentiation of languages in bilingual children. In K. Hyltenstam & L. Obler (Eds.), *Bilingualism across the life span: Aspects of acquisition, maturity and loss.* Cambridge, UK: Cambridge University Press.

Meltzoff, A. N., & Moore, M. K. (1977). Imitation of facial & manual gestures by human neonates. *Science, 198,* 75–78.

Meltzoff, A. N., & Moore, M. K. (1983). Newborn infants imitate adult facial gestures. *Child Development, 54,* 702–709.

Melzi, G., Schick, A., & Kennedy, J. (2011). Narrative participation and elaboration: Two dimensions of maternal elicitation style. *Child Development, 82*(4), 1282–1296.

Menga, H., Smith, S. D., Hagerc, K., Helda, M., Liud, J., Olsone, R. K., et al. (2005). DCDC2 is associated with reading disability and modulates neuronal development in the brain. *Proceedings of the National Academy of Sciences, 102*(47), 17053–17058.

Mercer, C. D., & Pullen, P. C. (2005). *Students with learning disabilities* (6th ed.). Upper Saddle River, NJ: Merrill-Prentice Hall.

Meringer, R. (1908). *Aus dem Leben der Sprache*. Berlin: B. Behr.

Meringer, R., & Mayer, C. (1895). *Versprechen und Verlesen: Eine psychologisch-linguistische Studie.* Stuttgart: G. J. Goschen. (Reprint with an introduction by A. Cutler & D. A. Fay, Amsterdam: J. Benjamins, 1978.)

Mervis, C. B., & Becerra, A. M. (2007). Language and communicative development in Williams syndrome. *Mental Retardation and Developmental Disabilities Research Reviews, 13,* 3–15.

Meschyan, G., & Hernandez, A. (2002). Is native-language decoding skill related to second-language learning. *Journal of Educational Psychology, 94*(1), 14–22.

Messner, A. H., & Lalakea, M. L. (2002). The effect of ankyloglossia on speech in children. *Otolaryngology—Head and Neck Surgery, 127*(6), 539–545.

Meyer, A. S., & Bock, K. (1992). The tip-of-the-tongue phenomenon: Blocking or partial activation? *Memory & Cognition, 20,* 715–726.

Meyer, A. S., Sleiderink, A. M., & Levelt, W. J. M. (1998). Viewing and naming objects: Eye movements during noun phrase production. *Cognition, 66,* B25–B33.

Meyer, D. E., & Schvaneveldt, R. W. (1971). Facilitation in recognizing pairs of words: Evidence of a dependence between retrieval operations. *Journal of Experimental Psychology, 90,* 227–234.

Miles, L. K., Tan, L., Noble, G. D., Lumsden, J., & Macrae, C. N. (2011). Can a mind have two time lines? Exploring space-time mapping in Mandarin and English speakers. *Psychonomic Bulletin and Review, 18*(3), 598–604.

Miller, G. A. (1956). The magical number seven, plus or minus two: Some limits on our capacity for processing information. *Psychological Review, 63*(2), 81–97.

Miller, K. F., Smith, C. M., Zhu, J., & Zhang, H. (1995). Preschool origins of cross-national differences in mathematical competence: Role of number naming systems. *Psychological Science, 6*(1), 56–60.

Miller, R. L. (1968). *The linguistic relativity principle and Humboldtian ethnolinguistics: A history and appraisal.* The Hague: Mouton.

Millichap, J. G. (2008). Etiologic classification of attention-deficit/hyperactivity disorder. *Pediatrics, 121*(2), e358–365.

Minagar, A., Ragheb, J., & Kelley, R. E. (2003). The Edwin Smith surgical papyrus: Description & analysis of the earliest case of aphasia. *Journal of Medical Biography, 11*(2), 114–117.

Minami, M., & McCabe, A. (1995). Rice balls and bear hunts: Japanese and North American family narrative patterns. *Journal of Child Language, 22,* 423–445.

Miozzo, M., & Caramazza, A. (1999). The selection of determiners in noun phrase production. *Journal of Experimental Psychology: Learning, Memory, and Cognition, 25,* 907–922.

Mishkin, M., & Foroays, D. G. (1952). Word recognition as a function of retinal locus. *Journal of Experimental Psychology, 43,* 43–48.

Mithun, S. (2006). *The singing Neanderthals: The origins of music, language, mind, and body.* Cambridge, MA: Harvard University Press.

Mix, K. S. (1999). Similarity and numerical equivalence: Appearances count. *Cognitive Development, 14,* 269–297.

Moats, L. C. (1999). *Teaching reading is rocket science: What expert teachers of reading should know and be able to do.* Washington, DC: American Federation of Teachers.

Moerk, E. L. (1972). Principles of dyadic interaction in language learning. *Merrill-Palmer Quarterly, 18,* 229–257.

Moerk, E. L. (1980). Relationships between parental input frequencies and children's language acquisition—a reanalysis of Brown's data. *Journal of Child Language, 7,* 105–118.

Moerk, E. L. (1981). To attend or not to attend to unwelcome reanalyses? A reply to Pinker. *Journal of Child Language, 8*(3), 627–631.

Monetta, L., & Pell, M. D. (2007). Effects of verbal working memory deficits on metaphor comprehension in patients with Parkinson's disease. *Brain and Language, 101,* 80–89.

Moon, C., Cooper, R. P., & Fifer, W. P. (1993). Two day old infants prefer native language. *Infant Behavior and Development, 16,* 495–500.

Morgan, J., & Demuth, K. (Eds.). (1996). *Signal to syntax: Bootstrapping from speech to grammar in early acquisition.* Mahwah, NJ: Erlbaum.

Morrell, C. H., Gordon-Salant, S., Pearson, J. D., Brant, L. J., & Fozard, J. L. (1996). Age and gender-specific reference ranges for hearing level and longitudinal changes in hearing level. *Journal of the Acoustical Society America, 100,* 1949–1967.

Morris, C. H., Lenhoff, H., & Wang, P. (Eds.). (2006). *Williams-Beuren syndrome: Research, evaluation, and treatment.* Baltimore: The Johns Hopkins University Press.

Morrison, C. M., & Ellis, A. W. (1995). Roles of word frequency and age of acquisition in word naming and lexical decision. *Journal of Experimental Psychology: Learning, Memory and Cognition, 21,* 116–133.

Morrison, C. M., & Ellis, A. W. (2000). Real age of acquisition effects in word naming and lexical decision. *British Journal of Psychology, 91,* 167–180.

Morrison, C. M., Ellis, A. W., & Quinlan, P. T. (1992). Age of acquisition, not word frequency, affects object naming, not object recognition. *Memory & Cognition, 20,* 705–714.

Morton, J. (1969). The interaction of information in word recognition. *Psychological Review, 76,* 165–178.

Morton, J., & Patterson, K. E. (1980). A new attempt at an interpretation, or, an attempt at a new interpretation. In M. Coltheart, K. E. Patterson, & J. C. Marshall (Eds.), *Deep dyslexia* (pp. 91–118). London: Routledge and Kegan Paul.

Moser, H. W. (1992). Prevention of mental retardation (genetics). In L. Rowitz (Ed.), *Mental retardation in the year 2000* (pp. 140–148). New York: Springer-Verlag.

Motley, M. T. (1980). Verification of "Freudian slips" and semantic prearticulatory editing via laboratory-induced spoonerisms. In V. A. Fromkin (Ed.), *Errors in linguistic performance*. New York: Academic Press.

Motley, M. T., Camden, C. T., & Baars, B. J. (1981). Toward verifying the assumptions of laboratory-induced slips of the tongue: The output-error and editing issues. *Human Communication Research, 8,* 3–15.

Motley, M. T., Camden, C. T., & Baars, B. J. (1982). Covert formulation and editing of anomalies in speech production: Evidence from experimentally elicited slips of the tongue. *Journal of Verbal Learning and Verbal Behaviour, 21,* 578–594.

Motta, M., Bennati, E., Ferlito, L., Malaguarnera, M., & Motta, L. (2005). Successful aging in centenarians: Myths and reality. *Archives of Gerontology and Geriatrics, 40*(3), 241–251.

Mulac, A., Bradac, J. J., & Gibbons, P. (2001). Empirical support for the "gender as culture" hypothesis: An intercultural analysis of male/female language differences. *Human Communication Research, 27,* 121–152.

Multhaup, K. S., Johnson, M. D., & Tetirick, J. C. (2005). The wane of childhood amnesia for autobiographical and public event memories. *Memory, 13,* 161–173.

Murdoch, B. E., Chenery, H. J., Wilks, V., & Boyle, R. S. (1987). Language disorders in dementia of the Alzheimer type. *Brain and Language, 31,* 122–137.

Murfitt, N. (2008). "I was called Dumb Dog": Henry Winkler's happy days as The Fonz were blighted by condition undiagnosed for 35 years. *Daily Mail.* Retrieved January 31, 2013, from www.dailymail.co.uk/health/article-1092477/I-called-Dumb-Dog-Henry-Winklers-happy-days-The-Fonz-blighted-condition-undiagnosed-35-years.html

Myerson, J., Ferraro, F. R., Hale, S., & Lima, S. D. (1992). General slowing in semantic priming and word recognition. *Psychology and Aging, 7,* 257–270.

Myerson, J., Hale, S., Chen, J., & Lawrence, B. (1997). General lexical slowing and the semantic priming effect: The roles of age and ability. *Acta Psychologica, 96,* 83–101.

Naeve Velguth, S. (1996). Prelinguistic gestures in a deaf and a hearing infant. (Doctoral dissertation, University of Minnesota, 1996). *Dissertation Abstracts International: Section B: The Sciences & Engineering, 56,* 8-B.

Nagle, S. J., & Sanders, S. L. (2003). *English in the Southern United States*. Cambridge, UK: Cambridge University Press.

Naigles, L., Bavin, E., & Smith, M. (2005). Toddlers recognize verbs in novel situations and sentences. *Developmental Science, 8,* 424–431.

Nation, K., Clarke, P., Marshall, C. M., & Durand, M. (2004). Hidden language impairments in children: Parallels between poor reading comprehension and specific language impairment? *Journal of Speech Language & Hearing Research, 47,* 199–211.

Nation, K., & Snowling, M. J. (1997). Assessing reading difficulties: The validity and utility of current measures of reading skill. *British Journal of Educational Psychology, 67,* 359–370.

Nation, K., & Snowling, M. J. (1998). Semantic processing and the development of word recognition skills: Evidence from children with reading comprehension difficulties. *Journal of Memory & Language, 39,* 85–101.

Nation, K., & Snowling, M. J. (2000). Developmental differences in sensitivity to semantic relations among good and poor comprehenders: Evidence from semantic priming. *Cognition, 70,* B1–13.

Nation, P., & Waring, R. (1997). Vocabulary size, text coverage, and word lists. In N. Schmitt & M. McCarthy (Eds), *Vocabulary: Description, acquisition, pedagogy* (pp. 6–19). New York: Cambridge University Press.

National Aphasia Association. (n.d.). Aphasia frequently asked questions. Retrieved January 31, 2013, from www.aphasia.org/Aphasia%20Facts/aphasia_faq.html

National Center for Elder Abuse. (2012). Retrieved January 31, 2013, from www.ncea.aoa.gov/ncearoot/Main_Site/index.aspx

National Institute of Neurological Disorders and Stroke. (2012). Parkinson's disease information page. Retrieved January 31, 2013, from www.ninds.nih.gov/disorders/parkinsons_disease/parkinsons_disease.htm

The National Institute on Aging. (n.d.). *Why population aging matters: A global perspective.* Retrieved March 16, 2013, from www.nia.nih.gov/health/publication/why-population-aging-matters-global-perspective/trend-1-aging-population

National Institute on Deafness and Other Communication Disorders. (2011). *NIDCD fact sheet: Cochlear implants.* Bethesda, MD: NIDCD Information Clearinghouse. Availablve from www.nidcd.nih.gov/health/hearing/pages/coch.aspx#a

National Institute on Deafness and Other Communication Disorders. (n.d.). *Speech statistics.* Retrieved March 15, 2013, from www.nidcd.nih.gov/health/statistics/pages/vsl.aspx#2

National Joint Committee on Learning Disabilities. (1990). Learning disabilities: Issues on definition. In *Collective perspectives on issues affecting learning disabilities: Position papers and statements* (pp. 61–66). Austin, TX: Pro-Ed.

Natsopoulos, D., Grouios, G., Bostantzopoulou, S., Mentenopoulos, G., Katsarou, Z., & Logothetis, J. (1993). Algorithmic and heuristic strategies in comprehension of complement clauses by patients with Parkinson's disease. *Neuropsychologia, 31,* 951–964.

Natsopoulos, D., Mentenopoulos, G., Bostantzopoulou, S., Katsarou, Z., Grouios, G., & Logothetis, J. (1991). Understanding of relational time terms before and after in Parkinsonian patients. *Brain and Language, 40,* 444–458.

Naveh-Benjamin, M., & Ayres, T. J. (1986). Digit span, reading rate, and linguistic relativity. *Quarterly Journal of Experimental Psychology, 38A,* 739–751.

Nebes, R. D. (1989). Semantic memory in Alzheimer's disease. *Psychological Bulletin, 106,* 377–394.

Nelson, T. D. (Ed.). (2004). *Ageism: Stereotyping and prejudice against older persons*. Cambridge, MA: The MIT Press.

Nerhardt, G. (1970). Humor and inclination to laugh: Emotional reactions to stimuli of different divergence from a range of expectancy. *Scandinavian Journal of Psychology, 11–13,* 185–195.

Nevins, M. E., & Chute, P. M. (1996). *Children with cochlear implants in educational settings*. San Diego, CA: Singular.

Newkirk, D., Klima, E. S., Pedersen, C. C., & Bellugi, U. (1980). Linguistic evidence from slips of the hand. In V. Fromkin (Ed.), *Errors in linguistic performance: Slips of the tongue, ear, pen, and hand.* New York: Academic Press.

Newman, P., & Ratliff, M. (2001). *Linguistic fieldwork*. Cambridge, UK: Cambridge University Press.

Newport, E. L., Gleitman, H., & Gleitman, L. R. (1977). Mother, I'd rather do it myself: Some effects and noneffects of maternal speech style. In C. Snow & C. Ferguson (Eds.), *Talking to children: Language input and acquisition*. Cambridge, UK: Cambridge University Press.

Nicolson, R. I., & Fawcett, A. J. (2011). Dyslexia, dysgraphia, procedural learning and the cerebellum. *Cortex, 47,* 117–127.

Nielsen, H., Lolk, A., Andersen, K., Andersen, J., & Kragh-Sørensen, P. (1999). Characteristics of elderly who develop Alzheimer's disease during the next two years: A neuropsychological study using CAMCOG: The Odense Study. *International Journal of Geriatric Psychiatry, 14,* 957–963.

Nikolopoulos, T. P., Dyar, D., Archbold, S., & O'Donoghue, G. M. (2004). Development of spoken language grammar following cochlear implantation in prelingually deaf children. *Archives of Otolaryngology—Head and Neck Surgery, 130*(5), 629–633.

Ninio, A., & Snow, C. E. (1996). *Pragmatic development.* Boulder, CO: Westview.

Ninio, A., & Snow, C. (1999). The development of pragmatics: Learning to use language appropriately. In T. K. Bhatia & W. C. Ritchie (Eds.), *Handbook of language acquisition* (pp. 347–383). New York: Academic Press.

Nippold, M. A., Leonard, L. B., & Anastopoulos, A. (1982). Development in the use and understanding of polite forms in children. *Journal of Speech and Hearing Research, 25,* 193–202.

Nippold, M. A., & Rudzinski, M. (1993). Familiarity and transparency in idiom explanation: A developmental study of children and adolescents. *Journal of Speech and Hearing Research, 36,* 728–737.

Nixon, P., Lazarova, J., Hodinott-Hill, I., Gough, P., & Passingham, R. (2004). The inferior frontal gyrus & phonological processing: An investigation using rTMS. *Journal Cognitive Neuroscience, 16,* 289–300.

Noble, K. G., Norman, M. F., & Farah, M. J. (2005). Neurocognitive correlates of socioeconomic status in kindergarten children. *Developmental Science, 8*(1), 74–87

Nottebohm, F. (1969). The song of the chingolo, Zonotrichia capensis, in Argentina: Description and evaluation of a system of dialects. *Condor, 71*(3), 299–315.

Nooteboom, S. G. (1980). Speaking and unspeaking: Detection and correction of phonological and lexical errors of speech. In V. A. Fromkin (Ed.), *Errors in linguistic performance: Slips of the tongue, ear, pen, and hand* (pp. 87–96). New York: Academic Press.

Northern, J. L., & Downs, M. P. (2002). *Hearing in children* (5th ed.). Baltimore: Williams & Wilkins.

Oberecker, R., Friedrich, M., & Friederici, A. D. (2005). Neural correlates of syntactic processing in two-year-olds. *Journal of Cognitive Neuroscience, 17*(10), 1667–1678.

O'Connor, B. P., & Rigby, H. (1996). Perceptions of baby talk, frequency of receiving baby talk, and self-esteem among community and nursing home residents. *Psychology and Aging, 11,* 147–154.

Office of the Commissioner of Official Language. (n.d.). History of official languages. Retrieved March 13, 2013, from www.ocol-clo.gc.ca/html/history_histoire_e.php

Oldfield, R. (1971). The assessment and analysis of handedness: The Edinburgh inventory. *Neuropsychologia, 9,* 97–113.

Oldfield, R. C., & Wingfield, A. (1965). Response latencies in naming objects. *Quarterly Journal of Experimental Psychology, 4,* 272–281.

Oliver, B., & Buckley, S. (1994). The language development of children with Down's syndrome: First words to two-word phrases. *Down's Syndrome: Research and Practice, 2,* 71–75.

Oliver, B. R., Barker, E. D., Mandy, W. P., Skuse, D. H., & Maughan, B. (2011). Social cognition and conduct problems: A developmental approach. *Journal of the American Academy of Child Adolescent Psychiatry, 50*(4), 385–394.

Oller, D. K. (1980). The emergence of the sounds of speech in infancy. In G. Yeni-Komshian, J. F. Kavanagh, & C. A. Ferguson (Eds.), *Child phonology: Vol. 1. Production* (pp. 93–112). New York: Academic Press.

Oller, D. K., & Eilers, R. E. (2002). *Language and literacy in bilingual children.* Clevedon, UK: Multilingual Matters.

Olsen-Fulero, L., & Conforti, J. (1983). Child responsiveness to mother questions of varying type and presentation. *Journal of Child Language, 10,* 495–520.

O'Neil, R., Welsh, M., Parke, R. D., Wang, S., & Strand, C. (1997). A longitudinal assessment of the academic correlates of early peer acceptance and rejection. *Journal of Clinical Child Psychology, 26,* 290–303.

O'Neill, D. K. (1996). Two-year-old children's sensitivity to a parent's knowledge state when making requests. *Child Development, 67*(2), 659–677.

Onishi, K., & Baillargeon, R. (2005). Do 15-month-old infants understand false beliefs? *Science, 308,* 255–258.

Opie, I., & Opie, P. (1951/1997). *The Oxford dictionary of nursery rhymes* (2nd ed.). Oxford: Oxford University Press.

Organisation for Economic Cooperation and Development. (2000). *Literacy in the information age: Final report of the International Adult Literacy Survey.* Ottawa: Statistics Canada.

Orton, S. T. (1925). Word-blindness in school children. *Archives of Neurology and Psychiatry, 14,* 285–516.

Orwell, G. (1948). *1984. A novel.* New York: Harcourt, Brace & Co.

O'Seaghdha, P. G., Chen, J.-Y., & Chen, T.-M. (2010). Proximate units in word production: Phonological encoding begins with syllables in Mandarin Chinese but with segments in English. *Cognition, 115,* 282–302.

Osgood, C. E., Suci, G., & Tannenbaum, P. (1957). *The measurement of meaning.* Urbana: University of Illinois Press.

Ouellette, G. P. (2006). What's meaning got to do with it: The role of vocabulary in word reading and reading comprehension. *Journal of Educational Psychology, 98,* 554–566.

Özçaliskan, S., & Goldin-Meadow, S. (2006). X IS LIKE Y: The emergence of similarity mappings in children's early speech and gesture. In G. Kristianssen, M. Achard, R. Dirven, & F. Ruiz de Mendoza (Eds.), *Cognitive linguistics: Foundations and fields of application* (pp. 229–260). Berlin: Mouton de Gruyter.

Padden, C., & Humphries, T. (1988). *Deaf in America: Voices from a culture.* Cambridge, MA: Harvard University Press.

Paivio, A. (1971). *Imagery and verbal processes.* New York: Holt, Rinehart, and Winston.

Pallier, C., Dehaene, S., Poline, J.-B., LeBihan, D., Argenti, A.-M., Dupoux, E., et al. (2003). Brain imaging of language plasticity in adopted adults: Can a second language replace the first? *Cerebral Cortex, 13,* 155–161.

Palmore, E. (2004). Ageism in Canada and the United States? *Journal of Cross-Cultural Gerontology, 19,* 41–46.

Paracchini, S., Ang, Q. W., Stanley, F. J., Monaco, A. P., Pennell, C. E., & Whitehouse, A. J. O. (2011). Analysis of dyslexia candidate genes in the Raine cohort representing the general Australian population. *Genes, Brain and Behavior, 10*(2), 158–165.

Patterson, J. (2004). Comparing bilingual and monolingual toddlers' expressive vocabulary size: Revisiting Rescorla and Achenbach (2002). *Journal of Speech, Language and Hearing Research, 47,* 1213–1215.

Patterson, J. L., & Pearson, B. Z. (2004). Bilingual lexical development: Influences, contexts, and processes. In B. Goldstein (Ed.), *Bilingual language development and disorders in Spanish-English speakers* (pp. 77–104). Baltimore: Paul Brookes.

Paul, L. K., Van Lancker-Sidtis, D., Schieffer, B., Dietrich, R., & Brown, W. S. (2003). Communicative deficits in individuals with agenesis of the corpus callosum: Nonliteral language and affective prosody. *Brain & Language, 85,* 313–324.

Paveza, G. J., Cohen, D., Eisdorfer, C., & Freels, S. (1992). Severe family violence and Alzheimer's disease: Prevalence and risk factors. *Gerontologist, 32,* 493–497.

Pavlov, I. P. (1960). *Conditional reflexes.* New York: Dover Publications. (Original work published 1927)

Pearson, B. Z., & Fernández, S. (1994). Patterns of interaction in the lexical development in two languages of bilingual infants. *Language Learning, 44,* 617–653.

Pearson, B. Z., Fernández, S., & Oller, D. K. (1993). Lexical development in simultaneous bilingual infants: Comparison to monolinguals. *Language Learning, 43,* 93–120.

Pearson, B. Z., Fernández, S., & Oller, D. K. (1995). Cross-language synonyms in the lexicons of bilingual infants: One language or two? *Journal of Child Language, 22,* 345–368.

Peek, F. (1996). *The real Rain Man: Kim Peek.* Salt Lake City, UT: Harkness.

Pellegrini, A. D., Brody, G. H., & Stoneman, Z. (1987). Children's conversational competence with their parents. *Discourse Processes, 10,* 93–106.

Peña, M., Maki, A., Kovačić, D., Dehaene-Lambertz, G., Koizumi, H., Bouquet, F., et al. (2003). Sounds and silence: An optical topography study of language recognition at birth. *Proceedings of the National Academy of Sciences USA, 100*(20), 11702–11705.

Penfield, W. (1958). Some mechanisms of consciousness discovered during electrical stimulation of the brain. *Proceedings of the National Academy of Sciences U.S.A. 44,* 51–66.

Penfield, W. (1972). The electrode, the brain & the mind. *Journal of Neurology, 201,* 297–309.

Penfield, W., & Roberts, L. (1959). *Speech & brain mechanisms.* Princeton, NJ: Princeton University Press.

Penn, A. (Director). (1962). *The miracle worker* [Motion picture]. United States: United Artists.

Pennebaker, J. W., & King, L. A. (1999). Linguistic styles: Language use as an individual difference. *Journal of Personality and Social Psychology, 77,* 1296–1312.

Penny, S. M. (2006). Agenesis of the corpus callosum: Neonatal sonographic detection. *Radiologic Technology, 78*(1), 14–18.

Peters, R. (2006). Aging and the brain. *Postgraduate Medical Journal, 82*(964), 84–88.

Peterson, N. (2008). *Embracing the edge.* Bothell, WA: Book Publishers Network.

Peterson-Falzone, S. J., Hardin-Jones, M. A., & Karnell, M. P. (2001). *Cleft palate speech* (3rd ed.). St. Louis, MO: Mosby, Inc.

Petitjean C. (1989). *Une condition de l'audition foetale: la conduction sonore osseuse. Consequences cliniques et applications pratiques envisagees.* France: University of Besancon.

Petitto, L. A., Katerelos, M., Levy, B. G., Gauna, K., Tetreault, K., & Ferraro, V. (2001). Bilingual signed and spoken language acquisition from birth: Implications for the mechanisms underlying early bilingual language acquisition. *Journal of Child Language, 28,* 453–496.

Petitto, L. A., & Kovelman, I. (2003). The bilingual paradox: How signing-speaking bilingual children help us to resolve it and teach us about the brain's mechanisms underlying all language acquisition. *Learning Languages, 8*(3), 5–19.

Petitto, P. F., & Marentette, L. A. (1991). Babbling in the manual mode: Evidence for the ontogeny of language. *Science, 251*(5000), 1493–1496.

Petryshen T. L., Kaplan, B. J., Liu, M. F., & Field L. L. (2000). Absence of significant linkage between phonological coding dyslexia and chromosome 6p23-21.3, as determined by use of quantitative-trait methods: Confirmation of qualitative analyses. *American Journal of Medical Genetics, 66,* 708–714.

Philip, W., & Coopmans, P. (1996). The role of lexical feature acquisition in the development of pronominal anaphora. In W. Philip & F. Wijnen (Eds.), *Amsterdam series on child language development* (Vol. 5). Amsterdam: Instituut Algemene Taalwetenschap 68.

Piaget, J. (1926). *The language and thought of the child.* Oxford, UK: Harcourt, Brace.

Piaget, J. (1957). *Construction of reality in the child.* London: Routledge & Kegan Paul.

Piaget, J. (1959). *The language and thought of the child* (3rd ed.). London: Routledge and Kegan Paul.

Pica, P., Lemer, C., Izard, V., & Dehaene, S. (2004). Exact and approximate arithmetic in an amazonian indigene group. *Science, 306*(5695), 499–503.

Pillemer, K. A., & Suitor, J. J. (1992). Violence and violent feelings: What causes them among family caregivers? *Journal of Gerontology: Social Sciences, 47*(4), 165–172.

Pinker, S. (1981). On the acquisition of grammatical morphemes. *Journal of Child Language, 8,* 477–484.

Pinker, S. (1984). Visual cognition: An introduction. *Cognition, 18,* 1–63.

Pinker, S. (1994). *The language instinct.* New York: Morrow.

Pinker, S. (1999). *Words and rules.* New York: Harper Perennial.

Pinker, S. (2007). *The language instinct.* New York: Harper Perennial Modern Classics.

Pinker, S., Lebeaux, S., & Frost, L. A. (1987). Productivity and constraints in the acquisition of the passive. *Cognition, 26,* 195–267.

Pisoni, D. B. (1993). Long-term memory in speech perception: Some new findings on speaker variability, speaking rate, and perceptual learning. *Speech Communications, 13,* 109–125.

Pohl, M. E. D., Pope, K. O., & Von Nagy, C. (2002). Olmec origins of Mesoamerican writing. *Science, 298,* 1984–1986.

Polian, G., & Bohnemeyer, J. (2011). Uniformity and variation in Tseltal reference frame use. *Language Sciences, 33,* 868–891.

Poline, J. B., Vandenberghe, R., Holmes, A. P., Friston, K. J., & Frackowiak, R. S. J. (1996). Reproducibility of PET Activation Studies: Lessons from a multi-center European experiment: EU concerted action on functional imaging. *NeuroImage, 4,* 34–54.

Pomerleau, C. S., Downey, K. K., Snedecor, S. M., Mehringer, A. M., Marks, J. L., & Pomerleau, O. F. (2003). Smoking patterns and abstinence effects in smokers with no ADHD, childhood ADHD, and adult ADHD symptomatology. *Addictive Behaviors, 28,* 1149–1157.

Popham, L., Kennison, S. M., & Bradley, K. I. (2011a). Ageism and risk-taking in young adults: Evidence for a link between risk-taking and ageism. *Death Studies, 8,* 751–763.

Popham, L., Kennison, S. M., & Bradley, K. I. (2011b). Ageism, sensation-seeking, and risk-taking in young adults. *Current Psychology, 30,* 184–193.

Poplack, S. (1980). Sometimes I'll start a sentence in Spanish y termino en español: Toward a typology of code-switching. *Linguistics, 18*(7/8), 581–618.

Poplack, S. (1981). Syntactic structure and social function of code-switching. In R. Duran (Ed.), *Latino language and communicative behavior* (pp. 169–184). New Jersey: Ablex Publishing Corp.

Portocarrero, J. S., Burright, R. G., & Donovick, P. J. (2007). Vocabulary and verbal fluency of bilingual and monolingual college students. *Archives of Clinical Neuropsychology, 22,* 415–422.

Potter, B. (2012). *Salma Hayek: "I'm Dyslexic, Short and Chubby."* Retrieved February 7, 2013, from www.entertainmentwise.com/news/90155/Salma-Hayek-Im-Dyslexic-Short-And-Chubby.

Premack, D. G., & Woodruff, G. (1978). Does the chimpanzee have a theory of mind? *Behavioral and Brain Sciences, 1*(4), 515–526.

Presnell, L. (1973). Hearing-impaired children's comprehension and production of syntax in oral language. *Journal of Speech and Hearing Research, 16,* 12–21.

Price, J., Roberts, J., Hennon, E. A., Berni, M. C., Anderson, K. L., & Sideris, J. (1999). Syntactic complexity during conversation of boys with fragile X syndrome and Down syndrome. *Journal of Speech, Language, and Hearing Research, 51*(1), 3–15.

Pullum, G. K. (1991). *The great Eskimo vocabulary hoax and other irreverent essays on the study of language.* Chicago; University of Chicago Press.

Pulsifer, M. B., Brandt, J., Salorio, C. F., Vining, E. P. G., Carson, B. S., & Freeman, J. M. (2004). The cognitive outcome of hemispherectomy in 71 children. *Epilepsia, 45*(3), 243–254.

Quaireau, C. (1995). Effet d'amorçage sémantique en décision lexicale: une comparaison jeunes versus âgés. *Bulletin de Psychologie, 49,* 14–18.

Quaresima, V., Ferrari, M., van der Sluijs, M. C., Menssen, J., & Colier, W. N. (2002). Lateral frontal cortex oxygenation changes during translation and language switching revealed by non-invasive near-infrared multi-point measurements. *Brain Research Bulletin, 59*(3), 235–243.

Quigley, S., & King, C. (Eds.). (1982). *Reading milestones: Level 5.* Beaverton, OR: Dormac.

Racine, M. B., Majnemer, A., Shevell, M., & Snider, L. (2008). Handwriting performance in children with attention deficit hyperactivity disorder (ADHD). *Journal of Child Neurology, 23*(4), 399–406.

Raffaele, P. (2006). The smart & swinging bonobo. *Smithsonian, 37*(8), 66.

Rafferty, E. A. (1986). *Second language study and basic skills in Louisiana*. Baton Rouge: Louisiana Department of Education.

Rapp, A. M., & Wild, B. (2011). Nonliteral language in Alzheimer dementia: A review. *Journal of the International Neuropsychological Society, 17*(2), 207–218.

Rapp, B. C., & Caramazza, A. (1995). Disorders of lexical processing and the lexicon. In M. S. Gazzaniga (Ed.), *Cognitive neurosciences* (pp. 901–913). Cambridge, MA: The MIT Press.

Ratcliff, R., Thapar, A., Gomez, P., & McKoon, G. (2004). A diffusion model analysis of the effects of aging in the lexical-decision task. *Psychology and Aging, 19,* 278–289.

Rauscher, F. H., Krauss, R. M., & Chen, Y. (1996). Gesture, speech, and lexical access: The role of lexical movements in speech production. *Psychological Science, 7*(4), 226–231.

Raver, C. C., & Zigler, E. F. (1997). Social competence: An untapped dimension in evaluating Head Start's success. *Early Childhood Research Quarterly, 12*(4), 363–385.

Rayner, K., Carlson, M., & Frazier, L. (1983). The interaction of syntax and semantics during sentence processing. *Journal of Verbal Learning and Verbal Behavior, 22,* 358–374.

Rayner, K., & Duffy, S. A. (1986). Lexical complexity and fixation times in reading: Effects of word frequency, verb complexity, and lexical ambiguity. *Memory & Cognition, 14,* 191–201.

Rayner, K., Inhoff, A. W., Morrison, R. E., Slowiaczek, M. L., & Bertera, J. H. (1981). Masking of foveal and parafoveal vision during eye fixations in reading. *Journal of Experimental Psychology: Human Perception and Performance, 7,* 167–179.

Rayner, K., & Pollatsek, A. (1989). *The psychology of reading*. Mahwah, NJ: Erlbaum.

Rayner, K., Pollatsek, A., Ashby, J., & Clifton, C. (2012). *The psychology of reading* (2nd ed.). New York: Psychology Press.

Rayner, K., Sereno, S. C., Morris, R. K., Schmauder, A. R., & Clifton, C., Jr. (1989). Eye movements and on-line language comprehension processes. *Language and Cognitive Processes, 4,* SI21–SI49.

Raz, I. S., & Bryant, P. (1990). Social background, phonological awareness and children's reading. *British Journal of Developmental Psychology, 8*(3), 209–225.

Reed, A., & Kellogg, B. (1899). Graded lessons in English. New York: Effingham Maynard & Co. Retrieved March 15, 2013, from http://digital.library.pitt.edu/cgi-bin/t/text/text-idx?idno = 00ABY4888m;view = toc;c = nietz

Reich, P. (1986). *Language development*. Englewood Cliffs, NJ: Prentice Hall.

Reicher, G. M. (1969). Perceptual recognition as a function of meaningfulness of stimulus material. *Journal of Experimental Psychology, 81*(2), 275–280.

Reichle, E. D., & Perfetti, C. A. (2003). Morphology in word identification: A word experience model that accounts for morpheme frequency effects. *Scientific Studies of Reading, 7,* 219–237.

Rein, J., Goldwater, M. B., & Markman, A. B. (2010). What is typical about the typicality effect in category-based induction? *Memory & Cognition, 38*(3), 377–388.

Rescorla, L. (1989). The language development survey: A screening tool for delayed language in toddlers. *Journal of Speech and Hearing Disorders, 54,* 587–599.

Rescorla, L. (2009). Age 17 language and reading outcomes in late-talking toddlers: Support for a dimensional perspective on language delay. *Journal of Speech, Language, and Hearing Research, 52,* 16–30.

Rescorla, L., & Achenbach, T. (2002). Use of the Language Development Survey in a national probability sample of children from 18 to 35 months old. *Journal of Speech, Language, and Hearing Research, 45,* 733–743.

Ricciardelli, L. A. (1992). Bilingualism and cognitive development: A review of past and recent findings. *Journal of Creative Behavior, 26,* 242–254.

Rice, M. L. (1980). *Cognition to language*. Baltimore: University Park Press.

Rice, M. L. (1997). Specific language impairments: In search of diagnostic markers and genetic contributions. *Mental Retardation and Developmental Disabilities Research Reviews, 3,* 350–357.

Rice, M. L., & Wexler, L. (1996). Toward tense as a clinical marker of specific language impairment. *Journal of Speech and Hearing Research, 39,* 1239–1257.

Richards, J. C., & Rodgers, T. S. (1987). Through the looking glass: Trends and directions in language teaching. *RELC, 18,* 45–73.

Richards, J. C., & Rodgers, T. (2001). *Approaches and methods in language teaching* (2nd ed.). Cambridge, UK: Cambridge University Press.

Riley, J. R., Greggers, U., Smith, A. D., Reynolds, D. R., & Menzel, R. (2005). The flight paths of honeybees recruited by the waggle dance. *Nature, 435*(7039), 205–207.

Ripich, D., & Terrell, B. (1988). Patterns of discourse cohesion and coherence in Alzheimer's Disease. *Journal of Speech and Hearing Disorders, 53,* 8–15.

Rips, L. J. (1975). Inductive judgments about natural categories. *Journal of Verbal Learning and Verbal Behavior, 14,* 665–681.

Ritti, A. (1973). Social junctions of children's speech (Doctoral dissertation, Columbia University, Teachers College, 1973). *Dissertation Abstracts International, 34,* 2289B.

Roach, P. (2000). *English phonetics & phonology: A practical course*. Cambridge, UK: Cambridge University Press.

Roberson, D., Davidoff, J., Davies, I. J. L., & Shapiro, L. R. (2005). Color categories: Evidence for the cultural relativity hypothesis. *Cognitive Psychology, 50,* 378–411.

Roberson, D., Davies, I. R. L., & Davidoff, J. (2000). Color categories are not universal: Replications & new evidence from a Stone-Age culture. *Journal of Experimental Psychology: General, 129,* 369–398.

Roberson, D., Pak, H., & Hanley, J. R. (2008). Categorical perception of colour in the left and right visual field is verbally mediated: Evidence from Korean. *Cognition, 107,* 752–762.

Roberts, K. (1997). A preliminary account of the effect of otitis media on 15-month-olds categorization and some implications for early language learning. *Journal of Speech, Language, and Hearing Research, 40*(3), 508–518.

Roberts, P. M., Garcia, L. J., Desrochers, A., & Hernandez, D. (2002). English performance of proficient bilingual adults on the Boston Naming Test. *Aphasiology, 16,* 635–645.

Robinson, R. J., (1991). Causes and associations of severe and persistent specific speech and language disorders in children. *Developmental Medicine and Child Neurology, 33,* 943–962.

Robinson, S. (2006). The phoneme inventory of the Aita dialect of Rotokas. *Oceanic Linguistics, 45,* 206–209.

Robock, A., Ammann, C. M., Oman, L., Shindell, D., Levis, S., & Stenchikov, G. (2009). Did the Toba volcanic eruption of ~74k BP produce widespread glaciation? *Journal of Geophysical Research, 114,* 1–9.

Rodriguez-Fornells, A., Rotte, M., Heinze, H. J., Nösselt, T., & Münte, T. F. (2002). Brain potential and functional MRI evidence for how to handle two languages with one brain. *Nature, 415,* 1026–1029.

Rodriguez-Fornells, A., Van der Lugt, A., Rotte, M., Britti, B., Heinze, H. J., & Münte, T. F. (2005). Second language interferes with word production in fluent bilinguals: Brain potential and functional imaging evidence. *Journal of Cognitive Neuroscience, 17,* 422–433.

Rogers, C., Lister, J., Febo, D., Besing, J., & Abrams, H. (2006). Effects of bilingualism, noise, and reverberation on speech perception by listeners with normal hearing. *Applied Psycholinguistics, 27,* 465–485.

Rogers, H. (2005). *Writing systems: A linguistic approach.* Malden, MA: Blackwell Publishing.

Rogers, S. J., & Puchalski, C. B. (1986). Social smiles of visually impaired infants. *Journal of Visual Impairment & Blindness, 80*(7), 863–865.

Rollins, P. (1980). *Benjamin Lee Whorf: Lost generation theories of language, mind, and religion*. Ann Arbor, MI: University Microfilms International.

Rosenblum, S., Epsztein, L., & Josman, N. (2008). Handwriting performance of children with attention deficit hyperactive disorders: A pilot study. *Physical and Occupational Therapy in Pediatrics, 28,* 219–234.

Rosenblum, S., Weiss, P. L., & Parush, S. (2004). Handwriting evaluation for developmental dysgraphia: Process versus product. *Reading and Writing, 17*(5), 433–458.

Rosenblum, S., Werner, P., Dekel, T., Gurevitz, I., & Heinik, J. (2010). Handwriting process variables among elderly people with mild major depressive disorder: A preliminary study. *Aging: Clinical and Experimental Research, 22*(2).

Rosselli, M., Ardila, A., Araujo, K., Weekes, V. A., Caracciolo, V., Padilla, M., et al. (2000). Verbal fluency and repetition skills in healthy older Spanish-English bilinguals. *Applied Neuropsychology, 7*(1), 17–24.

Rothbart, M., & Lewis, S. (1988). Inferring category attributes from exemplar attributes: Geometric shapes and social categories. *Journal of Personality and Social Psychology, 55,* 861–872.

Rovee-Collier, C. (1997). Dissociations in infant memory: Rethinking the development of implicit & explicit memory. *Psychological Review, 104,* 467–498.

Rovee-Collier, C., Hayne, H., & Colombo, M. (2001). *The development of implicit & explicit memory.* Philadelphia: John Benjamins Publishing Co.

Rubin, D. C., Schrauf, R. W., Gulgoz, S., & Naka, M. (2007). On the cross-cultural variability of component processes in autobiographical remembering: Japan Turkey, and the U.S.A. *Memory, 15*(5), 536–547.

Rubin, E. H., Storandt, M., Miller, P., Kinscherf, D. A., Grant, E. A., Morris, J. C., et al. (1998). A prospective study of cognitive function and onset of dementia in cognitively healthy elders. *Archives of Neurology, 55,* 395–401.

Ruhlen, M. (1975). *A guide to the languages of the world. Language Universals Project*. Palo Alto, CA: Stanford University Press.

Rumelhart, D. E., McClelland, J. L., & the PDP Research Group. (1986). *Parallel distributed processing: Explorations in the microstructure of cognition* (Vol. 1). Cambridge, MA: MIT Press.

Ryan, E. B., Meredith, S., MacLean, M., & Orange, J. B. (1995). Changing the way we talk with elders: Promoting health using the Communication Enhancement Model. *International Journal on Aging and Human Development, 41*(2), 89–107.

Sabat, S. R. (1994). Recognizing and working with remaining abilities: Toward improving the care of Alzheimer's disease suffers. *The American Journal of Alzheimer's Care and Related Disorders & Research, 9,* 8–16.

Saffran, J. R. (2003). Statistical language learning: Mechanisms and constraints. *Current Directions in Psychological Science, 12,* 110–114.

Saffran, J. R., Aslin, R. N., & Newport, E. L. (1996). Statistical learning by 8-month-old infants. *Science, 274,* 1926–1928.

Salthouse, T. A. (1996). General and specific speed mediation of adult age differences in memory. *Journal of Gerontology: Psychological Sciences, 51B,* P30–P42

Sander, E. K. (1972). When are speech sounds learned? *Journal of Speech & Hearing Disorders, 37,* 55–61.

Sandhofer, C., & Smith, L. B. (1999). Learning color words involves learning multiple mappings. *Developmental Psychology, 35,* 668–679.

Santiago-Rivera, A. L., & Altarriba, J. (2002). The role of language in therapy with the Spanish-English bilingual client. *Professional Psychology: Research and Practice, 33,* 30–38.

Santiago-Rivera, A.L., Altarriba, J., Poll, N., Gonzalez-Miller, N., & Cargun, C. (2009). Therapists' views on counseling the bilingual English-Spanish client: A qualitative study. *Professional Psychology: Research and Practice, 20,* 436–443.

Santrock, J. W. (2007). *Essentials of life-span development*. New York: McGraw-Hill.

Sassenberg, U., & van der Meer, E. (2010). Do we really gesture more when it's more difficult? *Cognitive Science, 34,* 643–664.

Saunders, B. (2000). Revisiting basic color terms. *Journal of the Royal Anthropological Institute, 6,* 81–99.

Saunders, G. (1998). *Bilingual children: Birth to teens*. Philadelphia: Multilingual Matters Inc.

Saussure, F. de (1977). *Cours de linguistique générale* (W. Baskin, Trans.). Glasgow: Fontana/Collins. (Original work published 1916)

Savage-Rumbaugh, S., & Lewin, R. (1994). *Kanzi: The ape at the brink of the human mind*. Hoboken, NJ: Wiley.

Savage-Rumbaugh, S., Shanker, S., & Taylor, T. (1998). *Ape, language & the human mind*. New York: Oxford University Press.

Scarborough, H. S. (1989). Prediction of reading disability from familial and individual differences. *Journal of Educational Psychology, 81*, 101–108.

Scarborough, H. S., & Parker, J. L. (2003). Matthew effects in children with learning disabilities: Development of reading, IQ, and psychosocial problems from Grade 2 to Grade 8. *Annals of Dyslexia, 53,* 47–71.

Scarborough, H. S., Rescorla, L. R., Tager-Flusberg, H., Fowler, A. E., & Sudhalter, V. (1991). The relation of utterance length to grammatical complexity in normal and language-disordered samples. *Applied Psycholinguistics, 12*, 23–45.

Scarmeas, N., & Stern, Y. (2003). Cognitive reserve and lifestyle. *Journal of Clinical and Experimental Neuropsychology, 25,* 625–633.

Schachter, D., Gilbert, D., & Wegner, D. (2011). *Psychology* (2nd ed.). New York: Worth Publishers.

Schachter, S., Christenfeld, N., Ravina, B., & Bilous, F. (1991). Speech disfluency and the structure of knowledge. *Journal of Personality and Social Psychology, 60,* 362–367.

Schiller, F. (1998). *Paul Broca: Explorer of the brain*. Oxford, UK: Oxford University Press.

Schmelz, M., Call, J., & Tomasello, M. (2011). Chimpanzees know that others make inferences. *Proceedings of the National Academy of Sciences, 108,* 17284–17289.

Schmitt, A. (1994). *Brilliant idiot: An autobiography of a dyslexic*. Intercourse, PA: Good Books.

Schmitt, J. F., & Moore, J. R. (1989). Natural alteration of speaking rate: The effect on passage comprehension by listeners over 75 years of age. *Journal of Speech and Hearing Research, 32,* 445–450.

Schnadt, M. J., & Corley, M. (2006). The influence of lexical, conceptual and planning based factors on disfluency production. *Proceedings of the twenty-eighth meeting of the Cognitive Science Society.*

Schneider, B. A., & Pichora-Fuller, M. K. (2000). Implications of sensory deficits for cognitive aging. In F. I. M. Craik & T. Salthouse (Eds.), *The handbook of aging and cognition* (2nd ed., pp. 155–219). Mahwah, NJ: Erlbaum.

Schorr, E. A., Roth, F. P., & Fox, N. A. (2008). A comparison of the speech and language skills of children with cochlear implants and children with typical hearing. *Communication Disorders Quarterly, 29,* 195–210.

Schumacher, J., Anthoni, H., Dahdouh, F., Konig, I. R., Hillmer, A. M., Kluck, N., et al. (2006). Strong genetic evidence of DCDC2 as a susceptibility gene for dyslexia. *American Journal of Human Genetics, 78,* 52–62.

Schwanenflugel, P. (1991). Why are abstract concepts hard to understand? In P. Schwanenflugel (Ed.), *The psychology of word meanings* (pp. 223–250). Hillsdale, NJ: Erlbaum.

Schwartz, B. L. (2011). The effect of being in a tip-of-the-tongue state on subsequent items. *Memory & Cognition, 39,* 245–250.

Schwartz, M. F., Marin, O. S. M., & Saffran, E. M. (1979). Dissociations of language function in dementia: A case study. *Brain and Language, 7,* 277–306.

Sciutto, M. J., Nolfi, C. J., & Bluhm, C. A. (2004). Effects of child gender and symptom type on elementary school teachers' referrals for ADHD. *Journal of Emotional and Behavioral Disorders, 12,* 247–253.

Scott, A. (1989). The vertical dimension and time in Mandarin. *Australian Journal of Linguistics, 9,* 295–314.

Scripps National Spelling Bee. (n.d.-a). *About the bee.* Retrieved March 10, 2013, from www.spelling bee .com/about-the-bee

Scripps National Spelling Bee. (n.d.-b). *Champions and their winning words.* Retrieved March 10, 2013, from www.spellingbee.com/champions-and-their-winning-words

Searchinger, G., Male, M., & Wright, M. (Writers). (2005). *Human language series* [DVD]. United States: Equinox Films/Ways of Knowing Inc.

Searle, J. (1965). What is a speech act? In P. P. Giglioli (Ed.), *Language and social context* (pp. 136–154). Harmondsworth, UK: Penguin Books.

Searle, J. (1979). Metaphor. In A. Ortony (Ed.), *Metaphor and thought* (pp. 92–123). New York: Cambridge University Press.

Sedivy, J. C., Tanenhaus, M. K., Chambers, C. G., & Carlson, G. N. (1999). Achieving incremental semantic interpretation through contextual representation. *Cognition, 71,* 109–148.

Seidenberg, M. S. (1985). The time course of phonological code activation in two writing systems. *Cognition, 19,* 1–30.

Seidenberg, M. S., & McClelland, J. L. (1989). A distributed, developmental model of word recognition and naming. *Psychological Review, 96,* 523–568.

Seidenberg, M. S., Waters, G. S., Barnes, M. A., & Tanenhaus, M. K. (1984). When does irregular spelling or pronunciation influence word recognition? *Journal of Verbal Learning and Verbal Behavior, 23,* 383–404.

Seliger, H. W., & Vago, R. M. (1991). The study of first language attrition: An overview. In H. W. Seliger & R. M. Vago (Eds.), *First language attrition* (pp. 3–15). Cambridge, UK: Cambridge University Press.

Senghas, A. (2005). Language emergence: Clues from a new Bedouin sign language. *Current Biology, 15,* 463–465.

Sera, M. D., Berge, C., & del Castillo Pintado, J. (1994). Grammatical and conceptual forces in the attribution of gender by English and Spanish speakers. *Cognitive Development, 9,* 261–292.

Sera, M. D., Elieff, C., Burch, M. C., Forbes, J., & Rodríguez, W. (2002). When language affects cognition and when it does not: An analysis of grammatical gender and classification. *Journal of Experimental Psychology: General, 131,* 377–397.

Sereno, S. C., O'Donnell, P. J., & Rayner K. (2006). Eye movements and lexical ambiguity resolution: Investigating the subordinate bias effect. *Journal of Experimental Psychology: Human Perception and Performance, 32*(2), 335–350.

Serretti, A., Olgiati, P., & De Ronchi, D. (2007). Genetics of Alzheimer's disease: A rapidly evolving field. *Journal of Alzheimers Disease, 12,* 73–92.

Seyfarth, R. M., & Cheney, D. L. (1992). Meaning & mind in monkeys. *Scientific American, 267,* 122–129.

Seyfarth, R. M., Cheney, D. L., & Marler, P. (1980a). Monkey responses to three different alarm calls: Evidence of predator classification & semantic communication. *Science, 210*(4471), 801–803.

Seyfarth, R. M., Cheney, D. L., & Marler, P. (1980b). Vervet monkey alarm calls: Semantic communication in a free-ranging primate. *Animal Behavior, 28*(4), 1070–1094.

Shapiro, L. R., & Hudson, J. A. (1991). Tell me a make-believe story: Coherence and cohesion in young children's picture-elicited narratives. *Developmental Psychology, 27,* 960–974.

Shattuck-Hufnagel, S. (1979). Speech errors as evidence for a serial ordering mechanism in sentence production. In W. E. Cooper & E. C. T. Walker (Eds.), *Sentence processing* (pp. 295–342). Hillsdale, NJ: Erlbaum.

Shatz, M., & Gelman, R. (1977). Beyond syntax: The influence of conversational constraints on speech modifications. In C. Ferguson & C. Snow (Eds.), *Talking to children: Language input and acquisition* (pp. 7, 189–198). Cambridge, UK: Cambridge University Press.

Shaywitz, B. A., Shaywitz, S. E., Pugh, K. R., Constable, R. T., Skudlarski, P., Fulbright, R. K., et al. (1995). Sex differences in the functional organization of the brain for language. *Nature, 373,* 607–609.

Shaywitz, B. A., Shaywitz, S. E., Pugh, K. R., Mencl, W., Fulbright, R., Skudlarski, P., et al. (2002). Disruption of posterior brain systems for reading in children with developmental dyslexia. *Biological Psychiatry, 52,* 101–110.

Shaywitz, S. E., & Shaywitz, B. A. (2004). Neurobiological basis for reading and reading disability. In P. McCardle & V. Chhabra (Eds.), *The voice of evidence in reading research* (pp. 417–442). Baltimore: Brookes.

Shi, R., & Werker, J. F. (2001). Six-month-old infants' preference for lexical words. *Psychological Science, 12*(1), 70–75.

Shi, R., Werker, J. F., & Cutler, A. (2006). Recognition and representation of function words in English-learning infants. *Infancy, 10*(2), 187–198.

Shi, R., Werker, J. F., & Morgan, J. L. (1999). Newborn infants' sensitivity to perceptual cues to lexical and grammatical words. *Cognition, 72,* B11–B21.

Shinn, M. W. (1900). *The biography of a baby.* Boston: Houghton Mifflin.

Shinn, M. W. (1908). *The development of the senses in the first three years of childhood.* Berkeley: University of California Press.

Shwe, H. I., & Markman, E. M. (1997). Young children's appreciation of the mental impact of their communicative signals. *Developmental Psychology, 33,* 630–636.

Simon, T., Hespos, S. J., & Rochat, P. (1995). Do infants understand simple arithmetic? A replication of Wynn (1992). *Cognitive Development, 10*(2), 253–269.

Singleton, D. (2005). The critical period hypothesis: A coat of many colours. *International Review of Applied Linguistics, 43,* 269–285.

Skinner, B. F. (1953). *Science & human behavior*. Oxford, UK: Macmillan.

Skinner, B. F. (1957). *Verbal behavior*. Acton, MA: Copley Publishing Group.

Skuse, D. H., Mandy, W. P., & Scourfield, J. (2005). Measuring autistic traits: Heritability, reliability and validity of the Social and Communication Disorders Checklist. *British Journal of Psychiatry, 187,* 568–572.

Slobin, D. I. (1973). Cognitive prerequisites for the development of grammar. In C. A. Ferguson & D. I. Slobin (Eds.), *Studies of child language development* (pp. 175–208). New York: Holt, Rinehart & Winston.

Slowiaczek, M. L., & Clifton, C. (1980). Subvocalization and reading for meaning. *Journal of Verbal Learning and Verbal Behavior, 19,* 573–582.

Small, B. J., Herlitz, A., Fratiglioni, L., Almkvist, O., & Bäckman, L. (1997). Cognitive predictors of incident Alzheimer's disease: A prospective longitudinal study. *Neuropsychology, 11*(3), 413–420.

Smith, E. E., Rhee, J., Dennis, K., & Grossman, M. (2001). Inductive reasoning in Alzheimer's disease. *Brain and Cognition, 47*(3), 494–503.

Smith, J., Brooks-Gunn, J., & Klebanov, P. (1997). Consequences of living in poverty for young children's cognitive and verbal ability and early school achievement. In G. J. Duncan & J. Brooks-Gunn (Eds.), *Consequences of growing up poor* (pp. 132–189). New York: Russell Sage.

Smith, M., & Wheeldon, L. (1999). High level processing scope in spoken sentence production. *Cognition, 73,* 205–246.

Smith, S. D. (2007). Genes, language development, and language disorders. *Mental Retardation and Developmental Disabilities Research Review, 13*(1), 96–105.

Smith, S. D., Kimberling, W. J., Pennington, B. F., & Lubs, H. A. (1983). Specific reading disability: Identification of an inherited form through linkage analysis. *Science, 219,* 1345–1347.

Smits-Engelsman, B. C. M., & Van Galen, G. P. (1997). Dysgraphia in children: Lasting psychomotor deficiency or transient developmental delay? *Journal of Experimental Child Psychology, 67,* 164–184.

Snedeker, J., & Trueswell, J. C. (2004). The developing constraints on parsing decisions: The role of lexical-biases and referential scenes in child and adult sentence processing. *Cognitive Psychology, 49*(3), 238–299.

Snow, C. E. (1977). Mothers' speech research: From input to interaction. In C. H. Snow & C. Ferguson (Eds.), *Talking to children*. Cambridge, UK: Cambridge University Press.

Snow, C.E. (1989). Imitativeness: A trait or a skill? In G. Speidel & K. Nelson (Eds.), *The many faces of imitation* (pp. 73–90). New York: Springer Verlag.

Snow, C. E., & Dickinson, D. K. (1990). Social sources of narrative skills at home and at school. *First Language, 10,* 87–103.

Snow, C. E., Pan, B., Imbens-Bailey, A., & Herman, J. (1996). Learning how to say what one means: A longitudinal study of children's speech act use. *Social Development, 5,* 56–84.

Snow, C. E., Perlmann, R., Berko Gleason, J., & Hooshyar, N. (1990). Developmental perspectives on politeness: Sources of children's knowledge. *Journal of Pragmatics, 14,* 289–305.

Snowdon, D. A. (1997). Aging and Alzheimer's disease: Lessons from the Nun Study. *Gerontologist, 37,* 150–156.

Snowdon, D. A., Greiner, L. H., Mortimer, J. A., Riley, K. P., Greiner, P. A., & Markesbery, W. R. (1997). Brain infarction and the clinical expression of Alzheimer disease: The Nun Study. *Journal of the American Medical Association, 277,* 813–817.

Soja, N. N. (1994). Young children's concept of color and its relation to the acquisition of color words. *Child Development, 65,* 918–937.

Solomon, S., Greenberg, J., & Pyszczynski, T. (1991). A terror management theory of social behavior: The psychological functions of self-esteem and cultural worldviews. In M. Zanna (Ed.), *Advances in experimental social psychology* (Vol. 24, pp. 91–159). Orlando, FL: Academic Press.

Soskin, W. F., & John, V. P. (1963). The study of spontaneous talk. In R. Barker (Ed.), *The stream of behavior*. New York: Appleton-Century-Crofts.

Spalding, D. (1873). Instinct, with original observations on young animals. *MacMillan's Magazine, 27,* 282–293.

Spector, C. C. (1996). Children's comprehension of idioms in the context of humor. *Language, Speech and Hearing Services in School, 27*(4), 307–313.

Spellman, B. A., & Holyoak, K. J. (1992). If Saddam is Hitler then who is George Bush? Analogical mapping between systems of social roles. *Journal of Personality and Social Psychology, 62,* 913–933.

Spence, M. J., & DeCasper, A. J. (1987). Prenatal experience with low-frequency maternal voice sounds influences neonatal perception of maternal voice samples. *Infant Behavior and Development, 16,* 133–142.

Sperling, G. (1960). Negative afterimage without prior positive image. *Science, 131,* 1613–1614.

Sperry, R. W. (1964). The great cerebral commissure. *Scientific American, 210*, 42–52.

Spieler, D. H., & Griffin, Z. M. (2006). The influence of age on the time course of word preparation in multiword utterances. *Language and Cognitive Processes, 21,* 291–321.

Stadthagen-Gonzalez, H., Bowers, J. S., & Damian, M. F. (2004). Age of acquisition effects in visual word recognition: Evidence from expert vocabularies. *Cognition, 93,* B11–B26.

Stanovich, K. E. (1986). Matthew effects in reading: Some consequences of individual differences in the acquisition of literacy. *Reading Research Quarterly, 21,* 360–407.

Stanovich, K. E., & West, R. F. (1983). On priming by a sentence context. *Journal of Experimental Psychology: General, 112,* 1–36.

Stark, R. (1980). Stages of speech development in the first year of life. In G. H. Yeni-Komshian, J. F. Kavanaugh, & C. A. Ferguson (Eds.), *Child phonology: Vol 1. Production*. New York: Academic Press.

Stephens, D. (1983). *Hemispheric language dominance and gesture hand preference.* Unpublished doctoral dissertation, Department of Behavioral Sciences, University of Chicago.

Stevens, G. (1999). Age at immigration and second language proficiency among foreign born adults. *Language in Society, 28,* 555–578.

Stevens, G. (2004). Using census data to test the critical-period hypothesis for second-language acquisition. *Psychological Science, 15,* 215–216.

Stevenson, J., Pennington, B. F., Gilger, J. W., DeFries, J. C., & Gillis, J. J. (1993). Hyperactivity and spelling disability: Testing for shared genetic aetiology. *Journal of Child Psychology and Psychiatry, 34,* 1137–1152.

Stewner-Manzanares, G. (1988). The Bilingual Education Act: Twenty years later. *National Clearinghouse on Bilingual Education, 6*(1).

Stine, E. A. L., Wingfield, A., & Myers, S. D. (1990). Age differences in processing information from television news: The effects of bisensory augmentation. *Journal of Gerontology: Psychological Sciences, 45,* 1–8.

Stoel-Gammon, C. (1988). Prelinguistic vocalizations of hearing-impaired & normally hearing subjects: A comparison of consonantal inventories. *Journal of Speech and Hearing Disorders, 53,* 302–315.

Strawbridge, W. J., Wallhagen, M. I., Shema, S. J., & Kaplan, G. A. (2000). Negative consequences of hearing impairment in old age: A longitudinal analysis. *Gerontologist, 40,* 320–326.

Stromswold, K. (2004). Universal grammar is innate, but environmental factors affect linguistic ability. Paper presented at the third annual Innateness & the Structure of the Mind Conference, 2004. Proceedings of this conference will appear in P. Carruthers & S. Stich (Eds.), *The innate mind: Foundations and the future.* Oxford University Press.

Studdert-Kennedy, M., Shankweiler, D., & Pisoni, D. (1972). Auditory and phonetic processes in speech perception: Evidence from a dichotic study. *Journal of Cognitive Psychology, 2,* 455–466.

Sundara, M., & Polka, L. (2008). Discrimination of coronal stops by bilingual adults: The timing and nature of language interaction. *Cognition, 106*(1), 234–258.

Sundara, M., Polka, L., & Molnar, M. (2008). Development of coronal stop perception: Bilingual infants keep pace with their monolingual peers. *Cognition, 108,* 232–242.

Svensson, I. (2003). *Phonological dyslexia: Cognitive, behavioural and hereditary aspects.* Unpublished doctoral dissertation, Department of Psychology, The University of Gothenburg.

Svensson, I., Nilsson, S., Wahlström, J., Jernås, M., Carlsson, L., & Hjelmquist, E. (2011). Familial dyslexia in a large Swedish family: A whole genome linkage scan. *Behavioral Genetics, 41,* 43–49.

Svirsky, M. A., Robbins, A. M., Kirk, K. I., Pisoni, D. B., & Miyamoto, R. T. (2000). Language development in profoundly deaf children with cochlear implants. *Psychological Science, 11*(2), 153–158.

Swain, I. U., Zelazo, P. R., & Clifton, R. K. (1993). Newborn infants' memory for speech sounds retained over 24 hours. *Developmental Psychology, 29,* 312–323.

Swinney, D. (1979). Lexical access during sentence comprehension: (Re)consideration of context effects. *Journal of Verbal Learning and Verbal Behavior, 18,* 645–659.

Tabor, W., & Hutchins, S. (2004). Evidence for self-organized sentence processing: Digging in effects. *Journal of Experimental Psychology: Learning, Memory, and Cognition, 30*(2), 431–450.

Taft, M. (1981). Prefix stripping revisited. *Journal of Verbal Learning and Verbal Behavior, 20,* 289–297.

Taipale, M., Kaminen, N., Nopola-Hemmi, J., Haltia, T., Myllyluoma, B., Lyytinen, H., et al. (2003). A candidate gene for developmental dyslexia encodes a nuclear tetratricopeptide repeat domain protein dynamically regulated in brain. *Proceedings of the National Academy of Science, USA, 100,* 11553–11558.

Tallal, P., Miller, S. L., Bedi, G., Byma, G., Wang, X., Nagarajan, S. S., et al. (1996). Language comprehension in language-learning impaired children improved with acoustically modified speech. *Science, 271*(5245), 81–84.

Tallal, P., & Piercy, M. (1978). Defects of auditory perception in children with developmental aphasia. In M. A. Wyke (Ed.), *Developmental dysphasia.* New York: Academic Press.

Tallerman, M. (2005). *Language origins: Perspectives on evolution*. Oxford, UK: Oxford University Press.

Tanenhaus, M., Spivey-Knowlton, M. J., Eberhard, K. M., & Sedivy, J. C. (1995). Integration of visual and linguistic information in spoken language comprehension. *Science, 268,* 1632–1634.

Tang, Y., Zhang, W., Chen, K., Feng, S., Ji, Y., Shen, J., et al. (2006). Arithmetic processing in the brain shaped by cultures. *Proceedings of the National Academy of Sciences, 103,* 10775–10780.

Tannen, D. (2001). *You just don't understand: Women and men in conversation*. New York: Quill.

Tannenbaum, K. R., Torgesen, J. K., & Wagner, R. K. (2006). Relationships between word knowledge and reading comprehension in third-grade children. *Scientific Studies of Reading, 10,* 381–398.

Tardif, T., Fletcher, P., Liang, W., Zhang, Z., Kaciroti, N., & Marchman, V. A. (2008). Baby's first 10 words. *Developmental Psychology, 44*(4), 929–938.

Tardif, T., Gelman, S. A., & Xu, F. (1999). Putting the "noun bias" in context: A comparison of Mandarin and English. *Child Development, 70*(3), 620–635.

Tarnopol, L., & Tarnopol, M. (1981). *Comparative reading and learning difficulties*. Lexington, MA: Lexington Books.

Tass, N. (Director). (2001). *The miracle worker* [Motion picture]. United States: Walt Disney Video.

Taylor, I., & Taylor, M. M. (1995). *Writing and literacy in Chinese, Korean and Japanese*. Amsterdam: John Benjamins.

Teichmann, M., Dupoux, E., Cesaro, P., & Bachoud-Levi, A. C. (2008). The role of the striatum in sentence processing: Evidence from a priming study in early stages of Huntington's disease. *Neuropsychologia, 46,* 174–185.

Teichmann, M., Dupoux, E., Kouider, S., Brugieres, P., Boisse, M. F., Eaudic, S., et al. (2005). The role of the striatum in rule application: The model of Huntington's disease at early stage. *Brain, 128,* 1155–1167.

Templin, M. (1957). *Certain language skills in children*. Minneapolis: University of Minnesota Press.

Terrace, H. S. (1979). *Nim*. New York: Knopf.

Thierry, G., Athanasopoulos, P., Wiggett, A., Dering, B., & Kuipers, J. (2009). Unconscious effects of language-specific terminology on pre-attentive colour perception. *Proceedings of the National Academy of Sciences, 106,* 4567–4570.

Thiessen, E. D., Hill, E., & Saffran, J. R. (2005). Infant-directed speech facilitates word segmentation. *Infancy, 7,* 53–71.

Thompson, R. F. (2000). *The brain: A neuroscience primer*. New York: Worth Publishers.

Tierney, M. C., Szalai, J. P., Snow, W. G., Fisher, R. H., Nores, A., Nadon, G., et al. (1996). Prediction of probable Alzheimer's disease in memory-impaired patients: A prospective longitudinal study. *Neurology, 46,* 661–665.

Tilque, D. (2000). Borrowed English. *Word Ways: The Journal of Recreational English, 33*(2), 114–118.

Timm, L. (1975). Spanish-English code-switching: El porque y how-not-to. *Romance Philology, 28,* 473–482.

Todd, L. (1990). *Pidgins & Creoles*. London: Routledge.

Toga, A., & Mazziotta, J. C. (2000). *Brain mapping: The methods*. New York: Academic Press.

Tomasello, M. (1992). *First verbs: A case study of early grammatical development*. Cambridge, UK: Cambridge University Press.

Tomasello, M., & Call, I. (1997). *Primate cognition*. Oxford, UK: Oxford University Press.

Tomasello, M., Carpenter, M., & Liszkowski, U. (2007). A new look at infant pointing. *Child Development, 78,* 705–722.

Tomasello, M., Conti-Ramsden, G., & Ewert, B. (1990). Young children's conversations with their mothers and fathers: Differences in breakdown and repair. *Journal of Child Language, 17,* 115–130.

Tomblin, J. B. (1996). Genetic and environmental contributions to the risk for specific language impairment. In M. Rice (Ed.), *Genetics of specific language impairment* (pp. 191–210). Baltimore: Brooks.

Tomlin, R. (1986). *Basic word order: Functional principles*. London: Croom Helm.

Trask, L. (1996). *The history of Basque*. London: Routledge.

Traxler, M., Pickering, M., & Clifton, C., Jr. (1998). Adjunct attachment is not a form of lexical ambiguity resolution. *Journal of Memory and Language, 39,* 558–592.

Traxler, M. J., Williams, R. S., Blozis, S. A., & Morris, R. K. (2005). Working memory, animacy and verb class in the processing of relative clauses. *Journal of Memory & Language, 53,* 204–224.

Treffert, D. A. (2009). The savant syndrome: An extraordinary condition. A synopsis: Past, present, future. *Philosophical Transactions of the Royal Society B: Biological Sciences, 364*(1522), 1351.

Treiman, R. (1985). Onsets and rimes as units of spoken syllables: Evidence from children. *Journal of Experimental Child Psychology, 39,* 161–181.

Tremblay, K., Piskosz, M., & Souza, P. (2002). Aging alters the neural representation of speech-cues. *Neuroreport, 13,* 1865–1870.

Trueswell, J. C., Sekerina, I., Hill, N. M., & Logrip, M. L. (1999). The kindergartenpath effect: Studying on-line sentence processing in young children. *Cognition, 73,* 89–134.

Trueswell, J., & Tanenhaus, M. K. (1994). Toward a lexicalist framework for constraint-based syntactic ambiguity resolution. In C. Clifton, L. Frazier, & K. Rayner (Eds.), *Perspectives on sentence processing* (pp. 155–179). Hillsdale, NJ: Erlbaum.

Trueswell, J. C., Tanenhaus, M. K., & Kello, C. (1993). Verb-specific constraints in sentence processing: Separating effects of lexical preference from garden-paths. *Journal of Experimental Psychology: Learning, Memory, & Cognition, 19,* 528–553.

Truffaut, F. (Director). (1970). *L'enfant sauvage*. France: MGM.

Tse, C.-S., & Altarriba, J. (2008). Evidence against linguistic relativity in Chinese and English: A case study of spatial and temporal metaphors. *Journal of Cognition and Culture, 8,* 335–357.

Tseng, M. H., & Chow, S. M. K. (2000). Perceptual-motor function of school-age children with slow handwriting speed. *American Journal of Occupational Therapy, 54,* 83–88.

Tversky, B., Kugelmass, S., & Winter, A. (1991). Cross-cultural and developmental trends in graphic productions. *Cognitive Psychology, 23,* 515–557.

Uller, C., Carey, S., Huntley-Fenner, G., & Klatt, L. (1999). The representations underlying infant addition. *Cognitive Development, 14,* 1–36.

Ullman, M. T., Corkin, S., Coppola, M., Hickok, G., Growdon, J. H., Koroshetz, W. J., et al. (1997). A neural dissociation within language: Evidence that the mental dictionary is part of declarative memory, and that grammatical rules are processed by the procedural system. *Journal of Cognitive Neuroscience, 9,* 266–276.

United Nations, Department of Economic and Social Affairs, Population Division. (2002). *World population ageing 1950–2050.* New York: United Nations.

United Nations, Department of Economic and Social Affairs, Population Division. (2011). *World population prospects: The 2010 revision.* New York: United Nations.

U.S. Census Bureau. (2000). Language use and English-speaking ability: 2000. Retrieved February 7, 2013, from www.census.gov/prod/2003pubs/c2kbr-29.pdf

Usher, J., & Neisser, U. (1993). Childhood amnesia and the beginnings of memory for four early life events. *Journal of Experimental Psychology, 122*(2), 155–165.

Vagh, S. B., Pan, B. A., & Mancilla-Martinez, J. (2009). Measuring growth in bilingual and monolingual children's English productive vocabulary development: The utility of combining parent and teacher report. *Child Development, 80,* 1545–1563.

van der Lely, H. K. J. (1998). SLI in children: Movement, economy and deficits in the computational-syntactic system. *Language Acquisition, 7,* 161–192.

van der Lely, H. K. J., & Christian, V. (2000). Lexical word formation in grammatical SLI children: A grammar-specific or input-processing deficit? *Cognition, 75,* 33–63.

van Gompel, R. P. G., & Liversedge, S. P. (2003). The influence of morphological information on cataphoric pronoun assignment. *Journal of Experimental Psychology: Learning, Memory, & Cognition, 29,* 128–129.

van Gompel, R. P. G., Pickering, M., Pearson, J., & Liversedge, S. P. (2005). Evidence against competition during syntactic ambiguity resolution. *Journal of Memory and Language, 52,* 284–307.

van Gompel, R. P. G., Pickering, M., & Traxler, M. (2001). Reanalysis in sentence processing: Evidence against current constraint-based and two-stage models. *Journal of Memory and Language, 45,* 225–258.

van Patter, B. (n.d.). *Greetings from around the world.* Retrieved January 29, 2013, from www.brucevanpatter.com/world_greetings.html

Van Riper, C. (1982). *The treatment of stuttering.* Eaglewood Cliffs, NJ: Prentice Hall.

Vargha-Khadem, F., Gadian, D. G., Copp, A., & Mishkin, M. (2005). FOXP2 & the neuroanatomy of speech & language. *Nature Reviews Neuroscience, 6*(2), 131–137.

Verreyt, N., Nys, G. M., Santens, P., & Vingerhoets, G. (2011). Cognitive differences between patients with left-sided and right-sided Parkinson's disease: A review. *Neuropsychology Revew, 21*(4), 405–424.

Vigliocco, G., Antonini, T., & Garrett, M. F. (1997). Grammatical gender is on the tip of Italian tongues. *Psychological Science, 8,* 314–317.

Vigliocco, G., Vinson, D. P., Martin, R. C., & Garrett, M. F. (1999) Is "count" and "mass" information available when the noun is not? An investigation of tip of the tongue states and anomia. *Journal of Memory and Language, 40,* 534–558.

Viljanen, A., Kaprio, J., Pyykko, I., Sorri, M., Pajala, S., Kauppinen, M., et al. (2009). Hearing as a predictor of falls and postural balance in older female twins. *Journals of Gerontology: Series A (Biological Sciences and Medical Sciences), 64,* 312–317.

Vital-Durand, F., Atkinson, J., & Braddick, O. J. (Eds.). (1996). *Infant vision.* New York: Oxford University Press.

Volkow, N. D., Wang, G. J., Newcorn, J., Telang, F., Solanto, M. V., Fowler, J. S., et al. (2007). Depressed dopamine activity in caudate and preliminary evidence of limbic involvement in adults with attention-deficit/hyperactivity disorder. *Archives of General Psychiatry, 64,* 932–940.

von Hentig, W. O. (1962). *Mein leben eine dienstreise.* Göttingen: Vandenhoeck & Ruprecht.

von Humboldt, W. (1841–1852). *Gesammelte werke* (7 vols.). Berlin: G. Reimer. (Reprinted by C. Brandes, Ed., 1988, Berlin: De Gruyer)

Vox, V. (1993). *I can see your lips moving: The history and art of ventriloquism.* Houston, TX: Plato Publishing.

Vygotsky, L. S. (1978). *Mind in society.* Cambridge, MA: Harvard University Press.

Vygotsky, L. S. (1986). *Thought and language* (A. Kozulin, Ed. & Trans.). Cambridge, MA: MIT Press. (Original work published 1934)

Vygotsky, L. S. (1987). Thinking and speech. In R. W. Rieber & A. S. Carton (Eds.), *The collected works of L. S. Vygotsky: Vol. 1. Problems of general psychology* (N. Minick, Trans.). New York: Plenum.

Wada J. (1949). A new method for determination of the side of cerebral speech dominance: A preliminary report on the intracarotid injection of sodium amytal in man. *Igaku Seibutsugaku, 4,* 221–222.

Wada, J. (1997). Clinical experimental observations of carotid artery injections of sodium amytal. *Brain & Cognition, 33,* 11–13.

Wada, J., & Rasmussen, T. (1960). Intracarotid injection of sodium amytal for the lateralization of cerebral speech dominance: Experimental and clinical observations. *Journal of Neurosurgery, 17,* 266–282.

Wadsworth, S. J., Corley, R. P., Hewitt, J. K., Plomin, R., & DeFries, J. C. (2002). Parent-offspring resemblance for reading performance at 7, 12, and 16 years of age in the Colorado Adoption Project. *Journal of Child Psychology and Psychiatry, 43*(6), 769–774.

Wang, Y., Paramasivam, M., Thomas, A., Bai, J., Kaminen, N., Kere, J., et al. (2006). Dyx1c1 functions in neuronal migration in developing neocortex. *Neuroscience, 143,* 515–522.

Warburton, E., Wise, R. J. S., Price, C. J., Weiller, C., Hadar, U., Ramsay, S., et al. (1996). Noun and verb retrieval by normal subjects: Studies with PET. *Brain, 119,* 159–179.

Warner, J., & Glass, A. L. (1987). Context and distance-to-disambiguation effects in ambiguity resolution: Evidence from grammaticality judgment of garden path sentences. *Journal of Memory and Language, 26,* 714–738.

Warren, K. (2011). *The revival of ventriloquism in America.* New York: Createspace.

Warren, T., & Gibson, E. (2002). The influence of referential processing on sentence complexity. *Cognition, 85,* 79–112.

Wartenburger, I., Heekeren, H. R., Abutalebi, J., Cappa, S. F., Villringer, A., & Perani, D. (2003). Early setting of grammatical processing in the bilingual brain. *Neuron, 37,* 159–170.

Waschbusch, D. A. (2002). A meta-analytic evaluation of comorbid hyperactive-impulsive-inattention problems and conduct problems. *Psychological Bulletin, 128,* 118–150.

Watkins, K. E., Vargha-Khadem, F., Ashburner, J., Passingham, R. E., Connelly, A., Friston, K. J., et al. (2002). MRI analysis of an inherited speech and language disorder: Structural brain abnormalities. *Brain, 125,* 465–478.

Watson, J. B. (1913). Psychology as the behaviorist views it. *Psychological Review, 20,* 158–177.

Watson, J. B., & Raynor, R. (1920). Conditioned emotional reactions. *Journal of Experimental Psychology, 3*(1), 1–14.

Watson, J. D., & Crick, F. H. C. (1953). A structure for deoxyribose nucleic acid. *Nature, 171*(4356), 737–738.

Waxman, S. R., & Lidz, J. L. (2006). Early word learning. In W. Damon & R. M. Lerner (Series Eds.) & D. Kuhn & R. Siegler (Eds.), *Handbook of child psychology* (Vol. 2, 6th ed., pp. 299–335). Hoboken, NJ: Wiley.

Weber-Fox, C. M., & Neville, H. J. (1999). Functional neural subsystems are differentially affected by delays in second-language immersion: ERP and behavioral evidence in bilingual speakers. In D. Birdsong (Ed.), *New perspectives on the critical period for second language acquisition.* Hillsdale, NJ: Erlbaum.

Weber-Fox, C., & Neville, H. J. (2001). Sensitive periods differentiate processing for open and closed class words: An ERP study in bilinguals. *Journal of Speech, Language, and Hearing Research, 44,* 1338-1353.

Wechsler, A. S. (1977). Three senile dementias presenting in aphasia. *Journal of Neurology, Neurosurgery, and Psychiatry, 40,* 303–305.

Werker, J. F., Polka, L., & Pegg, J. E. (1997). The conditioned head turn procedures as a method for assessing infant speech perception. *Early Development & Parenting, 6*(3–4), 171–178.

Werker, J. F., & Tees, R. (1984). Cross-language speech perception: Evidence for perceptual reorganization during the first year of life. *Infant Behavior and Development, 7,* 49–63.

Werker, J. F., & Tees, R. C. (1999). Influences on infant speech processing: Toward a new synthesis. *Annual Review of Psychology, 50,* 509–535.

Werker, J. F., Weikum, W., & Yoshida, K. A. (2006). Bilingual speech processing in infants and adults. In P. McCardle & E. Hoff (Eds.), *Childhood bilingualism: Research on infancy through school age* (pp. 1–18). Clevedon: Multilingual Matters.

Wernicke, C. (1908). The symptom of complex aphasia. Translated & republished in A. Church (Ed.), *Diseases of the nervous system.* New York: Appleton-Century-Crofts. (Original work published 1874)

Wetherby, A., Cain, D., Yonclas, D., & Walker, V. (1988). Analysis of intentional communication of normal children from the prelinguistic to the multiword stage. *Journal of Speech and Hearing Research, 31,* 240–252.

Wetzel, P. J. (2004). *Keigo in Modern Japan: Polite language from Meiji to the present.* Honolulu: University of Hawai'i Press.

White, K. K., & Abrams, L. (2002). Does priming specific syllables during tip-of-the-tongue states facilitate word retrieval in older adults? *Psychology and Aging, 17,* 226–235.

White, K. K., & Abrams, L. (2004). Phonologically mediated priming of preexisting and new associations in young and older adults. *Journal of Experimental Psychology: Learning, Memory, and Cognition, 30,* 645–655.

White, L. (1982). *Grammatical theory and language acquisition*. Dordrecht: Foris.

Whitehurst, G. J. (1997). Language processes in context: Language learning in children reared in poverty. In L. B. Adamson & M. A. Romski (Eds.), *Research on communication and language disorders: Contribution to theories of language development* (pp. 233–266). Baltimore: Brookes.

Whitehurst, G. J., & Lonigan, C. J. (1998). Child development and emergent literacy. *Child Development, 69,* 848–872.

Whorf, B. L. (1956). *Language, thought and reality*. Cambridge, MA: MIT Press.

Wierzbicka, A. (2008). Why there are no "colour universals" in language and thought. *Journal of the Royal Anthropological Institute (N.S.), 14,* 407–425.

Wigg, K. B., Couto, J. M., Feng, Y., Anderson, B., Cate-Carter, T. D., Macciardi, F., et al. (2004). Support for EKN1 as the susceptibility locus for dyslexia on 15q21. *Molecular Psychiatry, 9,* 1111–1121.

Wilcox, M. J., & Webster, E. J. (1980). Early discourse behavior: An analysis of children's responses to listener feedback. *Child Development, 51,* 1120–1125.

Wild, E. M., Teschler-Nicola, M., Kutschera, W., Steier, P., Trinkaus, E., & Wanek, W. (2005). Direct dating of early Upper Palaeolithic human remains from Mladeč. *Nature, 435,* 332–335.

Wiley, A., Rose, A., Burger, L., & Miller, P. (1998). Constructing autonomous selves through narrative practices: A comparative study of working-class and middle-class families. *Child Development, 69,* 833–847.

Wiley, E., Bialystok, E., & Hakuta, K. (2005). New approaches to using census data to test the critical period hypothesis for second language acquisition. *Psychological Science, 16*(1), 341–343.

Willcox, D. C., Willcox, B. J., Wang, N. C., He, Q., Rosenbaum, M., & Suzuki, M. (2008). Life at the extreme limit: Phenotypic characteristics of supercentenarians in Okinawa. *Journals of Gerontology Series A: Biological Sciences and Medical Sciences, 63,* 1201–1208.

Williams, D. F. (2012). *Communication sciences and disorders: An introduction to the professions.* New York: Psychology Press.

Williams, K. N. (2006). Improving outcomes of nursing home interactions. *Research in Nursing & Health, 29*(2), 121–133.

Williams, K. N., Kemper, S., & Hummert, M. L. (2003). Improving nursing home communication: An intervention to reduce elderspeak. *The Gerontologist, 43,* 242–247.

Williams-Gray, C. H., Evans, J. R., Goris, A., Foltynie, T., Ban, M., & Robbins, T. W. (2009). The distinct cognitive syndromes of Parkinson's disease: 5 year follow-up of the CamPaIGN cohort. *Brain, 132,* 2958–2969.

Wimmer, H., & Perner, J. (1983). Beliefs about beliefs: Representation and constraining function of wrong beliefs in young children's understanding of deception. *Cognition, 13*(1), 103–128.

Wingfield, A. (1996). Cognitive factors in auditory performance: Context, speed of processing, and constraints of memory. *Journal of the American Academy of Audiology, 7,* 175–182.

Wingfield, A. (1998). Comprehending spoken questions: Effects of cognitive and sensory change in adult aging. In N. Schwarz & D. Park (Eds.), *Cognition and aging self-reports* (pp. 201–228). Philadelphia: Taylor & Francis.

Wingfield, A., & Stine-Morrow, E. A. L. (2000). Language and speech. In F. I. M. Craik & T. A. Salthouse (Eds.), *Handbook of cognitive aging* (2nd ed., pp. 359–416). Mahwah, NJ: Erlbaum.

Wingfield, A., Tun, P. A., Koh, C. K., & Rosen, M. J. (1999). Regaining lost time: Adult aging and the effect of time restoration on recall of time-compressed speech. *Psychology and Aging, 14,* 380–389.

Wingfield, A., Wayland, S. C., & Stine, E. A. L. (1992). Adult age differences in the use of prosody for syntactic parsing and recall of spoken sentences. *Journal of Gerontology: Psychological Sciences, 47,* 350–356.

Winner, E., & Gardner, H. (1977). The comprehension of metaphor in brain-damaged patients. *Brain, 100,* 719–727.

Winxemer, J. (1981). *A lexical-expectation model for children's comprehension of wh-questions.* Unpublished doctoral dissertation, CUNY Graduate Center, New York.

Wolfe, H. K. (1890). On the vocabulary of children. *Nebraska University Studies, 1,* 205–234.

Wolfenstein, M. (1954). *Children's humor.* Glencoe, IL: Free Press.

Wood, C. C. (1976). Discriminability, response bias, and phoneme categories in discrimination of voice onset time. *Journal of the Acoustic Society of America, 60*(6), 1381–1389.

Wood, G., Willmes, K., Nuerk, H.-C., & Fischer, M. (2008). On the cognitive link between space and number: A meta-analysis of the SNARC effect. *Psychology Science Quarterly, 59,* 489–525.

Woodcock, R. W., & Muñoz-Sandoval, A. F. (1993). *Comprehensive manual: Woodcock-Muñoz Language Survey: English Form.* Itasca, IL: Riverside Publishing.

Wooten, J., Merkin, S., Hood, L., & Bloom, L. (1979). *Wh-questions: Linguistic evidence to explain the sequence of acquisition.* Paper presented at the biennial meeting of the Society for Research in Child Development.

Wordsmith.org. (n.d.). *A.Word.A.Day.* Retrieved March 14, 2013, from http://wordsmith.org/awad/index.html

The World Factbook. (2009). Washington, DC: Central Intelligence Agency. Retrieved from www.cia.gov/library/publications/the-world-factbook/index.html

Wynn, K. (1992). Children's acquisition of the number words & the counting system. *Cognitive Psychology, 24,* 220–251.

Wysyznski, D. F. (2002). *Cleft lip and palate: From origin to treatment.* New York: Oxford University Press.

Yairi, E. (1993). Epidemiology and other considerations in treatment efficacy research with preschool-age children who stutter. *Journal of Fluency Disorders, 18,* 197–220.

Yan, S., & Nicoladis, E. (2009). Finding le mot juste: Differences between bilingual and monolingual children's lexical access in comprehension and production. *Bilingualism: Language and Cognition, 12,* 323–335.

Yonan, C. A., & Sommers, M. S. (2000). The effects of talker familiarity on spoken word identification in younger and older listeners. *Psychology of Aging, 15*(1), 88–99.

Yuill, N., & Oakhill, J. (1991). *Children's problems in text comprehension: An experimental investigation.* Cambridge, UK: Cambridge University Press.

Zacks, R., & Hasher, L. (1997). Cognitive gerontology and attentional inhibition: A reply to Burke and McDowd. *Journal of Gerontology: Psychological Sciences, 52B,* 274–283.

Zangwell, O. L. (1967). Speech and the minor hemisphere. *Acta Neuroligica Psychiatrica Belgica, 67*(11), 1013–1020.

Zebian, S. (2005). Linkages between number concepts, spatial thinking and directionality of writing: The SNARC effect and the REVERSE SNARC effect in English and in Arabic monoliterates, biliterates and illiterate Arabic speakers. *Journal of Cognition and Culture. Special Issue: Psychological and Cognitive Foundations of Religiosity, 5*(1–2), 165–190.

Zevin, J. D., & Seidenberg, M. S. (2002). Age of acquisition effects in reading and other tasks. *Journal of Memory and Language, 47,* 1–29.

Zhang, S., & Perfetti, C. A. (1993). The tongue-twister effect in reading Chinese. *Journal of Experimental Psychology: Learning, Memory, and Cognition, 19,* 1082–1093.

Zuckerman, M., Eysenck, S., & Eysenck, H. J. (1978). Sensation seeking in England and America: Cross-cultural, age, and sex comparisons. *Journal of Consulting and Clinical Psychology, 46,* 139–149.

Photo Credits

Chapter 1: Photo 1.1, page 2. Istock/Morgan Lane Studios; Photo 1.2, page 5. Istock/WieslawFila; Photo 1.3, page 8. Istock/Givaga; Photo 1.4, page 21. Thinkstock/Mel Curtis

Chapter 2: Photo 2.1, page 28. Istock/Guido Vrola; Photo 2.2, page 29. From the collection of Jack and Beverly Wilgus; Photo 2.3, page 30. Thinkstock/ getty images; Photo 2.4, page 35. Thinkstock/ Adrian Neal; Photo 2.5, page 36. Thinkstock/ Getty Images/Jupiterimages; Photo 2.6, page 50. Istock/ Robert Maxwell; Photo 2.7, page 51. Thinkstock/ Medioimages/Photodisc; Photo 2.8, page 52. Istock/ Sergey Andreev

Chapter 3: Photo 3.1, page 58. Istock; Photo 3.2, page 67. Thinkstock/ Photodisc; Photo 3.3, page 70. Thinkstock/ Plush Studios; Photo 3.4, page 79. Thinkstock/ Stockbyte; Photo 3.5, page 80. http:// en.wikipedia.org/wiki/Helen_Keller; Photo 3.6, page 80. http://en.wikipedia.org/wiki/Helen_Keller

Chapter 4: Photo 4.1, page 88. Istock/Chris Bernard; Photo 4.2, page 101. Thinkstock/Getty Images/ Jupiterimages; Photo 4.3, page 109. Istock/DenKuvaiev; Photo 4.4, page 112. Istock/Loretta Hostettler

Chapter 5: Photo 5.1, page 118. Istock/Grzegorz Kula; Photo 5.2, page 124. Thinkstock/Getty Images/ Jupiterimages; Photo 5.3, page 131. Thinkstock/ Leonard Mc Lane

Chapter 6: Photo 6.1, page 150. Istock/andipantz; Photo 6.2, page 160. http://en.wikipedia.org/wiki/ Kim_Peek; Photo 6.3, page 163. Thinkstock/Getty Images/Jupiterimages; Photo 6.4, page 169. Istock/ jean schweitzer; Photo 6.5, page 172. Thinkstock/ Hemera Technologies

Chapter 7: Photo 7.1, page 178. Istock/Andy Dean; Photo 7.2, page 182. Thinkstock/Jupiterimages; Photo 7.3, page 190. Thinkstock/Hemera Technologies; Photo 7.4, page 195. http://en. wikipedia.org/wiki/Emil_Krebs

Chapter 8: Photo 8.1, page 206. Thinkstock/Tom Brakefield; Photo 8.2, page 211. Thinkstock/Pixland; Photo 8.3, page 215. Istock/Giedrius Dagys; Photo 8.4, page 225. Istock/matka_Wariatka

Chapter 9: Photo 9.1, page 234. Istock/Anatoliy Samara; Photo 9.2, page 244. Thinkstock/Siri Stafford; Photo 9.3, page 248. Thinkstock/ Goodshoot; Photo 9.4, page 248. Thinkstock/Aidon; Photo 9.5, page 248. Istock/SergiyN

Chapter 10: Photo 10.1, page 262. Thinkstock/Ryan McVay; Photo 10.2, page 277. Thinkstock/Getty Images/Jupiterimages; Photo 10.3, page 277. Thinkstock/George Doyle; Photo 10.4, page 283. Thinkstock/Comstock

Chapter 11: Photo 11.1, page 290. Thinkstock/ Digital Vision; Photo 11.2, page 292. Thinkstock/ Photodisc; Photo 11.3, page 306. Istock/Scott Griessel; Photo 11.4, page 309. Istock/Jeremy Sale; Photo 11.5, page 313. Istock/Andres Peiro Palmer

Chapter 12: Photo 12.1, page 318. Thinkstock/ Noel Hendrickson; Photo 12.2, page 323. Thinkstock/Getty Images/Jupiterimages; Photo 12.3, page 325. Thinkstock/Getty Images/ Jupiterimages; Photo 12.4, page 330. Thinkstock/ Getty Images/Jupiterimages; Photo 12.5, page 339. Istock/Mikhail Kotov

Author Index

Subject Index